Ecological Scale

Complexity in Ecological Systems

Complexity in Ecological Systems

T. F. H. Allen and David W. Roberts, Editors
Robert V. O'Neill, Adviser

Ecological Scale

Theory and Applications

David L. Peterson and V. Thomas Parker, Editors

Columbia University Press
New York

Columbia University Press
Publishers Since 1893
New York   Chichester, West Sussex
Copyright © 1998 Columbia University Press
All rights reserved

Library of Congress Cataloging-in-Publication Data
Ecological scale : theory and applications / David L. Peterson and
    V. Thomas Parker, editors.
        p.   cm. — (Complexity in ecological systems)
    Includes bibliographical references (p.   ) and index.
    ISBN 978-0-231-10503-3
    alk. paper)
        1. Ecology—Methodology.   I. Peterson, David L. (David Lawrence),
    1954-   .   II. Parker, V. Thomas.   III. Series: Complexity in
    ecological systems series.
    QH541.28.E36   1998
    577—dc21   97-48512
                                                                    CIP

Casebound editions of Columbia University Press books are printed on
permanent and durable acid-free paper.

Printed in the United States of America

For Linda, Gerry, and my parents, for sharing their love of Nature and making all things seem possible.

D.L.P.

For my children, Brian and Emily, who expanded my extent and changed my resolution.

V.T.P.

CONTENTS

PART IV Incorporating Scale Concepts in Ecological Applications

FOREWORD

T. F. H. Allen and David W. Roberts, Series Editors

A central idea in the Complexity in Ecological Systems series is the issue of scale. In the first volume, *Life Itself* (1991), Robert Rosen points to models as scaling devices. The second volume in the series, *Toward a Unified Ecology,* by Tim Allen and Tom Hoekstra (1992), cleaves ecological observation into two parts. The first part distinguishes the system from its background on organizing criteria that give it a type: population, ecosystem, biome, and so on. The second part of observation gives certain spatiotemporal scale to the object of study. Robert Ulanowicz, in the third volume of the series, *Ecology, the Ascendant Perspective* (1997), addresses scaling issues in complex systems described in thermodynamic terms. His central notion of ascendency is rich in its scaling implications. The very notion of complexity, the central issue for the series, at least implies relationships across scales. It is when coarse, overarching events appear to be closely related to fine-grained considerations that the system requires treatment as a complex system. Some accounts of complex systems may not be explicit about scaling, but the tension between scales can always be seen in complex systems if one only looks for it. Because scale and complexity are so inextricably entwined, the present volume with its central concern for scale in ecology is right down the middle of the purview of the series.

Originally, Complexity in Ecological Systems was conceived as a series of authored volumes rather than of edited collections. Indeed, this may well be the first and last edited volume in the series, but it fits so well with its companions that we feel it clearly belongs. The decision to have only authored volumes was deliberate. We wanted books that would drive in particular directions with a certain relentlessness. Bob Rosen described his book as having that quality, and we borrow the turn of phrase from him. So the present volume has been constructed to behave as much like an authored book as an edited volume can.

Some of the authors of the chapters have only met in the development of this book, but others know each other well and have engaged in extended discussions in corners at professional meetings. So there are more ties among the authors than is common in edited volumes. The authors here are part of a loose network of scholars. They did not just turn up at a meeting and have their papers lumped together in a proceedings, because they have all been centrally concerned with the scale issue for some time. Of course some choices of vocabulary and terminology are different among the authors, but to an extent that reflects the variety of approaches to the scale issue that are available. Scale is, after all, the new frontier for ecological investigations, and the frontier is still wide. It is hard to get funded to do research at just one scale these days, but that has not always been the case. Most of the authors here have been chastised at some point for complicating an issue with translations between scales. Some of us were over the barricades earlier than others and hurled ourselves onto the spears, but most of us have had at least a good nick that drew blood. There is an esprit de corps that lends unity to a broad treatment.

The present volume has been written and edited to be more than a collection. As series editors, we requested that the book editors put a firm and unifying hand on the whole, and feel that our request has been more than met. If this volume in the series is not as relentless as some of the others (they are all quite relentless), it gains by containing eclectic views and a range of styles of investigation inside a strongly unifying theme. Although some readers may find this volume through a reference to a particular authored chapter, the effort has been made to present the whole book in a fashion that causes such focused readers to pick up on other chapters. If we have all done our job right, then many readers will read all the chapters. All readers may not start at the beginning and read the volume straight through, but it has been organized and presented so that it is easy to read as a whole. For example, given the choice between putting a references section at the end of each chapter and collecting all the references at the end in one grand bibliography, the editors explicitly decided to do the latter, which provides a resource that is close to being the definitive bibliography of scale in ecology. It is certainly a reflection of a wide range of favorite references from an army of researchers. With an index and single references section, this book invites reading as a whole unit.

As we have suggested, scale is the new frontier of ecology. Indeed, until ecologists become adept at addressing the scale issue, the discipline will remain stuck in complicated descriptions at one level. Without a proper treatment of scale, ecological research papers will remain in one of two categories: narrow and isolated in some small arena of discourse, or elaborate, even tortuous narratives that present large findings. An appreciation of scale allows identification of the relatively small amount of crucial information that links different levels of function. A complex system requires several levels for its adequate description. But treating a system as complex allows one to escape the unwieldy, complicated mess that arises when trying to deal with everything at one level. A proper appreciation of scale allows one to invoke a complex system description, which actually makes it simpler to handle. So a facile treatment of the complexity of scale has very practical implications. The tractability that comes with adept scaling means that difficult applied ecological problems can now be addressed as never before. By no means are all chapters in this book applied ecology, but it is one of the strong common threads.

Probably, any contributor from this book who tried to cover the whole issue of scale as a lone author would fail. The frontier is so long we need a small army, the size of the entire contributor list for this volume, to give scale in ecology the broad treatment it deserves.

PREFACE

The concept of this book for one of us (D.L.P.) originated in a special-topics course at the University of Washington College of Forest Resources during the spring of 1993. The course, which focused on scale issues in ecology, addressed a wide range of topics in terrestrial and aquatic sciences. The class resolved that scale issues are implicit in nearly all ecological studies and applications but that those issues are rarely articulated or quantified adequately. The students found that the emerging field of landscape ecology has had a leadership role in investigating scale concepts, but that the fundamentals of ecological scale are not necessarily well defined or agreed upon. For the other of us (V.T.P.), the motivation for this book came from several years of lab meetings and courses with graduate students interested in hierarchy theory and scale concepts. Countless arguments ensued about how to apply some of these concepts to specific research questions. The frustrations in these efforts led to discussions each summer with Tim Allen and Bob O'Neill at meetings of the Ecological Society of America. Although a number of books have been published on scale, there is no single source to which ecologists and natural resource managers can turn for guidance on how to deal with scale in their respective disciplines.

The ecological literature of the past decade is filled with studies that collect data at the scale of a hectare or less, and extrapolate conclusions to a global scale. Extrapolating across several orders of magnitude on various scales may or may not be valid. However, it is only reasonable to acknowledge the sources of error and variation involved, as well as the potential for environmental relationships to change across different scales. Processes underlying patterns at increasingly larger scales may be substantially different from those at smaller scales of space or time. These issues became apparent early in research concerning large marine ecosystems, air pollution and acid deposition, and other issues addressed at nearly global levels. A number of books have brought together ideas about scale concepts, among them *Spatial Pattern in Plankton Communties* (1978, J. H. Steele, ed.), *Hierarchy: Perspectives for Ecological Complexity* (1982, T. F. H. Allen and T. B. Starr, eds.), *A Hierarchical Concept of Ecosystems* (1986, R. V. O'Neill et al., eds.), and, more recently, a number of other books focusing on large marine ecosystems, physiological ecology, and other more specific topics.

Most of our critical problems in natural resources—endangered species, landscape fragmentation, disruption of natural processes, air and water pollution—are affected by large-scale (in human terms) phenomena. These are issues for which the smallest appropriate spatial scale is a watershed or larger. Relevant data for addressing these problems are mostly available at much smaller spatial scales over a limited period of time. Should we sit back and wait for better data to arrive? Definitely not. We need to assertively resolve these environmental issues, but it is critical to acknowledge the role of scale issues in the application of ecological information to natural resource management.

The objective of *Ecological Scale: Theory and Applications* is to create a greater awareness of the need to incorporate scale concepts in ecological studies and applications. The individual chapters address a wide range of disciplines, demonstrating that ecological scale is a fundamental issue that cuts across all ecological endeavors. The book is not intended to be a comprehensive treatise on scale; that would require several more books. Nor does it have a consensus among chapters with respect to specific approaches or terminology. Rather, it reflects a range of perspectives currently used to address ecological scale. Whether these perspectives coalesce into a coherent body of theory and applications or into a unified approach for more specific models is not yet clear. We hope this book will generate additional momentum for more ecologists to integrate scale concepts in their work.

We feel privileged to have worked with the authors whose chapters appear in this book. Their creativity and enthusiasm have resulted in a volume of unique contributions on the topic of ecological scale. The job of serving as editors for this distinguished and eclectic group of authors has reminded us that diverse ideas and approaches are needed to make ecology a dynamic science.

Many individuals contributed to the completion of this book. We especially appreciate the support of Ed Lugenbeel (Senior Editor), Ron Harris, and Alissa Bader of Columbia University Press. Their advice and professionalism were valuable throughout the publication process. Maureen Laura served as technical editor for the entire volume and carefully edited each chapter. Beth Rochefort redrafted all figures used in the book. Darci Horner assisted with various editing tasks. Technical reviews of chapters were provided by the following individuals: Loveday Conquest, John Emlen, Gregory Ettl, Daniel Fagre, Nancy Grulke, Douglas Houston, Kathy Jope, Donald McKenzie, Thomas Stohlgren, Monica Turner, John Wiens, Jianguo Wu, and Peter Yodzis. A number of the chapter authors also reviewed other chapters. We thank all these colleagues for their efforts. We particularly appreciate the encouragement and ideas of Tim Allen and Bob O'Neill, who have inspired ecologists to think creatively and focus greater attention on ecological scale. We are also grateful to students at the University of Washington and San Francisco State University, who provided ideas and insights throughout the preparation of the book—their contributions were fundamental to our explorations of ecological scale.

This book is for scientists, students, and resource managers. It is for both theoreticians and applied ecologists. We believe that we can increase the integrity of ecology and improve our ability to address environmental problems by making scale concepts a standard part of research and applications. We hope that the issues discussed in this book will contribute to an ongoing dialogue on ecological scale.

David L. Peterson
V. Thomas Parker

PART I

INTEGRATING PATTERN, PROCESS, AND SCALE

1

HOMAGE TO ST. MICHAEL; OR, WHY ARE THERE SO MANY BOOKS ON SCALE?

R. V. O'Neill and A. W. King

The importance of scale has been evident to ecologists (Stommel 1963) and geomorphologists (Schumm and Lichty 1965) for three decades. The earliest inquiries (Feibleman 1954; Simon 1962) noted discrete levels of organization, a discovery well known to ecologists (Rowe 1961). So why are we still publishing books on scale? Why are the principles not well established and accepted? What is left unsaid or is still in need of clarification? It seems that the quest for understanding must continue. In a medieval legend, St. Michael met the journeying soul with scales in his hand; he seems an appropriate patron for our inquiry.

One of the problems is the diverse and scattered literature. Theoreticians (Loehle and Wein 1994) and empiricists (Harris 1980) publish in different journals. There are parallel developments in aquatic ecology (Frost et al. 1988) and terrestrial ecology (Thorhallsdottir 1991). It is difficult to find the extensive insights developed in landscape ecology (Urban et al. 1987) and geomorphology (Phillips 1988a,b; Huggett 1991). Citations in this chapter and another recent effort (O'Neill 1996) come from many different sources. Almost no investigator reads across this breadth of literature, and there is a continuing need for synthesis. The required synthesis is beyond the scope of this single chapter, but we can at least begin by providing access to a broad literature.

Beyond scratching the surface of synthesis, we have identified three issues that appear to need clarification. The first deals with the distinction between scale (i.e., dimensions of space and time) (in Peterson and Parker, chapter 22) and levels that occur at distinct breaks in the scale continuum. In many cases, the levels are not extracted from the data but are imposed a priori upon the data. As our second issue, we would like to make an effort to shake the reader's confidence in these a priori levels. The third issue deals with the role of hierarchy theory in scale research. Several authors have noted limitations in the theory (Allen, chapter 3; Kolasa and Waltho, chapter 4; Parker and Pickett, chapter 8). We would like to attempt some clarification and make certain that the baby is not thrown out with the bath water.

The Inductive Approach to Scale

Scale phenomena are complex, and we continue to accumulate data and experience. Theoreticians often feel that research should proceed in a tidy manner from theory to hypothesis to test. But, like many issues in science, the messy process of collecting and analyzing data is often primary. This inductive process will ultimately lead to an understanding of scale. Because the literature is so diverse, we begin by drawing attention to some recent findings that are not covered elsewhere in this book.

A body of useful information has developed in geomorphology (Phillips 1987; Hammer, chapter 6), where hierarchy theory has had an important impact (Haigh 1987; De Boer 1992). The work addresses decisions about which of the many scaled geological and ecological processes are needed to explain specific landscape changes, such as erosional leveling (Brunsden and Thornes 1979; Brunsden 1980).

Phillips (1995) presents a series of scale criteria for determining the interplay between geological changes and vegetation dynamics. Some criteria deal with temporal scale. For example, if the interval between disturbances is equal to or less than the recovery time, successional or transient communities dominate a landscape. Other criteria deal with the spatial extent needed to study a specific range of temporal dynamics.

This connection between temporal and spatial scale is also evident in ecological processes (Gardner, chapter 2; Schneider, chapter 12). At small spatial scales, predator and prey dynamics may appear to be negatively correlated. But at large scales, the correlation is positive, as both respond to a set of background environmental conditions (Rose and Leggett 1990).

In a related result, Carpenter and Kitchell (1987) used an aquatic model to study the correlation structure among ecosystem components. If they looked at every third day, algal production was negatively correlated with zooplankton biomass. But if they looked at every sixth day, the seasonal dynamics of nutrients took over and algal production was *positively* correlated with zooplankton biomass.

A different aspect of scale is presented by Jones (1995), who studied the genetic composition of broad-ranging species of *Phrynosoma* (horned lizards), which have existed for a long period of geological time. Isolating mechanisms that restricted gene flow among subpopulations had occurred in very different time and space scales. He found differences in DNA among races that indicated two scales of divergence. Considering the rate at which genetic differences occur, one scale corresponded to geological uplifts, separating an eastern from a western race, and the second scale corresponded to glacial–interglacial periods and accounted for more local changes.

In stream ecosystems, theoretical (O'Neill et al. 1979) and inductive (Newbold et al. 1981) approaches were combined to develop the concept of spiraling length as a measure of scale. Spiraling length is basically the distance traveled downstream by a molecule of carbon (Newbold et al. 1982a) or nutrient (Newbold et al. 1982b) as it recycles through the ecosystem (e.g., from algae and back into algae). Radiophosphorus studies (Newbold et al. 1983) were critical to developing an understanding of how spiraling length shapes and scales the aquatic invertebrate and fish communities (Elwood et al. 1983).

Consideration of scale has changed how we look at basic ecosystem properties such as stability (Turner et al. 1993). If the disturbances are large and rapid, compared to the system of interest, the ecosystem is unstable. But, for the same set of disturbances, systems considered at larger spatial scales respond in a stable manner. Thus, individual forest plots may come and go, but total forest cover over a region remains relatively constant (Vande Castle, chapter 13).

What determines spatial distribution of net primary production? At continental scales, net primary production is distributed by climate, with temperature and moisture determining the dynamics. But within a region where temperature and evapotranspiration are relatively homogeneous, production is determined by aspect and soils (Thorhallsdottir 1991). Thus, considerations of scale provide a reconciliation of Clements' (1916) continental picture of plant communities constrained by climate and Gleason's (1926) analysis of individual plant communities, where competition is the critical process.

These studies, combined with results presented throughout this volume, lead to one inescapable conclusion: if you move far enough across scale, the dominant processes change. It is not just that things get bigger or smaller, but the phenomena themselves change. Unstable systems now seem stable (Dutilleul, chapters 17 and 18; Martinez and Dunne, chapter 10). Bottom-up control turns into top down control (Hinckley et al., chapter 15). Competition becomes less important and climate seems to dominate patterns. These same changes can be observed in aquatic ecosystems (Pahl-Wostl, chapter 7), terrestrial vegetation (Lertzman and Fall, chapter 16; Parker and Pickett, chapter 8), and geomorphological dynamics. It is this observation of changing dynamics with scale that formed the basis for the development of hierarchy theory in ecology. And it is this same inductive observation that forms the foundation for the rest of this chapter.

Scales and Levels

Why are there so many books on scale? Maybe some of the books ostensibly about scale are as much or more about levels of organization. Ecologists have made a habit of confusing the terms scale and level. How often have we seen references to the scale of the leaf, the leaf level, the canopy scale, or canopy-level phenomena? Similarly, one finds reference to the population scale and the population level, the ecosystem level and the scale of the ecosystem, or the landscape level and the landscape scale. More often than not, the terms scale and level are used loosely, imprecisely, and essentially interchangeably. The lack of precision in the use of these concepts has been pointed out elsewhere (Allen and Hoekstra 1992; Allen, chapter 3). Nevertheless, the problem continues.

For example, scaling-up is generally understood to refer to the translation of information from smaller to larger scales (Bradshaw, chapter 11), but discussions of scaling are often more accurately characterized as examples of translation between levels. For example, scaling-up from leaf physiology to regional CO_2 exchange with the atmosphere (Hinckley et al., chapter 15) is more about translation between levels of organization than about changes in scale. The increase in spatial scale—square millimeters to hectares or square kilometers—results in new interactions and relationships (Hobbs, chapter 20; Stern, chapter 14), and changes in organization are encountered. A change in scale often necessitates consideration of new levels of organization.

Some definitions are in order. *Scale* refers to physical dimensions of observed entities and phenomena. Scale is recorded as a quantity and involves (or at least implies) measurement and measurement units. Things, objects, processes, and events can be characterized and distinguished from others by their scale, such as the size of an object or the frequency of a process (Innes, chapter 19). Even St. Michael's scales involve measurement and physical dimension: relative mass under the influence of gravity. As a rule of thumb, when we use the term *scale,* we should be able to assign or identify dimensions and units of measurement.

What then is the leaf scale? Is it a reference to temporal dimensions or to spatial dimensions? What about mass? How big is a leaf? What are the appropriate units? To what quantity does the leaf scale refer? Because the entity we commonly identify as a leaf has discrete dimensions, the leaf scale is at best an implicit reference to spatial scales in the range of a millimeter to a meter. Similarly, the landscape scale is at best an implicit reference to larger spatial scales. But what of the population scale? How big is a population? How big is an ecosystem? There is no ecosystem scale. Scale is not a thing. Scale is the physical dimensions of a thing.

Scale also refers to the scale of observation, the temporal and spatial dimensions at which and over which phenomena are observed. Scale of observation has two parts: grain and extent. Grain refers to the smallest temporal or spatial intervals in an observation set. Extent is the total area or the length of time over which observations of a particular grain are made. Sampling frequency in time and space determines the grain of observation. Instrument and observer response time contributes to the grain of observation, as does the time and space a single measurement integrates (e.g., the size of a sampling frame or the duration of a flux measurement). For example, electing to count individual stems in a 1-m^2 sampling frame once every 100 m along a transect once a week fixes the grain of observation. The length of the transect, number of transects, and number of weeks of sampling define the extent. Calculating means coarsens the grain. Subsampling reduces the extent. The scale of observation is a fundamental determinant of our descriptions and explanations of the natural world.

Level refers to level of organization in a hierarchically organized system. Hierarchy theory is about process, flow, interaction, and rates. Levels in the hierarchy are distinguished by rates that differ by one or more orders of magnitude. Indeed, it is the difference in rates that *causes* the structure of levels. Subsystems with similar rates occupy the same level within the hierarchical system.

A level is characterized by its rank ordering in the hierarchically organized system. One level may be higher or lower than another level in the system. Within a hierarchy, higher levels have slower rates and lower levels react more quickly. Levels of the same rank in different hierarchies can have different rates. Moreover, they need not be defined by the same processes, flows, or interactions. The definitions of scale and level seem clear enough. Scale involves physical dimensions, measurement, and quantification. Level of organization is a relative characterization of system organization. Why, then, is there any confusion, and why should we be concerned about it?

First, the source of the confusion. A level of organization is not a scale, but a level of organization can have a scale. More precisely, a level of organization can be characterized by scale, although Allen (chapter 3) points out that including scale in the *definition* of a level has a certain cost. The rates that define a level characterize the temporal scale of that level. Moreover, the spatial extent over which these rate processes operate characterize the spatial scale of the level. For example, an individual organism, a tangible entity associated with a particular level, is readily characterized by its size, its physical dimensions in space, or its mass. Because organisms can come in many sizes, organism as a class is independent of scale, but particular examples of organisms can be characterized by their scale. With more difficulty, the level of organization that emerges from the interactions among individuals (or the "population") can be characterized by its scale, the spatial extent it occupies, or the time it takes for all members of the population to be replaced. Cycling rates of nutrients can be used to define the temporal scale of levels, and the area over which cycles occur can be used to define their spatial scale. Thus, there is a definable association between level and scale. Note again, however, that it is particular examples that are characterized by scale and not the general class associated with a level (e.g., population). Unfortunately, the terms *level* and *scale* are sometimes used interchangeably.

When we observe the natural world, the things we discern, the patterns of these things, and the relationships among these things are determined by the scale at which we observe—the scale of observation, both grain and extent. Therefore, the hierarchical organization, if any, and the levels of organization are determined by the scale of the observation set. If you change the scale of observation, you change the observation set, and the hierarchical organization can change or disappear.

Within an observation set, we can locate processes at different levels by finding breaks or discontinuities in the data. An observation set of fine grain and very large extent is needed. The data are then analyzed at increasing scale, every second point is correlated, then every third point, etc. A number of statistical methods exist (Carpenter and Chaney 1983; Turner et al. 1991; O'Neill et al. 1991b; Cullinan et al., unpublished manuscript) to detect discrete levels. The methods include (1) graphing variance as a function of scale (Levin and Buttel 1986; O'Neill et al. 1991b), (2) wavelet transform (Li and Loehle 1995), (3) spectral analysis (Ripley 1978), and (4) detecting changes in fractal dimension (Krummel et al. 1987).

T. F. H. Allen has pioneered the use of multivariate statistical methods for detecting distinct levels in phytoplankton data (Allen and Koonce 1973; Allen et al. 1977). Others have followed his lead in developing multivariate methods for other systems (Burrough 1983; Oliver and Webster 1986; Kolasa 1989; Ver Hoef and Glenn-Lewis 1989). Discontinuities in the data suggest a change in level of organization. An association between scale and level is formed because each level is associated with a range of scales between the breaks. Scale can again be incorrectly considered a synonym for level.

Why should we care? Why should we be concerned that the concepts and terms of scale and level of organization are used imprecisely and interchangeably? Most of our audience really does know what we mean when we say "landscape level" or "landscape scale." Why confuse the issue further by carrying on about the error or inaccuracy of this usage? First, precision is one of the objectives of science. Thus, one can argue that careful use of terminology is part of the discipline required of ecologists as scientists. Think of it as a necessary cost of doing science.

Beyond a desire for precision, the confusion of level and scale can cause real problems. When you change scale (i.e., grain and extent), you eventually move across a discontinuity in scale and can change the level of observed organization. The general phenomenon, called transmutation or aggregation error, has been known for about two decades (O'Neill 1979). Basically, transmutation deals with extrapolation errors when you move across breaks in the scale continuum. If you stay on the same level, scaling can correct for scale differences (Schneider 1994a). Dimensional analysis is one approach. Similarly, you can expect self-similarity and correct for scale differences by the fractal dimension. But if you extrapolate across a break in the continuum, an error occurs. A significant literature has developed on this topic (table 1.1). Particularly important are methods for scaling-up that avoid or minimize this type of error. These methods are not generally known to ecologists because most presentations have focused on the problem in the context of mathematical modeling.

Traditional "Levels"

Most ecologists started as biologists exposed to a range of phenomena from DNA mechanisms to the geological record of evolution, from the microscope to global carbon cycling. Ecologists adopted hierarchy theory because of familiarity with phenomena segregated into discrete levels. This same heritage also led us to believe that the significant

TABLE 1.1
Papers describing transmutation or aggregation error occurring when information is moved to larger scales.*

Reference	Title, or abbreviated title
Cale and Odell (1979)	Aggregation in ecosystem modelling
O'Neill (1979)	Transmutation across hierarchical levels
Cale and Odell (1980)	Behavior of aggregate state variables
Gardner et al. (1982)	Robust analysis of aggregation
Cale et al. (1983)	Aggregation . . . in nonlinear models
Luckyanov et al. (1983)	Aggregation . . . water models
Iwasa et al. (1987)	I. Perfect aggregation
King et al. (1987)	Site-specific models to regional
Bartell et al. (1988b)	Relevant model structure
Iwasa et al. (1989)	II. Approximate aggregation
King et al. (1989)	Regional CO_2 exchange
King et al. (1990)	Tree physiology . . . with forest
King (1991)	Translating models across scales in the landscape
King et al. (1991)	Heterogeneous landscapes
Luxmoore et al. (1991)	Scaling up models of soil-plant
Yeakley (1991)	Organizational levels analysis
Wessman (1992)	Spatial scales and global change
Turner and O'Neill (1994)	Exploring aggregation in space and time
Cale (1995)	Model aggregation: ecological perspectives
Luckyanov (1995)	Model aggregation: mathematical perspectives

*The error occurs whenever the extrapolation moves across breaks in the scale continuum.

levels could be named a priori: organism, population, ecosystem, landscape, etc.

The levels of explanation must be extracted from data, not preimposed. To date, the empirical evidence shows that the levels extracted from the data do not correspond in any simple way to traditional levels of biological organization. But the heritage is so deeply ingrained that we feel it is necessary to take some space to shake the reader's confidence in the traditional concept of biological hierarchy.

Let us attack the tradition at its strongest point. Surely we can say that a population is at a lower level than, and therefore constrained by, the ecosystem? Not necessarily. Remember the old nursery song, "the bear came over the mountain." For the last 30 years, we have spatially delimited the energy and nutrient dynamics of ecosystems by the watershed, but the bear never read the book. In fact, most vertebrates cannot read. There are a number of studies showing that foraging strategies of large herbivores depend on moving across spatial scales usually identified with ecosystems (Senft 1987; Wallace et al. 1995; Goodwin and Fahrig, chapter 9; Stern, chapter 14). To which ecosystem does a population of migratory birds "belong"? Is *Ursus arctos* (Alaskan brown bear) a terrestrial or an aquatic organism? It may depend on whether you are a *Vaccinium* (huckleberry) or an *Oncorhynchus* (salmon). And is *Oncorhynchus* marine or freshwater?

So maybe there are some spatial exceptions, but surely the ecosystem dominates or constrains temporal rate processes of the populations. This is true for many observation sets. But precisely because consumers can integrate across space, the population will sometimes determine rate processes in an ecosystem. Consider that migratory duck populations probably determine production rates in flyway wetlands. Large ungulates, such as *Bison bison* (bison) and *Conochaetes taurinus* (wildebeest) (Hatton and Smart 1984), determine rates of production in grassland ecosystems. *Alces alces* (moose) certainly change plant composition and ecosystem properties (Pastor et al. 1988). It is very difficult today to guess what rate processes were like in forests of eastern North America when they contained *Ectopistes migratorius* (passenger pigeon) and *B. bison.* We are certainly impressed by how rate processes, such as decomposition, can be changed by an introduced consumer, such as *Sus scrofa* (wild boar) in Hawaii and forests of the southeastern United Stares (Bratton 1975).

Still not convinced? Then let's talk about *Castor canadensis* (beaver). Let's talk about rate processes in *C. canadensis* streams (Naiman et al. 1986, 1988). Let's talk about rates of nitrogen cycling (Naiman and

Melillo 1984) and rates of nitrogen fixation (Francis et al. 1985). Let's talk about rate processes in the riparian zone (Johnson and Naiman 1987). Let's talk about spatial heterogeneity across the ecosystem (Remillard et al. 1987; Broschart et al. 1989). And if you really want to know who is in control, consider the forest that went from 1 percent aquatic (1940) to 13 percent aquatic (1986) when the toothy engineers moved in (Johnson and Naiman 1990).

The same type of ambiguity occurs when we consider the individual species within the "higher level" of the community. In many observation sets, the species operates on a faster and smaller scale than the community; for example, competitive interactions can determine whether a species will occur in the community. But change the observation set, and the species is seen as the larger spatiotemporal entity, and the community is simply the "momentary" juxtaposition of a collection of species (Margalef 1995). The genome study of Jones (1995), mentioned earlier, documents that the space and time scales of *Phrynosoma* are larger than the present-day communities. Paleoecological studies (Davis 1976; Delcourt and Delcourt 1987; Schoonmaker, chapter 5) document that species combinations in communities during glacial–interglacial periods changed rapidly. It is was the species that remained stable, while the community was the "rapidly" changing, transient phenomenon.

When you ask a different question or simply change the grain and extent of your observation set, the levels in that observation set can change. Consider, for example, the study of Bartell et al. (1988a), in which variance was added to the parameters of a pelagic trophic model, and correlations in the output were analyzed. During periods of rapid algal growth, production is constrained by nutrients and algal growth rates—the control is bottom-up. However, these periods of rapid growth are followed by periods when the zooplankton populations have caught up with the algal growth. Now production is constrained by zooplankton parameters. The control is top-down. The traditional trophic levels do not correspond in any simple way to the levels of rate processes in the system.

Published studies that identify levels from breaks in the data certainly do not indicate that the levels correspond to traditional biological levels of organization. The clumped body weights of vertebrates (Holling 1992) do not correspond to traditional a priori levels. Vegetation transects (O'Neill et al. 1991a,b; Cullinan et al., unpublished manuscript) also reveal levels that do not correspond to traditional a priori levels.

Logical consistency is needed. The intrinsic rate structure of complex systems can assist us in explanation because components with similar rates interact; components with sufficiently different rates cannot interact but can only constrain. If we wish to take advantage of this important observation, then we must establish the rate structure—the levels—from data. The empirical evidence is clear: When you change the observation set, you can change the levels. Therefore, you cannot logically take a set of a priori levels established from other observation sets in the past, and impose them on a new observation set. The imposition leads to confusion at best and to serious errors at worst. The conclusion is straightforward: You cannot assume that the levels in your observation set are going to correspond to the chapter titles of your introductory ecology textbook!

What Is Wrong with Hierarchy Theory?

The empirical observation that changes in the observation set lead to changes in the levels of explanation, creates some problems for ecologists. The theory does not validate a priori, established theoretical concepts. A logical application of hierarchy theory leads to the conclusion that there may not be any *the* hierarchy. As a result, the hierarchies established by the theory seem arbitrary. Therefore, the theory itself may seem arbitrary or merely heuristic. It does provide objective methods for extracting levels of explanation from observed data. The extracted levels are not arbitrary, but they may not survive a change in the observation set.

When you change the questions you ask and what you observe about the system, you can completely change the explanatory principles required for explanation (Rykiel, chapter 21). You can go from the finest physical scale (e.g., quarks) to the largest physical scale (e.g., the universe). If you ask questions about gravitation, you never find "organism" or "ecosystem" as coherent levels that help explain it. You can ask questions about evolution and end up studying species. But you can also ask questions about CO_2 exchange or nitrogen cycling and never mention a Latin binomial. In one set of questions, the species is a sublevel within the community. But change the scale sufficiently, and the species appears as the stable spatiotemporal entity, and the community is simply the momentary juxtaposition of a collection of species.

We must not lose sight of the fact that hierarchy theory was offered as a solution to the dilemma of middle-number systems. You can predict a planetary orbit because this is a small-number system of sun and planet. You can predict the weight of a volume of air because this is a large-number system of very many particles. But you cannot predict the weather next week, much less next year, because this is a complex, middle-number system. Therefore, volcanoes, tornadoes, and ecosystems are phenomena whose behaviors are difficult to predict. There are many interacting parts, too many to analyze one at a time, too few to just average.

Hierarchy theory offers an escape from the middle-number dilemma. The escape is possible *if* you take a specific observation set and locate within that observation set distinct scales that allow you to divide the many parts into levels. Within a level, the parts interact; between levels, no interactions are possible. You can develop a small-number explanation within the level.

The basis for abstracting the small-number system is proven by Schaffer (1981). Within certain restrictions, the majority of interactions in a complex mathematical system minimally influence any specific eigenvalue (i.e., any single fundamental scale of rates). You can abstract the system to those components that have similar rates and differ by an order of magnitude from the rest of the system. In other words, you can ignore the dynamics of some components and aggregate others, but you have found a small-number explanation that is valid only for that observation set.

The theory does not maintain that you will necessarily find separate levels in every observation set. But the theory *does* advise you to find another approach if you do not locate distinct levels. It warns that you will be trying to predict volcanoes, earthquakes, and next month's weather.

There can be significant conflicts between hierarchy theory and traditional views of the ecosystem. Watson and Loucks (1979) analyzed rate processes in a lake ecosystem. They divided the ecosystem into the traditional biotic and abiotic components. They found that rates among these components differed by over three orders of magnitude. These components cannot *interact* in any meaningful sense of the word. In fact, most ecosystem models have this same property and violate a basic tenet of scale theory.

So the conflicts are real. If a hierarchy is limited to an observation set on a normal operating system, what can we predict when the system undergoes a radical change? The pathology may not be predictable from the explanatory structure built for the normal system. If the pathology involves the loss of a dominant component, or the introduction of a new dominant, the rate structure of the system can change and new levels of organization are required. Predictions are permissible only if the rate structure remains intact; this is a stringent requirement.

Our established theoretical constructs, such as organism, population, and community, are nested hierarchies, but there is nothing about the levels extracted from an observation set that requires them to be nested. In fact, the studies conducted to date seem to suggest unnested hierarchies. The most serious limitation is that 50 different observation sets may yield 50 different explanatory hierarchies. The only real escape from the middle-number dilemma may be to abstract the system into a large number of very specific hierarchies. This may explain specific phenomena but at the cost of theoretical chaos. Any specific hierarchy may seem arbitrary and (as stated) the theory that spawned them merely heuristic.

In fact, to date, no one has undertaken the difficult test needed to settle this issue. The test would involve a number of quite different observation sets on the same physical system, observation sets that ask different kinds of questions. Each observation set would have to be of sufficiently fine grain and sufficiently large extent to permit analysis of system organization. If, and only if, analysis of the different observation sets leads to the same breaks in the rate structure, can we maintain that hierarchy theory leads to a single, nonarbitrary hierarchy underlying all of the observations. If each observation set leads to a different rate structure, then hierarchy theory may turn out to be merely heuristic in the sense of explaining very specific problems. The theory would have to be evaluated as less fundamental and less synthetic.

Until such an ambitious test of the theory is conducted, hierarchy theory will remain the only available approach to middle-number systems. So our advice remains, don't throw away your hammer until you have invented the screwdriver. Even then, you may wish to keep your hammer in the bottom drawer for special jobs.

2

PATTERN, PROCESS, AND THE ANALYSIS
OF SPATIAL SCALES

Robert H. Gardner

The relationships among pattern, process, and scale have long been recognized as a central issue in ecology (Greig-Smith 1952; Pielou 1969; Usher 1969; Iwao 1972; McMahon 1973). Concerns about the detection of scale-dependent phenomena and the prediction of ecosystem dynamics across scales have increased significantly in recent years. Our inability to determine scale-dependent effects results in conflicting interpretations of data and an inability to reliably extrapolate results. Recognizing the seriousness of these problems, the United States Environmental Protection Agency established in 1992 a Multiscale Experimental Ecosystem Research Center (MEERC), whose goals are as follows: (1) to construct a series of experimental ecosystems differing in size, shape, and complexity of biota, (2) to quantify and parameterize their scale-dependent responses to nutrient and contaminant inputs, and (3) to construct predictive models that allow results to be extrapolated from experimental systems to the larger scale of ecosystems in nature. Ultimately, a combination of empirical measurements, experimental manipulation, and verification of predictions is required to resolve issues associated with the measurement and prediction of scale-dependent effects.

The failure to account for scale-dependent changes in pattern and process has confused and confounded the synthesis of ecological data and the extrapolation of ecosystem effects from sample plots to landscape scales (Dale et al. 1989; Wiens 1989; O'Neill et al. 1991b; Costanza and Maxwell 1994). The ability to detect changes in pattern and make predictions at more than one level of resolution would, therefore, seem to be a fundamental requirement of ecosystem science. Achieving this goal requires identifying the physical and biological processes of interest, estimating the variables and parameters that affect these processes at different scales, and developing rules to translate information across scales. The development of this capability will ultimately allow scale-dependent predictions to be tested and hypothesized relationships to be verified. Although some excellent examples have been published that illustrate how this process works (Wiens 1989), such cases have not produced a general methodology that provides the basic tools required for analysis and extrapolation of scale-dependent relationships (Bradshaw, chapter 11).

Compounding the difficulty of developing general and reliable tools for detecting scale is the fact that ecosystem data rarely produce a single scale that can be regarded as correct or optimal for measurement and prediction (Allen and Starr 1982; Meentemeyer and Box 1987; Steele 1989a; Wiens 1989; O'Neill et al. 1991b; Levin 1992). Although some authors assert that ecosystem data may reveal fundamental scales of dynamics (e.g., Carlile et al. 1989), the disagreement may be more semantic than real. The problem seems to be that temporal dynamics of ecosystems may decouple them from the direct effect of the physical world, producing a web of indirect effects that cause temporal or spatial lags in the patterns of interest (Wiens 1989; Holling 1992). Although measurements of these systems may show distinct changes in pattern with scales, the measured changes may be unique to that data set, introducing the important but difficult issue of pseudoreplication (Hurlbert 1984; Hargrove and Pickering 1992).

This chapter is an overview of some of the more familiar methods of detecting change in spatial pattern over time and space. It also discusses the limitations inherent in these techniques and identifies some guiding principles in the determination of scale-dependent relationships. Scale is defined as a change in pattern as determined by the spatial and/or temporal extent of measurements necessary to detect significant differences in the variability of the quantity of interest. Although the emphasis here is on characterizing spatial variability, this definition is consistent with the common usage of the term *scale* (Powell 1989; O'Neill et al. 1991b; Levin 1992; Schneider 1994a).

Detecting Spatial Scales

A variety of methods have been developed for detecting scale-dependent patterns. Each method has inherent strengths and weaknesses, with no single method being clearly superior to all others. Methods based on analytical procedures (Dutilleul, chapters 17 and 18), such as dimensional analysis (Platt 1981; Legendre and Legendre 1983) and consideration of allometric relationships (Calder 1983; Brown and Nicoletto 1991), are extremely powerful because they can solve complex problems by direct consideration of the equations describing the relationship between pattern and processes. Because these methods require an a priori specification of the equations describing causal relationships, their application is often beyond the scope of most ecosystem studies. Legendre and Legendre (1983) give several ecological examples of the use of dimensional analysis. Considering simple geometric relationships by dimensional analysis has recently shown that nutrient removal rates for different MEERC mesocosms can be directly related to differences in depth, volume, and the surface-area-to-volume ratio (L. Sanford, personal communication). More commonly, one must analyze spatial or temporal data without sufficient information to mathematically describe hypothesized relationships between pattern and process. Under these circumstances, a suite of possible causes can be posited without hope of identifying the subset of variables that will allow pattern to be predicted from process alone.

Summary of Basic Techniques

The inadequacies inherent in most data sets, and the availability of abundant data describing spatial patterns, often mean that the first step in an analysis will be an empirical description of changes in pattern with changes in scale (Schneider, chapter 12). A variety of quantitative methods are available to accomplish this objective. One of the simplest methods is to plot the change in variance with changes in spatial extent (Levin and Buttel 1986; O'Neill et al. 1991b). Estimates of the variance (S^2) are inversely proportional to sample size, $S^2 \approx 1/n$, where n is the number of samples or, for spatial data, the size of a sample quadrat. Taking the logarithms and introducing a proportionality constant yields: $\ln S^2 = a - \ln n$. When $\ln S^2$ is plotted as a function of $\ln n$, the result is a straight line with a slope of -1.0 (figure 2.1A). This relationship holds whenever the sample is randomly and independently distributed in space. However, if spatial correlations exist, then samples will not be independent, and the slope of the ln–ln plot will lie between -1.0 and 0.0 for correlations of 0.0 and 1.0, respectively. Wiegert (1962) used this approach to study vegetation on

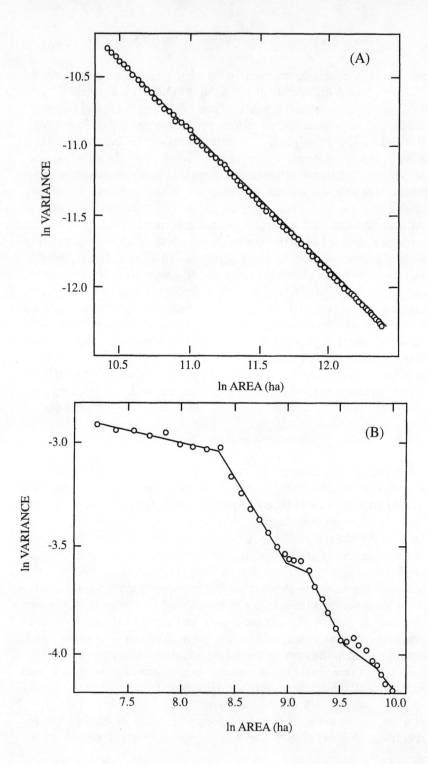

quadrats of different scales, and Levin and Buttel (1968) proposed that the deviation of the slope from -1.0 is a measure of the spatial scale or patchiness of the landscape. O'Neill et al. (1991b) used this technique to analyze a number of landscapes (see figure 2.1) and were able to demonstrate that land-cover patterns often display discrete levels of organization, although the analysis neither revealed the underlying factors nor predicted when one might expect these patterns to occur.

Methods based on geostatistical techniques (Deutsch and Journel 1992; Rossi et al. 1992) use estimates of the semivariogram to define the spatial scales over which patterns are dependent. The semivariogram,

$$\hat{\gamma} = \frac{1}{2N(h)} \sum_{i=1}^{N(h)} (x_i - x_{i+h})^2$$

is calculated by rearranging the data into i data pairs (x_i and x_{i+h}) separated by distance h, with $N(h)$ equal to the number of such pairs. The summation is made over all possible pairs and lagged distances, h. In practice, at least 50 data pairs are needed to adequately estimate the semivariance for each h. Assuming that the two-dimensional data set is isotropic, data pairs can be constructed by taking transects in all directions. Alternatively, if anisotropic properties of the spatial data are of interest, then values of the semivariance for specific directions can be compared. A variogram is used to visualize the results by plotting the semivariance on the ordinate against h on the abscissa. A relatively flat variogram indicates a pattern that lacks spatial dependencies (i.e., a random pattern). If a spatial data set is nonrandom and has been adequately sampled (i.e., the spatial extent of the data provides an adequate representation of the pattern of interest), then the variogram will ascend from an initial value at $h = 0$ to an asymptotic value. In the geostatistical jargon, these two values are the nugget and the sill, respectively. The difference between the nugget and the sill reflects the proportion of the total variance due to spatial dependencies within

FIGURE 2.1
Logarithm of variance in percent land cover for **(A)** barren ground and **(B)** grassland for Goodland, Kansas plotted over a range of sample sizes. The slope of the line for barren ground is equal to -1.0, whereas grassland areas show a hierarchical structuring within regions where slopes are greater than -1.0.

the data set. The value of h at which the semivariogram asymptotes indicates the spatial extent, or range, over which spatial dependencies can be detected.

A number of statistical measures of dependency have been employed to detect changes in pattern with changes in scale that are based on the change in variability between points at different values of h (Rossi et al. 1992). These statistics can all be shown to be closely related to the semivariance (Deutsch and Journel 1992). However, the semivariance, or the variogram, assumes a stationary process (no change in mean or variance with distance) and does not lend itself to statistical testing (Legendre and Fortin 1989), making the use of autocorrelations and correlograms a more preferred technique (Legendre and Fortin 1989; Legendre 1993). Estimating the autocorrelation requires the calculation of the covariance between points separated by h:

$$(3) \quad \text{regional dummies} + C_j\left(P_i - \overline{P}_j\right)$$

where x_i and x_{i+h} are, respectively, the tail and the head of the data pair, and m_i and m_{i+h} are the corrections for mean values of the tail and head, respectively. The estimated autocorrelation is then:

$$\hat{\rho}(h) = \frac{\hat{C}(h)}{S_i S_{i+h}},$$

where S_i and S_{i+h} are the standard deviations of the data points comprising the tail and the head, respectively. A correlogram, like a variogram, provides a visualization of the change in spatial relationships by plotting the autocorrelation values on the ordinate, and h on the abscissa. Indications of ecological scale can be verified by statistically testing the peak values of the correlogram (both positive and negative) for significant differences from zero (Carlile et al. 1989). The conditions for valid tests for the significance of these peaks are restrictive, requiring that the data be equally spaced, that gradients of change or trends in the data be removed [see Legendre (1993) for other restrictions in the analysis of gradients and autocorrelated data], and that the residuals be normally distributed (Legendre and Legendre 1983).

Power spectral analysis is frequently employed as a scale-detection technique for time series data. Spectral analysis can also be applied to spatial data by taking transects through the data set (as is done to construct a one-dimensional variogram and correlogram). Two-dimensional spectral analysis has been used to analyze ecological systems (Legendre and Fortin 1989). Power spectral analysis partitions the variance of the data series into contributions at frequencies that are harmonics of the length of the data set (Platt and Denman 1975). As a result, the highest frequency that can be estimated is equal to two times the sampling interval, and the lowest frequency that can be estimated is equal to half the total extent of the data series. The periodogram of a power spectral analysis is produced by the simultaneous least squares fit of a finite number of sine and cosine functions of sets of frequencies characteristic of the data series. The power spectrum (or, in statistical jargon, the variance spectrum) produced from the periodogram provides the necessary partitioning of the variance of the series. The power spectrum is a function of a range of frequencies, providing an evaluation of changes in pattern with changes in scale. The periodogram is very sensitive to repeatabilities in the data, but insensitive to other types of spatial patterns (Legendre and Fortin 1989). The statistical requirements of data to be subjected to a power spectral analysis are similar to those for the estimation of the autocorrelations, given that the data are normally distributed (i.e., subsets of the data set will have the same mean and variance) and the series is stationary (Platt and Denman 1975). However, it is the complexity of a spectral analysis, the difficulties in interpreting the results and the need for large data sets that have impeded the widespread use of spectral analysis for the detection of scale-dependent changes in spatial patterns.

A sampling procedure that compares the variance among paired quadrats (PQ) of different size has a long history of use in ecology. The methods are based on the principle that peaks in variance (or variance–mean ratios) between different plot sizes suggest intrinsic scales of patchiness (Greig-Smith 1952, 1983). The essential problem with PQ methods is the inefficiency of the sampling technique; each pair of quadrats results in a single data point for the estimation of means and variance. Because independent and nonoverlapping quadrats are sampled from the spatial data set, the range of scales that can be sampled reliably is severely limited, even when the data set being analyzed is quite large. For instance, if we analyze a data set with a spatial extent equal to 1000 cells (i.e., a grid with a total of 1,000,000 sites) and require that the estimated variance be sampled with at least 50 data points (an assumption similar to that required for estimation of the variogram and correlogram), then the maximum scale that can be reliably determined is equal to $1000/(50)^{0.5} \approx 141$. Thus, as a general rule, determination of scales by PQ methods will be reliable only when the scales of interest are less than or equal to 15 percent of the total extent of the data. Hurlbert (1990) evaluated the variance–mean ratio and determined that it "is useless as a measure of departure from randomness" and, therefore, an unreliable indicator of scale-dependent changes in pattern. Instead, Hurlbert (1990) found that a family of indices based on Morisita's index of aggregation, which measures the probability of two randomly selected individuals being found in the same quadrat (Morisita 1964, 1971), provides a more useful measure of spatial aggregation.

Distance sampling for determining scale-dependent changes in pattern is based on compared points in space (i.e., the location of individual plants or animals) rather than quadrat sampling (Campbell and Clarke 1971). The statistic of interest is usually based on the frequency distribution or the variance of co-occurrences of individuals at different distances (Pielou 1969; Getis and Franklin 1987). Statistical tests for the deviation from a random distribution have been established for a variety of indices based on distance sampling (Pielou 1969; Getis and Franklin 1987; Solow 1989). However, distance statistics sensitive to the truncation effects of spatial boundaries and analysis may require the assumption that pattern does not vary beyond the spatial extent of the data set (Getis and Franklin 1987), an uncomfortable assumption when the purpose of the analysis is to determine the scale of the pattern. The effect of this assumption on the reliability of the method, and a systematic comparison with quadrat sampling that does not require this assumption, have not been established.

Fractals

Fractals can be used to describe the complexity of natural patterns and the changes in these patterns with changes in scale. A fractal can be defined simply as any object that "looks the same whatever the scale" (Feder 1988). That is, a fractal need not be a geometrical object but might be a time series, a transect of points taken through space, a semivariogram, or any series of numbers. The property of sameness, or self-similarity, has been defined as invariance with respect to translation and scaling (Feder 1988). This means that a small portion of a fractal can be made to cover the original object simply by stretching or rotating an arbitrarily small subsample of the fractal object. Burrough (1981) was the first to suggest that changes in fractal dimension with changes in scale provide an appropriate and useful measure of self-similarity of pattern and the scales at which the self-similarity ceases. This is easily visualized by considering fractal objects occupying a two-dimensional surface (e.g., a cloud shadow or a habitat patch). The dimension, D, for such a fractal is estimated as: $D \approx \ln(A)/\ln(L)$, where A is the area and L is the perimeter of the object. Estimates of D can be made directly for each object by using the above ratio, or by regression where the slope of the line of $\ln(A)$ regressed on $\ln(P)$ statistically estimates the fractal dimension. A constant value of D taken at different spatial scales (i.e., changes in grain or extent) means that the pattern is self-similar, and any portion of the sample provides an adequate statistical representation of the whole. Similarly, when the value of D shifts with changes in scale, the pattern is no longer independent of the scale of measurement.

Sampling questions are important in estimating the fractal dimension, especially when the data are obtained from a gridded surface. Small objects (those represented by only a few sites on the grid) should not be included in the regression because they will result in an overestimate of D. This is easy to visualize when one realizes that the dimension of a single site will be exactly equal to the dimension of the grid. As a general rule, only objects represented by at least 50 data points (or grid sites) should be included in the analysis (Milne 1991b). It is common in the literature of the physical and material sciences to see estimates of the fractal dimension of objects when the system being studied is near the critical thresholds and, consequently, a single object is likely to dominate the pattern. In these cases, the preferred method for estimating the fractal dimension of the map is by direct calculation from the ratio of the natural logarithm of the area to that of the perimeter for the single largest cluster on the map (Stauffer and Aharony 1992). It is important to remember that even though an estimated dimension is fractal (i.e., not an integer), it is not necessarily self-similar.

One of the early uses of the fractal dimension for detecting scale-dependent changes in pattern was published by Krummel et al. (1987), who found a discontinuity in the fractal dimension of deciduous forest patches at scales of 60 to 70 hectares, corresponding to changes in shapes of wooded areas by conversion of large dendritic patches to smaller, more regular woodlots. Fractals have also been used to model patterns of spatial dependence in landscapes (Palmer 1992; With et al., in press), by describing the properties of home range size and dispersal ability (Milne 1992) and characterizing time series (Burrough 1981; Plotnick and Prestegaard 1993). The variety of applications of fractals, and the variety of fractal models, make the subject a confusing one. Excellent reviews by Sugihara and May (1990), Milne (1991a) and Plotnick and Prestegaard (1993) provide valuable insight into ecological applications of fractals.

One of the more recent uses of fractal geometry for the detection of scale is lacunarity analysis, a procedure originally suggested by Mandelbrot (1983). Lacunarity uses the first and second moments of the statistical distribution of "gaps" or "holes" within a spatial data set to describe changes in pattern with scale. The algorithm for calculating lacunarity employs a series of "gliding boxes" of increasing size to sample the gridded data. A gliding box with linear dimension of r and area of r^2 is placed at the upper left corner of the grid, and the number of sites occupied by the object of interest (e.g., land cover, habitat type, species abundances) are counted. Then the box is moved one column to the right, and the count is repeated. This process continues until all independent starting points for the gliding box have been sampled and the count summarizes as a frequency distribution, $QN(r)$ (the number of gliding boxes of size r that can be placed on a grid with linear dimension of M) will be equal to $(M - r + 1)^2$, and the intervals of Q will range from 1 to r. Then Q is converted to a probability distribution, q, by dividing each frequency by $N(r)$ and the first and second moments ($Z_{(1)}$ and $Z_{(2)}$, respectively) are calculated as

$$Z^{(1)} = \sum_{i=1}^{r} i q_i,$$

and

$$Z^{(2)} = \sum_{i=1}^{r} i^2 q_i.$$

The value of lacunarity, $\Lambda(r)$, for box size r is estimated by the ratio $Z^{(2)}/(Z^{(1)})^2$. Lacunarity values are similar to the variance–mean ratio, which can be calculated as $Z^{(2)} - (Z^{(1)})^2/Z^{(1)}$. The gliding-box sampling procedure can be used to estimate lacunarity values for one-dimensional transect data, two-dimensional mapped data, or even three-dimensional data sets. Additional details concerning the calculation of the lacunarity index and statistical tests of the pattern against either random or scale-dependent patterns can be found in Plotnick et al. (1993). The rate of change in lacunarity values with increases in spatial scale can be compared with the rapid decline of lacunarity values for random spatial patterns. If the change in lacunarity values is linear, then the pattern is self-similar (Plotnick et al. 1993). The ability to distinguish among different pattern types makes this index especially useful for characterizing the texture and pattern of remote-sensing images (Henebry and Kux 1995; Vande Castle 1997, chapter 13).

Gliding-box sampling is the most intensive, nonredundant sampling possible and differs substantially from methods that strive for independent, nonoverlapping samples of spatial data. The gliding-box sampling procedure is not unique to lacunarity analysis but has been used by a number of authors to estimate the change in variance with scale (Levin and Buttel 1986; O'Neill et al. 1991a) and to estimate scale-dependent changes in the fractal index of landscape patches (Krummel et al. 1987). Figure 2.2 illustrates the relative performance of these two sampling methods by comparing the estimated variance for a simple random map with the gliding box and nonoverlapping, independent

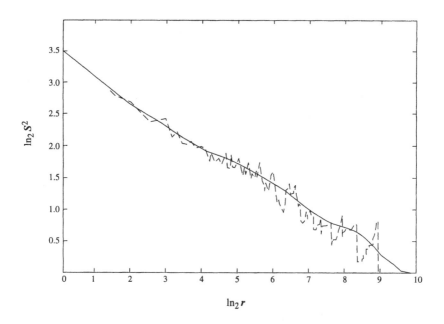

FIGURE 2.2

Two estimates of the change in ln transformed variance, S^2 (base 2), with increasing ln of quadrat size, r (base 2), measured on a random grid of 0s and 1s (grid of 100 rows and columns = 10,000 sites). The two sampling estimates are the gliding box (*solid line*) or independent quadrat (*dashed line*).

quadrats of increasing size. Both lines produce the expected slope of -1.0 when quadrat sizes are small and sample sizes are large. However, the reliability of the variance estimates of the independent quadrats declines rapidly when quadrat sizes are greater than $r = 16$. The gliding-box method produces 7225 samples when $r = 16$ on a 100×100 grid, whereas independent, nonoverlapping samples, such as those used by the PQ methods (Dale and MacIsaac 1989; Ludwig and Reynolds 1989), produce only 36 quadrat samples. Above $r = 16$, spurious peaks and valleys in the variance plots appear, indicating that the use of variance as an indicator of scale dependencies should not be attempted if sample sizes are less than 50.

Practical Difficulties in Determining Scale

This overview of scale detection methods is certainly not an exhaustive one. Many indices have been suggested for measuring changes in one-dimensional and two-dimensional spatial and temporal data sets, and more are certain to be developed. All these techniques are closely related, with similar assumptions regarding sampling and analysis of data. Therefore, the requirements and restrictions of each method must be carefully noted when attempting to measure changes in pattern with changes in scale. For instance, it is unlikely that any technique will be able to exceed the limits of spectral analysis, which measures frequencies within the range of half the extent to two times the grain of the data. All techniques are likely to be sensitive to sampling intervals (i.e., the grain), especially if this interval corresponds to some unmeasured frequency. One can also expect that all methods will be sensitive to sample size, and changes in statistics indicating a change in scale should not generally be based on fewer than 50 samples. Statistical tests account for the problem of small sample size, but graphical methods often ignore this problem. Additional assumptions, including stationarity within the data set, may be necessary before the mean and variance can be reliably estimated (but see Legendre 1993). Finally, it should be remembered that after an analysis has been performed, only a single data set has been characterized. Predicting results from the analysis of a single data set, no matter how intensively the data set has been sampled, is a classical case of pseudoreplication (Hurlbert 1984).

A significant factor affecting the reliability of all scale-detection procedures is the errors present in spatial data as the result of aggregating fine-grained information into coarser scales of resolution. The loss of information that results from aggregation of spatial data is well studied (Henderson-Sellers et al. 1985; Meentemeyer and Box 1987; Turner et al. 1989; Costanza and Maxwell 1994; Pierce and Running 1995), and analytical equations have been developed to quantify aggregation effects (Turner et al. 1989; Peitgen et al. 1992). Because the change in information with change in scale of resolution depends on the aggregation rule, insight provided by these analytical results allows alternative aggregation schemes to be evaluated. Suppose that the resolution of a square grid composed of randomly placed cells of two types, *a* and *b*, will be changed by aggregated 2 × 2 subsets of the grid into a single cell, thus reducing the number of cells in the aggregated grid by one fourth. If the proportions of cells of type *a* and *b* are equal to p and $1.0 - p$, respectively, then the proportion of cells of type *a* in the aggregated map, p', using the majority rule (at least three cells must be of type *a*), will be $p' = p^4 + 4p^3(1.0 - p)$. If a 50 percent rule for aggregation is invoked (at least two cells must be of type *a*), then $p' = p^4 + 4p^3(1.0 - p) + 6p^2(1.0 - p)^2$. Figure 2.3 plots the changes in p' for different values of p for these two aggregation rules. Both rules result in substantial aggregation errors, shifting the proportion of cells of type *a* by as much as 22 percent. However, the values of p that produce the greatest error differs between rules; in both cases, the error is zero when $p = 0.0$ and when $p = 1.0$ (an uninteresting result), but it is also zero at $p = 0.76$ for the majority rule and at $p = 0.24$ for the 50 percent rule.

The process of aggregation is similar to renormalization theory, which has been used to determine the parameters of physical systems at phase transition points (Milne and Johnson 1993). Renormalization theory for large random grids predicts three points where aggregation errors will be zero: at $p = 0.0$, at $p = 1.0$, and at the percolation threshold (Peitgen et al. 1992). Because the bond percolation threshold for a square grid is 0.5 (Stauffer and Aharony 1992), neither the majority rule nor the 50 percent aggregation rule is consistent with renormalization theory. A third rule, which produces zero error at the critical threshold (see figure 2.3), can be formulated as an average of the majority rule and the 50 percent rule. This averaging rule is equivalent to assigning the aggregated cell to type *a* or *b* at random when there are exactly two of each type in the 2 × 2 grid unit, and using the majority rule otherwise. Figure 2.3 shows that the averaging rule produces zero error in successive levels of aggregation when p is exactly equal to the critical threshold. Although the majority and 50 percent rules may be more useful when p is near 0.76 and 0.24, respectively, the maximum error produced will be a factor of two larger than the averaging rule (see figure 2.3). If p is not exactly equal to these points, then successive levels of aggregation will result in increasing levels of error (Turner et al. 1989).

The real world is far more complex than a simple two-phase random map. However, the results derived from this simple case provide a reasonable estimate of the expected errors when complex spatial data are aggregated. Analysis of actual landscape data by Turner et al. (1989)

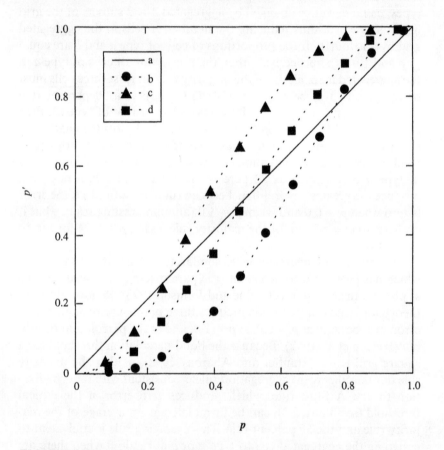

FIGURE 2.3

The fraction of occupied sites, p', that result from the aggregation of a square grid with the fraction of randomly occupied sites equal to p. Line (a) represents a hypothetical aggregation rule with no error; line (b) represents the majority rule; line (c) represents the 50 percent rule; line (d) is the "averaging rule." See text for further explanation of the equations used to generate these lines.

shows that rare land-cover types ($P < 0.5$) are rapidly lost as successive aggregations are performed. However, land-cover types that are clumped disappear slowly with increasing grain, whereas cover types that are dispersed are lost most rapidly. The "random aggregation" rule (Costanza and Maxwell 1994), which assigns the land-cover type of the aggregated cell at random to one of the four cells comprising the grid unit to be aggregated, preserves the fraction of rare land cover with increasing grain size. Mathematically combining terms for the random aggregation rule gives $p' = p^4 + 3p^3(1.0 - p) + 3p^2(1.0 - p) + p(1.0 - p)^3$. This rule is globally accurate because it preserves the value of p of all land-cover types, even over successive aggregations. Local errors (the accuracy in the representation of the amount and physical appearance of dominant land-cover types at fine scales) will increase, and the effect of the random aggregation rule on these errors has not been thoroughly investigated.

Recently, Sandra Lavorel, Roy Plotnick, and I have been experimenting with neutral models to determine if measures of scale correspond to the known frequencies of pattern used to generate maps of land cover. Preliminary results (Lavorel et al., unpublished data) indicate that methods quickly deteriorate when P, the proportion of the map occupied by the land-cover of interest, is high or low. The reason appears to be that the patterns of maps with low and high values of P are inadequately sampled unless the maps are extremely large (i.e., the extent is greater than or equal to 1000 grid sites). Maps with low values of P provide a sparse representation of the pattern. Although maps with high values of P are very dense, the pattern is actually determined by the gaps, and these are, of course, very sparse when P is high. The algorithms for identifying scale-dependent changes in pattern reviewed here are empirical estimates that do not reveal cause-and-effect relationships, so the possibility of spurious results is always possible. However, if predictions from models that are a parsimonious representation of the actual system demonstrate scale-dependent patterns similar to those observed for the actual system, then our confidence in the cause and reality of these patterns is greatly increased (Levin 1992).

The Practical Importance of Scale Detection

The identification of changes in pattern with scale—and an understanding of the ecological processes that effect these changes and are, in turn, affected by them—is of considerable importance for practical and theoretical reasons. It is evident that identifying and understanding scale-dependent changes in pattern and process must be a prerequisite for predicting the consequences of changes in ecological systems induced by natural disturbances and human alterations of the environment. Therefore, it is likely that scale-identification techniques will be increasingly employed as a means of observing and measuring natural and disturbed systems, designing and analyzing experiments, and setting management objectives and policies.

The practical importance of the systematic application of scale-identification methods includes the following:

1. Early detection of the spatial and temporal scales at which ecological changes occur is essential for the design of management strategies to monitor changes and minimize undesirable effects.

2. Recognition of the importance of predicting across scales, and familiarity with the techniques used to identify scale-dependent change, will encourage sampling designs that are sufficient to measure shifts in patterns across spatial and temporal scales.

3. Experiments testing effects of alternative management practices may be suggested by scale-detection techniques. For instance, management schemes designed to mimic natural disturbances may never capture the natural variation in frequency, intensity, and extent of disturbances. The significance of differences between the natural and managed system, and the temporal and spatial scales at which these differences become significant, can best be quantified by scale-detection techniques.

4. The critical test of our understanding of complex systems is our ability to make reliable predictions. The systematic comparison of model predictions against empirical results can be effected by scale-detection methods. Such comparisons will identify the specific temporal and spatial scales for which predictions differ significantly from observed dynamics.

Summary

The identification of appropriate scales for analysis and prediction remains an interesting and challenging problem. Although the factors producing scale-dependent patterns may not be clearly understood, we still require accurate and reliable descriptions of scale-dependent patterns and processes to design data sampling procedures and test the accuracy and reliability of methods of prediction. If we are to properly measure, understand, and predict, a number of issues must be addressed, including:

1. It is now recognized that the spatial and temporal scales set for each new research project will result in the observation of different, and perhaps unique, ecological patterns (Dayton et al. 1992). This result is acceptable if the a priori definition of scales is based on known relationships between pattern and process, and if the assumptions associated with the sampling procedures are taken into account when comparing results with other research projects.

2. Although a shift to mathematically complex sampling by the gliding-box method deviates significantly from more traditional ecological sampling, the superiority of this method for detecting scale-dependent patterns is evident. This is not surprising when one realizes that the gliding box is the quadrat complement of lag-distance sampling used in parametric statistics to estimate semivariances and autocorrelations.

3. It has always been difficult to obtain sufficient data to adequately characterize scale dependencies, because this requires the simultaneous estimation of relevant physical and biological variables, as well as data for describing the resulting patterns. Difficult as this information may be to obtain, increasing efforts need to be made if cause-and-effect relationships are to be revealed. The oceanic sciences have had many successes in this regard (Steele 1989b), resulting in the characterization of biotic changes as a result of the frequency and severity of changes in the physical environment. Empirical descriptions of the distribution of natural resources and biotic processes are not sufficient for predicting resulting patterns, because spurious correlations will always exist in any single data set, and the problem of pseudoreplication may mask true relationships. In addition, interactions and feedbacks between physical and biotic components of ecosystems often dominate patterns masking underlying relationships, thereby introducing strong temporal correlations for nearly all spatial patterns.

4. A number of research needs are obvious, including the testing of models against data to determine if correspondence in patterns is produced. Several methods, such as cross-correlation analysis or bootstrapping techniques, lend themselves to the statistical testing of results to verify patterns. Such information would provide strong support that potential cause-and-effect relationships have been adequately understood and described. Systematic comparisons, such as that begun by Hurlbert (1990), need to be performed for a variety of scale-detection techniques with spatial data generated with known patterns [see Gardner and O'Neill (1991) for a discussion of neutral models of spatial patterns].

Until we develop clear and reliable rules for spatial analysis and verify our ability to extrapolate temporal and spatial data across scales, we will be limited in our ability to predict to the arbitrary boundaries set by current research projects. Because we know that patterns are not uniform beyond the extent of our data (Wiens 1989), the broader implications of ecological studies must be extended beyond current limits.

Acknowledgments

I appreciate the stimulating discussions and helpful suggestions made by Tom Burns, Deborah Hart, Tony King, Jeff Klopatek, Sandra Lavorel, Bruce Milne, Bob O'Neill, Scott Pearson, Roy Plotnick, William Romme, Kenneth Rose, and Monica Turner concerning a variety of scale-related issues.

This research was supported in part by funding from the U.S. Environmental Protection Agency under contract R819640 to the Center for Environmental and Estuarine Studies, University of Maryland System, and by a grant from the National Science Foundation (EAR-9506606). This is publication number 2750 of the Appalachian Environmental Laboratory, Center for Environmental and Estuarine Studies, University of Maryland System.

3

THE LANDSCAPE "LEVEL" IS DEAD: PERSUADING THE FAMILY TO TAKE IT OFF THE RESPIRATOR

T. F. H. Allen

Fear not! Despite the title of this chapter, landscapes, landscape ecology, and spatially based ecology are all alive and well. The sickly members of the family are the specific term *landscape level* and its more moribund cousin *landscape scale*.

The essential flaw in *landscape level* as a technical term is the confusion created by mixing scalar relationships with relationships that derive from definitions. (Throughout this chapter, the word *scalar* is used as an adjective to mean pertaining to scale, not as the technical noun that appears in matrix algebra.) Levels of organization are definitional as opposed to scalar. Scalar relationships have a certain necessity, in that bigger entities must be physically contextual to smaller entities, no matter what decisions or definitions are employed by the observer. Relationships that derive from definitions come solely from assertions of the observer, and they follow from the material system only to the extent of the observed criteria raised in the definition. For example, most definitions of biological populations assert that populations are composed of organisms. There is nothing necessary about organisms in aggregate being called populations; it is merely what ecologists generally agree to mean when they use the word *population*. For example, a slime mold is a collection of organisms that is not generally addressed in population terms. It is also an arbitrary assertion that terrestrial ecosystems occur on landscapes that give ecosystems their context, so that landscapes are the level above ecosystems. One could also argue that the ground is only the soil and the geological component of eco-systems, putting landscapes inside ecosystems at a lower level. Neither is correct, because each relationship is the choice of the person making one assertion as opposed to another. In contrast, if the size of a pond skater is increased by a factor of 10, the insect will sink because it has a new relationship to the surface tension of water. There is nothing to be done about scalar relationships: one must accept them as they are, whereas definitional relationships come from what one chooses the definition to be.

Although the examples above distinguish between scalar and defi-nitional relationships, the confusion between these different relation-ships is a knotty problem in the new area of complex systems (Ahl and Allen 1996). The difficulty arises as specialists in an established area of ecological investigation, whose strategy is reductionist simplification, enter the realm of complexity where a new class of difficulties over-whelm the standard approach. The study of complex systems in ecol-ogy is a more general area of discourse. It is distinct enough to have its own specialists, some of whom are authors in this book, which, as a whole, addresses complexity as it arises from matters of scale.

This chapter has two objectives: (1) to clarify the issue of relation-ships between levels in spatial ecology and (2) to use landscapes as a means for advancing arguments about scaling in complex systems and the role of models in ecology. Complexity presses itself on all modern ecologists who wish to put their models and material systems formally in context, as well as to reduce their systems to a set of lower-level explanatory principles. The landscape-specific purpose of this chapter should interest landscape specialists, as well as those who have avoided embracing landscape ecology because they intuited a muddle underly-ing that specialist area. The more general intention of this chapter is to advance the central topic of this entire book—the issue of scale as it arises in ecology.

Our conclusion is that *landscape level* and *landscape scale* should defer to a terminology that avoids those phrases in favor of scale-specific and scale-independent usages. The distinction between types of things and the scale of things is fundamental. Scale-dependence and scale-independence are muddled in contemporary ecological parlance, so a revised vocabulary to clarify the issues surrounding the landscape approach is required. A discussion of landscapes explicitly will be pre-sented later, but first an extended discussion of entities and levels is in order.

Entities and Levels

Ecologists deal with things. The philosophy used here asserts that things may exist in the external world, but not as things. The attribution of "thingness" comes from the observer. Behind things are models that assign the thing in question to a type. A type implies relationships between parts that can be generalized across all exemplars of the type. In observation, a type is used to distinguish the whole (with boundaries) in the foreground from the background. A type is the decision of the observer, a decision to categorize to which class of entity a particular thing belongs (Allen et al. 1987). A type is a tool that often helps an observer in recognizing things, but it is important to note that the type exists in the mind of the observer even before the observation.

Helpful as types are, they are only part of the recognition of things in observations. Beyond the type to which a thing belongs, there are also matters of spatiotemporal scale. Everyone agrees on what constitutes one centimeter, and that scale, like type, exists before an observation is made. However, unlike type, scale becomes meaningful with regard to things only when it is operationalized in its use in a measure in the act of observation. Type is rich and fully formed before observation, whereas scale is tied more tightly to the act of observing. When one sees an entity, it is assigned to a type that existed in the mind of the observer before the observation, but its scalar properties are tied directly to the act of observing whatever in particular is observed. The type of an entity—such as organism—is independent of scale, whereas observed entities—such as a particular organism—belong to a type as well as exist at a particular scale. The theory of scaling that emerges from hierarchy theory emphasizes the need to distinguish between the spatiotemporal scale of a structure and its type; the scale that reflects the material dynamics of the system should not be confused with the arbitrary criterion for observation embodied in definitions, expectations, and recognition (Allen and Hoekstra 1992).

The distinction between prescribed, definitional things that are scale-independent, as opposed to observed scale-based things, translates into the need to distinguish between type-based levels of organization and scale-based levels of observation (Ahl and Allen 1996). The following illustrates the distinction between an entity and its level. A level is a class to which entities belong, but not all classes are levels. A class becomes the special case of a level when there are ordered relationships among classes. Scale need not be involved here, because the ordering that makes a set of classes a set of levels can take many forms, only some of which are relevant to issues of scale. For instance, in relation to entities at lower levels, entities that belong to upper levels may (1) be the context of, (2) be the controller of, (3) be the constraint of, (4) behave slower than, (5) contain and so consist of, (6) have less integrity than, or (7) exist at a larger scale than lower-level entities. These ordering principles can be applied one at a time, or in combination, depending on the purpose of the investigator and the material system at hand. For the most part, these ordering principles can be related one to another (Ahl and Allen 1996), but such an exploration is beyond the scope of this chapter.

Beyond the ordering principles given above, it is essential to divide levels into two major groups, one group based on typology and the other dependent on some manifestation of scale. Type-based levels are called levels of organization, and they are a matter of definition. If there is a relationship between levels of organization, then that relationship is asserted within the definitions of the respective levels. There is nothing coming from the material world that indicates anything necessary about the relationships between levels of organization (Allen and Hoekstra 1992). In such definitional levels, the one who states the definition for a given level of organization must take responsibility for its fixed characteristics. Levels of organization may be posited based on past experience, but the observer is still responsible for choosing the characteristics that make an entity a member of a level (Ahl and Allen 1996).

An example might be helpful at this point. There may be evolutionary reasons for organizing mammals into entities that meet the generally accepted criteria for being an organism. However, the concept of organism is still a human device for dealing with experience, not a necessity of nature. For instance, plants commonly fail to fit unambiguously inside most definitions of organism. Plants regularly fail to meet the organismal criteria of physical discreteness and genetic identity, with their runners, stolons, chimeras, and natural root grafts. Hoekstra et al. (1991) have argued at length that the organism is a deeply anthropomorphic concept. The most convincingly organismal entities are suspiciously closely related to humans. Note that the charge of deep anthropocentrism is directed at the most easily argued and tangible biological level of organization—the organism. Therefore, it will be even more difficult to consider other more obviously abstract levels of organization, such as the ecological community, as natural necessities.

Type-based levels of organization contrast with scale-based levels, which Ahl and Allen (1996) call levels of observation. The scale of an observation is determined by the observation protocol, and that is entirely a matter of subjective choice. However, if one changes the observation protocol, one also changes the spatiotemporal scale of the observed system and its necessary physical properties. If the spatiotemporal scale is enlarged, then one might observe a *larger* organism. In a larger organism, certain necessities emerge that are above and beyond the subjectivity of the observer. It would have been very convenient for Hannibal if his elephants could have skated like insects across the Mediterranean to Rome. However, there was nothing he could do to achieve pond-skating pachyderms, even if he had conceived such an idea.

There are necessary relationships between mass of an organism and the surface tension of water that have nothing to do with human decisions. When one deals with a larger organism, the size and connectivity of the molecules of water are constant, making the relationship of water to the larger organism necessarily different from the relationship of water to a smaller organism, such as a pond skater. The level of organization here, the organism, has not changed, because insects and elephants both fit the bill. However, there are important differences in scale, with material consequences. Level of organization has not changed, but the scale, identified by the changes in scale required to observe whole elephants, has changed radically. Therefore, we need to separate levels of organization related to type, from spatiotemporal levels of observation.

Levels of observation are tied to necessary conditions of the material system in a manner that cannot apply to definition-dependent levels of organization. The fluxes that are part of and surround biological systems are a matter of scaled material flows; they are captured in levels of observation. Levels of observation expose relationships that are presumed to relate to the observed material system above and beyond observer decisions. One might need to invoke levels of organization to define the entities used in a level of observation, but the two types of level (levels of organization and levels of observation) are different and have distinctive properties and consequences (Ahl and Allen 1996).

In summary, higher levels of observation are materially larger, whereas levels of organization cannot be assigned to any particular spatiotemporal size. Some of the worst confusion in ecological discussion arises when scale-based and type-based levels are mixed, precisely the point this chapter discusses. Biological levels of organization are not intrinsically scalar, but the material flows between and around particular biological entities *are* scalar. The scaled characters of a particular organism can be demonstrated from bacteria to dinosaurs, but the scale of organism as a level of organization is meaningless, even in relation to other levels of organization, such as the generalized notion of population or landscape. If organism as a class has no meaningful scale, then *landscape scale,* as an unqualified extension of landscape level, is meaningless too.

It is not the objective of this chapter to force a particular definition of a landscape on everyone; choose your own definition, if you wish. There is a more fundamental concern, namely that there is a serious problem with assuming that scale is covered by an assignment to type. Of course, one could assign a scale as part of the definition of a type. Allen and Hoekstra (1992) specifically identify that to be a restriction that offers little in return. Even so, the term *biome* is sometimes meant to apply only to large entities. Usually, scale is not mentioned at all in the definition of the type-based level of organization, and, if it is mentioned, general descriptors such as *big* are employed. There may be limits to the size of a type belonging to a certain level of organization, but that is different from assigning the level a particular scale used to describe the size of its occupants. Returning to the organism level of organization, the range of scales associated with that particular level lies between the size of the smallest virus and the size of the largest clonal plants. At the largest size, clonal plants win by mass, but fungi win by area occupied by a genotype presumed to arise from a single spore. If bacteria are the lowest form of organism, then the lower end of the range is increased to around a micron. From a *Populus tremuloides* (quaking aspen) clone to bacterium embodies a wide scalar range, thus denying the organism concept a particular scale (figure 3.1). As they are used in ecology, definitional levels of organization are independent of particular scales.

The independence of levels of organization from scale is with regard to particular assertions of size a priori. There may be important relative scaling considerations inside the definition of a type, but that is a different matter. For instance, a parasite is usually smaller and shorter-lived than its host, but that is inherent to the definition of the class parasite. The distinction between levels of organization and levels of observation pertain to the way the levels are ordered one to another. Only levels of observation are ordered by explicit scaling, which derives from the values of particular grains and extents in observation protocols and the values of particular measurements. Levels of organization are not explicitly scaled, although levels may be related by general reference to relative size. For example, populations contain and consist of entities from the organism level of organization, implying that the population is, by definition, bigger than organism. But note that there is no absolute statement of size, and organisms and populations occur across a wide range of overlapping sizes. In levels of observation, the order is fixed by the scale in all cases, with no apparent

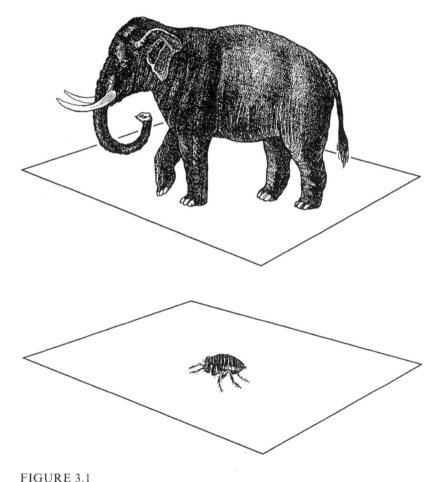

FIGURE 3.1

Both the insect and the elephant are organisms, so the class of things called organisms cannot be defined as existing at a particular scale.

From *Toward a Unified Ecology* by Timothy F. H. Allen and Thomas W. Hoekstra (1992). Copyright © 1995 by Columbia University Press. Reprinted with permission of the publisher.

contradictions such as some organisms being larger than some populations. Between levels of observation, higher always means bigger.

The Character of the Landscape Concept

Having made the distinction between level of organization and level of
observation, and having identified where and the degree to which scale
applies to levels, landscape can be addressed as a concept with scalar
and typological implications. A good place to start is to contrast usage
of landscape with that of the common vernacular. Because the term
landscape level is used in ecology, we need to identify the manner in
which landscapes are a level. In the scheme described that separates
levels of organization from levels of observation, the landscape level is
a level of organization, a type of organization, not a scalar concept.
Like the organism type of organization, landscape is a matter of definition. The organization of the parts of a landscape is usually defined
by spatial proximity that gives patterns in spatial terms. In fact, spatial
contiguity as a statement of proximity is nearly the only character that
is universal to all except the most fanciful definitions of landscape.

Having identified the essential spatial character of landscapes, it is
possible to identify what is level-like about the landscape concept. In
the hierarchy of life from cell to biosphere, landscapes are usually
placed above ecosystems and below biomes on the basis of relative
scale. All the levels in the hierarchy of life, as it occurs in textbooks,
are levels of organization, not scalar levels of observation. Of course,
much as particular organisms are a particular size, so are particular
ecosystems, biomes, and populations. But that does not make any of
those levels scalar, any more than a two-meter-high, one-ton elephant
makes organism a scalar concept. The interlocking definitions that are
at least implied in the hierarchy of life are pure convention, and they
are in no way necessary. According to the general definition employed
here, a mite on a leaf is on a landscape, and so is a *Sequoia sempervir-
ens* (coast redwood) tree on a mountainside (Allen and Hoekstra
1992). The concept of landscape could even apply to the red spot on
Jupiter, a landscape that dwarfs our entire planet. So why, one must
ask, are ecologists so insistent on using "the landscape level" to indicate a certain scale of investigation? Perhaps they mean something that
corresponds roughly to a Constable landscape painting, an area less
than a county, but more than a small field.

The sloppy use of the term *level* is by no means restricted to just the landscape level. In almost any ecological journal, *level* is tossed gratuitously into the middle of sentences where it has no particular meaning. R. V. O'Neill notes that a common usage of *level* is as a word tacked on behind some noun, when actually the noun stands alone, with no particular levelar implications. There is no need to embarrass any particular authors by making explicit citations; just look in the journals. So why do so many authors use the word *level* gratuitously? As ecological problems move up-scale, there is a need to be expansive in one's view, and in how one asks questions. This emphasis on being more inclusive and outward looking may well be responsible for the modern use of the word *level*. Tacking the word *level* after *ecosystem,* when the author means only ecosystem, not ecosystem level, is probably a flag that says, "No, really. I am thinking in expansive terms: what I am talking about is generally related to the rest of ecology, and particularly to large-scale systems such as the biosphere."

In the case of landscapes, there appears to be a strong inclination to stick *level* behind many uses of the word *landscape*. The first reason that *landscape level* remains common parlance is that landscapes imply a certain point of view. We mean point of view literally here, one that makes landscapes much larger than the observer. The ideal place from which to consider a landscape is with an oblique view from above. The ideal nature of this perspective has little to do with the material system of landscapes. Rather, it is a reflection of the proclivities that come with being human, Western cultural humans in particular. Thus, the archetypical landscape is what can be seen from a comfortable vantage place. There may be some biological basis to humans favoring safe lookouts. For example, the Canadian police know to seek for young children lost in the wilderness at lookouts where there is shelter. Without being told to go there, the child comes across the place and stays, more comfortable in such a setting than anywhere else. More anecdotal, but still convincing, is the ideal nature of the Memorial Union Terrace at the University of Wisconsin in Madison. Attendees of the 1993 Ecological Society of America Meeting will remember the comfortable, contrived savanna view through trees down the steps with the Union building to our backs. The Terrace is a very appealing place for primates to watch primates (figure 3.2).

Another reason that ecologists prefer to not let go of the landscape level as a scalar descriptor is that the landscape is the default setting for the thing surrounding an ecological entity. R. V. O'Neill points out that when ecologists of earlier days used the term *environment,* and even when most modern ecologists use it, what they mean is *landscape setting.* Intuition, not logic or an explicit protocol, makes landscape something scaled relative to the observer and other organisms. Intuitively, a landscape should be bigger than the observer, and the context of other organisms and their processes. Ecosystems, the process-focused view of ecological systems, are seen as somehow in the context of a landscape. If the area is so large that the organisms in question do not have a chance to experience most of it, then ecologists usually give the area a different name, perhaps a biome. Note that none of these relationships with other levels of organization, such as ecosystem or biome, are an explicit part of the definition of a landscape. This is in contrast to the interlocking definitions of population and organism in which the relationship between levels is explicit.

In the case of landscape, a general type of thing is given a scale, merely to satisfy intuitive feelings that arise capriciously from the stature of the average person. Intuition is not a good enough reason for raising the confusion that always occurs when a generalized type, a level of organization, is asserted to belong to a scalar level of observation. In failing to distinguish between level of organization and level of observation, conception and observation get confused. At best, it puts a straitjacket on what could be a more flexible concept. At worst,

FIGURE 3.2
The Memorial Union Terrace at the University of Wisconsin-Madison. The ideal view of a landscape is obliquely from just above. Photo by T. F. H. Allen.

it opens the door to confusion when meaningless phrases like *landscape scale* appear in the literature without qualification, leaving the reader to guess what scale the writer means.

What Is a Model?

The benefits that come with avoiding the landscape scale pertain to an increased generality to the concept of landscape. Landscape scale robs the concept of landscape of the general properties of being a model, the stock-in-trade of science. Scientific models derive their power from keeping all significant aspects of the system constant, except scale (Rosen 1991). Thus, imposing a particular spatial scale on the concept of landscape denies the possibility for change in scale that occurs when the ecologist uses the model. A scale-locked model may still be a model, but it is a far less powerful model.

The freedom to move across scales is clearly part of using model airplanes and boats. However, constancy of relationships between parts across scales also applies to the *Pan troglodytes* (chimpanzee) model for AIDS or the rat models for drug addiction, to pick a couple of less obviously scalar models. All significant aspects of the system are kept constant in the change in scale between the model and that which is modeled. The critical phrase here is "all *significant* aspects of the system." Thus, the model and that for which it is a model are not the same in every regard; they are the same only for certain critical characteristics that are acceptable for the purposes of prediction and the question at hand. A model boat does not have the same relationship to the surface tension of water as does a full-sized yacht, but so long as the differences do not cause the model to behave differently from the full-sized boat with regard to sailing characteristics that concern the designer, then the model is a good model. Although rats have a tail and humans have only a coccyx, the difference is only scalar, and this is unimportant when using rat behavior to make predictions about human drug addiction (Allen et al. 1993). The benefit from keeping an unspoken scalar limitation out of landscape level is that it allows the concept of landscape to be a more powerful model.

Models defined in this way have wonderful properties of prediction, and when they fail, they do so in ways that highlight the anomaly. In physics, but less so in ecology, anomalies are the device whereby the scientist learns. Rosen (1991) uses the notion of entailment to describe the model as well as the material system. The entailment in the material system is material causality. The entailment in the formal model of the material system is a logical consequence. A model is based on the assertion that, "If the world works in such and such a fashion, then the consequence of this (whatever *this* is) will be that" (figure 3.3).

A wrong model would be one in which the internal logic is flawed. The more usual notion of rightness and wrongness of a model does not relate to internal inconsistencies, but to how the model relates to the material system. No model that is other than a replica of the material system itself can be a complete representation of the material system for all situations. Therefore, all models are wrong in the sense of wrong being different from observation. Clearly, that sense of right and wrong has no power, so rightness and wrongness of a model that is logically consistent are beside the point.

The point of a model is the investigation of its assumptions. All assumptions are false at some level, so the correctness of assumptions is again beside the point. Science is not about finding the answers to

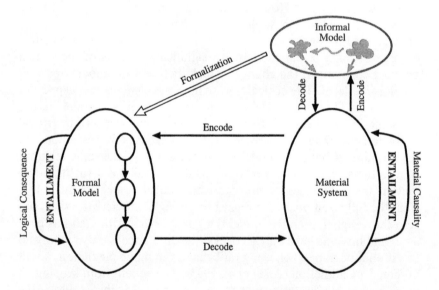

FIGURE 3.3

The material system appears coherent because it is entailed by material causality. The formal model is coherent through entailment by logical sequences. In ecology, we appear also to use informal models that are eventually formalized for testing and experimentation. The material system maps to the model through encoding. The formal model maps to the material system by decoding.

After *Life Itself* by R. Rosen. Copyright © 1991 by Columbia University Press, with permission of the publisher.

everything; it is about finding the situations in which one can get away with an assumption—and still have the model give the desired behavior. Because one cannot control everything, one wants to know what one must control to get by. Thus, the philosophically sophisticated scientist is not interested in the model being correct but is keenly interested in where and the extent to which the model maps to the material system. Does the model map widely enough to the material system in the area in which one is interested in predicting the observed behavior of the material system? The answer to that question is what matters.

If one needs to be further convinced on this point, consider one of the central models of physics. There is no such physical thing as a frictionless pendulum, but such pendula are as useful as any model in all of science. The insights come from the manner in which the material system does and does not map to the impossible frictionless model system. This verity in physics makes a nonsense of ecological statements such as "Lotka-Volterra and the monoclimax theory of Clements are wrong." No they are not, for they are both logically, internally consistent, and both capture some aspects of the material system. Even if they did not capture any aspect of the material system, they would still not be wrong. In that case, the models might not be so useful, except to say that one cannot afford to make that particular set of assumptions, if one cares about the outcome. Models that do not map at all to the material system generally do not apply.

Another feature of models, cast in Rosen's terms, is the way that model systems can become models of each other. Consider a formal model such as the laws of allometry in animals. They are remarkable in the way that the mass of so many animals can be placed on a straight line with respect to heart rate, running speed, metabolic rate, and many other features (figure 3.4). The definitive works are from Schmidt-Nielsen (1972, 1984), and the fit is usually spectacular.

The trick to getting such generalized predictions in biology is to employ a formal model, such as allometry, such that an exemplar can be cast in the terms of the model. The material system must be translatable both to and from the formal model. In Rosen's terminology, the material system can be encoded to and decoded from the model. If one can encode and decode another material system into and out of the same formal model, then the two material systems become models of each other. Both mice and dinosaurs can be encoded into and decoded out of an allometric formal model. The formal model relates body mass to factors such as the cross-sectional area of the legs and musculature that must bear that mass. In terms that are stated in the formal model, the mouse and the dinosaur can be models of each other, or of any other creature that can be similarly encoded and decoded. The formal model allows for continuous rescaling from one animal into another (figure 3.5). This is the reason that one wants a minimum of scaling fixed inside the definition of landscapes; if scale is fixed in some capricious fashion, then all landscape-like things that are not equivalently scaled to the scaling in the definition cannot be encoded and decoded with respect to the model implied in the term *landscape;* therefore, we cannot use them as analogs.

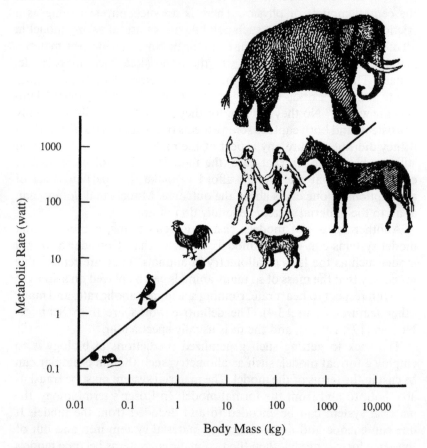

FIGURE 3.4

Allometric equations allow a simple logarithmic relationship to capture many of the differences in animals ordered by mass.

From *Toward a Unified Ecology* by T. F. H. Allen and T. W. Hoekstra. Copyright © 1992 by Columbia University Press. Reprinted with permission of the publisher.

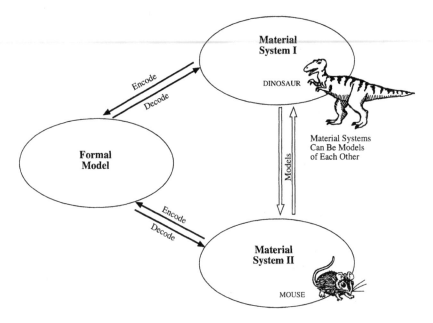

FIGURE 3.5

Two material systems are encoded into and decoded from the formal model. If such decoding is possible, then the two material systems become models of each other.

Much can be learned from things that cannot be encoded and decoded into and out of a formal model. Consider again a mouse and dinosaur. This time the formal model contains specification with respect to egg laying and the number of bones in the lower jaw. Mice do not lay eggs but use a placenta. Dinosaurs have three bones in their lower jaws, unlike mammals. With respect to this new and more demanding formal model, either the mouse or the dinosaur will fail to survive encoding and decoding, and so we have uncovered a critical instability in the application of the formal model. The instability emerges when a continuous scale change appears to give discontinuous behavior. The instability, in this case, is the separation of the mammal line from the dinosaur line; the discontinuity is the absence of individuals that are intermediate between mammals and reptiles after the separation. In a formal model that identifies that separation, even a mastodon cannot be a model for a dinosaur. Apparently out of nowhere, a new set of constraints involving nurture of young has emerged. These constraints also line up with constraints in the formation of the lower jaw, and with all the other constraints that separate dinosaurs from mammals.

Even when a model breaks down in part of the variable space, there is continuous behavior in other parts of the space of the formal model (figure 3.6). Systems that map to those areas can be treated as models of each other. However, if the model is stretched by employing extreme changes of scale, one enters the area of instability, a folded region of the model's behavior that gives discontinuous behavior. If one wishes to relate systems scaled so differently that they exist on different sides of the folded region of instability, then one must identify the critical features of the discontinuity and normalize around those critical features (Rosen 1989).

All this can apply to landscape issues. In more concrete terms, one rescales a landscape process either bigger or faster, and eventually the process will not only cover a larger area at a greater rate, but it will develop qualitatively different characteristics. Consider movement of propagules across a landscape. There comes a point where regular population dispersal becomes continental long-distance dispersal. In other words, movement of propagules across a landscape changes from establishment of new areas within the domain of a metapopulation to movement of a species to a new domain. This is the distinction between the constraints on multiple establishment events at the edge of a species range and constraints of survival at an enormous distance, perhaps on a new continent. These days, expansion to a new continent relates to human transportation, an issue that may be unimportant to movement of genes across a local landscape in the process of regular metapopulation dynamics.

The Consequences for *Landscape Level* as a Term

Organisms, as a concept, have a relationship between levels stated in their definition that places them below populations *of the same species.* This is not strictly scalar, because populations of other species may exist over shorter time frames and be spatially much smaller than organisms of the species in question. On the other hand, landscapes do not have obvious relationships to other biological levels of organization, despite the implicit juxtaposition to ecosystem and biome. Landscape level gives no advantage and appears only to carry a gratuitous

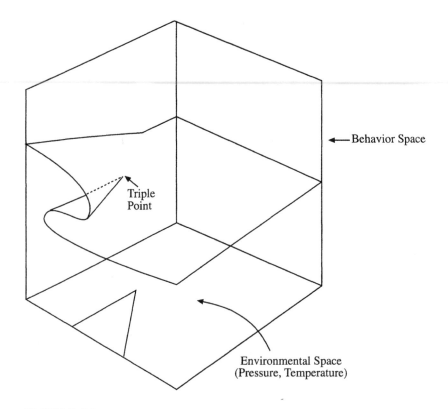

Behavior Space

Triple
Point

Environmental Space
(Pressure, Temperature)

FIGURE 3.6

A formal model gives a continuous surface that relates all systems that can be encoded and decoded into and from it. If one stretches the scale of a model too far, then the constraints that apply to the continuous part of the surface suddenly become irrelevant as a new set of constraints emerge. At that point, the surface derived from the model has folded in an instability. An example is the region in a temperature, pressure, and volume space where a gas changes into a liquid. Suddenly, the constraints on the particles change to those pertinent to the liquid phase, such as noncompression. Stay away from the fold, and a gas is a gas is a gas, and the model holds. Once the fold is crossed, a normalization around the essential character of the instability is required in a new metamodel that relates all gases to each other through all phase changes. In the case of the gas model used by Rosen (1989), the normalization is around the base of the pleat, the triple point of the Van der Waals equation. Pressure, temperature, and volume are expressed for all gases relative to the pressures, temperatures, and volumes at the base of their respective pleats. After such normalization, the instability and its manifestation in the fold disappear, and all gases then map to a series of unfolded sets of corresponding states. In the case of gases, there is only one instability, but in biological systems there are degrees of instability.

Redrawn from *Landscape Ecology*, vol. 3, pp. 207–216, by R. Rosen. Copyright © 1989 by Academic Publishing. Reprinted with permission of the publisher.

restriction that is not explicit enough to be useful. If landscape level, without qualification, is to be useful, it must mean some sort of default setting. It must imply a relationship to other classes that is as generally understood as the relationship between the class "population" and the class "organism." If not, what can the inclusion of *level* in *landscape level* mean? The conventional default setting appears to be the class landscape as a level between the ecosystem level and the biome level. Nevertheless, we have never seen an explanation in the literature of that juxtaposition, although ecosystem to landscape to biome appears often in textbooks as a capricious assertion. Without explanation, the juxtaposition gives little insight. The cost is that unqualified use of landscape level precludes other relationships to other classes such as community, population, and organism.

Allen and Hoekstra (1992) used the term *criterion* instead of level to mean level of organization. There is some advantage to using the term *landscape criterion* instead of *landscape level,* because it allows one to emphasize that landscape is a criterion for observation that has the general properties of a model. The landscape criterion means that one is considering relative spatial relationships on a plane, or a loose equivalent of a plane, as the organizing principle for proximity of parts of the observed system (in contrast, proximate vegetation in species space may well not be physically close on the ground). This definition is purposefully general, to avoid violating the definitional preferences of others. Landscape criterion is an optional definition, but it may be preferred because imposing a covert band of scales on top of those generally planar relationships gives little benefit, and only limits the model. Landscape level, with an underpinning of scale, is like insisting on the white rat model as opposed to the brown rat model, or the big rat model as opposed to the small rat model of human drug addiction. It is not so much incorrect as it sacrifices power, by gratuitously excluding data from the other color or size of rat, to no advantage.

The term *landscape level,* as it is commonly used, imposes unnecessary restrictions on the landscape model, and so it limits the model's ability to relate phenomena at more disparate scales. We need to stretch or shrink notions such as dispersal, migration, and predation across spatial frameworks, so that when they fail, we can learn. The point is to identify what was present, but only implicit, in our former definitions of terms such as dispersal, migration, and predation. Ecologists need to see what happens when we scale the use of such terms outside their normal range of competence. How do the notions of dispersal, migration, predation, and other common terms that pertain to landscape studies require fresh specifications when the landscape is a microscope field or the surface of a whole planet? Science learns through failure, but it learns much faster if the failure is engineered in a systematic fashion. Landscape level and landscape scale stand as obstacles to this systematization of the landscape investigation. It is time to break with stubborn past practices and embrace a more sound theory of landscape ecology.

Rosen (1989) gives a protocol for identifying when a scale change is fundamental and qualitative as opposed to a mere quantitative rescaling. He suggests stretching the model across ever-wider-scaled differences until a discontinuity occurs in the data and the predictions of the formal model. At that point, the change is not just a difference, it represents a dissimilarity, the emergence of a new set of constraints. Rosen explains how to get to the root of the discontinuity or instability, and how to normalize across to connect the two sides of the discontinuity in a coherent fashion. The separate conditions on either side of the discontinuity are linked in a metamodel that emerges from the normalization. Such relationships between models in biology could be discontinuity between species but a certain level of homogeneity in the genus. In ecology, the model-to-metamodel relationship could be discontinuity between short- and mixed-grass prairies, but a certain unity in the concept of prairie in general. Using Rosen's protocol, it could be possible to probe the basis of the conventional wisdom criticized here.

Perhaps we are wrong to assume that the scalar restrictions conventionally imposed on the meaning of *landscape* are only anthropocentrism. There may be critical relationships of a spatial sort that are indeed locked inside the band of scales that are intuitively asserted as being the landscape scale. If that is the case, then there should be critical instabilities between types of relationships at the conventional scale and types of relationship at scales that fall outside the realm of a landscape painting. Something can be clearly anthropogenic and human in its character but may still have a relationship to more general material system considerations. Rosen's protocol should allow us to see if predation, invasion, competition for space, and many other biological processes seen in spatial terms have a qualitatively different character at the scale of a leaf or that of a continent. If so, then the prevailing view that *landscape* should mean landscapes at only a human scale of occupancy is validated. However, that is all in the future, because at this time we have no particular reason to suppose that the band of scales that apply to the vernacular usage of landscape are special and distinctive. The scale-limited meaning of landscape would appear gratuitously restrictive. If there is a material basis to conventional wisdom, then it needs and deserves to be better articulated. Mere assertions that landscape should mean only a certain range of scales will not suffice. "Because we landscape ecologists say so" is an unconvincing reason that a landscape should range only from a small field to the size of a county.

4

A HIERARCHICAL VIEW OF HABITAT AND ITS
RELATIONSHIP TO SPECIES ABUNDANCE

Jurek Kolasa and Nigel Waltho

Ecologists often study communities by sampling and analyzing richness and abundance of species grouped by evolutionary taxonomy or
appearance. Abundances of such species usually show two general patterns (figure 4.1). One relates to the shape of rank-abundance curves
and the other to the relationship between the ecological range and
abundance of species. The rank-abundance curves are generally concave or hollow in shape when plotted on the arithmetic scale (Magurran 1988; Tokeshi 1993). Such curves have been shown to fit a variety
of related statistical distributions (May 1975; Sugihara 1980; Magurran 1988; Tokeshi 1993) and occasionally to have mysterious properties such as the canonical lognormal form (Preston 1948).

Niche-oriented models (MacArthur 1957; Whittaker 1965; Sugihara
1980; Tokeshi 1990) concentrate on resource subdivision, but they are
weak on other ecological factors known to affect abundances. The resource subdivision approach itself has serious conceptual inconsistencies when it is cast in terms of niche. For example, niche dimensions
describe different resource types, but they cannot capture the resource
amounts (if they did, then the niche of a species would change dramatically from year to year depending on resource supply). In addition,
niche dimensions have no spatial component, which may be crucial to
species density (Leibold 1995; but see Silvert 1994). The focus of much
previous research on fitting particular distributions to rank-abundance
data has been a limited research strategy. Few sound ecological reasons have been offered (MacArthur 1957; Engen 1978; Frontier 1985)
to support the view that the data should show any specific fit at all

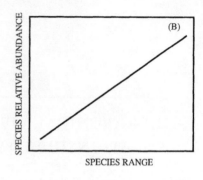

FIGURE 4.1

Species abundance patterns. **(A)** A generic pattern of species abundance distribution from a local or regional collection. **(B)** A typical relationship between species abundance and size of geographical range (as measured by area, number of sites, or latitude).

(Tokeshi 1993). When such reasons were provided (Sugihara 1980), the tests were faulty. For example, Nee et al. (1991) used a geographical distribution to test community level fit of data to the sequential break-age model whose ecological motivation is local or at best regional. However, when ecologists proposed ecological mechanisms under which species would exhibit regular patterns in distribution of abundances, they were successful in predicting quantitative trends (Brown 1984; Hughes 1986) and qualitative characteristics, such as the bimodal distribution of abundances (Hanski 1982), but not in predicting specific fits. The mechanism-focused approach is a better route to pattern detection and interpretation.

Ecological range and abundance have often been found to be positively linked (Hanski 1982; Bock and Ricklefs 1983; Brown 1984; Kolasa 1989; Gaston and Lawton 1990; Maurer 1990; Collins and Glenn 1991; Lawton et al. 1994). This means that habitat generalists are locally more abundant than habitat specialists (Kolasa and Drake 1997). However, sometimes researchers do not state clearly whether they mean a total count of a species in a collection or its density only in the habitats it occupies, or its density over the area of interest, including empty sites. The total observed abundance of habitat generalists, at the same density, will almost inevitably be higher because they occupy more habitat space. However, it is not obvious why the density of a habitat generalist should be higher than that of a specialist. It is thus useful to make a distinction between abundance and density (often termed local abundance), and we will interpret the pattern with a focus only on density. Gaston's (1994) considerations of rarity and scale are instructive in this context.

According to Hanski et al. (1993), there are three explanations of the range–abundance relationship. First, the sampling model attributes the relationship to the underestimation of distribution of rare species. This model is essentially correct in that it has a strong rationale and a detectable effect. It may constitute an adequate explanation in some situations in which the sampling methodology is exceptionally weak, but it is insufficient in other situations (see later). Second, the ecological specialization model (Brown 1984) assumes that species density follows a bell-shaped form over its geographical range. If the form is kept constant, the smaller ranges will produce lower peaks (corresponding to densities) in the respective curves (figure 4.2a). The area under the curve, which represents total abundance, will also be smaller.

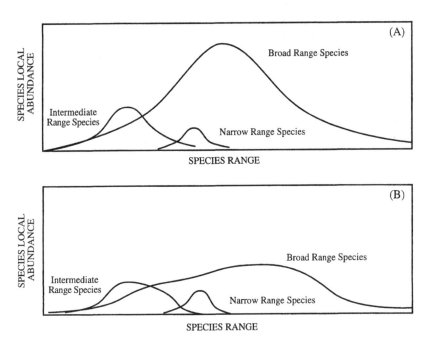

FIGURE 4.2
Species density over a geographical range. **(A)** According to Brown's model of specialization. **(B)** A more realistic form with hypothetical plateaus.

For this model to work, one must assume the same general shape for the bell curve for most species under consideration. If one sampled a pool of similar bell-shaped curves, those with broader bases would, on average, have higher peaks and higher mean densities. The assumption of curve similarity lacks an ecological justification. In contrast, curves may display a variety of shapes, with distinct plateau regions where densities would neither increase nor decrease significantly (see figure 4.2b). Furthermore, actual data do not show that widespread species have enormous densities in the center of their geographical range. The value of the Brown model lies in its emphasis on species properties (see later). Third, the pattern of abundance could be produced by metapopulation dynamics (Hanski 1982; Hanski et al. 1993). Occupancy of separate patches is linked to mean population density via probabilities of immigration and extinction. A net result is that species of limited distribution also maintain low population density.

This chapter reconciles the above propositions in one scale-independent model. We explore a conceptual approach to the problem of community organization. This approach appears (1) to solve the problem of explanation of species abundance and related patterns, and (2) to emphasize integration of previously unrelated phenomena and processes. The approach builds on previous attempts to deal with these patterns. Specifically, it combines the attributes of and differentiation among species, as well as the attributes and structure of the habitat in which they live. In other words, it combines the perspectives of classical community ecology, island biogeography, patch dynamics, meta-community, and landscape ecology. The rest of the chapter discusses this specific approach expressed in the form of a conceptual model. Although we elaborate the foundations of the model in greater detail than in previous papers, we also emphasize the operational aspects. In particular, we examine various field and analytical situations that ecologists are likely to encounter, and we provide suggestions and examples relevant to implementing the model.

Habitat-Based Model: General Approach

Unlike the niche-apportionment models, the habitat-based model (HBM) treats species and their habitat as separate variables. This approach is designed to be consistent with the logic of evolution, in which the interaction of the phenotypic pool of individuals with the environment determines their success in the next generation. In the HBM, by analogy, the anticipated patterns of abundance are viewed as the product of the interaction between species attributes and habitat attributes such that

$$\text{Abundance}_i = k \; (\text{specialization}_i \times \; \text{habitat}_i)$$

where k is a species–habitat adjustment factor that may take into account such factors as habitat quality. Specifically, the habitat of species i is expressed as a fraction of the parameter range with which the entire set of species interacts. Specialization is expressed as a ratio of ecological range of species i to that of the broadest species in the habitat unit under investigation, and it is calculated in units suitable for a particular system.

The model uses assumptions about (1) species attributes, (2) habitat attributes, and (3) the relation between species and habitat. It incorporates only those assumptions that apply universally to all species and all habitats, making it a general model. These assumptions are discussed next.

Assumptions: Species

Assumptions concerning species attributes relevant to the model interpretation apply only to habitats or regions whose species show one of the patterns described in the introduction. The assumptions are (1) that species differ in their ecological requirements and, thus, there will be some species with narrow and others with broad requirements (specialists and generalists, respectively) and (2) that any habitat may host a set of species representing a range of requirements from narrow to broad. Explanation of the assumptions about species will help avoid ambiguities and misunderstandings associated with textbook stereotypes. Different ecological requirements can imply many things, such as tolerance limits, resource needs, and conditions for successful reproduction and dispersal. For example, presence of certain pollinators may have a critical impact on the performance of a species in a particular habitat, and absence of wind may make patch recolonization difficult. Such a species may end up in few patches and will thus be considered a habitat specialist. Indeed, the terms *specialist* and *generalist* have a variety of meanings in ecology. The meaning here is general but precise: a specialist is a species restricted to a small volume of the multidimensional habitat space. In contrast, a generalist is a species that is not so restricted. Note that a species is evaluated by its relation to the habitat, and it is defined as specialist or generalist relative to other species in the community, which jointly define the multidimensional habitat space (see the next section on habitat). In a different habitat, classification might reverse, although this is not very likely. For these distinctions to be meaningful, one should evaluate species at a habitat scale at which species attributes are likely to make a difference in their performance as populations. At very small scales, immigration or emigration of individuals may cloud the relationships. Because abundance is a function of both habitat and species attributes (table 4.1), this flexibility of specialization concept is a strength rather than a liability.

The concept of specialization used here implies little about ecological efficiency. First, we wish to clarify the meaning in which the term is used. Specialization comes in two logically indistinguishable extremes: evolutionary specialization (genetically constrained) and ecological specialization (constrained only by immediately acting factors). All transitions between the two types are possible. Only the latter type responds dynamically to the change of conditions at a short time scale; this is the one of interest for the model we present. Although specialists may be more or less efficient in using accessible resources, many ecologists believe that specialists are generally more efficient. This advantage does not translate well into abundance patterns in the context of the proposed model, because consequences of habitat fragmentation

TABLE 4.1

Details of interpretation of species attributes

---

1. A species has ecological characteristics that can be expressed in an n-dimensional hyperspace, and when so expressed they will form a species ecological hypervolume.

2. These characteristics include variables permitting the species to exist and variables describing species impact on other components of the environment and other individuals of the same species.

3. Only those variables that define a species' ability to exist in an environment (thus, a portion of the hypervolume) are considered in connection with the habitat concept and approximate (but are not equivalent to) the concept of niche (the disclaimer is necessary primarily because of the lack of precise niche definition). Amount of resource is not a component of this hypervolume. If it were, the assessment of the habitat structure would be impossible. If quantity of resource were a part of a species hyperspace, then niche volume would be determined almost entirely by the number of individuals, N, thus rendering the concept of the niche useless, no matter how imprecise it is. The amount of resource determines N of a species but not its ecological characteristics.

---

and attendant size reduction (figure 4.3) are much greater (Kolasa and Strayer 1988). However, greater efficiency of resource utilization may make a difference between survival and extinction in a fragmented habitat.

We avoid using *niche* for describing ecological properties of species despite the fact that the concepts presented here have some affinity to the concept of the niche. There are many different interpretations of niche. Silvert (1994) and Leibold (1995) clarified important aspects of the niche concept, but until their ideas are more broadly accepted and supersede a plethora of earlier uses, the term will keep generating confusion. We limit ourselves to citing Silvert (1994) who stated that niche can be thought of as a generalization of range. Silvert views the niche as a collection of points in a multidimensional space where the species

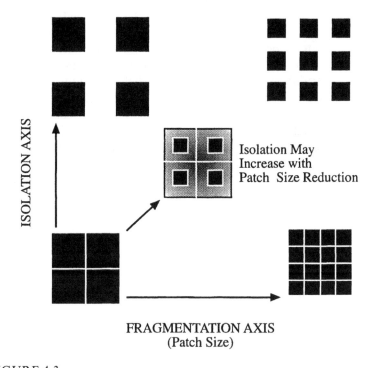

FIGURE 4.3

Fragmentation, isolation, and habitat size reduction. In most cases, increasing fragmentation is likely to correlate with the reduction of habitat size, although this may occur at the expense of the total habitat available or through emergence of interpatch barriers.

is observed. The difference between our use of niche and that of Silvert's is that observation of distribution is only a subset of species properties already defined by the available habitat template (see later).

Some species appear as specialists because of evolutionary constraints, whereas others may be products of novel or unusual circumstances. For example, many species requiring oligotrophic and cold lakes are currently habitat specialists, even though they may have appeared as habitat generalists 20,000 years ago. Some species may appear as specialists because of one particular limitation such as diet, whereas others may appear so because of a synergistic effect of several milder limitations. The reasons for the limitation may also vary greatly from immediate ecological factors such as trampling or seedling grazing to evolutionary constraints such as the inability by most mammals to chase insects in the air. For example, Tilman's (1982) resource ratio hypothesis illustrates how presence of competitors may limit the range and number of species with higher threshold levels of nutrient concentrations. Hanski's (1982; Hanski et al. 1993) metapopulation models call attention to dispersal abilities and their consequences. Natural history provides hundreds of other examples for different types of habitat limitation. Predation, ontogenetic shifts in habitat requirements, habitat connectivity (Milne 1992), magnitude of rescue effects (Stevens 1989), and heterogeneity of energy supply at regional scale (Wright et al. 1993a) all address one or several mechanisms limiting species in habitat space. These limitations may be fairly permanent or dynamic and variable. In all these cases, the HBM is concerned only with the actual status of a species in the community under consideration. This status affects a species' current performance, which, in turn, is subject to predictive deductions by the model.

Assumptions: Habitat

Before model assumptions can be presented, we need to clarify what we mean by *habitat*. Despite a long history of the habitat concept (Whittaker et al. 1973; Rejmanek and Jenik 1975; Wiens 1984; Addicott et al. 1987; McCoy and Bell 1991), the meaning and usage of the term are far from clear. This lack of clarity is intricately linked to the niche concept, which also has several inconsistent interpretations (Hutchinson 1957; Whittaker at al. 1973; Pianka 1976; Haefner 1980; Giller 1984; Herbold and Moyle 1986; see table 4.2 for several definitions and brief comments).

Species attributes and habitat attributes are two different things. Although they differ, they cannot be separated fully; they co-define each other as a key and lock are co-defined (in a hypothetical world without locks, the concept of key would have no meaning). We base further discussion on the following assumptions and definitions, some of which may sound trivial but are actually necessary for the completeness of the argument:

1. The physical world has an infinite number of descriptive dimensions.

2. Some of these dimensions form a subset relevant to ecology. This subset can be treated as, and used to characterize, an environmental template.

3. The subset is spatially and temporarily heterogeneous: values in each dimension may vary from place to place and time to time.

4. A combination of dimensions and a subset of dimension values (i.e., a portion of the template) that interact with a species, constitute the species habitat or multidimensional habitat space.

5. The portion of the template that interacts with a species is determined by a subset of species attributes. For example, lakes differ in phosphorus (P) concentration, and if a species requires a particular concentration of P (a dimension), interaction is likely to occur. (See table 4.1 for further explanation and comments on the concept of habitat.)

6. Portions of the physical environment match or fail to match the "niche" dimension values because they vary quantitatively (e.g., the amounts of resources: no resource—no match, correct concentration—perfect match).

7. Arrangement of those portions in space and time is habitat structure (McCoy and Bell 1991).

8. The totality of all species habitat spaces is the community habitat space.

A more precise definition of species attributes helps in identifying habitat structure. Consider the case of *Myrmecophaga tridactyla* (anteater). Let us use only a few dimensions, such as the habitat elevation and diet spectrum. *M. tridactyla* cannot feed under conditions of prolonged flooding because it normally requires high abundance of insects such as ants and termites. These dimensions constitute a subset of ecological attributes of *M. tridactyla* and roughly correspond to some interpretations of the niche. Other, nonniche attributes of *M. tridactyla* may include its reproductive strategy, destructive impact on termite nests (but see Leibold 1995), or intelligence. South American grasslands are gently undulating and host a variety of insects, and nonflooded portions of the grassland with their termite nests constitute the

TABLE 4.2

Definitions of *niche* and examples of conceptual diversity

Definitions and examples	Reference	Additional comments
A niche is an evolved, multidimensional attribute of a particular species population.	Whittaker et al. (1973)	This definition is compatible with the use of ecological characteristics of a species. It also precludes the amount of resource from being a part of the hyperspace.
A niche can be defined by the relations of a species with other species of the community.	Herbold and Moyle (1986)	Cited view.
A niche is an *N*-dimensional hypervolume enclosing the complete range of conditions under which an organism can successfully replace itself; every point of hypervolume corresponds to a set of values of the variables permitting the organism to exist.	Hutchinson (1957)	Vague on the resource amounts; it is easy to imagine that it sets a minimum resource level for a viable population, but because there is no upper limit to the amount of resource available, one should not interpret this definition as fully permitting resource amounts. Indeed, adding the resource amount as one of the axes would render the niche volume infinite from a mathematical perspective.
Niches are preexisting properties of the communities and their environment.	Giller (1984)	Cited view. It corresponds more to the HBM concept of habitat than to species attributes.
The range of values of environmental factors necessary and sufficient to allow a species to carry out its life history.	James et al. (1984), citing Grinnell's ideas	Niche defined by species requirements.
The representation of the population niche would consist of a cloud of points in phase, the more favored parts of the niche being represented by an increased density of points.	Wangersky (1972)	Niche defined by species performance on an environmental template. If points (individual organisms) are described by values of habitat axes, then this concept corresponds closely to the realized niche.

TABLE 4.2
Continued

Definitions and examples	Reference	Additional comments
The set of all observations of an organism in a multidimensional hyperspace can be thought of as the observed niche (with a probability function suggested as the mathematical approach to niche description).	Silvert (1994)	This view is close to the concept of realized niche. The fundamental niche, although addressable via evolutionary perspective, is not always definable through observation because a species may never be tested over the potential range of environmental values.
No explicit definition given. In the simplest terms, Leibold conceptualizes niche as two interacting sets of dimensions: those pertaining to species requirements and those describing species impact on the environment.	Leibold (1995)	This approach attempts to deal with the quantity of resources (and related factors), which was missing from other conceptualizations of the niche. It is not clear if the niche is a characteristic of species or a new conceptual entity incorporating "impacts."

anteater habitat. Landscapes of flood-free grasslands, and the density, distribution, and quality of termite nests constitute habitat structure.

The community habitat space is likely to coincide to a substantial extent with portions of the environmental template that are homogeneous (homogeneously diverse sensu Hutchinson 1957) relative to the surrounding matrix. This is not because the two are conceptually the same but because of the evolutionary history and adaptive responses of species. A lake is a relatively homogeneous patch of environmental template by most criteria. All fish species are likely to find their habitat spaces (or, simply, habitats) within that patch. Consequently, the habitat as identified by species is different from the full complement of variables the environmental template would require for description; a species "carves out" a habitat from the overall volume. Despite this difference, the habitats selected by species should generally coincide with the basic subdivisions of the environmental template, subject to empirical verification. This permits us to present views on the structure of habitat without constant distinctions between the species perspective and that of the external observer (the ecologist).

Model Structure and Requirements

The model requires that a habitat unit—a community habitat space—be identified. By defining a unit using all the species in a community, we create a reference level for habitat subdivision. Thus, the habitat unit is the highest level. There are practical reasons underlying the decision to have a large unit. The most important characteristic of this space is that it spans the full range of ecological conditions experienced by the species in question. The model permits the use of arbitrary habitat units, that is, units defined without a prior evaluation of the relative homogeneity of habitat. Such units may, however, introduce biases resulting from improper scaling (Addicott et al. 1987). The biases would emerge because the model relies on determination of ecological ranges of species, which would likely be incorrectly represented when using arbitrary habitat units.

Any habitat unit, whether a pile of dung, a lake, or a mountain range, is composed of subunits, which can be split into even smaller subunits. The criteria for distinguishing subunits are provided by species perception of boundaries in their environmental template which, as suggested earlier, are likely to correspond to the natural subdivisions of that template. The resulting subdivisions are a function of the perceptual resolution (whether by species or investigator). Low resolution yields coarse, high-level subdivisions, while high resolution identifies fine, low-level habitat subdivisions (figure 4.4). Criteria for subunit demarcation need not be invariant, and they can change from level to level. For example, at the highest level, the distinction may be made between water and land. When we analyze aquatic habitat (Pahl-Wostl, chapter 7), the criterion may be temperature, which separates the epilimnion from the hypolimnion zone of a lake. Such criteria are clearly scale dependent but fairly easy to identify and use. Scale-independent criteria exist and involve the magnitude of unit integration (Kolasa and Pickett 1989). However, these criteria are difficult to apply because methods for making them operational have not been sufficiently developed.

One striking feature of the model of habitat structure is nestedness. Smaller, low-level units are nested within larger, high-level units. This nestedness has potentially important consequences for constraining species activities and interactions and for creating differential species links to the environment. Consequences of nestedness on species performance will be discussed later.

Assumptions: Relationship Between Species and Habitat

The general assumption is that species interact with habitat and that this interaction has a number of detectable consequences for species abundance, selection of life strategies, probabilities of extinction, distribution patterns, and other ecological phenomena. The interaction of species with habitat is primarily via species tolerance limits, habitat variation, and habitat structure. Tolerance limits determine whether a species can live in a particular habitat. Habitat variation may influence species persistence through generation of temporarily adverse conditions, and the habitat structure interacts most visibly with population cohesion and individual mobility. Habitat structure and variation may have equilibrium and nonequilibrium components. For example, monkeys in seasonally flooded forest of the Amazon River drainage utilize different food resources during the flood and dry seasons. This seasonality determines the range of some species. Within the range, food availability is subject to asynchronous fruit-bearing by trees and to a patch disturbance regime. This nonequilibrium component of habitat

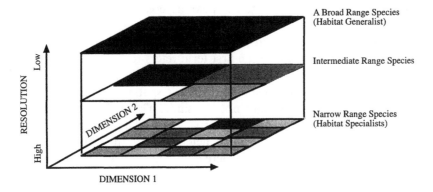

FIGURE 4.4

A conceptual model of habitat structure in which homogeneity and heterogeneity depend on the resolution with which species perceive their environment [modified from Kolasa (1989)]. The top level appears homogeneous and can be occupied by only one species. Levels below, which represent successively higher resolutions and smaller habitat grain, reveal mosaics of different microhabitat types and can, correspondingly, be inhabited by more species. Habitat of species using lower levels is more fragmented. In the HBM, specialization is correlated with the degree of habitat fragmentation, with the correlation strength dependent on variation in the actual arrangement of patches.

structure may be as important as the seasonality itself to the success of some species and the failure of other species.

Species perceive their habitat at different scales of resolution and respond to spatial scales at different levels (Kotliar and Wiens 1990; Milne 1991a; Sagova and Adams 1993). These differences make species that occur in the same locality live in different universes and interact with each other under different rules (Waltho and Kolasa 1996; Allen, chapter 3; Martinez and Dunne, chapter 10). The relations among species and their habitat include both static and dynamic links.

Differences in ecological requirements define the kind and hierarchical level of habitat unit a species can use. The view of specialization outlined earlier may now be combined with the hierarchical structure of habitat. Specialists use small subdivisions of the habitat and can be pictured as using lower levels of the hierarchy (see figure 4.4). Generalists use composite and thus larger fragments of the same structure and appear at higher levels. Size of the habitat portion is relative to the ecologically broadest species and should not be confused with the absolute size of two-dimensional patches (Forman 1995). This separation corresponds with the opposite levels of perceptual resolution. Specialists are high-resolution species, whereas generalists are low-resolution. Consequently, generalists often appear as coarse-grain species and specialists as fine-grain (Kotliar and Wiens 1990). However, the terminological distinction between coarse-grain species and habitat generalists is still required, because the first term refers to spatial fragmentation and patch size, whereas the second term refers to the number of habitat categories used, irrespective of their spatial arrangement; the two aspects are correlated but not the same. The difference between habitat generalists and coarse-grain species is related to the scale of perception and the scale of habitat heterogeneity. In simplest terms, one may consider a community where only one species is permitted to occupy a particular habitat subunit distinguished at a given hierarchical level. By definition, a subunit is homogeneous at that level, and thus, any two species using it would have to use identical resources and face severe competition. The competitive exclusion principle does not allow more than one species to persist in such an environment (Murray 1986a), although this is an idealization.

In natural systems, one would not expect perfect correspondence between habitat units and their respective species. First, species are not "perfect," because species associations with some habitat dimensions are probabilistic rather than deterministic, and even unsuitable areas of habitat space may be subject to exploratory colonization or establishment (Wu and Vankat 1995). Second, the habitat unit is assumed to be homogeneous at a particular scale of resolution, although scales of resolution may be incompletely isolated from each other from the species perspective. Some individuals may perceive more local (unit) heterogeneity than others, which may affect habitat use and result in variable competitive ability. Third, the species–habitat relationship may itself be far from equilibrium, especially at some scales or in some habitat units. In these situations, the model may perform inconsistently unless factors interfering with the dynamic equilibrium are taken into account (Wu and Loucks 1995).

Species' associations with their respective habitat units form through a variety of mechanisms and their combinations. In the terminology of the HBM, the habitat resolution of a species is a function of the species' association with a particular habitat subunit. Prohibited habitat is invisible habitat. As some limiting conditions are removed, species may move to habitat units previously unavailable, which produces a change in the level of habitat resolution. Finally, we must address additional complexity stemming from ontogenic habitat shifts. Many species change their requirements and habitat resolution as they develop from seedlings, eggs, and propagules into adults. While a full characterization of a species implies taking into consideration all these stages and their relationships to habitat, a number of ecological questions that can be addressed by the HBM need not involve all life stages for all analyses. The HBM is most effective when applied to a life stage that is most limiting to the population size. If the limiting life stage is survival of the young, then one should concentrate on the habitat structure as perceived by that life stage. If the limiting life stage is the habitat needs of the mature stages, then the autecology of juveniles can be ignored. Empirical work is needed to determine which is relevant before a general model is applied. The model itself could be used to determine the relative importance of the life stages in determining population size and distribution; its predictive power can be calculated independently for different life stages.

Predictions and Explanations

The strength of the HBM is its large range of mutually supporting and reinforcing applications. The HBM (1) explains patterns other models do not address, (2) allows a number of unique predictions, (3) agrees with successful explanations provided by other models, and (4) explains why various models apply at various scales (Tokeshi 1993). Other models cover but a fraction of the issues the HBM solves and identifies. Some examples will be discussed.

Patterns Not Addressed by Other Models

Rank-abundance curves range from a straight line to an S-shaped curve. These empirical distributions have stimulated the development of models based on fitting various mathematical functions. The two extreme forms involve the geometrical distribution and lognormal distribution. Biological motivation for these distributions is weak or nonexistent, depending on the model (Motomura 1932; Sugihara 1980; Ugland and Gray 1982; Magurran 1988). The HBM predicts the geometric distribution in simple (low dimensional) habitat units, such as in a conifer plantation (Magurran 1988) or those created by major disturbance; a lognormal-like distribution can be appropriate in more complex habitats or at large spatial scales. In complex habitats (see figure 4.4), species abundances are determined solely by scaling of one criterion. If space is the criterion, then the simplest subdivision of a habitat unit will result in two, then four, then eight subunits, and so on, with species abundances following a similar pattern.

In high-dimensional habitats, with many criteria involved and varying across levels, lognormal distribution of abundances is expected because random variation in many factors will tend to result in a normal distribution of abundances (May 1975; Magurran 1988). For example, a model that assumes seven species, with one species at the highest level, two at the intermediate level, and four at the lowest level, and that assumes (1) equal access to resources and (2) a three-to-one ratio of species occupying the same habitat unit (Sugihara 1980), produces a nearly straight linear fit ($r^2 = 0.9428$, $P = 0.0003$, $N = 7$) (table 4.3). This interpretation agrees with concepts proposed by Tokeshi (1993), who linked the pattern gradient from geometric series to lognormal series to the scale of sampling. Similarly, the HBM predicts the lognormal-like distribution of abundances at larger organizational scales in contrast to scales allowing analysis of single, small habitat units, because large-scale patterns are likely to involve a greater range of causal factors. Finally, in homogeneous habitats, one would expect species abundances to be determined by priority effects (niche-preemption model; Whittaker 1965), or by random processes if no priority effects apply (broken stick model; MacArthur 1957). Thus, the abundance distribution is not just a statistical product of multiple interacting variables (May 1975) but an explicit product of hierarchical habitat structure and unit size (see figure 4.4). The model predicts that simple structures produce straight rank-abundance lines, whereas rich structures produce S-shaped curves. In this sense, the model agrees with extensive ecological evidence that patterns of abundance differ among habitats, sampling scales, and patterns of disturbance.

Unique Predictions

We identify two unique predictions of the model: one about the discontinuities of community structure and the other about the effects of wrong scale on model accuracy. These predictions are a powerful testing ground for the conceptual foundations of the model. To the extent they have been verified, they provide substantial supporting evidence for those foundations. Consider two adjacent hierarchical levels of habitat. A species has a choice of using either the higher-level unit or

TABLE 4.3
Example of calculations for a simple, seven-species model*

Species 1 (1)					
Species 2 (.75)				Species 3 (.25)	
Species 4 (.56)			Species 5 (.19)	Species 6 (.19)	Sp. 7 (.06)

Species†	Proportion of habitat	Specialization	Expected abundance	Relative abundance
1	1.00	1.00	1.00	49.58
2	0.75	0.75	0.563	27.91
3	0.25	0.25	0.063	3.12
4	0.563	0.563	0.317	15.72
5	0.188	0.188	0.035	1.74
6	0.188	0.188	0.035	1.74
7	0.063	0.063	0.004	0.20

*Consider a case of seven species constituting the community shown. Size and position of the boxes represent relative ecological ranges and position in the habitat mosaic for each species. Proportions of the ecological range are shown for each species (e.g., the proportion for species 7 is 0.06). The size of the boxes is the mean random value attained by a species if it consistently acquires either more or fewer resources (Sugihara 1980). In this simple case, specialization assumes values identical to those of the range, but this is not necessarily true for data from natural communities unless the broadest species occurs throughout the whole measured range of values.

†Specific calculations for this part of the table assume equal division of resources among hierarchical levels. For communities in which data suggest a different division, the proportion of habitat values should be multiplied by a coefficient associated with a particular level.

one of its subdivisions at the lower level. Depending on which level the species is able to choose, the amount of available habitat differs dramatically. Because several species face similar choices at approximately the same levels of resolution, a natural division is likely to occur. Some species will occupy higher-level units and others the lower-level units. This should result in an emergence of species groups characterized by similar position in the structural hierarchy of habitat. Because the level at which a species operates predictably affects its abundance, species sharing a level should also fall into the same abundance classes. Thus, in addition to a positive relationship between ecological range and abundance, one could also postulate distinct discontinuities in parameter values (figure 4.5).

The second prediction is that if the scale of sampling is too small relative to the extent of ecological range required by the model for the analysis of a particular community, then the relative abundance of generalist species will be underestimated and the abundance of special-

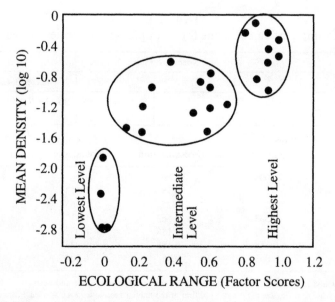

FIGURE 4.5

An example of discontinuities in species ecological range: each circle is an invertebrate species inhabiting a system of small rock pools on the northern coast of Jamaica. Ecological range is based on the maximum ranges of oxygen, salinity, pH, and light experienced by a species over 49 pools.

ist species will be overestimated. This poses a problem for study design, because it is not immediately obvious which scale is appropriate, at least not before the data are available (see also chapters 12 and 18 of this volume). The test of this prediction can be conducted in communities in which the ecological range of the top habitat generalist is known or possible to assess prior to the test. Alternatively, one might keep expanding the observation set until the range of conditions sampled is equal to or broader than the range observed for the broadest of species (Waltho and Kolasa 1994). In such a situation, one should initially observe an increasing trend in the estimated relative generalist density. This trend should stop after the expanding data set includes or exceeds the ecological range of the broadest species. The breakpoint in a regression could be used to identify the proper scale for a community in question, and this scale should coincide with the range of the broadest species. To the best of our knowledge, no other model predicts the above patterns.

Study Design and Methodological Concerns

Operational Approaches to Habitat Units

The preceding paragraphs imply that knowledge of habitat structure is important for the analysis of interactions among species and their habitat (unit). We envision an extensive set of rules on how habitat structure can be determined and evaluated. Common ecological sense may be the best guide. Given the definition of the habitat, however, one logical approach would include the following steps:

1. Identify dimensions of relevance to the species present (e.g., predators, resources, physical conditions).
2. Create a multilayer map representing distribution of the values for each dimension on the landscape of interest.
3. Superimpose individual species requirements ("niches") on the map to assess the configuration (fragmentation, amount, distribution) of its habitat space and nestedness relative to other species.

A simpler, acceptable approach is to measure the distribution of species over the habitat space and deduce habitat structure from it. Distribution of species over the habitat space can be measured along one or few dominant variables, or, even more simply, in physical space as long as the physical space reflects some gradient of habitat qualities. For aquatic invertebrates occupying rock pools, distribution in space correlates strongly ($r^2 = 0.88$) with composite measures of habitat variables (Kolasa and Drake 1997). For example, one can assess species distribution in a two-dimensional space of phosphorus and nitrogen concentrations, which might be appropriate for aquatic algae or along mountain slope sites that encompass a multivariable gradient. A species present at many sites defines a larger habitat unit than a species living only in the subalpine zone, a smaller habitat subunit.

Although the definitions of both habitat and species hypervolume refer to and rely strongly on multiple dimensions, the task of quantifying these concepts in the field is less complicated than the concepts might imply. This is because of two premises. First, any dimension relevant to species performance will map in space. Even such esoteric variables as predation risk will have spatial representation. Second, most variables are spatially integrated. For example, a clump of shrubs in a savanna has different food resources, risk of insect bites and infections, thermal and humidity regime, mate encounter, detection probability, and sun exposure than an adjacent patch of grass. It is possible to use only a limited number of measurements for assessing habitat structure, because many of these variables are correlated.

Scaling of Sampling Protocol

Because the primary application of the HBM is to examine differential performance of species over their habitat, it is important that the analysis does not underestimate or overestimate the habitat space of any of the species being compared. This requires that sampling should cover the range of values (for the variables one decides to analyze) that equal or exceed the range of the ecologically broadest species. It also requires that the sampling resolution is equal to or higher than the habitat grain to which the ecologically narrowest species responds.

In addition to the need for adequate scale for testing the model, other considerations may become important under some circumstances. As mentioned earlier, different scaling will result in different rank-abundance patterns (Tokeshi 1993). Thus, application of the model to areas differing in size or heterogeneity will result in data of different usefulness for testing specific predictions.

Summary

The HBM overlaps with other models to a varying degree. For example, the habitat specialization aspect is shared with Brown's (1984) model. Aspects of habitat fragmentation and patchiness are similar to the core–satellite model (Hanski 1982). In fact, each resolution level could be considered a landscape of patches in which species play out their patch occupancy game. The hierarchical subdivision of resources is conceptually similar to the sequential breakage model (Sugihara 1980) and is consistent with other resource-partitioning algorithms. Putnam (1994) supports Ugland and Gray's contention (1982) that the canonical lognormal curve arguments of Sugihara are wrong and that data actually refute them. If this is so, then this particular model of niche partitioning should not be legitimately compared to HBM. Nevertheless, other niche-partitioning models might be viewed as special cases of habitat structure perspective (Tokeshi 1993) in which resource availability declines with the size of the habitat unit used by a species. However, there is more than niche space to determining of species abundances. Indeed, a fully quantitative version of HBM would require formulae that combine scaling effects of increasing habitat resolution (specialization), such as reduction in available habitat, penalties of fragmentation, and allometric trends (Nee et al. 1991) associated with habitat size and differential patch dynamics (Collins and Glenn 1991). Pagel et al. (1991) analyzed 72 natural communities and found that relative abundances of species do not necessarily reflect division of resources. They further found that smaller species use substantially fewer resources than anticipated from allometric considerations alone. Coincidentally, Brown and Maurer (1986) estimated that habitat generalists, despite being generally larger and fewer in numbers, command more resources on a habitat-unit basis (thus reinforcing the notion of greater success of generalists). These two studies jointly suggest another factor—habitat structure—that determines the performance of species on an environmental template. Tests of the model should thus involve either a complete account of all these aspects or create conditions that permit controlling for some of them.

Although the HBM shares different assumptions and mechanisms with several other models, it is not a specialization hypothesis, a version of the core–satellite model, a heterogeneity-based explanation, or a niche-partitioning model alone. None of these assumptions and hypotheses alone sufficiently represents the consequences generated by the HBM. The specialization hypothesis does not consider scaling of habitat fragmentation, the heterogeneity perspective is incomplete without multiple levels of resolution and patch dynamics, and the niche-partitioning models have little to say about consequences of habitat structure.

Gaston and Lawton (1990) note that the distribution–abundance relationship may be negative when the habitat considered is rare in the region but positive when the habitat is common. In the HBM, the species ecological range is measured as a function of its use of the multidimensional habitat space. If a species has a broad ecological range, it will occur in most habitats of the region and thus cannot be affected by the habitat rarity unless an extremely common habitat type also has extreme attributes. In the HBM, the broad range correlates positively with abundance. If one defines ecological range at a scale different than the data used to evaluate it, or in some other arbitrary way, then spatial fragmentation must be considered separately and is likely to lead to the observation made by Gaston and Lawton (1990). In the HBM, a species should not be considered a generalist (broadly distributed) when it lives in a small fraction of available habitat types. Consequently, the observation is precisely anticipated by and consistent with the HBM: the geographical or spatial aspect of habitat fragmentation is subsumed in the definition of specialization and thus needs no explicit attention.

The HBM brings together important features of various models and hypotheses and places them within a single framework of the species–habitat relationship. By doing so, it frees those models from their scale blindness, permits addition of new mechanisms, and provides a context for evaluating the relative importance of various processes (Kolasa and Drake 1997). The species–habitat relationship is a more general type of relationship than the species-to-species relationship, which can be realized only in the context of the species–habitat relationship.

PART II

INTERPRETING MULTIPLE SCALES IN
ECOLOGICAL SYSTEMS

5

PALEOECOLOGICAL PERSPECTIVES ON ECOLOGICAL SCALE

Peter K. Schoonmaker

Ecological patterns and processes operate along a broad continuum of temporal and spatial scales, yet most ecologists focus on organismal to ecosystem questions, for which data are collected at temporal scales of seconds to decades, and spatial scales of square meters to square kilometers (figure 5.1). Paleoecology, the study of past environments and their biota, addresses patterns and processes at larger temporal and spatial scales—generally hundreds to thousands of years and hectares to biome—but has the potential to focus on higher-resolution questions. In general, paleoecologists want to know about the presence, abundance, location, and temporal occurrence of taxa, communities, and environments.

Ecologists' awareness of long-term paleoecological insights have waxed and waned during the past century. Early ecological thought was firmly grounded in earth sciences and paleontology, with evolutionary theory being a focus of controversy and speculation (Mayr 1984). Despite groundbreaking palynological studies of von Post (1917) and the advent of radiocarbon dating in the 1950s, awareness of the paleoecological perspective seemed to decrease for much of the twentieth century. However, in the past few decades, more ecologists have begun to recognize the importance of long-term processes, and more importantly, to factor long-term paleoecological considerations into their world view and research design, even if their focus is short-term phenomena (Prentice 1992; Richie 1994).

The conceptual framework of most ecologists now embraces paleo-ecological insights about the dynamic nature of communities, the per-

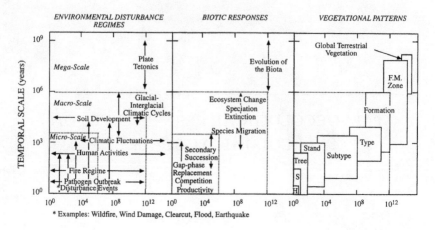

FIGURE 5.1

Disturbance regimes, biotic responses, and vegetation patterns operate at multiple time and space scales. Paleoecological data are useful at macroscales and all but the finest microscale.

Reprinted from *Quaternary Ecology: A Paleoecological Perspective* by H. R. Delcourt and P. A. Delcourt. Copyright © 1991 by Chapman and Hall. Reprinted with permission of the publisher.

vasive role of disturbance, the role of humans as agents of disturbance, and implications for resource management. This chapter provides an overview of recent (Quaternary) global environmental change, abiotic and biotic evidence for this change, an examination of temporal and spatial scales that paleoecologists can address, some paleoecological insights for contemporary ecologists, and some suggestions for future paleo-research relevant to ecological and social issues.

Environmental Change During the Quaternary

During the last 1.6 million years (the Quaternary Era), the earth's environment has been cold (compared to that of the warmer Tertiary Era), with 50,000- to 100,000-year glacial periods interrupted by fairly regular intervals of short, warm, 10,000-year interglacial intervals. The most recent interglacial, the Holocene, began some 10,000 years ago, and, based on observation of past glacial–interglacial cycles, it may draw to a close in a few hundred to a thousand years (Imbrie and Imbrie 1986).

The Quaternary's glacial–interglacial pulse is generally thought to be driven by changes in insolation as the geometry of the earth's orbit varies around the sun. Eccentricity, tilt, and precession fluctuate on 100,000-, 41,000-, and 22,000-year cycles. These cycles interfere with and amplify one another, they have different effects at different latitudes, and they cause time-transgressive events such as glaciation, ocean heating and cooling, and atmospheric chemistry change (e.g., CO_2 changes).

Viewed over a 1-million-year time frame, there is a regular global alternation of warm and cool climatic conditions that drive most ecosystem, community, and population processes. Viewed over shorter periods of time (centuries and decades), global climatic change can appear more stochastic: the onset of a glacial–interglacial change may be gradual (thousands of years), sudden (a few decades), or spasmodic (figure 5.2). The onset of the Holocene was characterized by all three of these properties, starting with a gradual warming 16,000 years ago, followed by significant temperature fluctuations 11,000 years ago (the older and younger Dryas intervals), and ending with rapid warming soon afterwards. Additionally, the onset of the Holocene was not globally simultaneous. Some past climates during the late glacial and Holocene appear to have no modern analogue; the corollary to this is that parts of the earth may experience future climates that have no present analogue. Global climate has been gradually cooling for the last 2,000 years. Recent human-driven atmospheric changes appear to be rapidly reversing that trend, at least in the short term.

Abiotic and Biotic Evidence of Environmental Change

The timing and spatial extent of glacial–interglacial periods and consequent biotic responses have been well documented by the study of glacial geomorphologic features, fluctuations in ocean and lake levels, stable isotope ratios in ocean sediments and ice cores, planktonic assemblages in ocean and lake sediments, and changes in sedimentation and chemistry in peat and lake sediments (Bradley 1985). Paleoecologists have concentrated on vegetation as a sensor of environmental change. Fossil plant parts, such as seeds, twigs, leaves, and especially pollen preserved in lake muds and bog peats, have been the primary evidence for inferring past vegetation changes (Birks and Birks 1980; Berglund 1986). In drier environments, pack-rat middens preserve large plant parts and other materials (Betancourt et al. 1990).

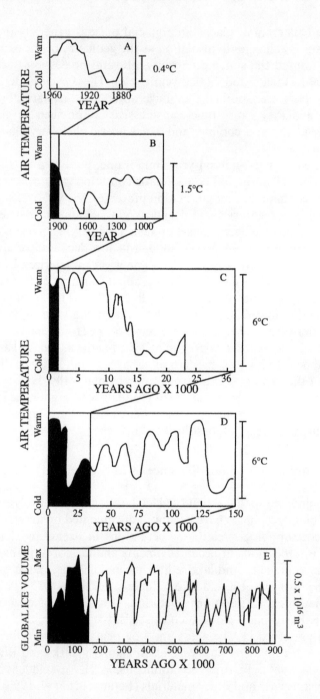

Pollen analysis has been the primary method for tracking vegetation change through time. Most late-glacial and Holocene vegetation data come from studies of fossil pollen in which samples are taken from lake and peat sediments. The pollen in these sediment samples is concentrated, subsampled, counted, and displayed using a pollen diagram. Macrofossils such as seeds, leaves, and twigs offer more taxonomic and spatial resolution than pollen, but they are less abundant in most sediments, so that problems of sampling and temporal resolution can arise. Information derived from fossil pollen lacks the spatial, temporal, and taxonomic resolution of living material, but it offers a long-term perspective unavailable to ecologists who study living organisms.

Pollen analysts have made great strides in understanding differential pollen production, dispersal, deposition, and preservation (Jackson 1994; Jackson and Wong 1994). Pollen studies, improved by advances in sampling, numerical analysis, and multivariate techniques, have helped define the taxonomic, numerical, spatial, and temporal confidence limits of pollen data. They also have provided methods for reducing the complexity of the data so that pollen assemblages can be interpreted more clearly with regard to stratigraphic relationships, environmental gradients, and rates of change (Gordon and Birks 1985; Berglund 1986; Jacobson and Grimm 1986).

Thus, a microscopic plant fossil, minutely subsampled, is our primary source of information about vegetation and environmental change, on scales of kilometers to continents, over time spans of thousands of years. Most pollen assemblages lack the taxonomic resolution of living vegetation in three respects: First, some plants produce more pollen (e.g., wind-pollinated trees) than others. Second, some pollen taxa tend to degrade rapidly. Third, most pollen taxa can be identified only to genus, some only to family. The first two factors may mask the presence of a taxon, and the third prevents species-level identification.

FIGURE 5.2

Global temperatures **A–D** (top four graphs); ice volume **E** (at bottom). Environmental phenomena exhibit different periodicities and trends depending on the scale of observation. For example, the rise in temperature from 1860 to 1960 appears to be a continuation of warming begun at the close of the Little Ice Age (B), which in turn appears to be a brief reversal in a slow cooling trend begun 6000 years ago when global temperatures during the current interglacial reached their maximum.

Reprinted from *Vegetatio*, vol. 67, pp. 75–91, by T. Webb. Copyright © 1986 by Kluwer Academic Publishers. Reprinted with kind permission from the publisher.

Focusing the Paleoecological Lens

Most pollen analyses investigate "landscape-level" phenomena by selecting small-to-medium-sized basins (which collect pollen from the surrounding 10 to 100 km^2), separated by distances sufficient to prevent extensive overlap of pollen source areas. Brubaker (1975) analyzed pollen from lake sediments on three different soil types in Michigan and found that the same regional climatic change resulted in different communities on different soils. Gaudreau (1986) compared two sites less than 15 km apart but with an elevational difference of 440 m in the northeastern United States and found that physiographic position affected the arrival dates of migrating plant species. Clark (1990) combined a stratigraphic analysis of charcoal on petrographic thin sections from varied sediments from three small (5-ha) lakes having adjacent catchments, with a study of fire scars on red pine trees. This combination of techniques produced a detailed picture of fire history and environmental conditions for a 1-km^2 area.

Paleoclimatic data—derived from radiocarbon-dated pollen samples, pack-rat-midden samples, lake-level reconstructions, and marine sediments—have been analyzed for thousands of locations throughout the world in the last few decades (COHMAP 1988). Analyses of late-glacial and Holocene paleoecological sites have resulted in detailed maps of the distribution of taxa through time on a regional or subcontinental scale in Europe and North America. For example, Huntley and Birks (1983) reconstructed past distribution and abundance of taxa by connecting similar pollen percentages with contour lines of pollen sampled from 423 sites in Europe, covering a period from 13,000 to 0 years before present (BP), with a resolution of 500 years. Davis (1981) was the first to map migrational fronts of taxa at different times using pollen data (figure 5.3). Paleoecologists have also mapped inferred vegetation types using modern analogues, an ambitious methodology given that some communities in the past had no modern analogues. Jacobson et al. (1987) mapped different taxa simultaneously using colored isopolls to reveal population centers and overlap of different taxa at 500-year intervals over the past 14,000 years. This latter method may offer the most insight into species interactions. Computer-enhanced interpolation with a continuous video output could provide additional insight.

Paleoecologists are not limited to these large scales. They can address various ecological problems along a broad continuum of spatial and temporal scales through the careful selection of study sites, organ-

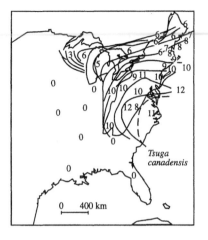
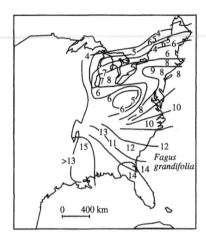
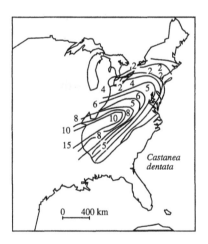
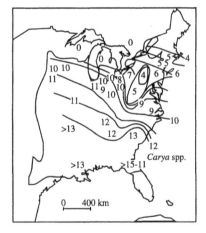

FIGURE 5.3

Pollen migration maps show that tree species moved at different rates and directions through time. These maps show migration fronts for *Tsuga canadensis*, *Fagus grandifolia*, *Castanea dentata*, and *Carya* species (hickory), from 15,000 years BP to the present, at 1000-year intervals.

From *Forest Succession*. D. C. West, H. H. Shugart, and D. B. Botkin, eds. Copyright © 1981 by Springer-Verlag. Reprinted with permission of the publisher.

isms, and field, laboratory, and analytical techniques. Some studies combine pollen analysis with analyses of plant and animal macrofossils, charcoal, sediment lithology and chemistry, human artifacts, tree-rings, historical records, and modern vegetation. These studies reveal patterns of climatic change, natural and anthropogenic disturbance, soil development, erosion, community composition and diversity, migration, invasion, and succession (Birks and Birks 1980; Berglund 1986; Huntley and Webb 1988).

One characteristic of retrospective studies is that the farther back in time one focuses, the fuzzier the view. This attenuation of resolution is more severe with some kinds of evidence than others. For example, human historical records and time series of tree-ring growth taper off within a few generations of living individuals, whereas microfossil records tend to persist for thousands of years. Pollen records are often shorter than those of marine microfossils (diatoms, zooplankton) because of the deterioration or destruction of their depositional environment (e.g., by glaciers).

A few examples illustrate the spatial, temporal, and ecological resolution available at various scales (figure 5.4). Tree-ring analysis, vegetation reconstruction, and historical records provide very fine resolution, often comparable to data from permanent plots. Henry and Swan (1974) and Oliver and Stephens (1978) analyzed the growth rates of live and dead stems on small plots (0.04 and 0.36 ha, respectively) in conjunction with historical records. They detected disturbances such as fire, windthrow, and tree-cutting, and subsequent changes in vegetation composition as far back as A.D. 1665. Tree-rings also provide paleoclimatic information at the local, regional, and global levels (Fritts 1976) by detecting phenomena such as the Little Ice Age (A.D. 1450 to 1850).

Annual laminations in lake sediments provide yearly records of vegetation and environmental change (Clark 1990). Swain (1978) analyzed varved sediments from Hell's Kitchen Lake in north-central Wisconsin for pollen, charcoal, and seeds. He found a successional sequence of *Populus* (aspen), *Betula* (birch), *Pinus strobus* (eastern white pine), and *Tsuga canadensis* (eastern hemlock) following fires, the frequency of which (100 to 140 years) was affected by available moisture of the prevailing climate. He concluded that major vegetation change occurred on two different time scales. Short-term changes related to individual fires were superimposed on longer-term changes resulting from increases or decreases in the frequency of those fires. We should bear in mind that the depositional dynamics of charcoal, which are incompletely understood, limit our ability to detect the frequency, severity, and extent of fires (Whitlock and Millspaugh 1996). The time-intensive nature of sampling sediments at close intervals (total processing time for pollen and charcoal may average from 4 to 6 hours per sample) places a practical limit on fine-temporal-resolution pollen analysis, but the results of such studies allow paleoecologists to address questions at scales of contemporary ecological interest (Green 1988) (figure 5.5).

Basin size (in addition to morphometry, inflow/outflow, and surrounding vegetation) may determine to a large degree the source area of pollen falling onto the surface of a lake or pond. Moderate-sized lakes (10 to 100 ha) receive pollen mainly from within a 20- to 30-km radius. Finer spatial resolution may be available through the analysis

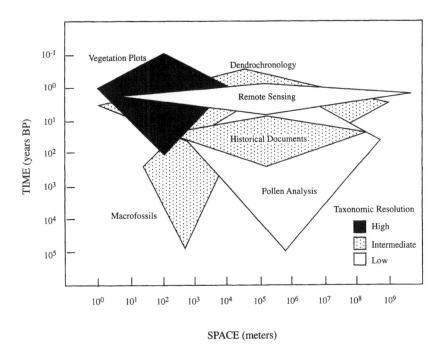

FIGURE 5.4

Although unable to detect environmental phenomena at the finest spatial and temporal scales, paleoecological data can address ecological phenomena at scales approaching those addressed by historical data and field studies, and they can extend our time frame farther back through time. Taxonomic resolution is also a consideration.

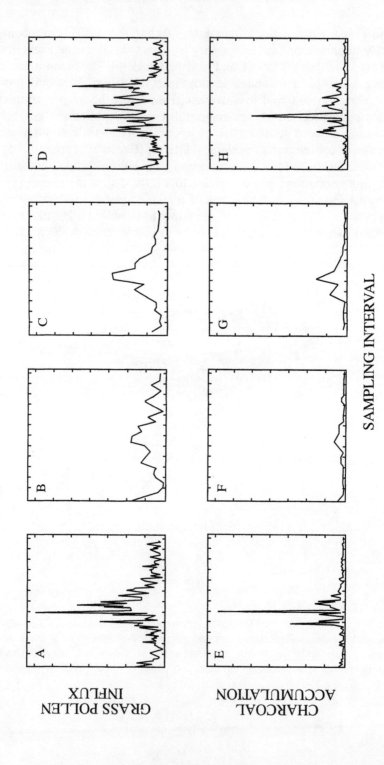

SAMPLING INTERVAL

GRASS POLLEN INFLUX

CHARCOAL ACCUMULATION

of humus and sediments from small forest hollows (Bradshaw 1988) (figure 5.6). These sites, overhung by a tree canopy that contributes abundant pollen and filters the regional pollen rain, collect pollen from a smaller source area than ponds and lakes. Small hollow sites have been used for stand-level interpretations of vegetation change over thousands of years in some cases (Andersen 1984). Analysis of pollen from small hollows and humus is a well-known technique in northern Europe, but North American paleoecologists have only recently adopted it (Heide 1984; Schoonmaker 1992; Davis et al. 1994).

Although humus and small forest hollows have been thought to collect most of their pollen from a radius of 20 to 50 m (Jacobson and Bradshaw 1981; Prentice 1985), recent evidence suggests that the pollen collection properties of these sites vary considerably, and that extralocal and regional pollen may contribute a substantial amount of pollen input (Schoonmaker 1992; Jackson and Wong 1994). Davis et al. (1994) found that 50 to 60 percent of pollen falling on small hollows in northern Michigan came from within 50 meters. But Jackson and Wong (1994) found that sites in New England may receive too much regional pollen to be useful for stand-level interpretation.

Results of a study in southwestern New Hampshire illustrate the caution that must be used in pushing the scale limits of paleoecological data. Pollen from two mor humus sites was analyzed and then compared to reconstructed vegetation plot records and the regional timber harvesting history. The pollen data are in clear disagreement with local stand history (the virtual elimination of *P. strobus* and *T. canadensis* by a hurricane in 1938); locally produced pollen was swamped by extralocal and regional pollen in response to large-scale logging in surrounding Winchester County and in response to the regional demise of *Castanea dentata* (American chestnut) by a fungal blight around 1915 (figure 5.7). Indeed, if the pollen profiles from the two humus cores were interpreted without access to any other information, the resulting "stand history" would be highly inaccurate. This study indi-

FIGURE 5.5

Long-term data offer the opportunity to sample various intervals. Graphs A through D show grass pollen influx sampled for the years 1862 to 1982 at Lake Parramatta (Sydney, Australia): **(A)** yearly, **(B)** 1-year samples spaced 5 years apart, **(C)** contiguous 5-year samples, and **(D)** contiguous thin sections. Graphs E through H show the same sampling intervals for charcoal.

From *Journal of Biogeography*, vol. 15, pp. 685–701, by D. G. Green and G. S. Dolman. Copyright © 1988 by Blackwell Science Ltd. Reprinted with permission of the publisher.

cates that one can not assume all closed canopy sites will provide a clear record of stand-level vegetation change, and that pollen analysts must be careful to test and acknowledge the spatial and temporal limits of data from small hollows and humus.

Because seeds, leaves, and woody plant parts are less mobile than pollen, macrofossil analysis is another alternative for studies whose aim is high spatial resolution (Jackson 1989, 1990). Macrofossils also provide more taxonomic resolution, often to the species level. They are less abundant than pollen, however, resulting in lower temporal resolution; large sediment samples (100 to 200 cm³) may be necessary to acquire a sufficient sample. More empirical work, such as that of Davis and her colleagues in the upper midwestern United States, is needed to define source areas for a range of vegetation types and collection basins. One promising development is the emergence of can-

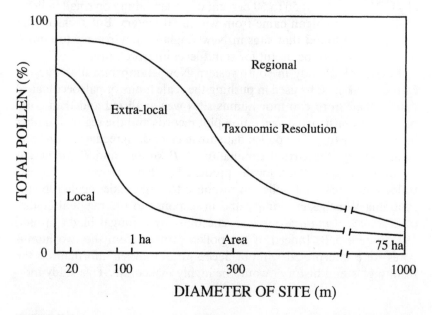

FIGURE 5.6

Hypothesized relationship between size of a site with no inflowing stream and the relative proportion of pollen originating from different areas around the site. Local = 25- to 50-m radius; extra-local = 50- to 500-m radius; regional is greater than 500-m radius. Empirical work is needed to test this hypothetical relationship in a variety of environments.

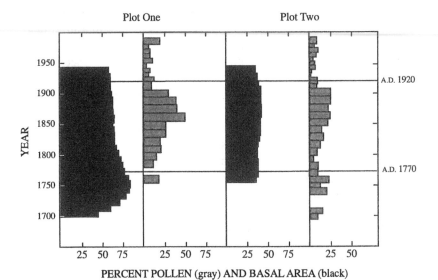

FIGURE 5.7

Pollen data may not provide the spatial resolution that vegetation and macrofossil data do. Pollen percentages of *Pinus strobus* from soil humus under closed canopy in two 400-m² plots, 90 m apart (shown in *gray*), show little relationship to relative basal area reconstructed from live and dead trees (shown in *black*) that records the elimination of pine by a hurricane in 1938. Rather, the pollen data appear to reflect regional pine harvesting, which began about 1850. Thus, the history of this stand in southwestern New Hampshire would be misinterpreted if pollen data alone were used.

From Schoonmaker (1992).

opy-access systems, from which controlled releases of marked pollen grains could further our understanding of pollen dispersal.

The Nature of Biotic Communities

The concept of the community as an organismic and cohesive unit was contentious for much of the twentieth century and has been out of favor for the last few decades. Paleoecological evidence has supported the individualistic interpretation of species roles within communities or assemblages. Much of the evidence has come from the construction of Holocene migration maps, which show that species migrate at different rates in different directions (see figure 5.3). Holocene pollen distribution maps also show that species assemblages have been highly dynamic, with species joining and leaving assemblages in a fluid fashion. For example, *C. dentata* arrived in central New England only 2000 years ago, thousands of years after the other major tree species, and rapidly rose to dominance in parts of its northern range. Other examples of the fluid nature of biotic assemblages include the disappearance of *T. canadensis* in the eastern United States 4800 years ago, probably because of a pathogen, and its reemergence as a dominant species only 500 to 1000 years later (Davis 1983).

Hebda and Whitlock (1996) have coined the term *biogeochron* to address the observation that assemblages of species have a finite existence in a place during an interval of time. Based on Krajina's (1965) term *biogeoclimatic zones* for regional ecosystem units, *biogeochron* fuses the biotic, geological, and time (chronos) elements of the ecosystem concept but includes climate as part of *geo.* The concept can be applied on a continuum of spatial and temporal scales from forest stand to continent, and from disturbance phase to glacial–interglacial cycle.

The biogeochron concept and the ephemeral nature of biotic assemblages imply that disequilibrium processes have characterized communities during the last several million years, and possibly far longer. Species that currently occur together, such as *T. canadensis* and *Fagus grandifolia* (beech), have different sensitivities to climate (on scales of microseconds to thousands of years), and therefore have different migration rates and patterns. Vegetation may appear to closely track climate if sampled every 1000 to 2000 years; vegetation often appears to lag climate if sampled at intervals of less than 500 years.

We need to consider processes on a continuum of spatial, temporal, and taxonomic scales to understand questions of equilibrium, but how are we to quantify rates of change over long periods of time? One approach used by Jacobson and Grimm (1986) was to use detrended correspondence analysis (DCA) to track directions and rates of change in fossil assemblages in Minnesota (figure 5.8). They showed that vegetation change was most rapid at two distinct times: soon after deglaciation in response to climate change, and during the last 200 years in response to widespread disturbance by European settlers. Their findings, which indicate that rates of ecosystem change vary widely through time, emphasize the importance of temporal scale for understanding community dynamics.

If disequilibrium conditions have been characteristic of communities throughout the Holocene, we need to reconsider theoretical and

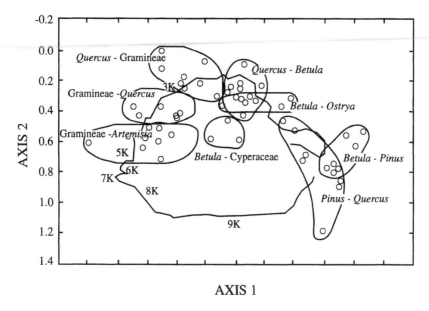

FIGURE 5.8

A five-sample running average of fossil pollen from Bily's Lake (Minnesota) [*irregular line* with time designations 6K, 7K, etc. (K = 1000 years)], plotted on top of analog pollen assemblages (*rough ellipses with open circles*) produced by various vegetation types, tracks the rate and direction of vegetation change at an ecologically meaningful temporal scale of 100-year intervals over the course of 10,000 years. Contrast the rapid vegetation change in a *Pinus-Quercus* assemblage at 9700 years BP to an assemblage with no modern equivalent at 8000 years BP, with the subsequent stasis dominated by grasses and shrubs until ca. 3800 years BP.

From *Ecology*, vol. 67, pp. 958–966, by G. L. Jacobson and E. C. Grimm. Copyright © 1986 by Ecological Society of America. Reprinted with permission of the publisher.

applied population and ecosystem models based on equilibrium conditions. Without factoring in short-term and long-term migration, models of population growth, competition, and predation may be heuristically useful but misleading. This also implies that models developed for short-term time scales do not extrapolate to long-term time scales.

The Role of Disturbance in Communities and Ecosystems

Extensive and severe disturbances are generally more uncommon than small, low-impact disturbances. A comprehensive picture of an area's disturbance regime may not be observable over a human lifetime; it may not be observable over the resident species' lifespans. Thus, we need to look at disturbance regimes over long periods of time to sample the full-range frequency, severity, and extent of disturbance at a site. In many ecosystems, this means we need environmental histories thousands of years in length (Lertzman and Fall, chapter 16). For example, the fire frequency in a 13,000-ha forest in southwestern New Hampshire is approximately 600 years, based on charcoal in a 7000-year sediment sample. However, if the sample period is reduced to 1000-year time slices, the temporal variation of disturbance could lead one to conclude either that fire does not occur in this region, or that these forests burn every 330 years, depending on the time period sampled (Schoonmaker 1992). The use of the long paleoecological record enables ecologists to avoid the space-for-time substitution that may be highly misleading because of the difficulty of assuring between-site uniformity.

Why should we care about disturbance? How do we know that the large-scale, severe disturbances we have documented during this century are not anomalous? Because the paleoecological record tells us so: it shows overwhelmingly that disturbance has been an important environmental driver in most regions of the world. Charcoal in sediment layers records fire events; geomorphologic features and sediment lithology record erosional events; and tephra layers document volcanic eruptions. Sudden changes in pollen assemblages often correlate with catastrophic events or anthropogenic activity (Brugam 1978; Green 1982). Patterns of disturbance change with environmental conditions. Grimm (1983) showed that mesic tree species invaded oak woodland in southern Minnesota when the cooler, moister climate of the Little Ice Age decreased fire frequency. He suggested that climate, physiography, vegetation pattern, and fire interact to cause thresholds for vegetation change to vary in space and time (figure 5.9). One result is that different vegetation types can persist in neighboring identical environments, with their locations being historically contingent on past disturbances (Parker and Pickett, chapter 8).

Recent and paleoecological evidence confirms that the regional vegetation mosaic in communities with fire-dependent species appears to have been maintained by naturally recurring fires (Heinselman 1973; Swain 1978). In these communities, fire is responsible for long-term stability; the lack of fire caused by climatic change might be considered more of a disturbance to the overall system, causing widespread species replacement (Grimm 1983). Conversely, during a time of climatic change, a community showing some inertia may be opened up to invasion by fire. Green (1982) documented rapid and permanent species replacement following fire in the early Holocene in Nova Scotia.

Disturbance in tropical areas has been characterized as less severe and less extensive than that in temperate regions (Whitmore 1975), with many studies focusing on tree- and stand-level disturbance (Denslow 1980; Uhl et al. 1988). But we know that hurricanes, droughts, and El Niño have widespread and severe effects on tropical ecosystems. The paleoecological record supports these observations. The

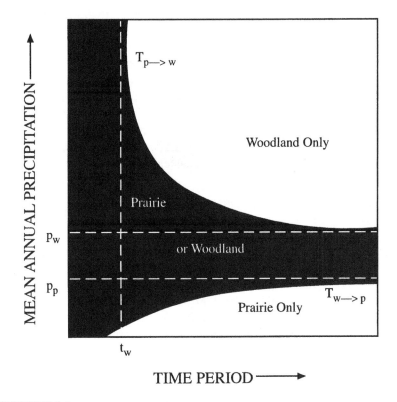

TIME PERIOD ⟶

FIGURE 5.9

Disturbance events and climatic change, unobservable over a lifetime, can interact with vegetation to create thresholds of vegetation change. The paleoecological record of vegetation change from the Big Woods (Minnesota) was used to develop this model of climate, vegetation, disturbance (fire), and physiography interaction. The curves are the temporal and climatic thresholds for changes from *Quercus* woodland to prairie and conversely, from prairie to *Quercus* woodland.

1983 El Niño that caused catastrophic rains in Peru's coastal desert was a 100-year event; the 1000-year events detected by archaeologists in pre-Colombian settlements are even more severe (Nials et al. 1979). Catastrophic flooding has occurred in the Amazon basin during the Holocene (Campbell et al. 1984). A quarter of the upper Manu region of the Amazon basin has been disturbed by flooding, probably within the last few hundred years (Salo et al. 1986). Charcoal dated back to 6000 years BP shows that fire has long been part of the Amazon environment (Saldarriaga and West 1986).

Prehistoric humans have had major impacts on their environments. Neolithic humans likely caused the sudden decline of *Ulmus* (elm) in Europe around 5000 years BP by girdling trees and using the foliage for animal fodder (Iversen 1973). North American Indians had a sophisticated understanding of the use of fire as a tool for hunting and clearing land (Pyne 1982; Boyd, in press). Human agricultural practices drastically altered vegetation in parts of central America during Mayan times (Binford and Leyden 1987). The clearance of forests by Euro-Americans resulted in a distinct settlement horizon of *Ambrosia* (ragweed) in most pollen records in eastern North America.

Focusing Deep: Quaternary Phenomena on an Evolutionary Time Scale

The temporal scales that interest most paleoecologists occupy an intermediate evolutionary time frame that can be organized into four categories: (1) ecological time (thousands of years), (2) orbitally forced climatic changes (20,000 to 100,000 years), (3) geological time (1 million years), and (4) mass extinction time (about 25 million years) (Bennett 1990). The overlap of these time frames offers exciting opportunities for fresh insights. Paleoecologists have accumulated considerable information about species migration behavior and refugial habitats relevant to evolutionary ecology. The rapid migration of individual species offers the best evidence for proposing and testing mechanisms of speciation. The individualistic nature of species movements suggests that, among plants in northern latitudes, long-term coevolutionary relationships are rare. Rather, a successful species is one that has enough inherent variation to adapt to both rapidly changing environmental conditions (at the transitions between glacial and interglacial periods) and to compete during longer periods of relative stability (10,000-year interglacials and 100,000-year glacial periods).

Species assemblages can be affected by historical contingencies, such as disturbance, and by past migrational episodes that constrain them to a limited range. These constrained ranges are, in effect, refugia. Numerous contemporary refugia can be observed, such as (1) the cloud forest located in the western Andes but with Amazonian affinities, (2) the range of *Gymnogyps californianus* (California condor) and (3) the range of *Pinus aristata* (bristlecone pine). Evolutionary ecologists often speculate on past refugia based on endemism and draw conclusions about evolutionary mechanisms of speciation. Paleoecological evidence for refugia is scarce; for example, few tropical rainforest refugia have been documented. This is not necessarily because they were rare in the past, but because they are difficult to find in the paleoecological record. One approach to examining past refugial dynamics might be to extrapolate past refugia based on high resolution paleoclimatic models and species environmental tolerances.

From Theory to Applications

The paleoecological perspective extends knowledge of three-dimensional spatial patterns deep into the fourth dimension of time. Patterns observed for extended periods of time can help us identify processes that operate at these scales and understand their interaction with shorter-term processes. Long-term patterns and processes can be compared with current phenomena to interpret and perhaps predict environmental and biotic change.

Paleoecology primarily has been a descriptive science, but Davis (1994) calls for more hypothesis testing and modeling. She identifies several areas in which paleoecological data can be used to provide a long-term perspective, as well as to test hypotheses and models: global change research, general circulation models (GCMs), models of lake chemistry and sedimentation, and competition and patch dynamics at several scales. Ecologists and paleoecologists must demand that paleoecological studies rigorously test contemporary ecological theory. Yet, paleoecological data will continue to be important for retrospective descriptions that detect long-term trends in local, regional, and global environments. Most ecologists are aware of the geology and recent history of their study sites, but few take full advantage of the available paleoecological data to better understand disturbance regimes and other historical constraints that may operate at several spatial and temporal scales (Hamburg and Sanford 1986).

Comparing paleoecological and current data not only allows the community ecologist and evolutionary biologist to place current phenomena in the context of a longer time frame, but also it can be helpful to environmental managers. We know soils have become increasingly leached and nutrient poor in northern temperate areas during the latter half of the Holocene (Engstrom and Hansen 1985). Long-term ecosystem acidification driven by climatic and vegetation change should have important implications for foresters and agronomists. On a shorter time scale, tree-ring chronologies and historical records may help us discern what changes constitute a long-term trend (such as global warming) versus a recent cycle (such as the 20-year precipitation pattern across much of Saharan Africa) versus random short-term variation (a 2-year drought in parts of North America).

Species responses to past environmental change can help predict future behavior. Most atmospheric scientists agree that the earth's atmosphere is warming because of anthropogenically produced pollutants. Most ecologists agree that species will respond at physiological, organismal, and population levels. One of the more obvious outcomes will be species migration. Past vegetation changes integrated with GCMs that include anthropogenically generated global warming may help predict species migration patterns. This is a difficult undertaking because (1) current and past climatic changes have occurred at various rates, (2) current changes are atmospherically driven rather than orbitally induced as in the past, (3) physiological effects of greenhouse gases may influence species responses, and (4) the earth's predicted future orbital geometry indicates global cooling, a trend reflected in global climate and vegetation over the last 2000 years up to the nineteenth century. Furthermore, GCMs and paleoecological evidence suggest that disturbance regimes will likely change with changing climate and vegetation (Rind et al. 1989; Mooney et al. 1993; Wright et al. 1993b). Predictions of biotic responses to global warming must account for all these factors and must recognize that future climate may not simply be warmer or drier but entirely different across a continuum of spatial scales.

In light of the difficulties of predicting global climatic change, what generalizations can we make about species responses? First, species will respond individualistically, but they will generally move poleward or upward in altitude. Second, species limited to high latitude and alpine environments will be the most likely to become extinct. Third, responses will vary locally with the severity of climatic change. Fourth, where climatic change is severe, response rates will be at or near their maximum, but they may still lag behind climatic change. Fifth, species that are more phenotypically and genotypically flexible in their responses will be favored over less flexible ones. Sixth, assemblages of species will change as species migrate at different rates.

How should we design wilderness preserves and refuges that protect endangered species in light of potential global warming? Past species migration and the variable nature of past Quaternary refugia suggest that planning refuges is a more complicated task than most conservationists realize. Conservation advocates and planners must seriously commit to understanding a fourth, temporal, dimension. Communities as we presently classify them will add and lose species, perhaps producing assemblages that have never occurred in the earth's history. Hunter et al. (1988) emphasize continental-scale migration corridors and large reserves encompassing a variety of physical environments. This contrasts with the present strategy of preserving specific communities or ecosystems.

Paleoecological techniques should be employed more widely to document an area's history before a decision is made to preserve it in its assumed natural state. Backman and Patterson (1984) found that a *Quercus* (scrub oak) and *Pinus rigida* (pitch pine) forest in southeastern Massachusetts, which has been managed to maintain that unique habitat, had been a *P. strobus*-hardwoods forest in precolonial times and was likely altered by land-clearing associated with settlement and extensive fire. Such information may not deter efforts to preserve an area, but it might greatly alter management procedures. Disturbance histories can be reconstructed to help (1) land managers understand natural disturbance cycles and plan accordingly (Heinselman 1973; Morrison and Swanson 1990; Lertzman and Fall, chapter 16), (2) implement national park management strategies (Romme 1982), and (3) provide a scientific basis for regional resource management plans (FEMAT 1993).

Ecosystem restoration is an emerging discipline for which paleoecological data are particularly useful in identifying baseline characteristics of predisturbance communities, and for documenting environmental changes caused by humans (Engstrom and Wright 1984). Past abundance of organisms sensitive to acidity, toxic compounds, and eutrophication can give us an idea of natural conditions in lakes that are currently polluted. For example, Stockner and Benson (1967) studied diatom assemblages and detected changes in eutrophication in Lake Washington associated with the development of and sewage input from the city of Seattle. After sewage was diverted from the lake, Shapiro et al. (1971) recorded improvements in water quality by comparing the chemistry of pre- and postsewage sediment cores. Davis (1987) describes an extensive effort to document changes in lake acidification by analyzing past diatom populations from a network of lakes in the northeastern United States.

Another arena in which long-term data are desperately needed is in fisheries management and restoration. Analyses of *Sardinops caerulea* (Pacific sardine) and *Engraulis mordax* (northern anchovy) show that populations of these species fluctuated greatly over the last 2000 years; such long-term records provide the context for planning a fishery that adapts to these fluctuations (Baumgartner et al. 1992) (figure 5.10).

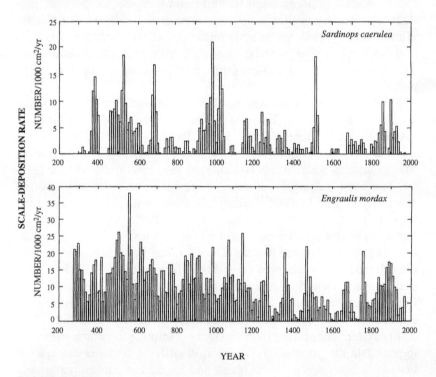

FIGURE 5.10

Sardinops caerulea and *Engraulis mordax* scales from annually varved sediments from the Santa Barbara Basin (California) show approximately 9 and 21 major recoveries and collapses for the two species, respectively, over the last 1700 years. These high-resolution data closely track the peak and collapse of the *S. caerulea* catch from 1920 to 1950. The same challenges that overwhelmed this fishery earlier in the twentieth century are faced by the *Oncorhynchus* fishery in the north Pacific, yet current management paradigms are limited to a time frame of 100 years or less.

From Baumgartner et al. (1992). Reprinted with permission of the author.

One of the most vexing problems in understanding the decline of the *Oncorhynchus* species (Pacific salmon) stocks is that there are few consistent, long-term population data. Most stock estimates are patched together from a variety of sources that go back only 100 years at best, and usually less than 50 years. Many fisheries biologists and managers assume that the early part of this period (1870 to 1900) represents the long-term productivity of *Oncorhynchus,* but there is no evidence of this over the long term. In fact, several observations suggest the opposite. By 1850, populations of the principal natural predator of pacific salmon, Northwest Coast peoples, had been decimated. The Little Ice Age may have enhanced habitat, especially in the southern portion of the range of *Oncorhynchus.* Furthermore, archaeological evidence suggests that Northwest Coast human populations increased dramatically about 2000 to 3000 years BP, possibly in response to an increase in *Oncorhynchus* abundance (Chatters 1996). On a glacial–interglacial time scale, *Oncorhynchus* has shown remarkable resilience, considering that over half its present range has been occupied by glaciers dozens of times (Groot and Margolis 1991). One must conclude that we are uninformed about long- and short-term fish population trends, which makes it difficult to develop management objectives that lead to long-term sustainability of *Oncorhynchus.* What should we do? Clearly we need long-term paleoecological data on salmon population dynamics. A network of lakes should be sampled throughout the range of *Oncorhynchus;* salmon scales and stable isotopes should be used to determine how *Oncorhynchus* populations have varied on a scale of millennia.

Related to the fate of Pacific salmon is a better understanding of the long-term dynamics of river channels and riparian areas. Historical studies show us that pre-European river channels in North America were highly dynamic and structurally complex. The complexity and connectivity of river systems have been simplified, especially in urban and agricultural settings (Maser and Sedell 1994). Late twentieth century flood events in the Pacific Northwest (United States and British Columbia) and along the Mississippi River watershed demonstrate that current management practices show little understanding of long-term river system dynamics. Our knowledge of pre-European channel morphology and subsequent riparian history provides a base for understanding longer-term dynamics, but Holocene-scale studies of channel dynamics, coarse woody debris, and riparian habitats are needed to fully inform aquatic ecosystem managers and land-use planners.

Forest, fisheries, and river system management is now evolving with a greater focus on sustainability of natural resources. Given this goal, the jobs of ecologists and resource managers have become increasingly complex because multiple spatial and temporal scales must be considered (Hobbs 1997, chapter 20). Any discussion of sustainability is meaningless without embracing long time scales and broad spatial scales. Forest harvesting in the Mediterranean region of Europe thousands of years ago may have appeared sustainable at the time; the long-term result was a legacy of wholesale vegetation change (Perlin 1989). The same could be said of European forestry systems in the sixteenth through twentieth centuries. Longer-term climatic and landscape dynamics, perhaps in conjunction with pollution effects, have shown that a paleoecological perspective can be valuable for long-term natural resource management.

Summary

Most humans perceive time on a diurnal to annual time frame (Rykiel, chapter 21). Businesses run on quarterly to several-year time scales. Some of us see and act on a decade planning cycle (career goals, family plans, retirement). Very few of us put much energy into multigenerational goals (e.g., estate planning for great-grandchildren). Yet, a paleoecological perspective shows that it is reasonable to identify the potential effects of a particular management regime 500 to 1000 years hence. Natural systems operate on these scales, and we also must think and plan with long-term perspectives if we wish to understand ecological phenomena and manage natural resources sustainably.

Long-term paleoecological data show that ecological studies limited to years and decades often overlook phenomena that operate over longer time periods. Few ecologists have a world view that embraces the notion that plants move, and that tree populations migrate often and continuously on a generational basis. Plant and animal populations and communities have often been out of phase, biotically and abiotically, during the last 10,000 years. Few ecologists realize that paleoecological techniques can probe issues at "ecologically relevant" scales of hectares and years, and that some fossil evidence cannot be pushed to these finer resolutions.

The paleoecological perspective offers great opportunities for testing current ecological hypotheses. The usefulness of hypothesis testing enhances, but does not replace, the value of historical inquiry. Just as we are years away from documenting the diversity of life on the earth through systematic studies, there are many decades of work to be done simply documenting past environments in remote, undersampled regions of the world.

Both approaches—historical documentation and hypothesis testing—will help scientists and managers expand their conceptual understanding of environmental issues by stretching the spatial and temporal vistas from which they observe living systems.

6

SPACE AND TIME IN THE SOIL LANDSCAPE: THE ILL-DEFINED ECOLOGICAL UNIVERSE

R. David Hammer

Soil and water permit terrestrial life. Soil—earth's accrued net balance of rock weathering—is the legacy of historical plant populations. Soil is held in place by plants, whose lives depend on both the soil and the waters that constantly seek to move soils to the seas. Synergisms among biota, soils, and terrestrial waters are ancient, complex, and so intertwined that distributions and processes of one can be understood only in context with the others.

Soils, landscapes, and their associated biota coevolve. Climatic changes affect biota, soils, and hydrology, which, in turn, modify geomorphic processes. Geomorphic changes affect soils and hydrology, causing the biota to adjust, and then soils and hydrology respond to the biota. Change begets change, and systems are in constant flux. The biotic and abiotic system components respond at different temporal and spatial rates and scales. Some soil and landscape changes are irreversible, others are not. The soil may retain consequences of certain processes after the causal influence is removed.

Organic matter accumulation to depths sufficient to form a mollic epipedon (dark surface horizon with high base saturation) may occur within centuries. Clay accumulation of adequate quantity and thickness to form a diagnostic argillic horizon may require millennia. Many northern Missouri soils contain both mollic epipedons bequeathed by prairie and argillic horizons that evolved under forests. Presence of both features in soils indicates that the forest–prairie ecotone shifted during the Pleistocene (Hammer et al. 1995).

Pipestems in upland Iowa soils (Ruhe 1956) are a second example of soil memory (the retention in soils of past soil-forming processes). Pipestems are cylindrical accumulations of iron oxides around plant roots. They form in wet soils when reduced iron, which is mobile in soil solution, is oxidized and precipitated in root rhizospheres. Pipestems are relicts in today's (now better-drained) soils, and are evidence of drainageway incision and accompanying landscape hydrologic changes after Wisconsin glacial retreat.

The important question in examining dynamics among the living and nonliving landscape components is: How profoundly and at what rates are the reactions of one system component compared to changes in another component or to perturbations? The research challenge lies in observing and measuring response to a particular event or process in the midst of natural dynamics (variance). This is the point at which scale concepts become relevant to, and perhaps inseparable from, individual or societal perspectives.

Spatial and temporal scales in nature can be matters of individual perceptions, which are dependent on one's belief system. Two people gazing at an ecosystem often will see different things, and when looking at the same entity, they may place it in different contexts or assign different importance to its role. Rowe (1984b) illustrates this by quoting a young geologist, who, when introduced to plate tectonic theory, stated, "I wouldn't have seen it if I hadn't believed it." Discussions by Rowe (1984a,b), Peters (1991), Rigler and Peters (1995), and Allen and Hoekstra (1992) on the impacts of belief systems in ecological studies are relevant to scale issues.

Unfortunately, many ecologists and other scientists have trivialized the soil landscape. Soil and site descriptions in ecological literature often are useless for those who might wish to replicate the described experiment or to extrapolate results. Scientific soil classifications are seldom presented in ecological literature, although Soil Taxonomy (Soil Survey Staff 1975) paints a vivid mental picture of the classified soil and its soil-forming environment. What would be the chances of publishing in the scientific literature a manuscript in which vegetation was described as "large trees with many branches and flat, green leaves"? This seems ridiculous, but it is no more so than describing soils in one of the following ways:

1. "basaltic andesite with a coarse sandy loam texture" (Pantastico-Caldas and Venable 1993);
2. "a pair of adjacent soil types: one derived from country rock and

one derived from hydrothermally altered parent materials" (Schlesinger et al. 1989);
3. *No description* of site or soil (Bowman et al. 1993);
4. "the two sites are located on somewhat different soils" (Burke et al. 1995a).

Many ecologists' assumptions of soil-landscape homogeneity would serve well as testable research hypotheses. Unfortunately, some soil scientists also do not view landscapes as interdependent, coevolving populations of geomorphic materials, biological communities, and soils. Much forest soil research is based on sampling the few centimeters of surface soil. Published research from many ecological and agricultural studies does not consider the entire plant-rooting volume or the role of hydrology in the distributions of minerals, chemicals, and plants.

Many pedologists have become so enamored of taxonomy that they have neglected depositional environments and soil-landscape relationships (Daniels and Hammer 1992). Soil scientists have been successful at dealing with management history, fertility, and increasing crop yields, but they have often neglected water quality, the importance of soil microorganisms, and the long-term effects of cultivation on drainage systems and microclimate. Society's environmental problems are serious, complex, and global. Solutions depend on scientists of all disciplines working together to study functional interactions of ecosystem components.

Variance can be measured and placed in a relevant context only if its existence is recognized. A priori assumptions of homogeneity compromise sampling strategies and data interpretations. Relevant questions in studying variance in natural systems are: How much, where, and when does it vary? What causes the variation? and What are the consequences of the variation for the intended use of the resource?

This chapter examines spatial and temporal soil-landscape relationships. The initial plan was to discuss soil variability within macro- (regional), meso- (watershed), and micro- (landform) scales, while partitioning variance by causal mechanisms as dictated by scale constraints. The discussion was to be framed in the context of current ecological issues. However, recent literature indicates that many ecologists' conceptual soil science models are dated and incomplete. Their understanding of soil systems is inadequate for substantive discussion of patterns and processes in soil systems. Thus, realizing that remedial discussion is necessary, the initial plan was modified.

This chapter assumes that ecologists (1) want to use sound scientific methods to solve problems important to sustaining the health of the earth and (2) will respond positively to well-intended criticism. Hydrology, geomorphology, and stratigraphy will be introduced as determinants of soil variability. The discussion will focus on the abiotic—not to trivialize the biota, but to emphasize landscape fabric (geomorphic and soil patterns and processes) to an audience that has conceptually homogenized it. Much of the soil science literature cited here is old but not antiquated. These citations are the origins of concepts that have endured through acquisition of new knowledge. Soils will be presented as multivariate systems and placed into a landscape context. A specific soil system will be used to illustrate ideas, and published literature will provide examples for discussion of soil geomorphology in an ecological framework. It is hoped that the concepts presented here can be used to guide more effective integration of soil science with ecology.

The Soil Landscape

The Landscape Defined

A landscape is a population of geomorphically related landforms. A landform is a landsurface unit defined by its surface shape, its location with respect to other landforms, the underlying geologic materials, and the kinds of soils that have formed on it. A landform may contain more than one geomorphic surface. For example, a backslope (the area between an upland shoulder and a lower alluvial bottom) may contain coalescing convex and concave surfaces. Similar landforms recur within the landscape but with degrees of variation.

Landforms cannot be classified by surficial features alone because polygenetic pathways can produce similar surface shapes through different geomorphic processes. For example, glacially modified landscapes may contain outwash plains and glacial lake beds. Both deposits will have nearly level surfaces, but textural patterns will differ. Lacustrine deposits will be fine-textured, and will fine upward with a narrow range of particle sizes. Outwash deposits can be very heterogeneous, with textures ranging from silts to coarse sands and gravels over short distances, both vertically and horizontally.

Macro features that affect the distribution of water in space and time are important landscape attributes (Daniels and Hammer 1992). Landscapes are partially defined by these features, which include the distance between interfluve summits, the shape and width of interfluve summits, the relief between summit and streams of specified order, and the lengths, steepnesses, and complexities of backslopes.

Merging Earth Sciences

Hillslope hydrology, pedology, and geomorphology are inseparable (Gerrard 1990; Daniels and Hammer 1992). Surface conditions, including shape, slope length, aspect, soil texture and structure, organic matter content, position in the landscape, and existing vegetation affect infiltration of water into the soil. Percolation (subsurface water movement) through the soil landscape is controlled by subsurface soil structure and textures of soils, sediments, and strata. Sediment distributions contribute to landscape heterogeneity. Many parent materials can be found in a cross section of terrain from loess hills to the river bank adjacent to the Missouri River in central Missouri (figure 6.1). At a finer scale, the river floodplain contains scores of deposits. Identifying and properly interpreting these materials is essential to understanding structure and functions of landscapes. Key questions in analyzing geomorphic processes affected by surface hydrology include the following: (1) How rapidly, when, where, and in what forms (snow, runon, etc.) does water enter the system? (2) Where does surface runoff go? and (3) What does it carry and either reorganize, relocate, or remove in transit?

The Role of Subsurface Water

Synergisms among landforms and soils derive from water movement on and within the system. Water shapes the landscape, participates in chemical and physical weathering, carries dissolved and suspended materials to different places in the soil profile, landscape, or drainage network, and is the solution within which new compounds are formed from the products of weathering. Mean annual precipitation is important to these soil-forming processes, but it is secondary to temporal water distribution (Hillel 1991). A landscape that annually receives 100 cm of water distributed primarily in winter has different vegetation, soils, and geomorphology than a landscape that receives the same amount of water distributed throughout the year. Temperature affects chemical reactions, which in turn affect dissolutions and control syntheses. Winter soil saturation and summer soil saturation affect soil development differently (Yaalon 1983).

Determinants of how water affects soil-landscape development are (1) how much water enters the soil, (2) what kinds of minerals the water contacts, (3) how long the water is in the soil, (4) the depth to which water penetrates, (5) how water is distributed internally in the soil profile and on the soil landscape, (6) the temperature when the water is in the soil, (7) whether the water is stationary or moving, and (8) the water oxygen content.

The Soil Survey Staff (1975) categorizes regional water regimes on the basis of plant-available water during the growing season. Annual

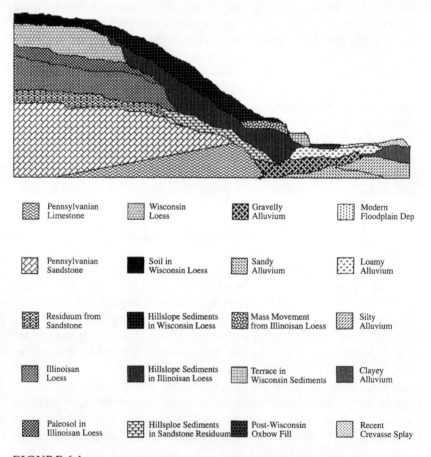

Pennsylvanian Limestone	Wisconsin Loess	Gravelly Alluvium	Modern Floodplain Dep
Pennsylvanian Sandstone	Soil in Wisconsin Loess	Sandy Alluvium	Loamy Alluvium
Residuum from Sandstone	Hillslope Sediments in Wisconsin Loess	Mass Movement from Illinoisan Loess	Silty Alluvium
Illinoisan Loess	Hillslope Sediments in Illinoisan Loess	Terrace in Wisconsin Sediments	Clayey Alluvium
Paleosol in Illinoisan Loess	Hillsploe Sediments in Sandstone Residuum	Post-Wisconsin Oxbow Fill	Recent Crevasse Splay

FIGURE 6.1

Possible parent materials in a cross section of central Missouri from loess bluffs to the Missouri River channel. Not all parent materials will be in a single transect.

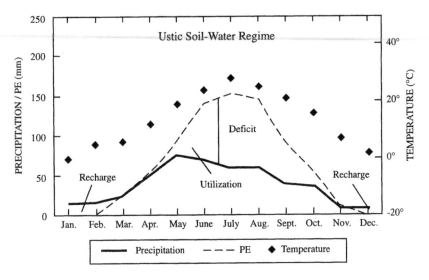

FIGURE 6.2

The ustic soil-water regime. Potential evapotranspiration (PE) is indicated. Adapted from Soil Survey Staff (1975).

distributions of precipitation and potential evapotranspiration are criteria. The ustic soil water regime (figure 6.2) has surplus water early in the growing season, but water becomes plant-limiting as soil reserves are withdrawn and summer rainfall is inadequate for soil water recharge. Soil water recharge occurs in most years after the growing season, so ustic soils are leached to varying depths depending on local combinations of soils, geomorphology, and stratigraphy.

Aridic, udic, and xeric are other regional water regimes. These classifications are modified locally by topography and soils. Aquic soil water regimes can exist in any landscape, and soils in the aridic soil water regime can have ustic attributes in some landscape settings. Crompton (1962) succinctly placed water and temperature as weathering and alteration agents into the contexts of richness of weathering and intensity of leaching.

Water Movement in the Soil: The Soil Profile as a Record

Water movement in the soil is temporally and spatially complex; it is site-specific and cannot be inferred on the basis of surficial features (shape and surface horizon texture) alone. Soil water storage and movement are controlled by sizes and distributions of soil pores. Structure (the aggregation of sand, silt, and clay separates into larger entities) may be more important than soil texture in partitioning soil water movement and storage, particularly for soils with fine textures. Sizes, shapes, and orientations of the structural aggregates are important porosity determinants. Parent material discontinuities (loess over till, hillslope sediments over preexisting surface, fluvial strata, etc.) create porosity and bulk density differences and often can cause seasonal zones of saturation.

Mineralogy is another overlooked porosity determinant in fine-textured soils. Montmorillonite expands and is sticky when wet and shrinks and becomes hard when dry. Soil water may move rapidly through large structural voids in dry montmorillonitic soils, but the voids will close and permeability is greatly reduced. Soils with high concentrations of kaolinite, gibbsite, or iron oxides are more structurally stable through changing wetness regimes.

Soils as Natural Bodies

Recognition of soils as natural bodies, and of the soil profile as a record of past and current processes in soil genesis (Coffey 1912; Marbut 1922) is one of the most significant principles of earth sciences (Ableiter 1949; Simonson 1951; Nikiforoff 1959). Horizonation is the systematic arrangement of the weathering products within soil profiles, and it produces attributes that are the keys to understanding a soil in its environment (Daniels and Hammer 1992; McSweeney et al. 1992; Wilding 1992). Anderson's (1972) perspective—more is different in biology—certainly pertains to soil horizons, soil profiles, landforms, and landscapes.

Soil Horizons

Soil horizons result from interactions of geomorphic and pedogenic processes and give each soil its unique character (Simonson 1959). Horizons are the arrangements of chemical and physical attributes into discernible volumes within soils and landforms. A soil volume may be leached of soluble constituents, some of which are lost in percolating waters, and others are synthesized into secondary clays that enrich deeper horizons in the profile and lower positions in the landscape. Elements and chemicals are differentially distributed within soil profiles and soil landscapes (Kurtz and Melsted 1973). Studying multiple soil pits along hillslope transects is an effective method of observing hydrologic and soil relationships, if the pits are sufficiently deep and wide.

Butler (1959) and Vreeken (1984) suggest that distributions of soil horizons in landscapes are the keys to inferring hillslope processes and distributions of surficial deposits. Bouma (1989) recognizes soil horizons as the key carriers of soil information. McSweeney et al. (1992) proposed that soil horizons should be used with three-dimensional computer simulation to model landscape behavior.

Chemical and physical (morphological) horizon features reflect temporal soil water distributions. Concretions, redoximorphic features (Vepraskis 1992), clay films, siltans, and ped orientations can be used with other soil and landscape features to determine relative distributions of soil water and the products of weathering. Interpretations must be verified by field measurements to establish a predictive model. Investigators who ignore soil horizonation lose a valuable source of information. Collectively, soil horizons indicate current soil processes important to distributions of organic carbon (C), water, and nutrients and can be valuable components of ecological studies. Ecologists do not necessarily need to become pedologists. However, multidisciplinary research benefits all earth and biological scientists who want to broaden their knowledge of ecosystem pattern and process.

Soil Sampling

Representative Samples

Studying soils is difficult, especially because exposing a soil profile may require considerable effort. More work is required to properly describe, sample, and analyze soils, and, unlike plants, birds, and insects, soils are not discrete, easily recognized individuals. Soils are complex natural bodies characterized by temporal and spatial variabilities of scores of chemical and physical attributes. Some of these attributes covary, others do not, and the relationships change within and among landscapes. Claims that an investigator has sampled a "representative" soil, site, or transect are rarely substantiated. Determining a representative portion of a multivariate continuum is a profound statistical challenge.

Figure 6.3 illustrates temporal and spatial variability of a single soil variable, organic C concentration, in the A horizon of fine-loamy,

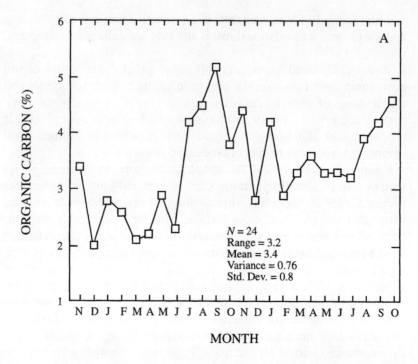

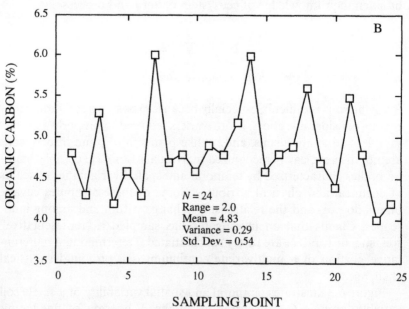

mixed, mesic Typic Hapludults under mixed mesophytic forest on the Cumberland Plateau (Tennessee) (Hammer et al. 1987). Data are from adjacent 2.5 m × 2.5 m plots divided into 0.5 m × 0.5 m subplots. Figure 6.3 shows temporal variability from 24 subplots in one plot sampled randomly, without replacement, at monthly intervals for 2 years, and spatial variability among 24 subplots sampled on a single day in the adjacent plot.

Which sample is "representative?" How many samples would be required to reduce the standard error to 10 percent of the mean? Can a single value represent organic C concentrations from this soil volume, or should one use some measure of the variance about a central tendency? A single value implies that no temporal or spatial variability exists.

The Soil Individual— Where Is It?

Soil properties change horizontally, vertically, and at different rates into different values or into different materials. Except for pedologists and soil mappers, most people do not view many soils or become familiar with the combinations of soil attributes that affect the biota and near-surface hydrology with which soils are intertwined. Therefore, important soil-landscape patterns are overlooked.

Soils are multivariate entities with many properties from each horizon. Most soils form within multiple parent materials, each with unique properties (Daniels and Hammer 1992; McSweeney et al. 1992; Wilding 1992). The lower limit of soil is the depth of rooting of native perennial plants (Soil Survey Staff 1975), which can penetrate soils and geologic materials to considerable depths (Stone and Kalisz 1991). Root pores may persist at depths for decades in soils converted to agriculture, so one usually can determine the lower soil boundary even in the absence of native plants. *Liriodendron tulipifera* (yellow poplar) roots extend 5 m deep on the Cumberland Plateau, and *Pseudotsuga menziesii* (Douglas-fir) roots extend 8 m deep in outwash sands

FIGURE 6.3

Temporal **(A)** and spatial **(B)** distributions of A-horizon soil organic C concentration on mid-Cumberland Plateau forested Ultisols. Samples are from twenty-four 0.25-m^2 cells in 0.5-m × 0.5-m grids. Temporal samples were collected monthly for 2 years and spatial samples were collected on a single day.

on the Olympic Peninsula (Washington). Determining soil nutrient pools and hydrologic relationships for these soils requires more information than is obtained from surface textures of haphazard grab samples.

The Soil as a Sampling Unit

Sampling, instrumenting, and interpreting soils requires observing and measuring numerous soil attributes for each horizon and developing a working model of the relationships of soil and landscape features within a particular weathering environment. Attributes that can be identified and measured in the field are (1) horizon thickness, (2) matrix soil color and size, abundance, colors, and contrasts of redoximorphic features, (3) horizon boundary thickness and pattern, (4) texture, (5) primary and secondary (and sometime tertiary) structure, (6) size, abundance, and orientations of roots and pores, (7) rock fragment content, (8) wet, dry, and moist consistencies, (9) abundance, thickness, and orientation of clay films, skeletons, and organic or other ped coatings or pore linings, and (10) size, color, shape, and abundance of concretions, nodules, and aggregations.

Laboratory analyses of certain physical and chemical attributes are necessary to interpret soil conditions, genesis, and response to perturbations. These include bulk density, pH, nutrient elements, extractable aluminum (Al), organic matter content, cation exchange capacity (CEC), base saturation, and extractable acidity. These data (requisites for other disciplines) can characterize soil processes and interactions, quantify nutrient pools, draw inferences about weathering, and accurately characterize plant rooting environments.

A Field Example

Structure and Soil Water in a Forested Ultisol

The Typic Hapludult from which the C data were presented illustrates synergisms among soil water, structure, and root distributions. The silty A horizon with granular structure overlies a loamy E horizon with weak, fine, subangular blocky structure, and an argillic horizon with moderate, coarse, prismatic structure parting to strong, medium, subangular blocks (figure 6.4). Illuviated clay in the argillic horizon welds silty loess to underlying sandstone residuum. Strong, medium, horizontal blocks are above the bedrock contact.

Clay film distribution in this soil is related to structural patterns and temporal soil water distributions. Wetting and drying cycles orient and deposit clay films in argillic horizons (Thorp et al. 1959). Thick, continuous, abrupt clay films on vertical upper prism faces (Bt1, Bt2) become moderately thick on vertical faces and thin on horizontal block faces within the prisms. This clay distributional change indicates less frequent wetting and drying cycles in prism interiors than on prism surfaces.

Clay films become thin and patchy on lower prism and interior block faces (2Bt3) but are very thick and continuous on block faces above the bedrock (2Bt1'). This pattern change seems to indicate a periodic, perched water table, which would inhibit clay deposition. Moderately abundant manganese concretions in deeper prisms (2Bc)

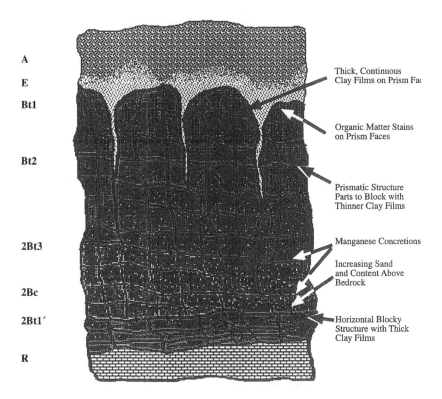

FIGURE 6.4

Representation of soil profile attributes for a fine-loamy, mixed, mesic Typic Hapludult on the mid-Cumberland Plateau in Tennessee.

enforce the perched water table theory because manganese can precipitate where periodic wetting and drying occur (Daniels et al. 1962). Thick clay films on horizontally oriented blocks above the bedrock suggest both periodic wetting and drying and a lateral vector of water movement over the bedrock.

Patchy organic matter coatings are on upper prism faces, whereas thin, continuous organic matter coatings are on the horizontal blocks, and no organic matter deposits are visible between these depths. Visible subsoil organic matter only with thick clay films supports Thorpe et al. (1959), who observed that organic acids enhance clay mobility in soil solution and may precipitate with the clay.

Inferences about soil-water distribution were tested by instrumenting the soil with fiberglass soil moisture cells. Cells were located below the A horizon, between prisms in the Bt1 horizon (labeled "high prism face"), in a Bt2 horizon prism interior, and between horizontal block faces in the 2Bt1' horizon (labeled "horizontal"). Readings were made at 3- to 4-week intervals for 2 years.

Interpretations were confirmed by field soil-water measurements (figure 6.5). The abscissa is soil-moisture cell readings in resistances, with higher resistances indicating relatively less soil water. The A horizon generally is drier than the subsurface except during dry summer periods. Excess water passes through the porous A horizon, and remaining A-horizon water is withdrawn by roots. The prism interior lags the prism face in response to wetting and drying. In the first (dry) summer, the prism interior was drier than prism faces, but in the second (relatively wetter) summer, the prism interior remained more moist than the upper prism face. In both years, the deeper horizons remained more moist than the upper argillic horizon. The periodicity of wetting and drying on the upper prism face during the second year, and the rapidity of deeper profile recharge in November, illustrate rapid water movement through voids to deeper horizons. This soil is hydraulically recharged from the bottom up. The soil-water profile is not a temporal or spatial gradient and can be inferred from soil morphology even when the soil is dry.

Soil Heterogeneity

Ped coatings differ texturally and chemically from ped interiors (Buol and Hole 1959; Heil and Buntley 1965). Roots can chemically modify their rhizospheres, and they have been measured to do so more profoundly for their lateral than for their vertical extensions (Yang et al. 1995; Zhang et al. 1996). This suggests that roots may have different functions according to their depth distributions. Do these root functions also change temporally? Do roots respond to differential soil conditions or help to cause them?

Clay mineralogy in this soil is complex. Hydroxy-interlayered vermiculites are the most abundant clay in the A horizon. Kaolinite is most abundant in the E and in argillic horizon prism interiors. Montmorillonite is abundant in clay films on upper argillic horizon ped faces. Gibbsite is in the 2Bt1'. Relative cation exchange capacity (CEC) is in the order montmorillonite > vermiculite > kaolinite > gibbsite, and depth distribution of CEC is controlled by a combination of organic matter and clay mineralogy (figure 6.6). The CEC decreases from the A horizon to the E, where its value is the lowest in the profile, although C content remains relatively high. The CEC then continually increases, even though C concentration remains relatively low. The CEC maximum is in the 2Bt1', and this seems to indicate that organic matter has a different form with higher CEC than in the A horizon. A

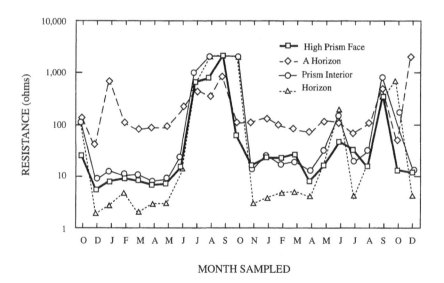

FIGURE 6.5

Resistance readings from cells placed in a Typic Hapludult on the mid-Cumberland Plateau in Tennessee. Observations were made monthly for 2 years. Higher resistances indicate relatively drier soil.

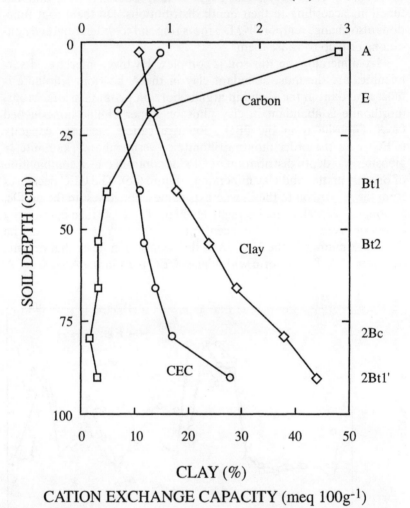

FIGURE 6.6

Depth distributions of soil organic C, clay, and cation exchange capacity (CEC) in a Typic Hapludult on the mid-Cumberland Plateau in Tennessee.

slight increase in C occurs in the 2Bt1', but the accompanying CEC change is disproportionate to the increase in clay or C alone.

Root distribution in this soil seems to be controlled by soil structure. A web of fine roots is distributed throughout the A horizon. Root density decreases abruptly and is confined to larger, unbranched, vertical roots that pass through the E horizon to vertically penetrate argillic horizons' prism faces. The vertical roots branch horizontally into the prisms along block faces. Root density increases above the bedrock, where horizontally oriented blocks are encased in fine roots as if wrapped in hair nets.

Interpreting Soil Attributes

This soil's morphology was described and interpreted from a carefully obtained monolith slice that contained complete prisms and retained the original orientations of structural aggregates. Sampling with a probe, with an auger, or from a pit face would not have revealed the organic matter stains on upper prism faces, or the horizontal blocky structure with its organic matter coats above the bedrock. Because the soil had been carefully described and sampled by horizons, and because bulk density samples also had been acquired, it was determined that most soil organic C is in the subsoil in this system. This discovery was a precursor for work that revealed that soil organic C may be concentrated in subsoils in much of the humid midwestern United States (Hammer et al. 1995).

Chemical analyses of bulk samples seem to indicate depth gradients of clay, CEC, and soil organic C. However, careful analyses showed that plant roots encounter abrupt chemical and textural discontinuities across ped faces. The subsoil contains structure within structure. Relative distributions of clay and C are influenced by location within the subsoil.

Soil Water Is Not a Gradient

Soil-water distributions are not in equilibrium (Jardine et al. 1990) and are more complex than indicated by figure 6.5. Perched water tables can occur above discontinuities. Macro-, meso-, and micropores, depending on their locations in the profile and on antecedent soil water, differentially contribute to volumes and distributions of water and solutes (Wilson et al. 1990). Soil water is distributed by macropores in larger volumes than suggested by their abundance (Wilson and Luxmoore 1988), but micropores control solute chemical concentrations (Luxmoore et al. 1990).

Most soil profiles are not a vertical gradient chemically, physically, mineralogically, or hydrologically. A profile is a single location in the landscape, and it grades laterally into different soils with different attributes. Figure 6.7 shows the complexity of temporal soil-water distributions in three profiles along a hillslope transect. A single profile is a "point in space" representation of a dynamic, differential distribution of a multivariate continuum. This concept is further illustrated by depth distributions of clay, iron (Fe), and C from shoulder, backslope, and first-order bottom profiles downslope from the previously described Ultisol (figure 6.8). The bottom soil is the most shallow. Clay maxima have different values and are at different depths in different profiles, and Fe has multiple maxima and minima within and among profiles. These distributions are from bulk analyses. Soil heterogeneity is more complex when ped coatings are considered.

Landscape Dynamics

It is not clear that *landscape* has been defined in a spatial context in the ecological literature (Allen 1997, chapter 3; Bradshaw 1997, chapter 11). Can meaningful field research be conducted in the absence of a sound conceptual framework for landscape scale sampling? Is expressing landscape as a "level" of research, rather than as a scale on which to frame pattern and process, an effort to convey importance to projects that have been defined in this context?

The idea that steady-state relationships drive hillslope processes is doubted by some pedologists (Wilding 1992), hillslope hydrologists (Anderson and Burt 1990), and geomorphologists (Butler 1959; Daniels and Hammer 1992). The steady-state theory is rooted in Milne's (1935) catena concept, which is not universally applicable and is often misunderstood and misapplied (Strahler 1952; Ruhe 1956; Hack and

FIGURE 6.7

Resistance readings from cells placed in Ultisols at three locations along a hillslope transect on the mid-Cumberland Plateau in Tennessee. Locations are (A) a shoulder slope, (B) a backslope, and (C) a first-order bottom with good surface drainage.

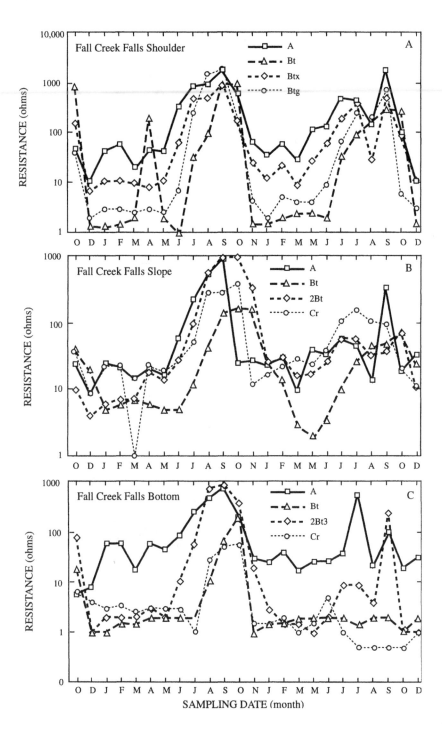

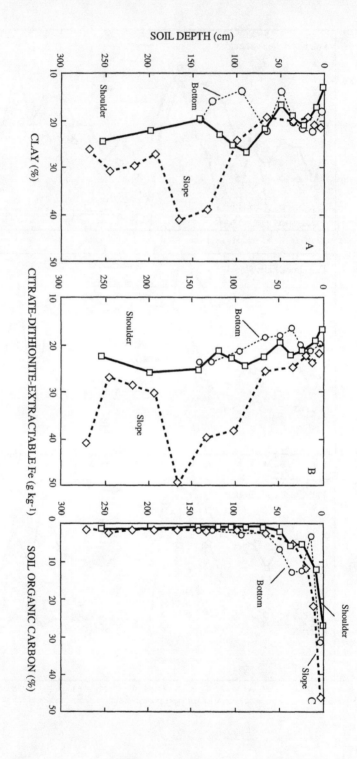

Goodlett 1960). A catena is a soil drainage sequence in uniform parent materials. Few hillslopes have a single parent material, drainage patterns can be complex, and hillslope dynamics are often episodic rather than steady-state. Persistence of the steady-state paradigm beyond pedology and geomorphology further illustrates the need for collaborative, multidisciplinary research.

For example, Vitousek (1992) discusses Jenny's (1941) "state factor" model of soil genesis as a template for ecological research. Vitousek presents the factors in the context of gradients and does not link the factor interactions. He discusses parent material only in the context of texture, which he says is the major controlling factor for soil N. Evidently, Vitousek thinks N determines availability of all other soil nutrients (Vitousek and Howarth 1991). Vitousek presents relief from the perspective of toposequences with no indication that subsurface factors control temporal and spatial distributions of water and nutrients. Vitousek concludes that Jenny's factors are a framework for gradient analyses to predict impacts of global climate change. In the same symposium, Wilding (1992) states that:

> One of the major limitations of the state factor model as viewed from breakthroughs in modern pedology is the recognition and general acceptance that soils are developed along polygenetic pathways, on dynamically evolving landforms under the influence of paleoclimates, in nonuniform parent materials and through combinations of processes. Most pedologists have strong reservations about the rigorous independence of state factors.

Jenny's model is an elegant paradigm for pedologic research. Like all good models, it set the template for its own demise (Cline 1961) as new data emerged.

Different spatial scales exist for different soil attributes, and not all soil attributes are related to topographic variables (Kachanoski 1988). Vitousek's (1992) presumption of soil-state factors as gradients is unsupported by pedologic literature and falls into that realm of specula-

FIGURE 6.8
Depth distributions of **(A)** clay, **(B)** citrate-dithionite-extractable iron, and **(C)** soil organic C from forested Ultisols on the mid-Cumberland Plateau in Tennessee. Soil profiles are on a shoulder, a backslope, and a first-order bottom with good surface drainage.

tion that Stone (1975) categorizes as bearing "a burden of supposition and lack of rigor."

Reductionism Manifested—Misperceptions and Illustrations

Mayr (1976) believes that Western society is culturally ill prepared to study natural variance because:

> Shaped by Plato's essentialism, the Western mind has tended to think in terms of unchanging types and essences and to regard variation as fleeting and unimportant. The thinking of dominant schools of philosophy has throughout been incompatible with an adequate consideration of the importance of variation.

Mayr (1976) proposed that biologists' failures to recognize variability might result from reductionism, a perspective shared by Peters (1991). Mayr (1976) attributes the roots of reductionism to physical sciences (chemistry, physics, and mathematics), which are older than biological sciences and for which certain laws are pervasive and relevant. Does the search for simple underlying truths promote reductionism in biological sciences?

Ecologists' reductionism of the soil resource can be categorized under three broad assumptions: (1) the soil landscape is comprised of univariate gradients, (2) the upper few centimeters of soil represent the temporal plant-rooting environment, and (3) N is a surrogate for soil fertility. These closely related misperceptions are issues of scale and perspective.

Reductionism in Ecological Teaching

The soils chapter in Colinvaux's (1986) introductory ecology text discusses regional soil color in the context of Oxisols and Spodosols. The following passage contains misconceptions, errors, gross generalizations, and outdated terminology. Oxisols are referred to as

> thick, red, and heavily weathered and altered strata of the tropical ground. Theirs is the red color of much of the wet tropics. . . . Whether we live in places of red dust, brown, or gray depends on the zonal soil of our region. . . . The north is gray or brown because the soils are podzols (spodosols); the tropics are red because soils are laterites (oxisols). The gray of a podzol is the gray of silica minerals that have been washed clean of iron and aluminum compounds. . . . A simple answer to why temperate and tropical

lands are of different colors, therefore, is that the local soils collect different minerals.

The zonal concept was part of the philosophy (Byers et al. 1938) on which the 1938 soil classification system (Baldwin and Kellogg 1938) was constructed. Shortcomings soon became apparent, and the zonal concept and its companion taxonomy were abandoned within a decade (Ableiter 1949; Cline 1949).

Most northern soils are not Spodosols, and not all Spodosols are in the north. Spodosols are in hyperthermic and isohyperthermic soil temperature regimes (Soil Survey Staff 1975) in sandy coastal plain soils from the Carolinas through Florida and into the tropics. The E horizon is not diagnostic for Spodosols, many of which do not contain one. Dark spodic horizons often grade to underlying reds and yellows, and an eroded Spodosol landscape can be a veritable rainbow of colors.

The term *soil type* was declared obsolete by the Soil Science Society of America nearly 30 years ago. The term originated near the turn of the century and was modified several times as early taxonomic concepts evolved (Simonson 1951). Soil type came to mean different things to different people, so it was dropped with the introduction of the current soil taxonomy.

Oxisols are intensely leached soils on old, stable landforms; few tropical soils are Oxisols. The erroneous perception that tropical soils are Oxisols is so pervasive that the Soil Science Society of America addressed the problem in a special symposium:

> The widespread use of the term *tropical soils* has led to the general misconception that soils of the tropics comprise a group of soils with common properties. These soils are often referred to erroneously as being synonymous with lateritic soils, laterites, or red soils. Although there are soils in the tropics that contain laterite (hardened ironstone) and red soils are common, most soils in the tropics do not have laterite nor are they red. Soil scientists with extensive experience in the tropics are convinced that there are more different kinds of soils in the tropics than in the temperate regions. (Drosdoff et al. 1978)

Soils generally do not "collect" minerals. Certain minerals are concentrated in leached soils as more soluble constituents are progressively removed. Simonson's (1959) model is an excellent primer on this topic.

Ignoring Information from a Cited Source

Drake and Mueller-Dombois (1993) studied forest composition on Mauna Loa in Hawaii. They said, "Moisture availability affects succession rates and rainfall changes rapidly with elevation." This was justification for confining the study area to a narrow elevational range. Citing a book on the geology of Hawaii, the authors said, "Hawaiian lavas are chemically uniform, but structurally variable" (MacDonald et al. 1984). The authors analyzed vegetation, with no mention of soils or landscape parameters on the site (as if soil texture, soil depth, mineralogy and soil structure, and slope attributes have no affect on plant-available soil water). Drake and Mueller-Dombois (1993) omitted part of the cited sentence and several important ensuing paragraphs:

> In general the mineralogic composition of Hawaiian volcanics is relatively uniform and *cannot account for the complex patterns of soil occurrence* [italics added for emphasis]. The wide range of conditions under which soils form in Hawaii has resulted in a complex pattern of soil. . . . Climatic conditions, which are extremely variable over short distances in Hawaii, greatly influence soil sequences. . . . Topography and drainage conditions significantly influence soil conditions in Hawaii. . . . Where topography causes internal seepage, complex patterns of well-drained red kaolinitic soils grade into poorly drained black montmorillonitic soils (MacDonald et al. 1984).

Hawaii's soils are a complex mosaic containing all 11 soil orders, and Hawaii may be the only area of comparable size containing all 11.

Information in the Public Domain

Soils of most of the United States and Canada have been classified and mapped in progressive soil surveys. Most surveys are at a scale too coarse to identify soils on small research plots. The surveys contain useful information about the kinds of soils in an area, the most important soil attributes, and the relationships of specific soils to one another, the landscape, and the climate. All states have a Natural Resource Conservation Service (formerly Soil Conservation Service) staff willing to help identify soils in the field. If resources are not available to classify soils, an acceptable alternative is to identify from soil surveys how the soils are mapped and to include the taxonomic classifications in the site description.

Sala et al. (1992) modeled water infiltration in the Ascalon soil, a "common rangeland soil in north-central Colorado," but no additional information about the soil was provided. Their hypothetical soil column was divided into 11 layers of unspecified texture. They said the surface was sandy loam, but it was not clear if they used this texture for all layers. Water infiltration was modeled on the basis of a several criteria, including precipitation, evaporation, and transpiration. Sala et al. concluded that "no deep percolation events were recorded."

Ascalon soils are fine-loamy, mixed, mesic aridic Argiustolls occupying 2.4 percent of Logan County (Amen et al. 1977) and 9.4 percent of Morgan County (Spears et al. 1968). These soils have argillic horizons containing between 18 percent and 35 percent clay. The soil is in the ustic soil moisture regime, but on landscape settings (ridgetops with up to 9 percent slopes) producing aridic conditions. The soil formed in loess over coarse-textured outwash or fluvial materials with high variability in gravel and sand content. Depth to massive calcium (Ca)-cemented horizons ranges from 30 to 75 cm depending on landscape position and slope steepness.

A model was not necessary to determine depth of wetting. Soil morphology indicates that the wetting front does not penetrate past the depth of Ca concretions. On the few occasions in which water reaches the cemented layer, it will move laterally across the discontinuity. Site-specific wetting depths can be determined by sampling soils to the depths of visible calcium concretions.

Surface Soil Nitrogen as a Surrogate for Soil Fertility

Ecologists seem to be fascinated with surface soil nitrogen (N): they seem convinced either that N and phosphorus (P) are the only plant nutrients of consequence, or that all other nutrients in soil systems are proportional to N and P and behave as they do. Soil fertility is classically defined as "the quality that enables a soil to provide the proper compounds, in the proper amounts, and in the proper balance, for the growth of specified plants when other factors are favorable" (Millar et al. 1965). Nutrient investigations based on this hypothesis likely would reveal more information about plant interactions with one another and with their environments.

Sixteen elements generally are essential to plants: the macronutrients including C, N, P, Ca, oxygen, hydrogen, potassium, magnesium, and sulfur (S), and the micronutrients Fe, manganese, zinc, copper, molybdenum, chlorine, and boron (B). Some plant species require sodium, silicon, nickel, and aluminum (Troeh and Thompson 1993).

An ecological focus on surface N and P probably can be traced to Vitousek, who has written that N limits productivity in forest ecosystems and that N mineralization controls the availability of other nutrients (Vitousek and Howarth 1991). The former is not always true, and the latter is not based on published research. Forest soils of the U.S. Eastern coastal plain are P deficient (Gholz and Fisher 1984), and many forested Ultisols of the southeastern United States are Ca deficient (Johnson et al. 1995a). Many New Zealand forest soils are both P and B deficient (Turner and Gessell 1990).

Does ecosystem response to N indicate an N deficiency that affects form and function? Ecosystems have evolved and functioned for thousands of years. Midwestern prairies in the United States respond differentially to fire regime, grazing, or any other disturbance (Collins and Barber 1985), but this in no way indicates that more of anything is necessary for their continued existence. In fact, if species diversity is a management goal, evidence exists that soil heterogeneity enhances diversity, and increased soil N favors some species at the expense of others (Tilman and Olff 1991).

Ecological reports of ecosystem N and C generally ignore the forest fertility literature, much of which is long term, addresses multiple inputs and disturbances, and includes well-defined hydrological and mineralogical parameters. Stevenson's (1982, 1986) lifetime of work with N and its interactions with C and S also is largely ignored. Many ecological studies include laboratory measurements of N mineralization from sampled soils, although increased N mineralization of disturbed soils is well documented (Johnson et al. 1995b).

The role of subsurface nutrients in plant nutrition deserves additional attention. The importance of subsoil fertility varies according to species, soil water distributions, and the relative nutrient distributions in the soil profile (Comerford et al. 1984). Hypotheses that subsurface nutrients play a role in surface organic-matter dynamics probably would provide more useful information regarding ecosystem function than assuming that N mineralization is the key to fertility.

Ratios of N to P to S under field situations and how these nutrients interact with other plant nutrients require clarification. Fisher (1995) suggests that soil organic matter research should focus on the ecological role of humus. The extent to which humus complexes increase nutrient availability is largely unknown. What is the role of humus in N, P, and S cycling? The role of water-soluble C in soil fertility is unknown. The role of the detrital trophic system in the decomposition and maintenance of soil organic matter is undefined (Fisher 1995).

How would the decomposers react to differing C to N to S ratios? What are the roles of micronutrients in the detrital system? Much remains to be learned.

Do Univariate Gradients Exist?

Gradient is undefined in the ecological literature, and untested assumptions of gradients are common (table 6.1). For example, Matson and Vitousek (1990) claim that sites in Brazil, Costa Rica, and Colombia represent a nutrient gradient. How is this assumed gradient identified among the other undefined system variables? Why is this called a *nutrient* gradient when only N is measured?

Schimel et al. (1991) examined prairie response to a "resource gradient" in lower landscape positions on the Kansas Konza Prairie. They presented an "idealized" valley cross section (figure 6.9), in which the soil was presented as homogeneous, and hydrology was not investigated. A maximum soil depth of 50 cm was reported after probing with a rod. This would be classified as a shallow soil (Soil Survey Staff 1975).

The soil survey in the Kansan study area is not complete, but the Shawnee County soil survey (east of Manhattan) (Abmeyer and Campbell 1970) and the Ellis County soil survey (west of Manhattan) (Glover et al. 1975) bracket the area. Parent materials in both counties include horizontally stratified limestones and shales with a loess veneer on ridgetops. Soil-water regimes are ustic.

One shallow soil on ridgetops was reported in the two counties. Most local ustic soils have Ca concretions or cemented layers and would seem shallow if probed. Many hillside soils are skeletal, and rock fragments also would give the impression of shallow soils. Soil pits are necessary to determine soil depths in rocky or cemented soils. Both surveys have complex soil patterns related to geologic strata on backslopes (figure 6.10). Different strata impart different properties to the soils depending on resistance to weathering, Ca content, texture, etc. Soil gradients do not exist on these hillslopes, which would more accurately be described as multivariate differential distributions.

Is the Surface a Surrogate for the Subsoil?

Sampling the upper few centimeters of a soil does not represent the temporal plant-rooting environment. The temporal plant-rooting environment, as used in this paper, *includes the complete soil volume occu-*

TABLE 6.1
Examples of assumed gradients in recent ecological literature

Author	Gradient	Assumptions necessary for a gradient, based on conditions presented by the authors
Aber et al. (1993)	Forest productivity	Surface texture and landscape position alone determine temporal and spatial distributions of soil water. No differences in water-holding capacities among sites. No reported soil depths or other soil attributes.
Burke et al. (1995a)	Soil	Surface 10 cm and point-in-time nutrient sampling represent the entire soil profile. No between-site variance in soil depth, texture, mineralogy, etc.
Callaway et al. (1994)	Soil and geology	No textural, mineralogical or chemical differences in bedrock at different sites. No soil or site differences important to tree growth within or among sites. Soil not described, analyzed or sampled below 5 cm. No bedrock samples, analyses, or descriptions.
Gosz (1992)	Altitude and elevation	Everything but the climate related to altitude and elevation is constant across sampling scales. No micro- or macro-scale differences in topography, soils, geology.
Hanson et al. (1994)	Hydrology and N dynamics in a floodplain	Soil textures, organic C, plant rooting, species composition, and anything else affecting water movement and N dynamics is constant in a fluvial system.
Lauenroth (1994)	Infiltration into and percolation through soils	Topography, soil texture, mineralogy, bulk density, porosity, organic matter, subsurface attributes, and evapotranspiration stresses are constant in the modeled landscape.
Matson and Vitousek (1990)	Nutrients	All nutrients in sites in Brazil, Costa Rica, and Hawaii behave as surface N does. No within- or among-site soil variability affecting nutrient supply or uptake. Vegetation in different locations utilizes N and nutrients in same way.
Pantastico-Caldas and Venable (1993)	Soil water and nutrients	Soil to depth of 8 cm and single-time sampling represent hydrologic and nutrient distributions along a slope of undetermined attributes.
Pastor et al. (1984)	N mineralization	N mineralization in bags represents N dynamics in Entisols and Spodosols under different vegetation.

TABLE 6.1

Continued

Author	Gradient	Assumptions necessary for a gradient, based on conditions presented by the authors
Running (1994)	Hydrology, C, N	A few measurements can simulate ecological responses across the regional range of topography, geology, climate, and soils.
Sala et al. (1992)	Soil hydrology	Undefined attributes of a single soil represent regional responses to precipitation input. No local modification of hydrology due to soils, stratigraphy, or botanical variability.
Schimel et al. (1991)	N and soil water	Uniform hydrology, N, and other nutrient availability down a hillslope transect in complex geologic strata.
Schlesinger et al. (1989)	Soil and geology	Soils kilometers apart have no within- or among-site differences below surface 5 cm.

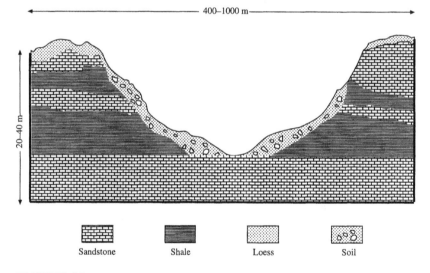

FIGURE 6.9

An "idealized" cross section of a Konza Prairie (Kansas) watershed.

From *Ecology*, vol. 72, pp. 672–684, by D. S. Schimel et al. Copyright © 1991 by Ecological Society of America. Reprinted with permission of the publisher.

pied by roots during the life of a plant and, over time, *the chemical, physical, mineralogical, and biological conditions* that determine plant-root water and nutrient utilization. Coile (1952) defines the ability of a site to support a particular botanical community and to permit a specific amount of productivity dependent on "the quantity and quality of the rooting volume." Coile's qualitative statement defines the relevant universe and provides a conceptual research framework. The productivity index concept (Kiniry et al. 1983; Gale and Grigal 1987) is a quantitative application of Coile's idea.

Many ecologists sample as if all soils are only a few centimeters deep or as if a few surface attributes—primarily texture and C—are correlated with all subsurface features. Soil horizons are largely ignored (Schlesinger et al. 1989; Sala et al. 1992; Bowman et al. 1993; Davidson et al. 1993; Pantastico-Caldas and Venable 1993; Pastor et al. 1993; Callaway et al. 1994; Hanson et al. 1994; Reiners et al. 1994; Burke et al. 1995a,b).

Testing the hypothesis that the soil surface is correlated to the sub-

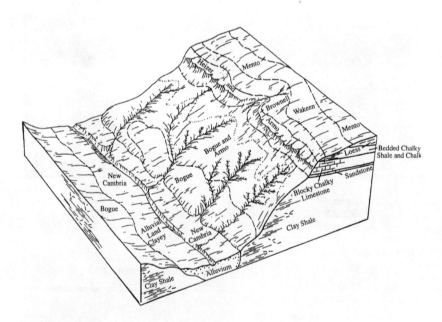

FIGURE 6.10

Block diagram of soil–stratigraphic relationships in Ellis County, Kansas.

From R. K. Glover et al. (1975).

surface is done by regressing A-horizon elemental concentrations with Bt-horizon concentrations within the same profiles. These data are from grids at two scales on the previously discussed Cumberland Plateau site. Correlations between surface and argillic horizon variables were uniformly low for texture, C, and all cations. Calcium is the example (figure 6.11). The hypothesis is rejected.

A Detailed Example from the Literature—An Undefined Universe

The Setting

The previously mentioned soil sampling problems all are illustrated in Burke et al. (1995a), who sampled soils from two locations along a reported "constant precipitation gradient but varying temperature gradient" in eastern Colorado. A hypothesis was not stated. The purpose was to "evaluate the influence of a suite of land management practices on soil organic matter turnover and nutrient supply." The sampled soils were neither described nor classified. Soil parent materials and geological substrate were neither described nor mentioned. Soil depths within and among treatment locations were not reported. Soil sampling was from three positions "across a landscape gradient" at both locations. Landscape positions from which soil samples were obtained were a "summit, midslope and toeslope." The only mention of soil conditions is that the soils between sites were "somewhat different." Slope steepness, slope length, slope shape, and relief from summit to toeslope are not given for either site, and the total sampling areas are unknown.

Both sites are reported to have "been managed for conventional tillage" for approximately 50 years prior to current treatments, which were initiated in 1985. No evidence is given that soils at these sites separated by 370 km were managed in similar ways for similar crops in the 50 years prior to the current, incompletely defined practices. Tillage traditionally is defined in the context of a specific crop, fertility, weed treatment, and residue management.

The sites are named Sterling and Walsh. Towns with these names are about 370 km apart on a north–south transect. Soil surveys of Baca (Walsh) and Logan (Sterling) Counties indicate that 11 of the 28 soils mapped in Baca County (Woodyard et al. 1973) were among the 55 soils mapped in Logan County (Amen et al. 1977). The soil surveys caution that wind erosion and local relief produce high local soil heterogeneity.

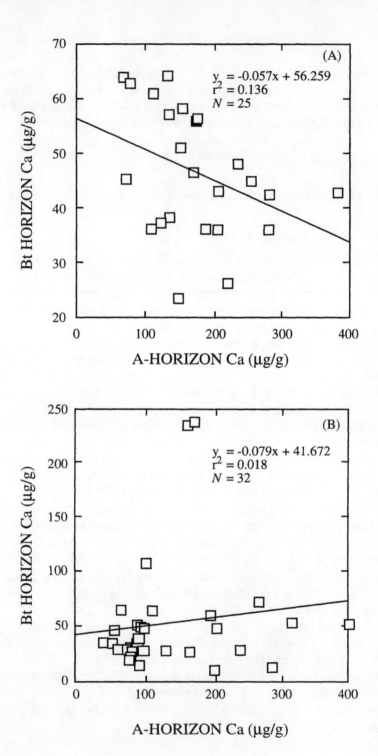

Sampling and Analyses

Soils were sampled with a probe (kind and size unspecified) from the 0- to 5-cm depths (Burke et al. 1995a). Laboratory analyses included "total N" by Kjeldahl, "total organic P," "total C" by wet oxidation, and particle size by hydrometer, and results were analyzed by analysis of variance.

Wet oxidation C determination produces analytical errors of ±40 percent. Oxidation by combustion furnace with errors of ±1 percent is now the laboratory standard. Particle size by hydrometer (±10 percent) is less precise than pipetting (±3 percent) (Kellogg 1961; Indorante et al. 1990). Of what value are statistical analyses when the data have error ranges of several times the measured statistical significance? Are statistical analyses of a complex universe of any value when based on one or two degrees of freedom? Can five composited surface samples taken from an unspecified area represent the spatial variation in a disturbed landscape?

Udawatta and Hammer (1995) examined organic C distributions in surfaces of upland Ozark forests. Twelve hillslope transects were established across concavities and convexities. Organic C concentrations increased downslope along concavities and decreased downslope along convex surfaces. However, the trends were apparent only when the means were determined across all transects. The measured values follow the trends of the means on 1 of the 12 transects. Over 180 samples were required to reduce the standard error to 10 percent of the measured mean.

The Reported Results

Burke et al. (1995a) state that, "soil texture, total soil C, N, and P, and potentially mineralizable N and C varied consistently across the sites and landscape gradients, with each variable showing the same pattern." They repeatedly refer to the data as "gradients" at "landscape scale."

FIGURE 6.11

Correlations between A-horizon and Bt-horizon calcium concentrations in a forested Typic Hapludult on the mid-Cumberland Plateau in Tennessee. Samples were collected from **(A)** 0.5-m and **(B)** 10-m grids.

Adapted from data in *Soil Science Society of America Journal*, vol. 51, pp. 1320–1326, by R. D. Hammer et al. Copyright © 1987 by Soil Science Society of America. Reprinted with permission of the publisher.

Graphs (reproduced from their table 1) indicate that their measured soil attributes do not vary consistently across sites and do not have the same patterns (figure 6.12). Are these gradients? Is "landscape scale" meaningful in the absence of a defined sampling area or replication of variance patterns?

Burke et al. (1995a) state in their abstract, "Total soil C, N, and P, microbial biomass, and C and N mineralization were highest at a northern site and on toeslope landscape positions." Total C, N, and P were not measured because only the upper 5 cm was sampled, even

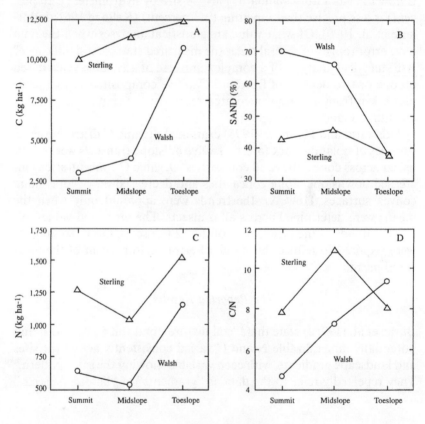

LANDSCAPE POSITION

FIGURE 6.12
Surface **(A)** C, **(B)** sand, **(C)** nitrogen, and **(D)** C/N ratios from three landscape positions on unspecified soils near Walsh and Sterling, Colorado.

Adapted from data in *Ecological Applications*, vol. 5, pp. 124–131, by I. C. Burke et al. Copyright © 1991 by Ecological Society of America. Reprinted with permission of the publisher.

though most soil organic matter is in subsoils (Hammer et al. 1995). Soil bulk density measurements, an attribute necessary to convert C concentrations to quantities, are not described. The low C to N ratios suggest the possibility of N fertilizer applications in the recent past (figure 6.12D).

Among their conclusions, Burke et al. (1995a) suggest "that the simple relationship between equilibrium soil organic matter content and nutrient availability found across natural gradients, such as regions or landscapes, does not apply to transient systems." They also discuss "nutrient supply capacity" as indexed by microbial biomass, and they say that the nutrient supply capacity "varied concomitantly with total soil pools."

What is the basis for assuming that a "simple relationship" exists between equilibrium soil organic matter content and "nutrient availability?" Neither was measured. Their assumption of gradients contradicts published soil inventories. Can a single tree, sampled without regard to species or age, represent a mixed mesophytic forest covering thousands of square kilometers? Can a single leaf, taken from a tree without regard to position on the tree or time during the growing season, represent the tree? Can one expect that a few soil samples, extracted from a pair of undefined hillslopes separated by 370 km, would represent the nutrient and organic matter dynamics of the soils of eastern Colorado?

Kellogg (1961) reviewed the accomplishments of 30 years of soil fertility research. He lauds increased crop yields but laments that a focus on the response of plants to nutrient inputs has superseded investigations of the interactions of soil properties and plants: "Think of the man-years that have been spent on elaborate field experiments set on an undefined universe, with results hopelessly confused by unknown and unmeasured variables!" (Kellogg 1961).

Summary

Many misperceptions have been created and sustained because the dynamic processes of soil landscapes have been ignored. Both ecology and soil science would benefit from cross-fertilization. The importance of truly collegial, multidisciplinary, long-term research cannot be overstated. Many complex, important environmental problems await solutions that will be found only in rigorous, systematic research based on hypotheses designed to quantify and predict nature's variation.

Solving complex problems requires that people know *what* to do rather than that they merely try harder to do *something* (Deming 1982). Knowing what to do in the landscape requires knowing something of the patterns, processes, and depositional environments. Assumptions of hydrologic and nutrient gradients must be rigorously tested. Finding answers to difficult questions requires careful testing of paradigms, rigorously quantifying entities under consideration, and admitting that more can be learned.

Acknowledgments

This chapter is manuscript 12475 from the University of Missouri Agricultural Experiment Station. Data from the Cumberland Plateau were obtained from research partially funded by research grant FS 19-81-34 from the USDA Forest Service, Southern Forest Research Station, New Orleans, Louisiana. I have discussed frequently with colleagues the topics presented in this chapter. Some ideas expressed herein originated with them. I am grateful for such scientific dialogue, particularly with Gray Henderson, Kevin McSweeney, Dick Fisher, Dale Johnson, and Ray Daniels. The chapter was improved by critical, constructive reviews by Gray Henderson, Kevin McSweeney, John Kabrick, and Alan Haney. I sincerely thank Richard Schlepp and Lewis Daniels, State Soil Scientists in Kansas and Colorado, respectively, for prompt, courteous responses to requests for soil surveys.

7

ECOSYSTEM ORGANIZATION ACROSS A CONTINUUM OF SCALES: A COMPARATIVE ANALYSIS OF LAKES AND RIVERS

Claudia Pahl-Wostl

Lakes and rivers are aquatic ecosystems that share habitat structure and patterns of environmental variation in time and space, which, in this case, are determined by water and its physical properties. However, lakes and rivers also have major differences in habitat diversity and ecosystem organization along the dimensions of space, time, and function. As a consequence of such differences, management practices and the impacts of human activities are quite distinct. A comparison of patterns of ecosystem organization and their differences can be performed only within an overall conceptual framework. The first part of this chapter is devoted to outlining such a framework, which depicts ecosystem organization across a continuum of scales. The reasoning behind this framework is based on knowledge of aquatic and particularly pelagic ecosystems. However, the basic ideas underlying the theoretical concepts have general relevance.

Ecosystems are assumed to organize themselves along a continuum of scales along space, time, and function, which is important for combining a high degree of adaptability and the integrity of functional properties. The resulting dynamic behavior of ecosystems is characterized by high spatiotemporal variability and a wide range of degrees of freedom that render futile any attempts at detailed prediction and directed control. Human management aims at making ecosystems more controllable by reducing degrees of freedom and variability wherever possible. The impacts for the dynamic organization of ecosystems and thus for ecological diversity are often disastrous. These considerations suggest a shift in our basic approach towards human–ecosystem interactions.

Spatiotemporal Organization of Ecological Networks

Traditionally, ecologists have claimed stability to be the prime attribute of ecosystems. It has been assumed that, in the absence of external perturbations, ecosystems settle into a time-invariant equilibrium state. But these assumptions have been challenged by an increasing awareness of the importance of variability and scale. Model simulations conclude that it is the dynamic nature of ecosystems, and the interplay between high variability of the components and constraints at the level of the system as a whole, that result in functional integrity combined with a high potential for change (Huberman and Hogg 1988; Kauffman 1993; Pahl-Wostl 1995; Perry 1995). These findings have convincing empirical support in work reported by Tilman and coworkers (Tilman and Downing 1994; Simon Moffat 1996; Tilman et al. 1996), which shows that species-rich grassland systems have higher resistance to disturbances with respect to functional properties at the system level, whereas the biomass of the individual species exhibit higher fluctuations than in species-poor systems. Understanding these findings requires theoretical concepts that focus on the relationship between levels of ecosystem organization.

Figure 7.1 provides a framework for addressing these issues from the perspective of ecological networks. To characterize networks, one may distinguish (1) the level of the network components and their individual properties, (2) the level of the network structure and its connections, and (3) global system properties at the level of the system as a whole. System properties emerge from lower-level dynamics, and at the same time they constrain these dynamics. Interactions between ecosystem properties and different elements of component diversity are propagated by the dynamic structure and organization of ecological networks. Diversity at the level of the network components is the source for degrees of freedom. In this context, degrees of freedom refer to

FIGURE 7.1

Hierarchical ecosystem concept emphasizing the mutual relationship between system-level properties and diversity at the level of system components. Further explanation in the text.

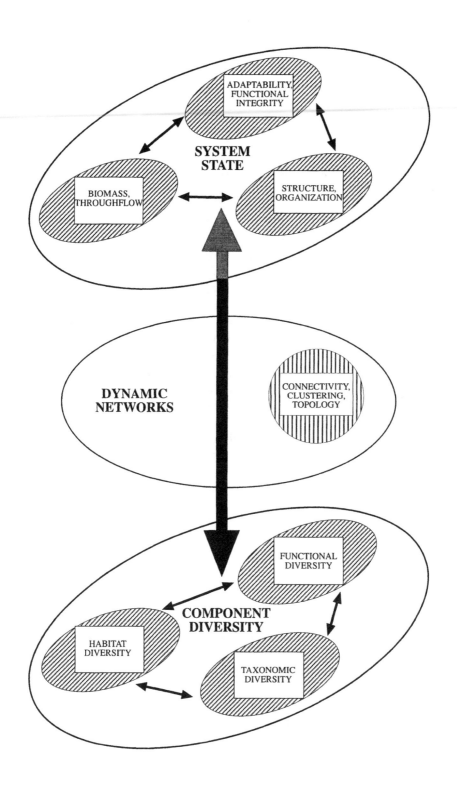

SYSTEM STATE

ADAPTABILITY, FUNCTIONAL INTEGRITY

BIOMASS, THROUGHFLOW

STRUCTURE, ORGANIZATION

DYNAMIC NETWORKS

CONNECTIVITY, CLUSTERING, TOPOLOGY

COMPONENT DIVERSITY

FUNCTIONAL DIVERSITY

HABITAT DIVERSITY

TAXONOMIC DIVERSITY

the flexibility of the network and its potential to adapt to changing environmental conditions and escape measures of external control.

The system as a whole can be characterized by quantitative properties such as total biomass or energy throughflow. Dynamic and structural characteristics such as adaptability or organization may be referred to as qualitative properties. Understanding the relationship between qualitative and quantitative system properties requires that we investigate the interdependence between system function, network structure, and the diversity of the network components. Network structure is characterized by properties such as connectivity or the clustering of interactions. The level of the network components is characterized by different elements of diversity. Notwithstanding the importance of taxonomic diversity, functional and habitat diversity will be emphasized here. Functional diversity refers here both to function in the network context and to dynamic properties of ensembles of component organisms. Network function may be determined by trophic function and interactions with the environment. Dynamic properties of network elements refer to the "typical" temporal and spatial scales of patterns of activity and response.

Rules have been derived for delineating the structure of ecological networks in a systematic fashion (Pahl-Wostl 1995). Figure 7.2 depicts an ecosystem as a network through which energy and matter are processed. Network components represent ensembles of organisms aggregated according to their dynamic and functional properties. The axis of dynamic characteristics refers to a component's rate and range of action and response. These dynamic properties may be identified with the biomass turnover time of the organisms and their spatial range of active or passive dispersal over that time period. For many organisms, body size constitutes a satisfactory substitute because of allometric relationships between body weight and physiological and ecological time scales. The axis of trophic function refers to a component's location along the trophic gradient and how it is embedded in the ecological network. The simplest possible structure is that of a linear food chain with defined trophic levels. However, a chain does not adequately represent the complex network of real systems that can be derived in a recursive fashion by starting with the primary producers at the interface with the environment (Pahl-Wostl 1995). One may thus derive the network structure for a specific system in terms of trophic dynamic modules that comprise organisms with similar dynamic and functional properties. The definition of *similar* depends on the resolution with respect to functional and dynamic properties.

The third axis of environmental niches (see figure 7.2) indicates the influence of spatiotemporal patterns in the abiotic environment (e.g., habitat diversity) on spatiotemporal patterns in biological organization. Organisms experience environmental patterns in relation to their own resolution of temporal and spatial heterogeneity (Allen, chapter 3;

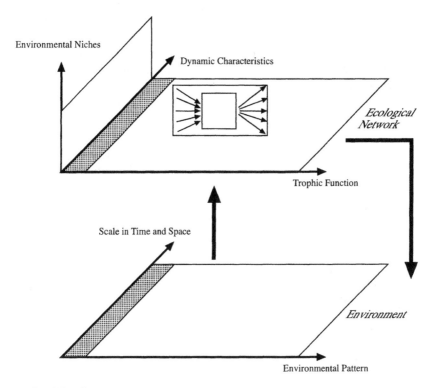

FIGURE 7.2
Ecological network in which energy and matter are processed along a trophic gradient. A network component is equivalent to an ensemble of organisms that are aggregated according to their dynamic and functional properties. Function may be expressed by a component's embedding in the ecological network, which may be characterized by an input- and an output-environment. Network components may be further distinguished by their niche with respect to specific environmental patterns. The *broad arrow* pointing up indicates that environmental variables and their dynamic patterns influence network dynamics on a wide range of temporal and spatial scales. Interactions with the environment are further given in terms of exchanges of flows of energy and resources. Imports enter the network at the base of the food web. In lakes, such imports comprise predominantly sunlight and nutrients. In rivers, the import of detrital material as energy source may be of major importance. Exports are more diffuse and are scattered over the total food web.

Kolasa and Waltho, chapter 4). Patterns that vary on scales larger than those experienced by the organisms constitute opportunities for niche differentiation [e.g., algal species may differ in their optimal conditions for light or temperature and thus may thrive during different phases of a seasonal cycle (Ward and Stanford 1982)]. Patterns that vary on the same scales may trigger life cycles of organisms (e.g., organisms responsive to environmental cues in light or temperature attune their own life cycle to the overall seasonal pattern). Patterns that occur on smaller scales must be compensated by a response from organisms (e.g., algae must cope with varying light regimes under conditions of turbulent mixing).

The broad arrow in figure 7.2 indicates that the organization of the network as a whole reflects the environmental template. Over evolutionary time scales, an organizational structure has developed that allows organisms to exploit patterns in the environment and to absorb disturbances in the dynamic network structure. The dynamic patterns of the environment have predictable components, such as the daily or seasonal cycles. They have stochastic components that may be experienced as disturbance regimes such as episodic flooding and droughts.

This space, which is composed of dynamic properties, trophic function, and environmental patterns, may be perceived as a continuous niche space that expresses the relevant properties regarding dynamic organization of ecological networks. The essential elements characterizing the organisms' niches in such a dynamic network are the following:

1. Functional properties in the network: organisms have different feeding preferences or resource requirements (e.g., food specialist or generalist).
2. Spatial and temporal niches in the environment: organisms are sensitive to seasonal patterns or spatial gradients of environmental variables (e.g., light- or shade-tolerant plants).
3. Dynamic properties: organisms grow fast or slow; they remain in place or disperse (e.g., phytoplankton or macrophytes).

These characteristics describe a flexible network organized along a continuum of functional and dynamic properties. Ecosystem organization along a continuum of scales results from the tendency of component populations to fill the envelope of the available niche space as fully as possible. The niche space is not a fixed rigid structure but co-evolves and changes in mutual interaction with the network components and the dynamic pattern of the environment (Kolasa and Waltho, chapter 4). Ecosystem adaptability is derived from the degrees of freedom extending over a wide range of dynamic and functional properties. Integrity of system function is achieved by high flexibility and diversity of the network components. Ecosystem dynamics thus continuously generate and relax tension on a continuum of scales.

Such behavior is not desirable from the point of view of traditional management practices. Human management schemes often aim at exerting control, at forcing dynamic patterns of interactions into a rigid framework that allows directed interference. To meet human perceptions, ecosystems should be stable, have little autonomy and variability. Variability that cannot be predicted is not desirable. To reduce an ecosystem's degrees of freedom, internal feedback cycles are disrupted by high loads of external inputs of energy and resources. Diversity is reduced by the selection of single scales along the dimensions of space, time, and function. Therefore, human management interferes with ecosystem dynamics over a wide range of spatiotemporal scales.

Natural systems are characterized by change on a continuum of scales (figure 7.3A). With increasing spatial scale, change takes place at increasingly longer time scales. Human influence speeds up change over large spatial scales. It reduces variability at small spatial scales by homogenizing and stabilizing undesirable variation. Human interference often leads to a fragmentation of the continuum by selecting single scales. In natural systems, disturbances occur over a wide range of scales. The repetition interval of a disturbance generally increases with increasing intensity; large disturbances are rare, but they constitute the extremes of a continuous spectrum (figure 7.3B). Although human management attempts to reduce disturbance, homogenizing patterns and reducing ecosystem potential to relax tension on a continuum of scales may cause catastrophic disturbance to become more common. Examples of this are extreme floods, large fires, and the outbreak of pathogens in monocultures.

This theoretical framework will now guide a more focused discussion on the importance of temporal and spatial scales for understanding the dynamics of lakes and rivers and the impacts of human management practices. Special emphasis will be given to pelagic ecosystems and floodplain rivers and their management.

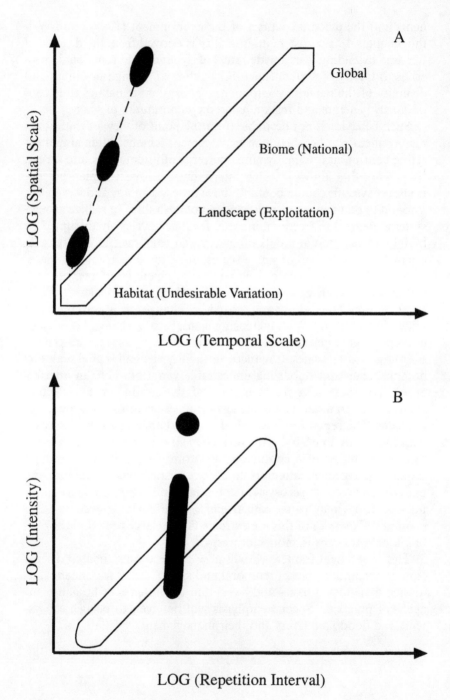

A

Global

Biome (National)

Landscape (Exploitation)

Habitat (Undesirable Variation)

LOG (Spatial Scale)

LOG (Temporal Scale)

B

LOG (Intensity)

LOG (Repetition Interval)

Lakes and the Continuous Organization of Pelagic Ecosystems

Unit of Investigation

Most of the knowledge about lacustrine ecosystems has been derived from pelagic systems. Pelagic ecosystems have a deep water body sufficiently large that the interactions (material exchanges) within this habitat dominate those across habitat boundaries, especially regarding the littoral zone and sediments. The difference between small ponds and large systems such as Lake Baikal may be greater than the difference between Lake Baikal and some marine systems. Considerations of scale are of vital importance in determining the characteristics of lacustrine habitats and their relationships (e.g., volume to surface area ratio, water retention time, and the ratio of epilimnion depth to total depth determine the influence of sediments and nutrient gradients).

The smaller the lake, the more benthic and pelagic food webs must be considered as a single unit of organization. However, this does not imply that links between the pelagial, benthos, and profundal are of no importance for large lakes (Lodge et al. 1988). The benthos is the major spawning ground for many pelagic fish. Resting stages that may remain in the sediments are of vital importance for the continued survival of many planktonic organisms. Phosphorus stored in the sediments provides a lake with a long-term memory for periods of eutrophication. The process of recovery to an oligotrophic state may be slowed down as a result of resuspension of nutrients from the sediment (internal loading). Thus, to understand the present state of a pelagic ecosystem, one must include information on the temporal and spatial scales of the whole catchment area. To understand the characteristics of community organization, it is important to focus on the characteristics of the pelagial as a habitat.

FIGURE 7.3

Pattern of (**A**) change and (**B**) disturbance in natural systems (light) and predicted influence of human activities (dark). (**A**) In natural systems, change occurs on a continuum of scales. Human influence fragments the continuum and leads to a shift towards a decrease in time scales. (**B**) In natural systems, disturbances are distributed over a continuum of scales. The typical interval between two disturbances of similar magnitude decreases with increasing intensity of the disturbance. Human influence in general reduces variability but may lead as well to unprecedented catastrophic events.

The Physical Environment

Pelagic ecosystems are often perceived as simple with respect to spatio-temporal habitat characteristics, and they are perhaps simple compared with the heterogeneous pattern in riverine systems. The large water body moderates daily environmental fluctuations in temperature or precipitation. However, numerous mixing processes on a wide range of spatiotemporal scales make this environment highly variable as well (Imboden and Wüest 1995). The pelagic habitat is characterized by a dynamic pattern of ephemeral patches on a continuum of scales. Spatial patches extend over the microscale of millimeters to the macroscale of kilometers, and they change on time scales of seconds to hours (Legendre and Demers 1984; Verhagen 1994; Imboden and Wüest 1995). Pelagic organisms have only one clear spatial frame of reference on the vertical dimension, along which strong environmental gradients are encountered. For example, phytoplankton photosynthesis is influenced by turbulent transport along large vertical gradients (Imboden 1990; Pahl-Wostl and Imboden 1990).

The seasonal pattern of stratification dominates the seasonal succession of the planktonic community. Mixing during winter results in a replenishment of nutrients in the photic zone of a lake. In spring, the water column stratifies as a result of temperature-dependent gradients of water density. As a consequence, the epilimnion becomes separated from the hypolimnion and thus from further nutrient supply. During periods of stratification, there is a stable spatial gradient that allows community organization in space. Strong episodic mixing results from wind storms and disturbs this vertical organization.

In temperate climates, the spectrum of variability of the pelagic habitat is characterized by strong periodic signals on the time scale of daily and annual changes. Irregular fluctuations are superimposed on these regular patterns. Because of the turbulent nature of the pelagic environment, the spectrum of variability extends over a continuum of spatial and temporal scales. Spatial heterogeneity is ephemeral, and spatial and temporal patterns are intimately linked. Organisms have evolved life history strategies to exploit regular patterns and cope with randomness. Indeed, Reynolds (1984a,b) reveals regularities in phytoplankton succession and their link to characteristics of the abiotic environment. The continuous and highly dynamic habitat is reflected in the organization and dynamic network structure of the biological community as a whole.

Biological Organization

The network concept depicted in figure 7.2 can be applied to delineate the food web of a pelagic community. The two axes of dynamic properties and trophic function are intimately related to body weight. Numerous investigations have revealed allometric relationships between body weight and ecologically relevant temporal and spatial scales, such as turnover rate, generation time, and habitat range. Body weight can thus be viewed as a substitute for dynamic properties (Pahl-Wostl 1995). In the overwhelming majority of predator–prey relationships, the size of the predator exceeds that of its prey. Energy flow is directed along a gradient of increasing body weight and along a gradient of increasing temporal and spatial scales (Legendre and Demers 1984; Powell 1989; Denman 1994).

The simplest model structure to express such relationships is given by a food chain with discrete trophic levels and dominant temporal and spatial scales (figure 7.4A). The niches related to trophic function are identified with discrete trophic levels. The body weight at a trophic level is confined to a narrow weight range.

However, the body weight of planktonic organisms is not confined to narrow modes, as numerous studies of biomass weight spectra reveal (Ahrens and Peters 1991; Gaedke 1992; Thiebaux and Dickie 1993). To obtain such spectra, organisms are aggregated into logarithmic weight classes, and the total biomass is determined for each weight class. Uniform distributions have been reported for different lakes (Rodríguez and Mullin 1986; Ahrens and Peters 1991; Gaedke 1992). Pelagic communities of large lakes and marine systems seem to be characterized by even biomass distributions, with an absence of major gaps in the spectrum. In these systems, the pelagic habitat is large enough to be considered an autonomous ecosystem. The continuous biomass distribution may be interpreted as a sign that the niche space related to dynamic properties is uniformly occupied. The continuous biomass distributions in pelagic systems seem to reflect the continuous nature of the physical environment.

In contrast, the size distributions of biomass and species numbers in benthic environments show pronounced modes corresponding with the sizes of benthic bacteria, meiofauna, and macrofauna. Such patterns may be explained by the presence of a discontinuous spatial habitat structure. Organisms may either colonize the particle surfaces, inhabit the interstices between particles, or regard the sediment as non-

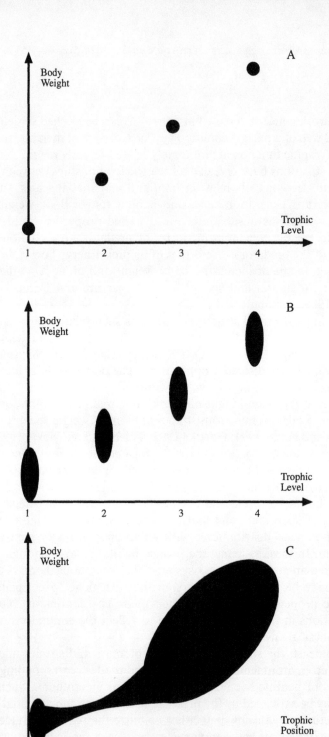

particulate (Schwinghamer 1981; Warwick 1984; Warwick and Joint 1987). We may thus discern a close relationship between the spatio-temporal organization of the biological community and the physical environment on both ecological and evolutionary time scales. In the absence of physical constraints and major external stress, there is a tendency to evenly fill the available niche space that is identified here with the body-weight axis (Pahl-Wostl 1995).

The continuum in body size results in a continuum in dynamic properties because of the close relationship between an organism's body size and its characteristic spatial and temporal scales. The trophic model depicted in figure 7.4B takes these empirical findings into account. Each trophic level is still well defined but comprises a wider range in body weight. However, there is also a continuum in trophic function (figure 7.4C). The notion of trophic level has been replaced by trophic position to emphasize that it is not possible to confine organisms to discrete trophic levels. For example, the trophic positions of a consumer feeding on primary producers and herbivores range between 2 and 3. The bottleneck between primary producers and herbivores (levels 1 and 2) indicates that there is not sufficient knowledge of the quantitative importance of mixotrophy (a switching between autotrophy and heterotrophy) to assume a continuum path.

A specific food-web model illustrates how a continuum of trophic positions may emerge. Figure 7.5 contrasts a food-chain model having discrete trophic levels in selected weight classes, with a food-web model having a continuum in trophic function and body weight. In the food-web model, a predator is characterized by a prey window along the body-weight axis. The width of a predator's prey window increases with increasing body weight (Hansen et al. 1994; Gaedke 1995). In

FIGURE 7.4

Hypothetical distributions of biomass in a two-dimensional framework of trophic level and body weight for different models describing the energy transfer along a gradient of increasing body weight. **(A)** Distribution discrete in body weight and trophic function. **(B)** Distribution with a continuum in body weight but still discrete in trophic function. **(C)** Distribution with a continuum in both body weight and trophic function. Trophic levels that can take only integer values correspond to the distinction of (1) primary producers, (2) herbivores, (3) primary carnivores, and (4) secondary carnivores. Trophic position, which may take any value, is determined as the weighted average over all feeding pathways of a consumer.

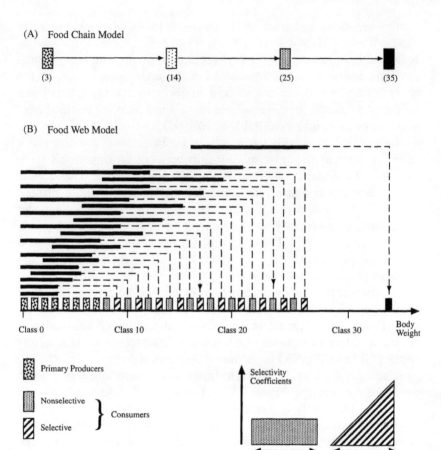

FIGURE 7.5

Energy transfer diagrams for two different simulation models: **(A)** a linear food chain with discrete trophic levels, corresponding to figure 7.4A; **(B)** a food web with a high diversity in dynamic and trophic function, corresponding to figure 7.4C. The body weight axis is divided into weight classes that are equally spaced on a logarithmic scale: $w_{i+1} = 2w_i$ where w_i denotes the average weight in class i. The different *shadings* denote different functional groups. In **(B)**, each consumer weight class is linked by a *dashed line* to a *dark bar* denoting the range of the consumer's prey window. The width of a predator's prey window increases with increasing body weight. Nonselective predators corresponding with filter feeders prey indiscriminately on all classes in their prey window; selective predators corresponding with raptorial feeders prey preferably on the higher classes in their prey window.

addition, two types of shapes are distinguished: nonselective and selective. Nonselective predators, corresponding with filter feeders, prey indiscriminately on all classes in their prey window. Selective predators, corresponding with raptorial feeders, prey preferably on the higher classes in their prey window.

Figure 7.6 compares the trophic positions derived for the chain and web models. In the web model, trophic function approaches a continuum. Because of the high diversity in its dynamic and functional properties, the web model is characterized by a dynamic and flexible network, which is in sharp contrast to the rigid structure of the chain model. Model simulations show that, unlike the chain model, the web model buffers structural changes and environmental fluctuations (Pahl-Wostl, unpublished manuscript). This gives the web a greater ca-

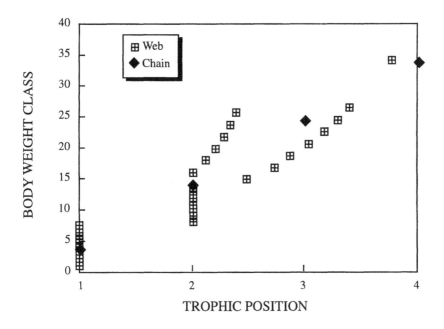

FIGURE 7.6

Comparison of theoretical trophic positions derived for the chain and the web model. To calculate trophic positions, it is assumed that the same amount of biomass is available in all weight classes. In this case, the energy transferred along a pathway is determined only by the selectivity of a predator. Because of their asymmetric prey window, raptorial feeders have a higher trophic position than similarly sized filter feeders.

pability to cope with the vagaries of a highly variable environment. It also renders the web less amenable to measures of direct control.

Model simulation further reveals that initial species biomasses at the beginning of the growth period have a major influence on the dynamics of seasonal succession. Initial biomasses reflect the presence of resting stages that preserve a memory for past conditions. A memory that extends over several years is also introduced by the dynamics of fish cohorts. It is evident that even the dynamics of short-lived planktonic species reflect the influence of a number of processes that range in time scale from days to years.

This modeling exercise provides an example of a systematic approach to investigating the structural and dynamic complexity of ecological networks. It also elucidates the problems of prediction and control in such diverse and dynamic systems.

Human Influence

Human activities may impact material exchanges and structural properties that influence habitat diversity and biological organization. The structural properties of the pelagic habitat cannot be controlled because the degrees of freedom in physical structure are inherent in the turbulent dynamics of the fluid medium. In this respect, lakes are less sensitive to impacts on physical structure. Structural changes mainly have been imposed onto littoral zones, but this pattern may change in the future. Climate change could seriously affect the overall seasonal dynamics of lakes and, in turn, biological organization. Climate change may also lead to qualitative and quantitative shifts in spatio-temporal environmental patterns over time scales that are too short for such overwhelming changes. Analysis of temperature data reveals an increase in the length of the stratification period for large lakes in Switzerland over the last decade (D.M. Livingstone, unpublished manuscript). Because the seasonal development of the planktonic community is closely linked to the seasonal cycle of stratification, major influences can be expected on seasonal successional patterns.

Until now, the major influence of humans on large lakes has been related to quantitative aspects, such as the input of nutrients and anthropogenic chemicals, or changes in the water balance. Lakes are very sensitive to changes in material exchanges with their environment and have turnover times that range from years to decades. This provides a lake with a large buffering capacity for short-term changes. However, long turnover times include a long-term memory for past changes. This has already led to considerable frustration with lake restoration efforts for which short-term success is an objective.

Lake management has been dominated by the management of eutrophication, the excessive growth of algae brought about by the input of sewage or fertilizers. The logical response is management efforts at the level of the catchment to reduce nutrient inputs. However, such measures may not always be politically feasible. Because of hysteresis effects, recovery—if possible at all—may be much slower than the progress towards a eutrophic state. Therefore, management efforts (e.g., artificial oxygenation) at the scale of the lake have increasingly gained importance in speeding up recovery processes.

Another method for controlling excessive algal growth is biomanipulation, which is based on the concept of trophic cascading interactions along a linear food chain (figure 7.7). It is assumed that increasing the number of herbivorous zooplankton directly leads to a decrease in algal abundance as a result of higher predation pressure. Decreasing the number of planktivorous fish indirectly leads to a decrease in algal abundance as a result of the decreased predation pressure on herbivores, which in turn leads to an increase in herbivore abundance and an increase in predation pressure on algae. The suitability of biomanipulation as a management tool is controversial (Carpenter and Kitchell 1992; DeMelo et al. 1992; Reynolds 1994), but it seems to be most successful in small lakes where spatial heterogeneity of large organisms is virtually absent and trophic structure corresponds to a linear food chain. However, from the perspective of ecosystem organiza-

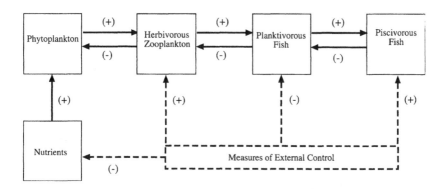

FIGURE 7.7
Simplified food chain and major pathways of eutrophication control by the reduction of nutrient input and measures of biomanipulation.

tion across a continuum of spatiotemporal scales and function, the long-term success of such management practices is questionable.

Fisheries management is another important issue. Post and Rudstam (1992) used model simulations to examine the influence of management practices on fish populations. Increasing rates of exploitation lead to higher amplitudes of irregular fluctuations in the dynamics of fish populations. Empirical studies have identified population cycles to be common in unexploited populations. However, intense exploitation may enhance such natural fluctuations to the point of catastrophic breakdowns of populations. Even intensive stocking efforts prove insufficient to overcome the impacts of exploitation in model simulations. The only sustainable strategy would be the reduction of exploitation rates.

These examples indicate the shortcomings of identifying management with direct control. Management cannot focus on single issues but must start at the catchment and proceed to the landscape level. It must include all levels of human activities. Interestingly, it seems to be much easier to raise money for lake restoration than to raise money for preserving pristine undisturbed lakes (Larson 1996). As in human health care, it is easier to motivate action when ecosystem health has already seriously declined; action seems to be warranted and effects are measurable, at least in principle. How are the effects of preventive action measured? First, there should be increased awareness at all scales, from managing a lake to managing climatic change. Preventive action is also critical for rivers, in which the negative impacts of human activities are not always as disturbing and evident as algal blooms and foaming water.

Riparian Ecosystems and the Diversity of Floodplain Rivers

Unit of Investigation

The larger and longer a river, the more diverse are the different types of ecosystems located along its course. A river is closely connected to its terrestrial environment. Aquatic habitats are rarely as uniform and isolated as the deep water body in the pelagial of a lake, but they must be defined in the context of their terrestrial surroundings. Because water movement in rivers, unlike that in lakes, is unidirectional, the horizontal dimensions are not equivalent. Ward (1989) emphasized the importance of three qualitatively different dimensions in space to characterize the dynamic and hierarchical nature of lotic ecosystems. Exchanges on the vertical dimension occur between the channel and the aquifer. The longitudinal dimension is characterized by upstream–downstream interactions and exchanges because of the directed motion of the stream. It can be viewed as a dimension of succession in space where different types of interactions prevail. The lateral dimension is characterized by exchanges with the terrestrial surroundings, and it is increasingly important as one moves from headwater streams to braided and meandering rivers. In floodplain rivers, the spatial unit of ecosystem investigation includes the river channel and its terrestrial surroundings—the floodplain.

The Physical Environment

Natural streams are characterized by high habitat diversity over a range of spatial scales. Frissell et al. (1986) introduced a hierarchical framework for stream habitat classification based on geomorphological and physical criteria. Figure 7.8 shows the distinction of different habitat types as a function of their typical extension in space and their typical period of persistence in time.

The catchment encompasses all surface waters and all terrestrial habitats in a drainage basin, and catchment features change on geological and climatological time scales. A segment is a portion of stream confined to a single bedrock type with tributary junctions as boundaries. A reach is a subunit of a segment with rather uniform characteristics regarding the embedding of the channel in the landscape. This is the typical scale at which channel migration is observed. A pool-riffle system is a reach subsystem with an alternation of shallow areas of higher velocity and coarse substrate (riffles) and deeper areas of slower velocity and finer substrate (pools). The distinction is brought about by topographic features in the river bed and channel sinuosity. Typical changes are caused by flooding-induced alterations in deposits of sediments or coarse material such as boulders. Microhabitat systems are defined as patches within pool-riffle systems that are uniform with respect to substrate type, water depth, and velocity.

Episodic flood events are important disturbance events in rivers, albeit on longer time scales, compared to mixing events in lakes. In both rivers and lakes, such events disrupt spatial structures and patterns of biological organization. The effects of flooding events may have quite different consequences depending on disturbance frequency and the type of river system.

Along the longitudinal dimension, one can distinguish three types of systems based on characteristics at the scale of the reach where the dominant feature is channel stability (Ward and Stanford 1995b). Mountain headwater reaches are typically constrained and channel stability is high. The longitudinal dimension is the dominant spatial dimension of interaction, and floodplain development is essentially absent. In the region of braided reaches, channel stability is very low because of the continuously shifting nature of the substratum. Despite extensive interactions along the lateral dimension, it is not a dimension of organization, because of the high rate of unpredictable change. The terrestrial surroundings are characterized by a pioneer community, and biodiversity is generally low. This pattern changes in the region of the meandering reaches, where floodplain development is extensive. The time scale of channel migration is slow enough to develop diverse ecosystems. The heterogeneity of habitat structure, community types, and successional stages extends over a continuum of temporal and spatial

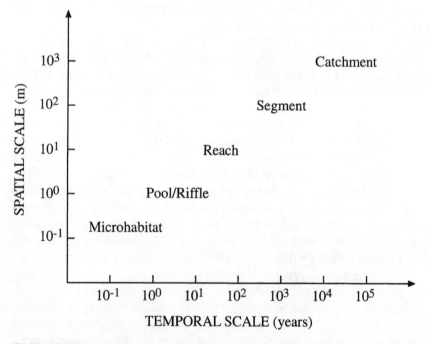

FIGURE 7.8

Hierarchical scheme of habitat classification in streams.

From *Environmental Management*, vol. 10, pp. 199–214, by C. A. Frissell et al. Copyright © 1986 by Springer-Verlag. Reprinted with permission of the publisher.

scales. The lateral dimension is the major dimension of spatiotemporal and functional organization. As in pelagic systems, strong seasonality dominates floodplain river dynamics. The seasonal flood pulse is essential for the spatial and temporal diversity of structural characteristics, as well as for nutrient and energy dynamics, especially the exchange of nutrients and biomass between a river and its wetlands (Ward 1989a,b).

Schumm (1988) pointed out that the complexity of fluvial habitat systems and their continuous change over a wide range of temporal and spatial scales render the assignment of single cause–effect relationships obsolete. He provided empirical examples for several problems associated with such variability. Patterns may be explained from a historical perspective, taking into account trajectories of change. They may also be explained from a process-oriented approach focusing on the present state of a system. Similar effects may be produced from different processes and causes. Similar processes and events may produce different effects. The specific response observed is context and scale dependent, and the existence of threshold conditions may lead to unexpected sudden change. These observations support the approach discussed here, which emphasizes overall patterns of organization rather than focusing on single mechanistic explanations.

Biological Organization

Rivers and streams are physically harsh environments. Therefore, a habitat-centered view has prevailed, stating that community organization is largely determined by abiotic forces. However, the relative importance of abiotic forces depends on scale and disturbance regimes. Poff and Ward (1989) predicted the relative contributions of abiotic and biotic processes to community structure for streams that differ in discharge regime. They state that abiotic forces prevail in systems where floods are intense and the overall pattern of the discharge regime is unpredictable; abiotic forces also prevail in intermittent streams that lack surface waters during part of the annual cycle. Seasonality increases the predictability of environmental patterns and thus the importance of biotic processes.

Townsend (1989) introduced the patch dynamics model, which states that communities are structured by frequent disturbances that remove organisms and generate opportunities for new colonizations. The concept of spatiotemporal organization discussed here complements the patch dynamics model to some extent. Whereas the patch dynamics model attributes change exclusively to abiotic disturbances such as flooding events, the concept of spatiotemporal organization emphasizes change resulting from the complex dynamics of biological interactions in diverse ecological networks. The above considerations suggest that the relative importance of abiotic and biotic factors depends on the characteristics of the system.

The seasonal flood pulse results in a predictable overall pattern of change, generating a high diversity of habitat patterns. The biological community has evolved a dynamic organization to exploit the ephemeral niches that emerge in dynamic flooding regimes. The dynamic mosaic of terrestrial and aquatic habitats encompasses a wide range of temporal and spatial scales. Water bodies in seasonally isolated side channels vary in age and thus in successional stage. On a smaller spatial scale, a variety of lotic and lentic habitats offer a wide range of environmental conditions.

Biological organization reflects this habitat diversity. Spatial and temporal heterogeneity at the base of the food web is far more pronounced than in pelagic systems, in which primary production is derived primarily from phytoplankton. In contrast with pelagic systems, there is not a uniform increase in temporal and spatial scales along the trophic gradient. Energy is channeled through the network along a number of pathways that differ in function and dynamics.

Allochthonous inputs constitute a major energy source. Because of the heterogeneity of detrital input, ranging from fine particulate organic matter to entire trees, consumption and decomposition encompass a wide range of time scales. The half-life of coarse detrital material may vary from days to decades, or even centuries for large logs (Allan 1995). The functional diversity of autotrophic producers— comprising periphyton, macrophytes, and phytoplankton—reflects the diversity in habitat. Phytoplankton is important in floodplain rivers during phases in which lentic habitats become available. Because of the dynamic nature of habitat, the relative importance of the different allochthonous and autochthonous pathways of energy and nutrient supply varies over time and space. Empirical observations suggest a complex spatiotemporal functional organization that is poorly reflected in the time-and-space averaged quantitative contributions of the various pathways. Heterogeneity at the base of the food web is propagated along the trophic gradient, making functional diversity for a given trophic position much larger than in pelagic systems.

As in pelagic systems, the body size of heterotrophic consumers increases with increasing trophic position. Size preferences in predator–

prey relationships have been observed, although body size receives less attention in rivers than in pelagic systems. Systematic investigation of body size spectra may be of interest, especially for habitat characteristics and disturbance regimes. The slope of biomass size spectra, a measure for the efficiency of energy transfer, seems to be a sensitive indicator for stress. For example, acidification in mountain streams results in a shift of the biomass distribution towards small opportunistic species (Seifried 1993), a finding that has been reported for lakes as well (Ahrens and Peters 1991). Uniform distributions of biomass across the body-weight gradient seem to indicate an undisturbed community that organizes itself on a continuum of dynamic properties and function.

Human Influence

In 1772, Benjamin Franklin noted, "Rivers are ungovernable things, especially in hilly countries. Canals are quiet and very manageable" (cited in Allan 1995). Since Franklin's time, rivers and streams have experienced major modifications by human activities, including input of material and chemicals as well as changes in structural characteristics at all scales of river habitats. Because of the directed nature of advective transport, rivers have been abused to dispose of all types of waste, ranging from household sewage to industrial effluents. In the 1970s, the Rhine river in its upper course in Germany could be described only as a stinking sludge. Public awareness about chemical pollution has increased over the past decades. Major improvements in water quality have been achieved by huge investments in waste-water treatment, at least in industrialized countries that can afford it. However, improved water quality alone is not sufficient to guarantee the integrity of riverine ecosystems.

Changes in structural characteristics have had major impacts on biodiversity in riverine ecosystems. Modifications have altered basic characteristics of lotic habits, and connections to the terrestrial and aquatic surroundings have been disrupted. Unmodified floodplains are a rarity in most industrialized countries. The spatiotemporal pattern of habitat variability has been fundamentally altered.

Figure 7.9 contrasts the catchment of a natural river in a natural landscape, with a highly exploited agricultural landscape where streams and rivers are reduced to straight, concrete channels. This comparison highlights differences in systems and is not intended to portray details in a realistic fashion. Natural landscapes are characterized by a fractal pattern (Gardner, chapter 2), indicating variability across a continuum of scales. In contrast, managed landscapes are characterized by straight geometric lines and uniform patches where variability is confined to selected scales. Empirical investigations supporting this overall pattern have been reported from both landscapes (Krummel et al. 1987) and river basins (Gordon et al. 1992). In addition, land use changes may affect runoff rates and input rates of sediment, woody debris, and chemicals. Clearing land for urbanization or agriculture generally results in increased peak discharges with reduced duration.

Variability of the reach and smaller scales has been homogenized by different activities. Channel modifications typically aim at improving conveyance and the predictability of hydraulic behavior. In extreme cases, the river bed may be reduced to a concrete trapezoid channel devoid of microhabitat structure. Channelization reduces structural diversity at different scales by eliminating meanders and by smoothing

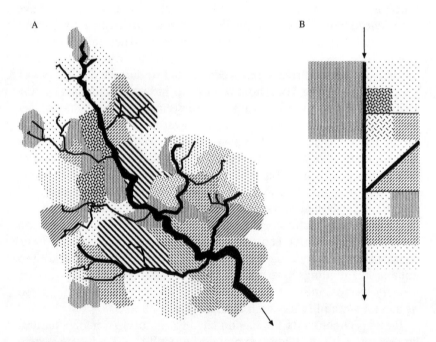

FIGURE 7.9

Highlights of the major structural differences between **(A)** a natural river in a natural landscape and **(B)** a highly exploited agricultural landscape where rivers are confined to straight concrete channels.

pools and riffles. It disrupts interactive pathways and isolates the river from its floodplain.

Rivers are major sources for the extraction of energy from hydropower, and dams exert their most profound influence on the river downstream. Several characteristics of the flow regime are altered, including total annual discharge, duration of flows of a given magnitude, and variability, frequency, and intensity of flood events. Flow regulation leads to reduced variability in discharge and temperature. Figure 7.10 schematically depicts the overall pattern of these changes. Flooding events that are essential for maintaining the morphological characteristics of river channels are often eliminated. Figure 7.10A shows typical seasonal variations in discharge in the Colorado River before and after impoundment of Lake Powell (Paulson and Baker 1981); the spring flood pulse has disappeared entirely. Figure 7.10B indicates the alteration of the seasonal thermal regime downstream of a deep-release reservoir. Because water comes from the hypolimnion with rather constant temperature all year round, summer temperatures decrease and winter temperatures increase relative to a natural regime. Seasonal temperature changes are important for many species whose life history characteristics respond to environmental cues of a seasonally changing environment.

Ward and Stanford (1995a) characterize the overall effect of impoundments as displacements along the longitudinal successional gradient. Specific effects depend on the type of stream under consideration and the location of a dam along the river continuum. The effects of different measures of regulation on floodplain rivers are generally disastrous. Anthropogenic impacts have severed major interactive pathways, thereby isolating rivers from their floodplains and disrupting the dynamic patterns and processes that provide structure for river–floodplain ecosystems (Ward and Stanford 1995a).

Despite its overall homogenizing influence, a dam may introduce extreme daily fluctuations in discharge as a result of peaking hydropower (see Figure 7.10C). Therefore, a dam can both reduce seasonal variability and increase diurnal variability. Dam construction can stabilize runoff, but other human activities can lead to instability. Land use changes and river channelization may cause extreme floods, the result of accelerated runoff rates and reduced buffering capacity of the floodplain for storing and releasing water. These floods may really be catastrophic events, because the regular disturbance regime and the biological organization adapted to it have been removed. In natural

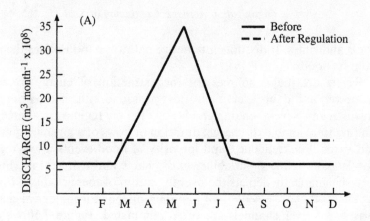

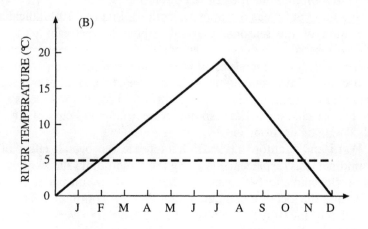

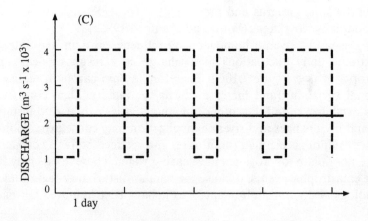

disturbance regimes, extremes are rare but are events along a continuum. Equally disastrous are no-flow conditions (absence of surface waters or dry channel segments), which are generally more frequent in managed than unregulated rivers.

Human management practices interfere strongly with the organization of riverine ecosystems along a spatiotemporal continuum. Single scales are selected along the dimensions of time, space, and function. The degrees of freedom and diversity at all scales are reduced. Such changes are the direct consequence of the strong human desire for managing ecosystems toward maximum yield and optimal control. A fundamental change is required in the pattern of human–ecosystem interactions.

Ecosystem Management and Sustainability

Ecosystem management and sustainable development have become popular concepts in recent years (Hobbs, chapter 20). Both concepts address the overall need to change human–ecosystem interactions. However, the vagueness of these terms leaves room for interpretation about what they mean. Of course, people can apply these terms any way they want, but the ambiguity points out the necessity for participants in a social discourse about environmental issues to address conflicting points of view. Such a dialogue must be viewed as an ongoing process in which the meaning of controversial concepts is defined and redefined. Social discourse should not be considered an impediment to preserving and restoring ecosystem integrity but as an essential step in dealing with the complex environmental problems facing us today.

Many natural resource problems of the past were reasonably well-defined, with clear objectives for resolving those problems. If lakes are anoxic as a result of eutrophication, if most organisms die because of

FIGURE 7.10

Influence of impoundments on the river downstream. **(A)** Typical seasonal variations in discharge in the Colorado River before and after impoundment of Lake Powell (Paulson and Baker 1981). **(B)** Alteration of the seasonal thermal regime downstream of a deep-release reservoir. **(C)** Extreme daily variations in discharge caused by peaking hydropower production.

From *Symposium on Surface Water Impoundments*, vol. II, by L. J. Paulson and J. R. Baker. Copyright © 1981 by American Society of Civil Engineers. Reprinted with permission of the publisher.

chemical pollution, or if human health is threatened, it is relatively easy to agree on goals for improving the situation. However, in the case of sustainable development, goals are more elusive. Restoring a managed river and preserving a pristine floodplain are issues that can be controversial if there are conflicting interests. Control by management is difficult, if not impossible, in natural ecosystems. In intensively managed systems, one tries to reduce a system's degrees of freedom by homogenizing the environment and the dynamics of the system in space, time, and function. Such landscapes are depauperate in species and life-forms, as well as in esthetic appeal. Restoring some of the lost degrees of freedom leads to some loss of control, and exploitation for human purposes may be less efficient from a short-term perspective. This may lead to conflict.

Even in situations in which the need for action has not been under dispute, conflicts have often been avoided by choosing "end-of-a-pipe" solutions. As in waste water treatment, such solutions are often energy intensive, expensive, and highly technical. Technical measures such as artificial oxygenation are designed to speed up the recovery of anoxic lakes. Management practice has been shaped by a view that is based predominantly on a mechanistic perception of nature. Defined pathways of interaction are needed to implement specific measures of control. Therefore, preference is given to mechanistic approaches and models that provide numbers and advice for targeted action. To fit into such a scheme, it is helpful to reduce environmental problems to specific cause–effect relationships.

Space–time windows are often determined by institutional and political constraints rather than scientific considerations. Such practices prevent a meaningful approach to basic questions dealing with environmental management and sustainable development. Environmental management should be defined as the management of anthropogenic activities with respect to interaction with environmental systems, rather than as the management of environmental systems to meet human perceptions. New concepts for human–ecological systems are required in which human activities are managed; flexibility is retained for ecosystems to choose their degrees of freedom themselves (e.g., leave woody debris in streams rather than removing it, and calculate appropriate measures of restoration later). Flexible management concepts are required that take into account the dynamic and diverse nature of ecosystems as well as socioeconomic systems. Management goals must be flexible enough to coevolve dynamically with our knowledge about specific ecosystems. Management cannot be limited to

TABLE 7.1
Management of the human–ecological system as a social process

	Prevailing philosophy	Advanced perspective
The overall goal is to	Search for an optimal solution	Design an adaptive process
	Reach a preconceived goal	Define and redefine goals
Participants involved are	Decision makers, scientists	Decision makers, scientists, citizens
Research	Reduces uncertainties	Reveals indeterminate factors
Models	Make quantitative predictions	Convey qualitative knowledge
Political measures	Are rigid, controlling	Are flexible, adaptive
	Enforce regulation	Foster innovation

planning in governmental offices based on advice from scientific experts. It must be a comprehensive societal process (table 7.1).

This distinction between prevailing philosophy and advanced perspectives (see table 7.1) should not convey the impression that a particular approach is inherently superior: the two approaches are complementary. Quantitative predictions may be possible for specific questions. However, consider the problem of the application of models in conservation biology. Wennergren et al. (1995) attribute only a minor role to spatial models, because simple models that represent metaphors rather than predictive tools risk being interpreted too literally. Complicated models that include increasing levels of detail are unwieldy, require excessive amounts of data, and are impractical to use. However, one should not conclude that simple models that convey qualitative insights are irrelevant in a social discourse. Models can serve as tools to increase and communicate our understanding about the dynamic nature of complex systems.

Summary

Models of complexity in ecological systems are not restricted to mathematical expressions. Every model reflects the perspective of its creator. This perspective determines how research problems are formulated, what type of scientific knowledge is highly valued, which strategies for solving environmental problems are chosen, how science and its social role are perceived, and how institutional settings are designed. The predominant perspective is still based on mechanistic perception, a belief in predictability and controllability and a search for instrumental knowledge. In such a world view, there is only one scientifically correct answer to a problem. Ecology has often been blamed for its inability to provide quantitative predictions and unique answers. However, our improved understanding of the behavior of complex systems allows us to conclude that the lack of general predictive theory cannot be perceived as a weakness of ecosystem science. Complementary perspectives are needed to cope with complexity. Collaboration between natural and social scientists is needed to address complexity and to develop new forms of valuing and disseminating scientific knowledge (Pahl-Wostl et al., in press).

Acknowledgments

James Ward provided helpful comments, particularly regarding scale issues in rivers.

8

HISTORICAL CONTINGENCY
AND MULTIPLE SCALES OF DYNAMICS
WITHIN PLANT COMMUNITIES

V. Thomas Parker and Steward T. A. Pickett

Community dynamics is defined as change at a variety of scales within an assemblage of organisms. A measured pattern of change is assumed to reflect relevant underlying processes. This relationship is the central model in community ecology, and it is reflected in the pairing of "pattern and process." Any pattern of change is believed to be scaled to a responsible process, although ecologists may not always explicitly recognize the scaling (Hoekstra et al. 1991). Examples of patterns that ecologists attempt to relate to underlying processes are shifts in richness or composition (Kolasa and Waltho 1997, chapter 4; Martinez and Dunne 1997, chapter 10), and change in life forms, productivity, and rates of transpiration or photosynthesis (Hinckley et al. 1997, chapter 15). These characteristics are different aspects of community response to any scalable feature (frequency, intensity, or area) of a process. Dynamics emphasizes temporal flux as well as spatial pattern, and a principal issue for community ecologists studying flux of whole assemblages is that organisms and their characteristics exist on a variety of scales, as do the processes that impact systems (Milne 1991a). How do we determine what processes underlie observed patterns? A number of models are available for examining spatial patterns within communities (e.g., several of the chapters in this book); recent theory emphasizes hierarchical approaches based on rates (O'Neill et al. 1986) or spatial extent of patches (Wu and Loucks 1995). What seems to be missing is an explicit model that attempts to integrate spatial and temporal patterns. This discussion introduces the development of such an integrated approach, and although the focus is on vegetation, the model is general.

One approach for studying communities assumes that dynamics can be understood best as a function of a limiting process (Tilman 1988). In this approach, a few processes, usually involving resources, are believed to constrain all the dynamics of the system, and models of competition are developed to explain the dynamics. Furthermore, if populations are experimentally found to be competing for a resource, then subsequent interpretations of community patterns are based on competition. Such models have been developed in the context of Hutchinson's (1957) multidimensional niche concept, but with the assumption that only one or a few dimensions are important. A key difficulty with this approach is the lack of criteria for determining limiting processes.

Currently, no explicit model of community processes reflects the broad implication of the Hutchinsonian multidimensional niche, that there are multiple processes that can be limiting in a system. Patterns of assemblage dynamics reflect a number of processes, such as droughts, canopy gaps, or fires, that can be limiting at some point in time but not necessarily continuously. Furthermore, organism characteristics such as resource acquisition, seedling recruitment, or reproductive success may each be limited by various environmental processes at different times. The significance of incorporating such a broad array of organism characteristics for explaining community patterns has recently received greater emphasis (Grubb 1977; Harper 1977; Murray 1986b; Leck et al. 1989; Wyatt 1992; Charnov 1993).

Multiple scales of processes can be seen from several perspectives. Processes that can be limiting exist on various scales (Damman 1979; Neilson and Wullstein 1983); for instance, soil moisture may fluctuate to an extreme with droughts every decade, whereas fires may return only once every hundred years. Each process itself has a number of characteristics such as frequency, extent, duration, and magnitude. Different processes may scale together in some combination of these characteristics but not in others. Finally, any community pattern reflects a limited range of each of these process characteristics, such as a particular assemblage conforming to some fire frequency. These ranges of different process characteristics occurring together have been described as a *process regime*—for example, a fire or disturbance regime (Lertzman and Fall, chapter 16).

The other side of the pattern-process model also can be multivariate and multiply scaled (Hoekstra et al. 1991). Members of the assemblage will differ considerably in the scale of any of their measured characteristics. Some plants may be much larger than others, and some exhibit greater longevity. Differentiation occurs among many characteristics, such as breeding system, seed production, dispersal, growth rates, etc. These differences illustrate variation in life histories of species and indicate some of the ways species are themselves multidimensional. Pairs of species may scale together at some life history stages and not at others. For example, obligate seeders in chaparral are morphologically similar to resprouting species, but they differ considerably in lifespan, population turnover rates, physiological rates, and other dimensions (Wells 1969; Keeley and Zedler 1978; Parker 1984; Parker and Kelly 1989). As another example, mature trees have considerable impact on understory herbaceous vegetation, and the environment shaped by the impact of large trees is generally interpreted as constraining the herbaceous understory.

Ecologists assume that species having similar scales of characteristics will respond similarly in comparison to species of different scale for that same characteristic. This is why we tend to investigate interactions among trees and ignore impacts of herbaceous species, and their patterns in the understory, on community dynamics. Yet we know understory species can influence environmental conditions (Muller and Bormann 1976; Peterson and Rolfe 1982) and recruitment rates of overstory populations. For example, ferns and other understory plants inhibit tree regeneration in the Allegheny (Pennsylvania) hardwoods forest (Horsley 1977). The animals with which plant species interact are also scaled differentially. For example, deer and molluscan herbivores both may impact seedlings, but the two groups scale differently in terms of spatial range and life history timing. Thus, disparate lifeforms influence each other through general processes such as biogeochemistry and can have size-specific interactions through the bottleneck of recruitment, when individuals of the larger species are smallest and most readily influenced by a variety of processes to which the adults are immune. To understand assemblage dynamics, by which we mean the dynamics of all the species found together, we need to incorporate all members of the assemblage, not just those that appear to act at similar rates.

The main point is that, when comparing organisms, similarity in response and scale may not be related. For example, species that appear to be of different scale may not act independently or respond to other processes independently despite their scale differences. Our goal is to develop a conceptual model of temporal community dynamics that includes the variety of scales exhibited by processes that affect whole communities and all the organisms they contain, and thus the patterns of dynamics exhibited at a variety of scales. Scale-focused models already developed are typically hierarchical models that differentially examine the structural aspects of scale of the system (Allen and Starr 1982; Allen and Wyleto 1983; O'Neill et al. 1986; Kolasa 1989; Pickett et al. 1989; O'Neill and King, chapter 1). The key aspects of these current hierarchical models will be considered, but if we emphasize the ongoing and historically contingent nature of community dynamics, then these types of models can be viewed as causally incomplete.

To focus on dynamics of complete assemblages, we propose a fundamentally different approach to these issues of scale. Spatial models emphasize the scaling of pattern to process, but because we emphasize temporal dynamics, we need to focus on the continuous impact of a number of processes on the dynamics of a single site. Because historical contingency is fundamental to community change (Morin 1984; Foster et al. 1990), we have developed an alternative conceptual model of community processes using a patch-dynamic approach consistent with historical contingency. This model also examines community patterns as a hierarchical model, yet incorporating species and processes of various scales at the same level of investigation. It is appropriate to use models that scale processes of different rates or impact onto different levels of a structural hierarchy, for independently investigating spatial or temporal scales of community characteristics. However, a model is needed that focuses at a local scale, explicitly considers dynamics, and yet incorporates all possible processes that may influence dynamics at a site (Pickett et al. 1987a,b), regardless of whether they are contained within the boundaries of the site.

Hierarchy Theory and Processes of Different Scales

How do we investigate systems with such a diverse array of scales of processes and responses? Some have proposed that developing models of hierarchical structure clarifies how ecologists investigate and understand these systems (Allen and Starr 1982; O'Neill et al. 1986, Kolasa 1989; Pickett et al. 1989). One way is to approach interactions hierarchically, arranging processes by rates. For some ecological questions, this approach appears effective (Allen and Starr 1982; Allen and Hoekstra 1992; O'Neill et al. 1986; O'Neill 1989; O'Neill and King, chapter 1). In a nested hierarchical system, levels of the system are recognized by having rates of processes that differ from others in the hierarchy. Within a level, subsystems differentiate by the strength of interactions among members of that level. Systems are investigated in hierarchy theory by considering three levels of organization or structure.

The middle level represents the scale of the investigation, and processes of slower rates act as the context and processes of faster rates reflect other mechanisms, initiating conditions or variance. Such a hierarchical structure leads to other assumptions about interactions between and within levels. Interactions between levels are considered asymmetric and the higher level acts as a constraint on the next level below. This just means that processes having slower rates limit the potential pattern generated by processes with faster rates. Within a level, subsystems are identified by a higher frequency or strength of interactions among members of the same level, with boundaries between these subsystems recognized by a lower strength or frequency of interaction. For plants or other sessile organisms, this interpretation of hierarchical system structure suggests that members of subsystems must be sufficiently close that they have greater frequency of interactions for the process under consideration.

In formal hierarchy theory, processes at the level above are slower and represent constraints on the dynamics of the level investigated. Interpreting exactly how slow a process must be for a contextual process versus one at the level of investigation has been difficult, because as yet there are no explicit criteria. Adding to this difficulty has been the recognition that constraints for an ecological system can be processes of any sort of rate. In other words, we need to clarify the distinction between contextual processes [slower processes that act as general system constraints (e.g., historical climate changes)], and ecological constraints (processes that characterize particular assemblages and limit other assemblages). For these types of processes, a range of rates may underlie recognizable patterns of dynamics, and thresholds can exist at both lower and higher rates of any process (figure 8.1). Vegetation that burns by means of canopy fire shows a clear example of these bounded thresholds in terms of fire frequency. For example, when chaparral burns with a short between-fires interval, obligate seeders and other dominants can be eliminated (Zedler et al. 1983; Haidinger and Keeley 1993); at long intervals between fires, chaparral can be converted into forest. In this example, both the slow rates of fire and the faster rates act as constraints, but they are still within the scale of the ecological level of vegetation dynamics regardless of their rates. Although all processes may not behave this way, thresholds exist for many processes at both slow and fast rates that constrain the dynamics of the system investigated. The model of dynamics is developed in the following sections based on this principle of ecological constraints sorting assemblages.

On ecological time scales, processes are multidimensional, so that any model based solely on rates, spatial extent, or other characteristics may capture only a portion of the dynamics. Processes may act as constraints along one dimension (e.g., in intensity) during some time intervals or for some sets of species, and other dimensions (e.g., frequency) may be limiting at other times or for other sets of species (Wiens 1989). Hierarchical models generally provide accurate frameworks of constraints, but they do not fully represent various dimensions of processes that occur at quite distinct rates. For example, forest vegetation patches of a range of sizes may yield predictable results that select for shade-intolerant species, but when rare large-scale events such as tornadoes create unusually large openings, results can vary greatly. At the same time, individual treefalls select for a different suite of species

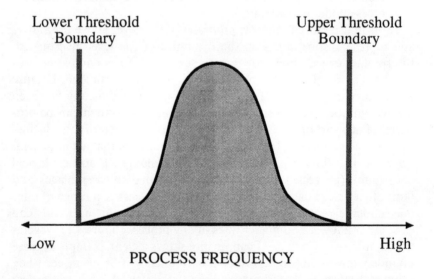

FIGURE 8.1

Illustration of how a system of interest may be characterized by a process (e.g., fire frequency) having an upper and lower threshold for frequency. The curve represents an arbitrary response for a community, such that thresholds for any dimension of a process (frequency, intensity, area, etc.) may exist for any defined assemblage.

than larger patches. Different aspects of any process such as canopy openings may end up on different levels in a hierarchical model because of their relative rates. Describing the overall dynamics of communities requires another kind of approach that can incorporate processes and characteristics of processes that occur at quite different scales.

Integrating Processes of Different Scales at a Single Level

At a particular site, abiotic and biotic processes of a variety of rates and scales continuously impact the dynamics of the assemblage. This observation suggests that analysis of community dynamics must incorporate processes of different scale and organisms of different scale at the same "level," otherwise we can only study pieces of an assemblage. On the other hand, hierarchical approaches that assign processes to different levels of organization because of contrasting rates have been an effective way to model some types of system interactions. Models such as hierarchy theory that combine processes of multiple scales are rare and often are not directed at community dynamics. How do we incorporate disparate scales of processes into a coherent model of assemblage dynamics, and can we do this in a way that meets assumptions of existing models such as hierarchy theory?

Historical pattern provides an insight into this issue. Here we accept the unique nature of specific sites based on a particular history of events and the composition, sequence, magnitude, frequency, or other characteristics of those events (Facelli and Pickett 1990). The historical contingency of sites results from the cumulative pattern of the impact of this diversity of processes at various scales. Vegetation may be constrained by composition and species availability at some scales, and by age structure on other scales. Because the history of any one site is unique compared to other locations, the assemblage is likely to be different (figure 8.2). Given historical contingency, what assumptions should we make to develop a model of dynamics? History suggests two critical features. One is that the model incorporate multiple processes and include all processes having a differential impact on the assemblage. It should also integrate the range of frequency, magnitude, and other characteristics (i.e., the environmental regime) appropriate for any particular site. Increasingly, current approaches to populations or communities have recognized multiple environmental processes (Brown 1984, Kolasa 1989).

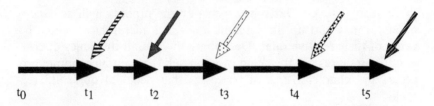

FIGURE 8.2

Representation of an assemblage at a site through time. At different intervals, events impact the site (represented by arrows). In reality, some processes are ongoing, and others occur sporadically. Through time, any site will have a unique history of such processes, each event constraining further dynamics at the site.

Second, historical contingency indicates that community dynamics reflects not just the tolerance or enhancement of species relative to particular processes, but the response of an assemblage at a site to a relatively unique set of processes in time and space. Because populations are interactive and because populations differentially respond to processes, whole community dynamics must reflect the ongoing sequence of processes as a set. Site dynamics are difficult to analyze because the ongoing flux in vegetation patterns results from an assemblage of processes, both biotic and abiotic.

Scaling the pattern of community dynamics to processes requires scales of analysis appropriate for each process. We can probe for process-specific patterns, but the pattern with which we are actually concerned is the dynamic changes of vegetation at a site, and that pattern is scaled not to a single process regime but to the combination of all processes at a site. This also suggests that forcing the scaling of pattern and process can be quite useful for some questions, but it also may restrict analysis of overall dynamics (May 1994). To integrate scale concepts in vegetation dynamics, one must interpret the combined impact of a large number of processes of different scale, not search out patterns determined by single processes.

Developing a Concept of Dynamics Using a Patch Model

How can we develop a model that yields historical contingency? First, we can consider how processes are spatially and temporally distributed across sites. A conceptual model of a system impacted by a number of processes can start with the impact of a process as a single patch (figure

8.3). Not all processes may express discrete boundaries or result from a discrete temporal event, but conceptually these are issues of the scale of measurement. Our assumption is that a process has spatial and temporal limits. Furthermore, any process varies spatially and temporally (figure 8.4), so the system may illustrate a patchwork of a process, each patch of different age, size, and other features (figure 8.5). This patchwork may result in a random pattern of patches, or locations within any area may be differentially impacted by various processes. For any location considered, a process can be described by frequency both spatially and temporally. A second process would add a similar patchwork that may differ in any of the other dimensions of processes, such as size or degree of impact. With subsequent processes also considered, the conceptual system illustrates an overlapping patchwork of these processes (figure 8.6).

This simple patch model incorporating multiple processes provides an effective starting point for considering community dynamics. Particular locations within the system may have a unique history of processes that satisfies one of the criteria expressed above. Additionally, the model provides an inclusive framework by considering all potential processes. The model also illustrates that processes must have spatio-

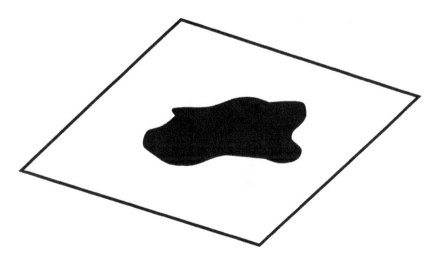

FIGURE 8.3
Representation of a site in which some process (e.g., disturbance) impacts a portion of the area.

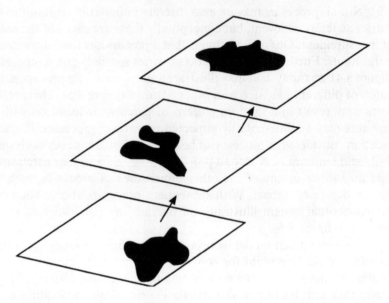

FIGURE 8.4

Representation of a process varying spatially and temporally at a site. Each layer represents the site at a different point in time, and each patch represents the same process impacting differing portions of the site at each time interval.

FIGURE 8.5

Representation of multiple events of the same process at a site. Each patch differs in age, representing an event of a process at a different time period.

Multiple Processes

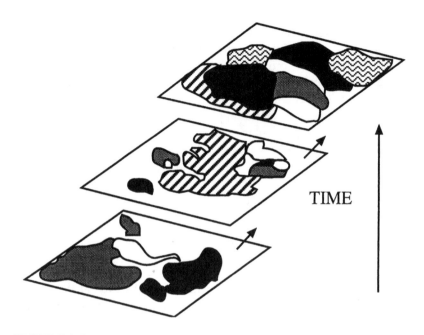

TIME

FIGURE 8.6

When viewed through time, a site illustrates a changing patchwork representing the impact of a number of processes of different ages, frequencies, intensity, and other characteristics.

temporal dimensions and be expressed in terms of a process regime (e.g., a fire regime). Processes also overlap in their impact (Addicott et al. 1987). This latter characteristic permits us to investigate the impact of any particular process in the context of another; in other words, it provides a nested structure for processes. Finally, this patch model is not restricted to particular spatial or temporal dimensions; it can apply at different scales depending on the question being asked.

Setting Scale Domains

Can we develop an explanatory, operational model from this graphic patch-process approach that can be applied at multiple levels of organization? One important point to notice in this conceptual model is that

we have considered only spatial and temporal characteristics of processes. Organisms respond to the sequence and extent of processes and
yield the observable community dynamics. To develop an operational
model, we have to be able to establish the domain of the model, the
boundaries inside of which it will work (Pickett et al. 1984). In doing
this, we set the scale of the system that allows us to develop criteria for
relevant processes. Two directions appear to be available at this point:
setting spatial boundaries or setting assemblage boundaries. Assemblage composition is considered in the following discussion; the arguments apply equally to spatial approaches.

Relevance of any particular process depends on the organisms investigated. We can assess how an assemblage sorts in interactions with a
collection of processes based on the characteristics of the organisms.
Even without knowing the relative magnitudes of processes, this approach can make meaningful predictions about assemblage structure.

Differential Impact of Processes

For any set of processes, what criteria would we use to determine which
processes are relevant for the organisms being observed at the level of
interest? Do we need to express the vegetation dynamic as a rate and
look for processes that occur at that rate? This does not seem practical,
because vegetation change is the dynamic we are interested in analyzing, and the processes that create change can occur at a variety of rates,
spatial extents, or duration. Processes that create change do not occur
at the same rate as system change, because system change reflects the
total collection of processes. The criteria for relevance, therefore, must
involve responses involving organisms in the assemblage, but the criteria also should be general enough to apply the model at a variety of
assemblage sizes or levels of organization. For these assumptions to
apply, one must focus on responses of organisms scaled at the level of
the total assemblage.

If a process can divide any analyzed species into two groups of contrasting responses, then that process would be an important differentiating force for the assemblage examined. For any process that influences the community, its relevancy is determined by whether it
differentially impacts the assemblage, that is, whether the process can
sort the assemblage. Processes that directly affect only a few members
play a role, but only if their consequences scale up to effect differential
response of the entire assemblage. For any process that is experienced

by the community, the criterion for including a process in the model is that it differentially impact the assemblage.

This model of system dynamics recognizes that processes must by definition have both positive and negative effects. This model views community assembly as a problem of assortment driven by factors that can influence the pool of potential community members jointly or simultaneously. The focus thus shifts from fine-scale, pairwise interactions to the role of coarse-scale and long-term processes. For any assemblage, one group is negatively impacted and another positively impacted. This indicates that for any relevant process there are thresholds of the magnitudes, frequencies, or other characteristics of these processes beyond which certain populations will be extirpated or suffer reduced recruitment or growth. These different levels of response are a critical feature of community dynamics. Threshold responses are how competition effects on individuals are judged, whether it is for rates of growth or changes in population size. If this is true, then the frequency and intensity of many different processes would be critical for understanding relative roles of processes. Thus, the next step is to analyze the dynamics of the processes.

Hierarchical Structure (Levels of Impact)

For a process to be important in structuring an assemblage, it must have a differential impact on all members of the assemblage being considered. Any process with a homogeneous impact would reside at a hierarchical level above the level of interest. Any process that has a differential impact only on a subset of species would reside at a level below. Thus, this model can be based on a structure that is hierarchical and has three levels of organization, but they are arranged differently from rate-based hierarchical structure such as that of O'Neill et al. (1986). In this case, at the level considered, processes that impact system structure would have a differential impact on the assemblage as a whole. That differentiation would provide the mechanism for shifts in composition, productivity, density, or any other measure of structure.

This is how a process is scaled to the assemblage being investigated. When a process has a differential impact on the assemblage observed, then the process has been scaled to that of the pattern. For example, shade tolerance is an important differentiating characteristic among herbaceous species living in a mosaic of forest and open meadow habitats, but less so within a forest or within a meadow. In the latter situation, other characteristics, such as soil conditions, tend to be more important. The criterion of differential impact on all members of the assemblage thus determines the scale of the process examined. *Scaling* processes to an assemblage, therefore, does not mean the processes have the same rates or magnitudes, only that they have a differential impact for the assemblage as a whole.

Toward a Predictive Model

We have argued that patterns in vegetation reflect a large number of processes, so what happens in this model if we look at more than one process? The criterion for considering a process is that species will be differentially impacted. Differential impact can be defined in a variety of ways, depending on the nature of the question. For community dynamics, we can define *differential impact* as having a net positive or net negative impact on the population dynamics of any species in particular. Those species with similar responses to the impact of a process (e.g., net increases in population size) will likely co-occur spatially and temporally. Subsequently, they will also be more likely to interact with one another. When these co-occurring species experience the next environmental process, they will sort again by their responses. Two important features emerge here: the set of differentiating processes and the sequence of those processes. The more species share similar responses to a variety of processes, the more likely they will co-occur spatially and temporally. Thus, a model of association can be based on the cumulative number of shared responses among species. We will call such a model a cumulative shared response.

A model of cumulative shared response would be based on only a few assumptions. System dynamics is assumed not only to reflect the impact of limiting processes, but also to be sensitive to the context of other processes, the sequence, magnitude, and other process characteristics. It is assumed that there will be differential responses among members of the assemblage before a process is "scaled" to the level of interest. Once an assortment has been accomplished by the relevant processes on a particular scale, within each cluster or component, further ecological processes may come into play. For example, assortment by tolerance to fire can be followed, within the tolerant and intolerant assemblages, by assortment based on competition. Of course, competition may be a coarse differentiation factor. Factors that act as coarse filters and those that act as fine filters are determined by the scale of analysis. For example, to model the assemblage structure of the flora of New Jersey, the relevant factors likely will be drawn from soil "quality" and fire frequency. In contrast, at the scale of a few meters in old fields, competition and herbivory may be the dominant differentiating factors.

These assumptions predict that the degree of association among species most likely relates probabilistically to their cumulative shared responses to processes. This is an extension of reasoning that individuals of a single-species population should exhibit the maximal proportion of shared responses. Species that are quite similar also should co-occur for the same reasons. Cumulative shared response thus becomes a statistic by which we should be able to predict patterns of co-occurrence. For the level of assemblage organization investigated, shared response also determines multispecies subsystems or "entities" within a level. Using shared responses also permits us to inclusively consider a variety of processes at the same level, both biotic and abiotic, rather than making assumptions about a hierarchical role of physical factors determining species presence, and biotic interactions determining subsequent pattern. The relative strength of processes and changes in the context of other processes will influence association patterns and in some cases increase the variance in these patterns. On the other hand, co-occurrence would be reinforced the "stronger" or more limiting the process is to the species having similar responses.

Investigating Spatiotemporal Dimensions for Dynamics

Community ecology must cope with organisms and processes of considerably different scale, which nonetheless interact directly or indirectly with one another. We have attempted to develop an alternative approach that is based on a conceptual description of spatial and temporal patterns of a community. Dynamics places a constraint on models because of the variety of processes that influence system dynamics. Hierchical models often solve this by using rates of processes to segregate systems into different hierarchical levels (Allen and Starr 1982; O'Neill et al. 1986; O'Neill 1989); in these cases, rates determine levels of organization such that families of processes of similar rates act on any one level. Other hierarchical models focus on structure, using interactions as a criterion for segregating into levels (Pickett et al. 1989; Kolasa and Pickett 1989). A difficulty of these models is the interpretation of the criteria. Segregating by rates may separate interacting species among different levels. Different life history or developmental stages of species can differ in rates of longevity, growth, or other characteristics by orders of magnitude; for some research questions, organisms may not fit well in rate-based models (O'Neill et al. 1986). Alternatively, processes of a variety of rates that can influence dynamics would also be segregated among levels of organization in some interpretations. Interaction models have interpretation problems as well, because interactions between specific pairs of species are usually not continuous; therefore, infrequent or rare interactions are difficult to evaluate, especially because the history of interactions is relatively short or incomplete for most communities.

Other models are explicitly spatial and attempt to cope with variously scaled events by nesting smaller, more frequent processes into larger, less frequent events (Wu and Loucks 1995). An important assumption of this approach is that there exists a direct correlation between spatial and temporal scales. This approach also assumes that processes of different spatial scales somehow vary in their impact on communities, such that smaller spatial processes are necessarily nested within larger spatial processes. Hierarchically separating these processes works for a spatial description of pattern; unfortunately, because dynamics is not the focus of the model, this approach separates processes influencing community dynamics among different hierarchical levels of organization.

Vegetation exhibits great variation in the scale of organisms and their characteristics, as well as in the scale of abiotic processes that influence the system (Addicott et al. 1987). Approaches to community dynamics have generally assumed the prominence of a particular process such as competition, herbivory, predation, or disturbance. Subsequent interpretation has unfortunately assumed the process studied is the limiting one for community dynamics, without specifying the spatial or temporal limits to the interpretation. However, dynamics for the assemblage as a whole is the consequence of a large and diverse array of processes. If a number of characteristics can be limiting at some temporal and spatial scale, and if these relevant processes constrain further dynamics, then the structure, composition, and dynamics of the site observed are a consequence of all relevant processes acting at the site. This is the fundamental basis for the historical uniqueness of locations. Pickett et al. (1987a,b) illustrated a successional framework for this array of processes, both biotic and abiotic.

We have built on interpretations of the successional framework for general community dynamics (Pickett and McDonnell 1989; Brand

and Parker 1995) to establish a spatiotemporal approach applicable to various assemblage sizes, levels of organization, and spatial extent. By using a patch-dynamic model for processes, we developed a conceptual model for environmental dynamics (both abiotic and biotic) applicable to spatial and temporal dimensions that can be applied to multiple scales of investigation. In this model, individual sites are shown to have a unique combination of processes, or a unique spatiotemporal pattern. This indicates concordance with what can be termed historical uniqueness or contingency (Gould 1989). The conceptual model presented provides an additional hierarchical approach to system dynamics that integrates processes and organisms of quite different scale at the same level of focus. Although this model differs in construction from rate-based models (Allen and Starr 1982; O'Neill et al. 1986) and from spatial-based models (Wu and Loucks 1995), it retains the features associated with hierarchy theory (as discussed later).

Historical contingency is based not just on the composition of processes that impact a site, but also on the characteristics of these processes, such as frequency, intensity, and area. To be inclusive of all pertinent processes, as well as to be able to generate historical uniqueness, concepts must see local pattern as resulting from the pattern of all processes simultaneously and in sequence (site history is sensitive to process context and sequence). In other words, it is not merely the process that is important but the spatiotemporal pattern of processes. The model indicates that processes occur in the context of all other processes, and their impact is not independent. Instead, by sequentially nesting within the context of other processes, each process [a mode of research illustrated by multisite analysis-of-variance (ANOVA) designs] can be investigated; spatial or temporal dimensions do not establish a hierarchical priority for dynamics. A historical context determines local system characteristics as each event constrains further dynamics by limiting species availability and influencing their performance. Local pattern results from the overall local regime of processes.

Adding to Hierarchical Approaches

One might assume that any scalable features should permit hierarchical organization of systems. The criterion of differential impact is a scalable characteristic that allows placing processes at appropriate hierarchical levels of organization in this scheme. As in other hierarchical models, there are three levels of organization that must be considered (Pattee 1973). At the level of interest, processes must exhibit a differential impact on the entire assemblage, allowing a comparative approach permitting prediction of subsequent trends. The higher level is represented by processes that have a similar impact on all the organisms being considered. At a lower level, processes also have differential impact on species populations but only on a subset of species considered in the assemblage. This makes the model applicable to multiple scales of investigation, in the sense that dynamics for any list of species chosen can be appropriately focused based on the preceeding criteria. An example of this is the invasion of *Pseudotsuga menziesii* (Douglas fir) into chaparral in the western United States. Within chaparral, important processes include soil nutrient and moisture gradients and the differential response of species to burns. With a larger scale of consideration (larger species list), including the adjacent mixed evergreen forest with which chaparral interacts, a different array of processes becomes important. At this level, fire frequency is an important differentiating process. Fire frequency can differentiate among any pair of species because of their individualistic responses, but threshold frequencies between chaparral and forest exhibit a different and more inclusive range than within either system alone. Similarly, dynamics between chaparral and forest is also influenced strongly by the type of mycorrhizal symbionts hosted by chaparral plants. Stands of *Arctostaphylos* (manzanita) species, which host arbutoid mycorrhizal fungi (mostly basidiomycetes), are differentially invaded by the dominants of the mixed evergreen forest, especially *P. menziesii,* which are principally ectomycorrhizal plants and can utilize the same species of fungi found on the shrubs. Other areas of chaparral dominated by arbuscular mycorrhizal plants are mutualistic with different groups of mycorrhizal fungi (zygomycetes) and do not facilitate the establishment of trees. At the level of the mosaic of chaparral and forest stands, the differentiating processes are now shifted in frequency, intensity, area, or other characteristics such as mycorrhizal symbionts. As a result, the general order of importance of processes has changed.

Shared cumulative response provides an index for probing structural association patterns resulting from dynamics. At the assemblage level, association patterns result from cumulative similarity among or between species for responses to processes that have differential impact on assemblage composition. This is simply a formalization of a general ecological understanding (Chesson 1991; Holling 1992; Busing and White 1993; Palmer and White 1994). It is also amenable to other types of analyses such as matrix or multivariate approaches (Dutilleul, chapters 17, 18; Gardner, chapter 2; Schneider, chapter 12). Matrix data can be analyzed by multivariate approaches to illustrate clustering of species by their cumulative shared responses. A matrix of species to processes includes only whether the process enhances or inhibits a species (+/−), or a numerical value indicating the relative strengths of processes for more sensitve analyses.

Several implications become apparent when using a patch-dynamic model of a multidimensional process regime. One is that we must incorporate processes of different rates together. Another is that we must focus on thresholds so we can differentiate between enhancement and inhibition. At the level of comparing among sites, each site ends up with a unique history in terms of its spatiotemporal sequence as we increase the time span or spatial extent of consideration. At the same time, another implication of the model is that sites impacted by the same processes of larger spatial scale then share historical events in the same sequence. A measure of the relative strengths of large-scale versus small-scale processes would be the shift in biotic similarity per unit distance. Variance in biotic similarity indicates the strength of local-scale processes on general trends (i.e., large-scale processes). Traditionally, we have thought of these types of results (large variance) as indicative of a weak large-scale process or influence. Yet this type of result can support more than one interpretation. In some cases, it may be accurate to think of the variance as indicating a weak influence. In other cases, the process may be strong, but other processes with a more local or smaller-scale impact also may be strong and have "diluted" the impact. To understand dynamics, we must be able to distinguish between these two interpretations. A resource manager who assesses a large-scale process at a site to be weak and then modifies the habitat, may discover the process was strong but "masked" by a large number of smaller-scale processes. Therefore, a fundamental error in interpretation can be made.

The North American prairie provides examples of how different scales of approaches yield different conclusions about the role and importance of different processes. Depending on the scale of focus, several processes sort species into different association patterns. At relatively large scales, climatic features such as precipitation and precipitation–soil interactions sort the pool of prairie species into different groupings (Transeau 1935; McMillan 1959; Sims et al. 1978). In the eastern reaches of the prairie, the presence or absence of fire sorts between prairie and forest dominance, but at finer scales throughout the biome, the frequency of fire sorts prairie species into different assemblages (Curtis 1959; Wright and Bailey 1982; Allen and Wyleto 1983). Local-scale differentiation among prairie species can be effected by large grazers (Anderson 1982; Axelrod 1985), small disturbances such as *Taxidea taxus* (badger) mounds (Platt 1975; Platt and Weiss 1977), differentiation in use by pollinators (Parrish and Bazzaz 1976), and competition (Werner and Platt 1976; Werner 1979). All the processes that sort prairie species are characterized by different scales of frequency, duration, and magnitude. Different sets or frequencies of the processes act to sort assemblages at the large scale, and guilds or other species groups on the fine scale. The processes can be arranged hierarchically based on the proportion of the total prairie species pool among which they differentiate, and on the spatial extent they affect. The resultant assemblages, from the biome to interacting species pairs, can likewise be arranged hierarchically.

Historical Contingency as an Integration of Spatiotemporal Processes

Our conceptual approach models systems as multidimensional in that species respond differentially along multiple dimensions of processes (Hutchinsonian niche). Species populations respond variously to processes and sequences of processes because of their own unique combination of life history and morphological and physiological characteristics. Species exhibit dimensionality in their life histories; each stage—seed, seedling, vegetative adult, reproduction, and dispersal —may scale with different processes or sets of species. Various morphological or physiological characteristics similarly will scale to different types of processes. For any process in particular, some members of the assemblage will be enhanced in their own dynamics, whereas others may be inhibited. Differentiation among species can occur on a number of dynamic dimensions because a variety of dimensions may be "limiting." When we focus on composition or proportion, or any scalable biological characteristic, the impact of any process of any rate or ultimate spatial extent may constrain further dynamics at a site.

This leads back to our original concept of community dynamics as change, but the change as a result of historical contingency, a particular spatiotemporal sequence of processes. Because we assume that pattern results from the combined impact and sequence of all processes on a site, for any scale of organisms considered or any assemblage size, the criteria developed earlier would work for determining relevant processes in this community dynamics model. The nature of questions being asked (e.g., concerning changes in growth rates, productivity, reproduction, recruitment, or composition) all change the relative importance of processes and of the spatiotemporal scale being investigated.

This means assemblages can be assessed from a number of characteristics of organisms, ranging from diversity to physiology. At one scale of investigation, inhibition of individual physiology that ultimately limits some plant response, such as rate of a metabolic process, growth rate, or rates of reproduction, might be one approach. At another scale, inhibition of population processes, such as rate of recruitment or change in population range or density, leads to a different assessment. Threshold changes in these categories lead to change in community composition, diversity, or species density. Thus, changes in scale of how we view dynamics can cause substantial changes in how we might view patterns of assemblage distribution.

Historical contingency suggests an approach in which processes of different spatial or temporal scale are combined at the level of investigation. This seems to be a requirement of investigating dynamics, and it also implies that all processes act in the context of other processes. Their temporal sequence may be critical. To work with community dynamics, rate and size of processes (spatiotemporal dimensions) must be independent characteristics of the model utilized. By using criteria based on shared impact of processes on organisms, one can build vertical hierarchical structure. Subsystems within a level are determined based on cumulative shared impact of processes. This model is consistent with our observations of historical uniqueness of sites, as well as with our understanding of spatiotemporal dynamics. Hierarchical approaches to ecological problems with scale problems are quite informative and appropriate, and this investigation suggests that there may be a number of criteria not yet explored that can be used to build useful hierarchical structures.

9

SPATIAL SCALING AND ANIMAL
POPULATION DYNAMICS

Brett J. Goodwin and Lenore Fahrig

Studies of changes in animal population size over time (population dynamics) are conducted primarily for three reasons: (1) to make predictions of population persistence (Lande 1988), (2) to investigate the mechanisms causing population fluctuations (Royama 1992), and (3) to calculate sustained yield harvests (Bailey 1984). The spatial scale of the investigation has important consequences in all cases. This chapter reviews our current understanding of how the spatial scale of a study of population dynamics affects the conclusions one is able to draw from it.

Spatial Structure of Populations

To understand how the spatial scale of an investigation affects the conclusions it engenders about population dynamics, it is first necessary to understand how populations themselves are scaled. First, we must distinguish between the spatial scale of an experiment or census, as opposed to the scaling of a population. The spatial scale of an experiment or census is chosen by the investigator and may or may not be appropriate for investigating the population dynamics of a species. The scaling of a population is the spatial structure of the population itself. Populations can be organized into different hierarchical levels (figure 9.1), each with its own typical scale. The lower and upper limits of a species' spatial scales are the scale of the individual and of the global population. The spatial scale of the individual is the range over which it moves during usual (nondispersal) activities; this has been referred to as the ecological neighborhood by Addicott et al. (1987). The spatial scale of the global population is the scale over which all movements of all individuals in the species occur, that is, the range of the species. Between these two limits there may be two additional population scales: the local population and the regional population.

A local population is a collection of individuals with a high probability of encountering each other during usual activities (Andrewartha and Birch 1984; Wiens et al. 1986; Addicott et al. 1987). The encounter probability must be high relative to the probability of dispersal, where dispersal refers to movements of individuals between local populations. Therefore, most individuals remain within the local population and only a small fraction (e.g., less than 5 percent) move between local populations. Examples of this type of population structure have been reported for a wide range of insects (Addicott 1978; den Boer 1981; Cappuccino and Kareiva 1985; Cappuccino 1987; Harrison et al. 1988;

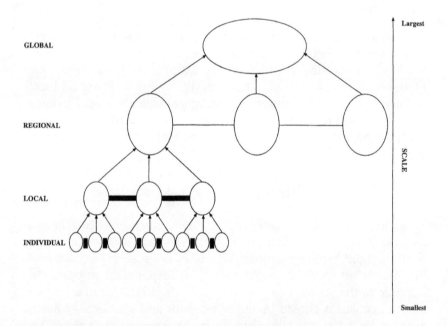

FIGURE 9.1

Schematic representation of the organization of a species. Levels higher in the hierarchy are at larger spatial scales. The thickness of the interconnecting lines indicates the degree of interaction within levels; thicker lines indicate a greater probability of interaction.

Solbreck and Sillén-Tullberg 1990), amphibians (Gill 1978; Laan and Verboom 1990; Sjögren 1991), birds (Saunders and Ingram 1987; Villard et al. 1992), and mammals (Henderson et al. 1985; Arnold et al. 1993). The spatial scale of the local population is defined by the scale over which all its individuals move during their usual (nondispersal) activities.

Of increasing interest to ecologists is the regional population scale, which consists of all the local populations within a particular region (Gurney and Nisbet 1978; Taylor 1988a). The scale of the regional population is arbitrarily determined by the total area (or region) of interest to the researcher. If the region includes the whole range of the species, then the regional and global populations are the same.

The spatial structure of a regional population is determined by the combined effects of the spatial pattern of the population and the rate of successful dispersal between local populations, which in turn depends on the interaction between the dispersal characteristics of the organism and the landscape structure (Merriam 1984; Burel 1989; Taylor et al. 1993; Fahrig and Merriam 1994). Population spatial structure is determined functionally by the set of probabilities of movement between all pairs of local populations. For example, a population of birds inhabiting 10 small forest patches could have a spatial structure similar to that of a population of dung beetles inhabiting 10 dung pats in a field, if their abilities to disperse relative to interpatch distances were equal. Also, a population of frogs inhabiting 10 ponds at 1-km spacing within a large forested area could have a spatial structure similar to that of a population of the same species inhabiting 10 ponds at 100-m spacing but separated from each other by roads or other barriers to movement, causing the probability of successful dispersal over 100 m to be the same as that over an unimpeded kilometer.

The spatial structure of a population is not static; it changes over time. It is necessary, therefore, to consider population dynamics within an appropriate temporal frame. The appropriate temporal scale is the generation time of the organism (McArdle et al. 1990). It has been suggested that populations should be censused once per generation (Connell and Sousa 1983); however, with overlapping generations, this concern can be relaxed (Gaston and McArdle 1993). Before making any inferences about population dynamics, it is necessary to study a population over an appropriately long time frame by gathering data on several generations of a species. Spatial and temporal scaling are difficult to separate, and the investigator needs to keep both in mind when designing a study. This chapter focuses on spatial scale; in our discussion, we assume that the population is sampled at an appropriate frequency and for an appropriate duration.

Scales of Population Studies

Two components of measurement scale can be identified: extent and grain (Wiens 1989, Gardner, chapter 2; Vande Castle, chapter 13). Extent refers to the range over which measurements are taken or how much area is covered in the sampling scheme. Grain often refers to the amount of area sampled per quadrat or the size of quadrats used (Wiens 1989). This definition of grain is most applicable to sessile organisms. Mobile organisms are usually censused using traps or visual counts. These methods produce estimates of abundance, but it is difficult to determine how much area they sample effectively. In the case of mobile animals, it may be more appropriate to think of grain as the number of samples per unit of length or area. Increasing the grain of a sampling scheme increases the resolution of the observations made. More sampling points must be added to maintain a constant grain and to increase extent. If the number of sampling points is restricted, the only way to increase extent is to decrease grain. In general, when population ecologists refer to the spatial scale of a study, they are in fact referring only to its extent (Kareiva and Andersen 1986; Taylor 1988; Rose and Legget 1990).

Studies of population dynamics are conducted at a variety of scales relative to the actual spatial structure of the population. An important problem in the interpretation of population dynamic studies is that the scale of observation often does not coincide with one of the population scales. For example, the samples might cover only part of a local population (Bach 1980), or only part of a regional population (Gill 1978).

Population ecologists use a variety of criteria when deciding on the sampling scale for a study. The most common is probably expediency and cost. Kareiva and Andersen (1986) reviewed 6 years of experimental community ecology studies published in *Ecology* and found that half of the studies used plots no larger than a meter in diameter. Although Kareiva and Andersen were not examining studies of population dynamics *per se,* their review demonstrates that many ecologists do investigate populations at scales smaller than the scale of a single local population. This inequity between the scales of most ecological investigations and population structure is mostly caused by high costs of and logistical constraints on larger-scale studies (Kareiva and Andersen 1986; May 1994). The scale of such studies may be too small to quantify properly, let alone understand the mechanisms driving population dynamics.

A second common criterion used is the human-centered view of habitat structure. Researchers often define the boundaries of populations based on landscape elements obvious to human observers, such as woodlots, clearcuts, agricultural fields, or urban areas (Opdam et al. 1984; Fahrig and Merriam 1985; Saunders et al. 1987; Fahrig and Paloheimo 1988; Szacki and Liro 1991; Villard et al. 1992). There is no guarantee that the population structure of the organism of interest coincides with these landscape elements. In such instances, a study of the local population dynamics of a particular species may in fact be a study of only a small part of a local population. For example, populations of *Peromyscus leucopus* (white-footed mouse) were thought to occur primarily in wooded areas (Graves et al. 1988), so studies of *P. leucopus* dynamics were rarely performed outside of forested areas (Wegner and Merriam 1990). It has recently been found that *P. leucopus* utilizes some crop types in agricultural areas (Wegner and Merriam 1990), and, therefore, the population structure does not match the spatial structure of forest. A large woodlot could contain a number of local populations, or a collection of small woodlots and fields could contain only one local population. For a small organism with limited dispersal, a study of a local population in a landscape element could actually represent a study of a regional population composed of several local populations. Such a species may divide a patch more finely than the observer would: a patch that appears homogeneous might actually be heterogeneous for the species and contain a number of local populations.

The third possible criterion for selecting the scale of a study is to match the scale of the study to the actual spatial structure of the population. As we will argue, this is the most useful criterion. Unfortunately, it is difficult to do because it requires two pieces of information that are seldom available: (1) information on the actual spatial distribution of the population, and (2) information on the rates of dispersal of the organism. The latter depends not only on the intrinsic dispersal characteristics of the species, but also on the structure of the particular landscape in question (Fahrig and Merriam 1994). Once this information is available, one can adjust the scale of observation to match a single local population or an entire regional population.

Sampling Scale and Population Predictability

The growing emphasis on scale issues in ecology can, in most cases, be reduced to a concern over the appropriate scale to investigate a particular ecological problem (Morris 1987; Hastings 1988; Smith and Urban 1988; Wiens 1989; Rose and Legget 1990; Taggart and Frank 1990). In the case of population dynamics, the "correct" scale is the scale that permits prediction of population persistence based on a statistical or mechanistic model of the population. But how can one select the scale of the study to maximize predictability?

Success of model predictions depends on the importance of factors internal to the system relative to the influence of external factors (Allen and Starr 1982; Wiens 1989; Levin 1992). In general, predictability is highest at the scale of the global population and lowest at the scale of the individual (Levin 1986). The decline in predictability from global to individual is not monotonic (figure 9.2). Relative increases in pre-

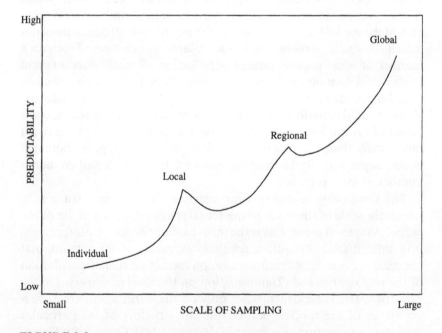

FIGURE 9.2
Hypothesized relationship between predictability of population persistence and scale of sampling. Increase in predictability with scale is not monotonic, and local maxima coincide with intrinsic scales of organization (see figure 9.1).

dictability will occur when the scale of sampling matches the scale of an entire regional population or an entire local population.

At any scale in the population hierarchy, population persistence depends on two factors: (1) the probability of extinction of the population given that it is extant, and (2) the probability of reestablishment of the population given that it is extinct. The first of these (extinction) depends primarily on factors internal to the population, and the second (colonization) depends primarily on external factors. Therefore, as the frequency of extinction and reestablishment increases, the importance of external factors relative to internal factors increases, leading to decreasing predictability.

At the scale of the individual territory, extinction probability is the product of mortality probability and the probability of movement out of the territory. However, once the individual dies or moves, the probability of reestablishment of the area by another individual depends on the density of the local population and the movement rate of individuals within it. Because extinction rate is very high, the probability of persistence of a population within an individual territory depends almost entirely on factors external to the territory.

Similarly, if only a portion of a local population is sampled, predictability will be low unless that area of the population happens to be representative of the whole local population. The greater the extent of the local population, and the greater the asynchrony over space in factors affecting the local population, the less likely a subset of the population is to represent the whole. For example, small portions of a local population could become extinct as a result of localized feeding of a predator. Recolonization of these extinctions depends on the density of the remaining population and the movement rates within it.

At the scale of the entire local population, there is a sharp increase in predictability. By definition, movement rates between local populations are much lower than movement rates within local populations. Extinction probability depends primarily on factors operating at the scale of the local population, with the possible exception of a rescue effect through immigration (Brown and Kodric-Brown 1977). Recolonization depends on the larger regional scale, so short-term persistence is relatively predictable, whereas long-term presence or absence is not. To gain the highest possible predictability for a local population, one should conduct sampling over the entire area of the local population.

Regional population persistence depends on the overall regional population density, which is affected by the processes of growth, extinction, and recolonization of local populations within the region.

Once the population in an entire region is extinct, recolonization depends on immigration from outside the region. However, it is not clear whether one can sample a subset of local populations to make predictions of regional population persistence.

A possible method for sampling local populations to predict regional persistence has been proposed by Hanski (1994a). He developed a method for estimating regional persistence based on incidence functions, which could be applied to a subset of patches. Using patterns of presence and absence and patch arrangement, simple models can be parameterized and used to predict the regional persistence of a species. Hanski (1993) has applied these models to *Sorex* species (shrews) on islands and found these predictions to compare favorably to empirical studies. However, his model assumes a large number of local populations (more than 50) and a balance between occupied and extinct patches (between 20 and 80 percent in one state). Thomas (1994) has also raised the concern that for this model, populations must be at a steady-state equilibrium, which probably does not hold for many natural systems. Therefore, the incidence-function approach is probably applicable only to a very limited number of systems.

In subsampling a regional population, we again have the same difficulty in making predictions about the whole population. The various local populations within the regional population may be controlled by different factors. For example, within one portion of the region, the local dynamics may be primarily density dependent, and in another portion they may be primarily density independent. To obtain the highest predictability of a regional population, one should, therefore, conduct stratified sampling: the sampled strata are the local populations. This ensures that all local populations are sampled. Similarly, predictions about a global population will be most accurate when they are based on observations stratified according to the population structure.

Finding the "Correct" Scale

Predicting Persistence

Based on the above description, it follows that the correct sampling scale for population dynamics should coincide with the intrinsic scales determined by the population structure, such that sampling is conducted over either an entire local population or an entire regional population. However, this approach requires knowledge of the population structure, which is often lacking. One way to find the scales of local populations is to determine what is usable breeding habitat through autecological studies or natural-history observations, and then search out that type of habitat in the study area. This could be facilitated through the use of aerial photographs or satellite images. Capture–recapture studies could then be used to determine the rate and extent of demographic flows between habitat areas, and to determine whether the different habitat areas represent different local populations (Stern, chapter 14). Note again that the population structure depends on the landscape structure, because movement through the landscape depends on both the structure of the landscape and the movement attributes of the species (Fahrig and Merriam 1994).

If it is not clear what demarcates a local population or where a local population might occur for a species of interest, then population structure could be determined by censusing in a grid pattern and looking for aggregations of individuals that could represent local populations. The size of grid chosen and the grain of sampling should ideally be based on information about the movement scale of the organism, and it may be beneficial to determine this empirically (e.g., mark-release, radio tracking) before conducting the census. Inability to detect aggregations could be the result of either the entire sampling area being within a single local population or the grain of the sampling design being too large to detect local populations within the area (figure 9.3). These have opposite implications for changes to the sampling scheme (increased extent versus decreased grain), and in the absence of additional information on movement scale, there is no way to decide which alternative to try.

If a sampling scheme is found to yield low predictability once the scale of the local or regional population is taken into account, then forces external to the populations sampled are more important than internal forces. This indicates that to increase predictability (approach the correct scale), it is necessary to increase the scale of investigation to incorporate external forces. This is opposite to the usual scientific tradition for dealing with uncertainty, which is to further reduce the system of interest and look at a finer scale (Allen and Starr 1982). However, in studies of population persistence, this approach may indicate that the population is structured at a larger scale and needs to be studied at this larger scale.

In many cases, the appropriate scale of study will be too large to census realistically. Sometimes, this problem can be overcome through a marriage of theory and empirical study (Kareiva and Andersen

1986). Autecological studies performed at smaller spatial scales can provide parameters (e.g., reproductive rates, dispersal ability) for either simulation or theoretical models. Results from detailed studies of both movement (Kolasa and Waltho, chapter 4; Stern, chapter 14) and habitat use can be incorporated into an individual-based, spatially explicit model of population dynamics. This model can be made to reflect a larger area than what was incorporated in the garnering of parameter values, by using aerial photographs or satellite images and geographic information systems (GIS) software. The model is run by incorporating this larger area and used to produce hypotheses (Fahrig 1991) that can be tested by using a few strategic experiments or observations (Fahrig and Freemark 1995).

For example, parameters for a species could be measured either directly or gathered from the literature. Fahrig and Merriam (1985) used

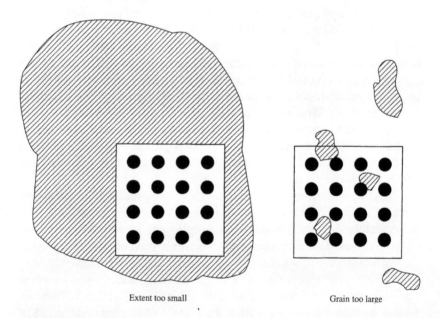

Extent too small   Grain too large

FIGURE 9.3

Two possible explanations for the inability of a sampling scheme to detect local populations (*shaded areas*). The sampling scheme is indicated by the square enclosing the area sampled and the dots indicating sampling points. **(A)** Extent of the sampling scheme is too small to detect the local population, because all samples are within the local population. **(B)** Grain of sample is too large because local populations fall between the samples.

literature values to develop a model of *P. leucopus* survival in woodlots. They tested the short-term predictions of the model regarding interconnection of woodlots and survival. Lefkovitch and Fahrig (1985) then used the same model to make longer-term predictions about the survival of mice in different configurations of woodlots. If it was then necessary to attempt to understand mouse population dynamics at a larger scale, it would be possible to set up the simulation to emulate a specific landscape and then use the model to make specific predictions about population dynamics (e.g., population levels in certain woodlots). These predictions could then be tested by sampling a subset of the area. If the predictions and observations match, then the model describes the dynamics well and can be used both to reduce the fieldwork required to estimate population sizes and to make predictions about the population. A similar approach can be taken using theoretical models. For example, Kareiva and Andersen (1986) created a theoretical model explaining *Coccinella septempunctata* (ladybug) aggregations in *Solidago canadensis* (goldenrod) habitats based on fine-scale studies of movement rates, turning rates, and prey densities. The theoretical model was then successfully scaled up to explain *C. septempunctata* aggregations in larger patches of *S. canadensis*. When the model failed to work in pea fields, it illustrated to Kareiva and Andersen that plant architecture played an important role by limiting the ability of *C. septempunctata* to disperse. This failure of their model did not lead Kareiva and Andersen to reject their model but instead to be aware of another parameter (plant architecture) that would need to be considered in larger-scale extrapolations.

Investigating Mechanisms

There is no single correct scale for understanding the mechanisms driving population dynamics (Levin 1992). Mechanisms could be operating at larger or smaller scales than the scale of the local or regional population. To understand these mechanisms and how they function, it may be necessary to sample for the mechanisms at larger or smaller scales. This should not be confused with resampling the population itself at another scale. For example, population levels of a common prey species may be largely determined by predation, which makes it necessary to understand the distribution and dynamics of the predator. If the predator's population is structured at a larger spatial scale than that of the prey, then sampling for the predator (but not necessarily for the prey) must be performed at that larger scale.

One should investigate a mechanism at both the appropriate spatial scale and the appropriate temporal scale. Wiens (1989) raised the concern that factors acting at larger spatial scales may take longer to become apparent. If one is looking for a relationship between a mechanism and a population, it is clearly necessary to investigate that mechanism at the appropriate temporal scale.

There are multiple mechanisms driving population dynamics. Understanding what drives the dynamics of a population requires investigating all possible mechanisms, each at its appropriate scale. If even a few of the possible mechanisms operate at a large spatial scale, then a systematic investigation of all possible mechanisms will be impossible. To determine which mechanisms are most likely to explain population dynamics and are therefore worth investigating empirically, one could: (1) identify possible mechanisms, (2) identify their appropriate scales, (3) construct a multiscale simulation model incorporating the possible mechanisms, (4) use sensitivity analysis to determine which mechanisms are most important, and (5) test the one or two most important mechanisms with appropriately scaled field tests. If the field tests fail to demonstrate that the mechanism has a large influence on population dynamics, then it will be necessary to repeat the process while reevaluating other possible mechanisms and the structure of the model.

Human Impacts

Human use of lands and resources changes landscape structure (Merriam 1988; Groom and Schumaker 1993), which can change population structure and thereby change the degree to which a particular sampling scheme coincides with population structure. Most natural habitats are reduced through human alterations, which in turn reduces local population sizes and increases the probability of local extinction at both local and regional scales. The possibility of extinction increases the relative importance of external factors affecting recolonization, which reduces predictability of population dynamics.

Loss of habitat is also often accompanied by fragmentation of habitat into a larger number of habitat patches. It is often assumed that habitat loss and fragmentation convert a single large (local) population into several smaller local populations (Robinson and Quinn 1992). However, it has also been found that habitat loss over large areas can lead to an increase in the scale of movement of some organisms. For example, it has been reported for small mammals [*Clethrionomys glareolus* (bank vole) and *Apodemus agrarius* (field mouse) (Szacki et al.

1993), and *P. leucopus* (Wegner and Merriam 1990)] that movement distances double in subdivided habitats compared to homogeneous habitat. *Calopteryx maculata* (forest damselfly) individuals in subdivided habitats have larger wings and greater thoracic mass (stronger flight muscles) than individuals in continuous forest, indicating an increased ability to move (Taylor and Merriam 1995). This switch to greater movement ability following habitat fragmentation can also change population structure, converting a regional population of several local populations into a single large, low-density local population. This change in population structure could lead to a decrease in predictability for a given sampling scheme and thus require an increase in observational scale.

In the case of harvested populations, such as ungulates and waterfowl, it is necessary to accurately estimate the population level to determine an appropriate harvest size. If the population estimate is too low, the amount harvested is less than it could be. More important, if the population estimate is too high, the amount harvested can be too great for the population to sustain, leading to the eventual extinction of the population. As we have seen above, to accurately census a population, the spatial scale of the census needs to coincide with a natural scale of the population. If sampling does not coincide with a natural scale of the population, it is possible that only part of a local population is censused, which results in low predictability. If predictability is low, then estimates of population size and subsequent calculations of harvest size will be poor. If a poor estimate of population level leads to an overestimate of the population, and subsequently an overestimate of the allowable harvest, the population could be overharvested. Therefore, it is necessary for managers to be aware of the spatial structure of the harvested population. Fisheries have long been concerned with the spatial structure of the harvested population, as evidenced by attempts to identify fish stocks (Garrod 1977) and movements between stocks (Fahrig 1993). Expansion of this spatial awareness into terrestrial populations would be beneficial.

Summary

Populations can be conceptually arranged into organizational levels including local, regional, and global populations. This spatial structure results from an interaction of dispersal ability and landscape structure. Each level has an associated spatial scale. Most studies of population dynamics are performed at scales determined either by expediency or anthropocentric views of landscape structure. These sampling scales may be inappropriate for some animal populations and can lead to poor predictive power. Maximal predictability can be accomplished by sampling at a scale that matches an intrinsic organizational scale of the population (local, regional, or global).

Choosing an appropriate sampling scale requires specific knowledge about the spatial pattern of the population and dispersal ability of individuals, but this information is not commonly available. Studies of population persistence should be conducted to determine the appropriate sampling scale by studying habitat use and dispersal ability.

If the population study is concerned with elucidating the mechanisms controlling the population dynamics, then it will probably be necessary to take a multiscale approach. Mechanisms could be occurring at smaller or larger scales than the scale of the population. Understanding the mechanism requires studying it at the appropriate scale (or scales) for the mechanism itself. This requires thinking about, proposing, and testing hypotheses at multiple scales. Testing hypotheses can be aided by the use of theory and modeling coupled with empirical tests.

As humans change landscape structure, we change the scales of both the population structure and the mechanisms driving it. The scale of population structure could be reduced in some cases and increased in others. Sampling schemes should be adjusted to compensate for anthropogenic changes in population structure, but this will be extremely difficult if the rate of landscape change is high.

Acknowledgments

We thank two anonymous reviewers who helped to improve the quality of this chapter. This work was supported by a National Science and Engineering Research Council of Canada grant to Lenore Fahrig.

10

TIME, SPACE, AND BEYOND: SCALE ISSUES IN FOOD-WEB RESEARCH

Neo D. Martinez and Jennifer A. Dunne

Ecologists almost exclusively restrict analyses of scale to spatiotemporal aspects. Although this helps focus analyses in important ways that we discuss, such restriction also limits the lessons learned from ecological analyses of scale. Here, we expand beyond these limits by illustrating how the classic concepts of extent (or range) and resolution (or grain) apply to nonspatiotemporal scales such as species richness and productivity, as well as space and time, within food-web research.

A *scale* is a metric and therefore a means of measuring or quantifying differences among observations. Once a scale has been employed in the observation of a phenomenon, the scale of observation can be specified by stating the variables observed, along with the extent and resolution of the observations. *Extent* describes the realized scope of the phenomenon or entity of interest, such as identification of the lowest and highest values observed for a particular variable. *Resolution* describes how finely observations within the extent are divided or distinguished. Resolution can be thought of as how similar two observations can be and still be recognized as two observations rather than one.

In ecology, the concept of scale is almost exclusively explained and applied in terms of time and space. The Division of Environmental Biology at the National Science Foundation, the primary source of funding for basic ecology in the United States, virtually always stipulates that temporal and spatial scales be addressed where scale is mentioned as a research topic in requests for proposals. Explanations of hierarchy theory, which is perhaps the main vehicle for promulgation of scale concepts in ecology, emerge out of definitions of scale, extent, and resolution (or grain) that are based on considerations of space and time (O'Neill and King, chapter 1). For example, Allen and Starr (1982) give the definition, "A scale is the period of time or space over which signals are integrated or smoothed to give message." Allen and Hoekstra (1992) state that extent "refers to both the spatial and temporal universe of discourse embodied in the data collection," and resolution or grain "determines the smallest and most ephemeral entities that can be found in the data." Allen (chapter 3) reiterates this spatiotemporal fixation by stating, "The theory of scaling that emerges from hierarchy theory emphasizes . . . the spatiotemporal scale of a structure." This emphasis follows from one of the central hypotheses on which hierarchy theory rests: larger, slower entities constrain smaller, faster entities (O'Neill et al. 1986). Following hierarchy theorists, many ecological subdisciplines are increasingly paying attention to spatiotemporal scaling issues. An example from the subdiscipline on which this chapter focuses is provided by a recent compilation of advances in food-web research (Polis and Winemiller 1996), in which the section devoted to scale issues is entitled "Spatial and Temporal Scales."

Many types of ecological analyses and data involve examination of spatial and temporal scales of observation. For example, consider a single aerial photograph of a vegetated landscape. The boundaries of the photograph, which establish the scope of the recorded image, determine the spatial extent of the observation. Spatial resolution is determined by the grain of the photograph, which fixes the size of the smallest unit that can be distinguished within the picture. A series of photographs of the same area taken at different times introduces temporal scaling. The beginning and end dates of the photographs determine the temporal extent of the set of observations. Temporal resolution is determined by how frequently pictures are taken between the beginning and end dates. The establishment of the spatial and temporal scales of these types of observations is critical for determining which hypotheses can be addressed using the observations.

Although the tendency in ecology is to limit the use of scale concepts to the consideration of spatial or temporal extent and resolution, this chapter explains how scale may more broadly refer to any metric that can quantitatively measure difference among observations in order to create data. For example, the use of a thermal scale to produce a set of observations will have an extent determined by the range of temperatures measured, and a resolution determined by the preciseness of measurements made with a particular thermometer. Depending on what kinds of hypotheses are being examined, analyses of the thermal scale of observation could be sufficient, or the analyses could be augmented by examining spatial and temporal scales of observation related to placement of temperature measurements in space and time. In ecology, nonspatiotemporal scales have already advanced food-web research through extent- and resolution-based sensitivity analyses of various food-web properties. Two scales that have proven most useful for such analyses are primary productivity and species richness.

This chapter describes how analyses of scale issues related to the metrics of primary productivity and species richness, as well as time and space, can help trophic ecology mature as an ecological subdiscipline through the reassessment and refinement of food-web generalities and predictions. First, we discuss analyses of ecological variation along species richness and productivity scales that employ concepts of extent and resolution. We then discuss the relevance of more traditional considerations of spatial and temporal scale issues in food-web research. The chapter concludes by proposing three general hypotheses whose examination may help to expand research on the alternative ecological scales discussed, as well as research concerning spatial and temporal aspects of scale in food-web research.

Scales of Food-Web Research Beyond Time and Space

The Species-Richness Scale

A nonspatiotemporal scale of substantial historical as well as contemporary ecological interest is species richness, or the number of species in an ecological system—one of the simplest measures of biodiversity (Gaston 1996a,b). Ecologists have long been interested in which conditions are responsible for different ecological systems having different numbers of species (Hutchinson 1959; May 1988). Interest in the impacts of species richness on the structure and function of ecological systems has also experienced a resurgence resulting from the accelerating loss of biodiversity caused by humans (Schultz and Mooney 1993). One of the early hypotheses relevant to this issue stated that increasing species richness increases the stability of populations because of the effects of diversity on food-web structure (MacArthur 1955). A later alternative hypothesis suggested that if several food-web parameters, including average interaction strength and connectance (the fraction of all possible interactions within a system that are realized), are held constant, increasing species richness leads to decreased stability of populations (May 1974). These two arguments are briefly summarized below.

MacArthur (1955), along with several others (Hutchinson 1959; Odum 1971), argued that diversity in terms of species richness enhances population stability as a result of the effects of increasing complexity that accompanies increasing diversity. Complexity is considered to increase with diversity because species in more species-rich systems typically have more kinds of trophic resources than species in more species-poor systems. If one of the resource populations crashes or goes extinct, the linked consumer population in the more diverse system is more likely to have an alternative resource population on which to feed. Paine (1966) contributed to this theory by pointing out that an increase in complexity, caused by an increase in predator species and predation interactions, can stabilize populations by preventing certain prey species from excluding other species from the system. In summary, diversity (in this case, species richness) begets stability through a process of increasing trophic complexity that accompanies increased diversity.

May (1974) challenged the diversity-begets-stability theory by arguing that increased complexity decreases stability instead of enhancing it. This argument is based on results from mathematical food-web models that demonstrate that increasing trophic complexity increases the probability of positive feedback loops, which drive populations to explode to infinity or to go extinct. Loosely translated, the result commonly known as "May's stability criterion" holds that the product of average interaction strength (i) and the square root of species richness (S) multiplied by connectance (C) should stay below 1 [$i(SC)^{1/2} < 1$] for the population processes of a multispecies system to be stable. Average interaction strength is the average intensity of all interactions among species within a web. Connectance is frequently measured (Warren 1994) in terms of directed connectance (Martinez 1992), which equals the number of trophic interactions or "links" in a food web (L) divided by the square of the number of species in the food web (S^2). This criterion assumes that i does not systematically vary with S, which leads to the assertion that C must decrease as S increases in order to maintain stability. In other words, if all else (i.e., i and C) is constant, stability is undermined by increasing S.

Although limitations of May's model have been illustrated by several researchers (Cohen and Newman 1985; Taylor 1988b; Law and Blackford 1992; Haydon 1994), it continues to play an active role in interpretations of food-web data (Moore et al. 1993; Warren 1994). Of particular interest is the extensive empirical testing of these theories within structural food-web research, the branch of trophic ecology that examines properties of food-web structure specified by descriptions of unweighted feeding links within a system. Two equivalent means of evaluating May's relationship are to investigate whether (1) C hyperbolically declines as S increases, such that the product SC remains constant, and (2) the links-per-species ratio (L/S) remains constant among food webs of varying S, because $SC = L/S$ follows from $C = L/S^2$. Several prominent investigations found these two forms of constancy to be empirically supported (Rejmanek and Stary 1979; Cohen and Briand 1984; Sugihara et al. 1989; Pimm et al. 1991), and constant L/S became known as a law of scale-invariance called the link-species scaling law (Cohen et al. 1990a). Links-per-species (L/S) was asserted to be scale-invariant with regard to both extent (Cohen and Briand 1984; Sugihara et al. 1989) and resolution (Sugihara et al. 1989). Resolution, in this case, refers to the degree of aggregation of different taxa within a food web (Martinez 1991).

Further analyses, centered around consideration of the species-richness scale of observation, have rejected the widely accepted link–species scaling law and related properties. Most of these analyses favor alternative relationships with different scaling properties leading to alternative interpretation of diversity, complexity, and stability theories. One set of analyses centers on the extent of S (Martinez 1991; Polis 1991). Early findings of constant SC and L/S were based on food webs with less than 50 species (Rejmanek and Stary 1979; Cohen and Briand 1984). The small extent of the species-richness scale of observation represented by those webs thus restricted findings to a small subset of ecological systems (Martinez 1991; Polis 1991). Other recent analyses that support the link-species scaling law (Sugihara et al. 1989; Havens 1992) do so under the methodologically suspect exclusion of webs of 10 species or fewer (Martinez 1993a, 1994), even though the original purveyors of the law assert that it applies to such small webs (Cohen et al. 1990a). Only after Havens (1992) was faced by the contradictions between his data and scale invariance did he try to exclude such small webs from analysis (Martinez 1993a; Murtaugh 1994). In other words, the extent of species richness to which the law applies appears very narrow.

Food webs that include many more than 50 species have been constructed to extend the species-richness scale of observation. These webs suggest that the links-per-species ratio (L/S) increases, rather than staying constant, with increasing S (Winemiller 1989; Martinez 1991, 1992). Contrary to the earlier results, the increase in the extent of web size considered produces the result that connectance C stays roughly constant among food webs with variable S (Warren 1990, 1994; Martinez 1992, 1993a; Deb 1995). The finding that L/S increases with species richness (S) is also supported by more rigorous and methodologically consistent reanalyses of earlier data with smaller extents of S (Schoener 1989; Martinez 1993a, 1994; Murtaugh 1994).

The constant connectance hypothesis, as corroborated by more extensive and highly resolved data, supports MacArthur's (1955) theory that diversity enhances stability by corroborating a key component of the theory: consumers in more species-rich systems appear to have more resources or links per species (L/S) than consumers in less diverse systems. However, if systems are constrained by May's stability criterion, i must decrease as S increases (Warren 1994; Martinez 1995). If i decreases relatively fast as S increases, then diversity could enhance stability and still be consistent with May's stability criterion (Martinez 1996). It should be noted that the relative constancy of L/S^2 could be related to mechanisms other than those suggested by MacArthur and May (Lawton 1995).

Another set of critiques of earlier food-web analyses brought attention to the lack of consistency in collection of trophic data (MacDonald 1979; May 1983; Paine 1988). Many of these critiques are concerned with the lack of even and consistent resolution of taxa within food webs. Such resolution problems have been hypothesized to lead to inaccurate descriptions of food-web attributes (Lawton 1989). The initial claims of scale-invariance were based on early webs containing a wide variety of differently resolved taxa, where taxa refers to any named group of organisms that appears in a food-web description. For example, webs can have taxa as variably resolved as "*Caenorhabditis elegans,*" "insects," "plankton" and "protozoa/bacteria" (Schoenly et al. 1991). In more recent webs, greater attention has generally been placed on more detailed resolution of taxa, down to taxonomic species (Havens 1992). Although this type of consistency is an important first step, it has been noted that "it is a fact of systematic biology that the vast majority of the species recognized are equivalent by designation only, not through their degree of evolutionary, genetic, ecological, or phenetic differentiation" (Heywood 1994). In other words, taxonomically consistent resolution provides little if any biologically consistent resolution.

A promising procedure for more consistently resolving taxa within food webs (Martinez 1995, 1996) is to first finely resolve taxa into taxonomic species and then to aggregate the taxa into functional groups called "trophic species" (Cohen and Briand 1984). Trophic species are groups of taxa within food webs that appear to share the same set of trophic consumers (e.g., predators, parasites) and resources (e.g., prey, hosts). The use of trophic species within food-web analyses standardizes the grain of resolution and is particularly appropriate for structural food-web analyses; trophic interactions provide a consistent basis for identifying both the components (trophic species) and their relationships (trophic consumption) within the system being studied (a food web).

Further considerations of the extent (Martinez 1993a) and resolution (Martinez 1993b) of S have shown that L/S^2 varies little as these two aspects of scale vary. For example, when S of a web is reduced to the small range between 2 and 10 species, either by limiting the extent to this range (Martinez 1992) or by reducing resolution through the aggregation of species in more species-rich systems, such that those systems are described by food webs with few species (Martinez 1991), L/S^2 increases as S decreases. However, the increase is so slight that constant connectance more accurately predicts L within these small webs than does the link-species scaling law (Martinez 1992). Among webs with more than 10 species, little if any systematic variation of connectance is observed among the more recent and higher quality (i.e., more rigorously and consistently collected) data (Martinez 1993a, 1995; Deb 1995).

The constant connectance hypothesis asserts only that connectance does not systematically vary with S (Martinez 1992). It does not assert that there is no variation in connectance. Warren (1990) hypothesizes that connectance varies with habitat type and organisms chosen. For example, connectance would be expected to be higher if trophic generalists were selectively included in food webs than if specialists were selectively included (Warren 1990). Also, connectance is increased when criteria for designating trophic links are made more permissive (Martinez 1991, 1993b). Taking into account such sources of variation, we suggest a "constrained connectance hypothesis" that more generally asserts that connectance varies roughly between 0.01 and 0.3. This range is only a fraction of the theoretically possible range of 0.0 to 1.0 and well below the null expectation of 0.5 (Kenny and Loehle 1991) when each species consumes a random fraction of the species in a food web. Of the several hundred food webs with which we are familiar, none fall outside the range from 0.01 to 0.3.

This discussion demonstrates that analyses of diversity–complexity–stability relationships have made substantial use of scale issues. In particular, these relationships are illustrated by considering the extent and resolution of the species-richness scale of observation within structural food-web research. Early empirical research supported laws of scale invariance that in turn supported the hypothesis that increasing diversity does not lead to increased stability. Later empirical research increased the extent and resolution of S and overturned the laws in favor of an alternative hypothesis of constant connectance, which corroborates the theory that diversity begets stability. Further consideration of the species-richness scale of observation is warranted because connectance is almost certainly not limited to the range of 0.01 to 0.3 at very large scales of observation. For example, if connectance is constrained to 0.1 at global scales of observation (Martinez and Lawton 1995), species on average would eat and be eaten by about 10 percent of the species on the planet. If there are 10 million species on the planet, this would mean that on average each species eats 1 million other species and is eaten by the same number as well. In other words, whereas constrained connectance appears to apply to local scales of species diversity (10 to 1000 species), the pattern should break down at larger extents of species richness such as those found at global and regional scales (Martinez and Lawton 1995). More specific identification of where, how, and why the breakdown occurs may help identify mechanisms that constrain connectance at more local scales.

Several other food-web properties that were originally thought to be scale invariant, based on small extent and low resolution of species richness, are also greatly affected when data that increase the extent and resolution of the observations are considered. These other properties include the fraction of species at the top, intermediate, and basal trophic levels (Briand and Cohen 1984). Top species have prey but no predators. Intermediate species have both prey and predators. Basal species have no prey, but do have predators. Earlier analyses held that these fractions did not systematically vary with species richness and were therefore scale invariant (Briand and Cohen 1984; Sugihara et al. 1989; Pimm et al. 1991; Havens 1992). Partially motivated by consideration of larger and more highly resolved food webs (Schoener 1989; Martinez 1991), later analyses assert that, instead, these fractions are scale dependent and vary with species richness (Schoener 1989; Martinez 1993a, 1994). Specifically, the fraction of intermediate species increases with species richness, whereas the fractions of top and basal species decrease in the range between 2 and 100 species (Martinez 1993a, 1994). Martinez and Lawton (1995) extend consideration of these fractions to regional and global scales and hypothesize that these fractions become scale invariant between 10,000 and 10,000,000 species (figure 10.1). The bulk of webs with 1 to 100 species appear consistent with scale dependence (figure 10.2).

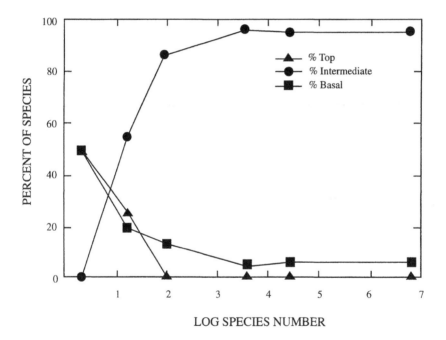

FIGURE 10.1

Effects of species richness on fractions of species at top, intermediate, and basal trophic levels. These effects were hypothesized for food webs with two species, to food webs with the total number of species on earth. These predicted effects are generally observed among the data (see figure 10.2) among food webs with less than 100 species.

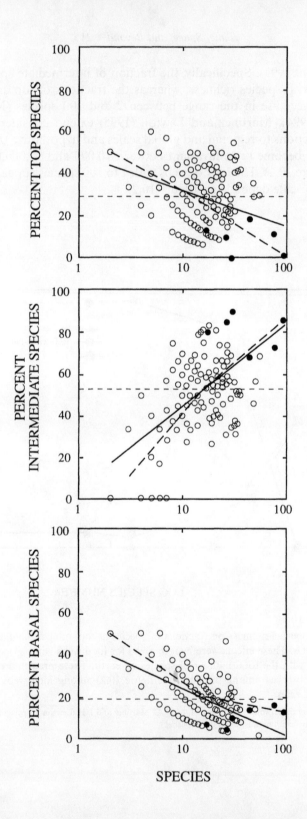

The Primary-Productivity Scale of Observation

Another more prominent debate in ecology concerns the effects of productivity on food-chain length (Lawton 1989). Hutchinson (1959) advanced the early hypothesis that the low thermodynamic efficiency of transferring energy from one trophic level to another limits food-chain length. Hutchinson argued that the amount of energy available to successively higher trophic levels is progressively more limited because only a small fraction (e.g., 10 percent) of the energy available at one trophic level is converted into energy in the form of biomass at the next trophic level. This suggests that at some relatively low trophic level (e.g., level 4 or 5), there is no longer enough energy to support another trophic level. Pimm and Lawton (1977) advanced the hypothesis that mathematical instabilities in the dynamics of long food chains, rather than energetics, limit food-chain length. They found that the time necessary for perturbed populations within mathematical models to return to their original sizes increases when the populations are embedded in longer food chains. Perturbed populations attain their original sizes more quickly when embedded in short food chains.

This debate relates two metrics by which ecological systems are measured: productivity and food-chain length. The energetic theory

FIGURE 10.2

Effects of species richness on fractions of species at top, intermediate, and basal trophic levels. *Open circles* represent trophic-species versions of Cohen et al.'s (1990) 113 community food webs and trophic-species versions of Sugihara et al.'s (1989; Martinez 1994) 60 community food webs. *Closed circles* represent community food webs constructed by Warren (1989), Winemiller (1990), Martinez (1991), Polis (1991), and Goldwasser and Roughgarden (1993). See Martinez (1994) for more details regarding these filled data points. The *horizontal dashed lines* represent the scale-invariant effects of species richness on the fractions of species as hypothesized by Cohen et al. (1990) and Pimm et al. (1991). The *solid lines* represent least-squares regressions of the fractions of species as a function of the log of species number. For top species, $r^2 = 0.12$. For intermediate species, $r^2 = 0.36$. For basal species, $r^2 = 0.34$. These regressions are generally very close to the trends hypothesized by Martinez and Lawton (1995) in figure 10.1 and redrawn here as *sloped dashed lines*. Nonparametric correlation tests reject ($P < .025$) the scale-invariant (no-correlation) hypotheses in favor of scale-dependent hypotheses that top species and basal species decrease, whereas intermediate species increase with species richness (Martinez 1993a, 1994).

(Hutchinson 1959; Oksanen et al. 1981; Yodzis 1984; Fretwell 1987) holds that these two variables are positively correlated, whereas the population stability theory (Pimm and Lawton 1977; Pimm 1982; Briand and Cohen 1987; Pimm et al. 1991) holds that the variables are uncorrelated. However, a consideration of the extent and resolution of the scale of observation of primary productivity suggests different possible empirical evaluations of the debated correlation. If the possible extent of variation in productivity includes systems with zero as well as nonzero levels of productivity (Power 1992), it can be argued that the data lead to rejection of the population stability theory in favor of the energetic theory: systems with zero productivity have food chains of zero length, whereas systems with positive productivity have longer than zero food-chain lengths. However, if extending the scale of observation to include systems with zero productivity is held to be unacceptable or trivial, then the energetic theory may not be supported in this manner. If the extent of productivity considered varies from the very low productivity observed in large hyperoligotrophic lakes [e.g., Lake Tahoe (California and Nevada)] to the high productivity found in hypereutrophic ponds with large amounts of inedible cyanobacteria, one could find a negative correlation between productivity and food-chain length. This is because large oligotrophic lakes often have large piscivorous fish that primarily consume organisms that are at trophic levels 3 and 4, whereas highly eutrophic ponds often lack piscivorous fish and have few herbivores.

Resolution considerations also have direct applicability to this research. Merely choosing to compare the extremes set by the extent of the scale, as described above, misses much information that may be revealed if productivity observations are more highly resolved so that many intermediate levels of productivity are also considered. Such information could include the identification of specific ranges within a wide extent of productivity values where food-chain length is positively, negatively, and not-at-all correlated. Moore et al. (1993) have recently proposed just such a solution to the food-chain-length debate by more finely examining levels of productivity between zero and the levels most commonly observed in nature. They found that the dynamic constraints on food-chain length hypothesized by Pimm and Lawton (1977) apply only at very low productivities, lower than that observed in most ecosystems that account for approximately 99 percent of global productivity. Above this low productivity threshold, ecological systems appear productive enough to not have their food-chain lengths constrained by the dynamics of food chains modeled after

Pimm and Lawton (1977). Consideration of more highly resolved food-chain lengths has historically contributed to another line of food-web research. Hutchinson (1959) restricted resolution of food-chain length to integers, and later research (Fretwell 1987) increased the level of resolution to include fractional chain lengths.

Generalities Among Scales

The previous discussions of scale issues related to species richness and productivity illustrate that scientifically productive consideration of scale is not limited to spatial and temporal aspects of observation. Indeed, analyses of scale, such as those focused on extent and resolution, may be usefully applied to any metric that can quantitatively measure differences among observations. Perhaps the general hierarchical theory of scale, that larger spatiotemporally scaled entities constrain smaller spatiotemporally scaled entities, is also more generally applicable. Contrary to hierarchy theory, the link-species-scaling law asserts that the number of links within a food web is constrained by the limitations of individual species, which restrict each species on average to consume two other species irrespective of the number of species in the web (Pimm 1982; Cohen et al. 1990a). Consistent with hierarchy theory, the constrained connectance hypothesis asserts that the number of links within a web is constrained by the number of species in the web to be a fraction of all possible links. This constraint means that the size of food webs, which are larger-scaled systems than their component organisms, constrains the number of embedded trophic interactions linking the smaller-scaled component organisms. In this way, empirical corroboration of constrained connectance (Warren 1989, 1994; Martinez 1992, 1993a, 1995; Deb 1995) appears to generalize the hierarchical theory of scale-related constraint beyond spatiotemporal scales of observation to the species-richness scale of observation within food webs.

Beyond extent and resolution, another general aspect of scale involves the dimensions of scale. Time appears to be the paradigmatic single-dimensional scale. A single sequence of numbers may refer to all of time—past, present, and future. Space, on the other hand, is three-dimensional. Whereas time may extend and be resolved along a single dimension, space extends and is resolved along three dimensions (i.e., length, width, and depth). High resolution or large extent along one dimension of space does not necessarily correspond with high resolution or large extent along another dimension of space. Consider a continuous set of remote-sensing images of the earth that follows narrow 1-m-wide longitudinal transects from north to south. Such transects could be spaced roughly 3200 km apart at the equator, allowing 12 transects for the entire planet. Although the continuous longitudinal transects highly resolve observations of the earth's surface (e.g., reflectance of a particular wavelength of light) from north to south (Vande Castle, chapter 13), the large spacing between the transects greatly limits the latitudinal resolution from east to west.

In the case of species richness, the concept of dimensions may appear unusual. However, consideration of dimensions brings to light important issues. Two dimensions of the species-richness scale appear most ecologically salient. One dimension is that of taxonomic extent and resolution (table 10.1). Taxonomic extent may refer to how many kingdoms, such as prokaryotes, protozoa, animals, plants, and fungi, are represented within a species list. Taxonomic resolution may refer to the level of identification of the taxa, whether it be at the species,

TABLE 10.1

Examples of variable levels of extent and resolution of the species-richness scale along a taxonomic dimension.*

Large extent		Small extent	
Low resolution	High resolution	Low resolution	High resolution
Prokaryotes	*Escherichia coli* Other prokaryotes identified to species level	Plants	Woody plant species 1 Woody plant species 2
Protozoa	*Paramecium* spp. Other protozoa identified to species level		Woody plant species 3 Woody plant species 4
Fungi	*Laccaria laccata* Other fungi identified to species level	Animals	Woody plant species 5 Vertebrate species 1
Plants	*Ceanothus cuneatus* Other plants identified to species level		Vertebrate species 2 Vertebrate species 3
Animals	*Caenorhabditis elegans* Other animals identified to species level		Vertebrate species 4 Vertebrate species 5

*Each column represents a different species list whose extent and resolution are characterized in the column heading. There is no correspondence between the left two columns and the right two columns.

genus, family, order, class, or kingdom level, from highest to lowest resolution. The functional dimension is another ecologically salient dimension of the species-richness scale. The function of a species may be generally considered to be its role with respect to some activity or the species' interaction with a particular process (Martinez 1996). Trophic processes described by structural food webs can be used to exemplify this dimension. In this case, trophic extent may refer to how many autotrophs, decomposers, and consumers are contained within a set of observed species. Trophic resolution may refer to the level of trophic similarity among groups of observed organisms. Trophic species may be considered highly resolved functional groups within structural food webs, because the organisms within a trophic species have the same set of predators and prey and are therefore completely similar for the purposes of structural analyses. Less trophically resolved groups could contain all organisms within a structural web that share, for example, only 50 percent of their predators and prey (Martinez 1993b).

It may be that the functional dimension is too broad to be considered a single dimension, and that practically any ecological process forms a base for a dimension of the species-richness scale. This would mean that there are as many different ecologically salient dimensions of the species-richness scale as there are different processes with which species interact. These dimensions of species richness represent a more complex view of scale than in the above discussion, where it is simply assumed that more species represent larger extents of species, and more precisely identified groups mean higher resolution. Nevertheless, it would not be surprising if the disposition of food-web hypotheses ends up being determined with these more complex criteria. In a closely related example, Tilman (1996) recently proposed a resolution to the diversity/stability debate by distinguishing between the stability of population processes involving taxonomic species and the stability of ecosystem processes involving biomass and nutrients.

Reconsideration of Spatial and Temporal Scales in Food-Web Research

Although consideration of scales beyond time and space is a productive aspect of structural food-web research, there are also many opportunities for spatial and temporal scales to receive more detailed consideration in trophic ecology. Within structural food-web research, the effects of spatiotemporal scale issues (e.g., the effect of changing the physical boundary of a system to include larger or smaller areas, or the impact of making annual, seasonal, or more frequent observations of trophic links in a system on structural food-web properties such as connectance) are being explored by various researchers (Warren 1989; Winemiller 1990; Schoenly and Cohen 1991; Closs and Lake 1994; Deb 1995; Martinez and Lawton 1995; Tavares-Cromar and Williams 1996).

Spatial and temporal scales of observation have been examined relatively little in functional food-web research, particularly in classic exclosure experiments that are conducted to analyze interaction strength within food webs (Paine 1966, 1980, 1992). Functional food-web research emphasizes quantitative estimates of trophic interactions such as interaction strength or energetics (Paine 1992; DeRuiter et al. 1995), whereas structural food webs emphasize binary designations of trophic interactions (Cohen et al. 1990a; Pimm et al. 1991). Interaction–strength experiments are conducted by comparing exclosures where a species is removed, to exclosures where the species has not been removed. The effect of excluding the species on other species populations is used to calculate the strength of its interactions on other species in the system. There are several issues of scale, extent, and resolution that may greatly affect the conclusions of such experimental results. The following discussion focuses on the spatial scale of observation.

One aspect of the spatial scale of observation in interaction-strength experiments is the size of the exclosures used, which are typically on the order of 1 m^2 or 1 m^3. This can be seen as the extent of the set of interaction-strength observations in the same way that the boundaries of an aerial photograph can be seen as the extent of visual observations. Just as results from a particular-sized photograph do not necessarily extend to larger-sized areas, the results from a particular-sized exclosure do not necessarily extend to larger-sized areas. Additionally, unlike our example photograph in which the pixels spatially resolve smaller areas that can usefully summarize what a smaller photograph may reveal, data from a particular-sized exclosure do not necessarily summarize what can be expected from smaller exclosures. Thus, in the absence of other information, the experimental results from a particular-sized exclosure may not apply to other sizes of exclosures. Although replicate exclosures within similar environmental conditions and treatments increase the statistical certainty of the results in certain-sized enclosures, the results are still keyed to the particular spatial extent defined by the size of the exclosure. Ongoing studies that examine effects of varying the size of exclosures will allow investigation of the hypothesis that different spatial extents of observation quantitatively and qualitatively affect experimental results (Ruesink 1996; Sarnelle 1997).

Spatial resolution within exclosures brings up another aspect of the spatial scale at which interaction-strength experiments are conducted. For example, attention to species within intertidal exclosures is restricted to species whose adult stages are on the order of centimeters in size. Smaller species may not be detected and analyzed. One of the more famous interaction-strength exclosure results that relates to the earlier diversity/stability discussion is that *Pisaster ochraceus* (seastar) can prevent intertidal mussels, such as *Mytilus californicus,* from excluding several other sessile organisms from the rocky substrates in the intertidal zone by preying on the mussels, thus apparently increasing diversity (Paine 1980). However, when the level of resolution is increased to the point that hundreds of subcentimeter-sized species are observed, the strong interactions of the *P. ochraceus* that prevent the attainment of a mussel monoculture also effectively exclude many small species that depend on mussel monocultures. Thus, instead of enhancing species diversity as originally claimed (Paine 1966), *P. ochraceus* may decrease diversity in terms of the species richness of a more highly spatially resolved community assemblage (Suchanek 1979). Again, changing spatial resolution may alter or reverse the conclusions of experiments conducted at a particular resolution. Alternatively, changing the spatial resolution may corroborate results among different scales and provide for greater generality of explanation and prediction.

The spatial scale of interaction-strength exclosure experiments can also be seen in a different manner. Instead of the size of the exclosure reflecting spatial extent, and the smallest species observed reflecting spatial resolution, the area over which the experiment is conducted may be thought of as reflecting spatial extent, whereas the size of the exclosure may reflect the spatial resolution. From this perspective, the concerns about exclosure size determining a particular spatial resolution are similar to the concerns about exclosure size determining a particular spatial extent, as discussed. However, the consideration of spatial extent as defined by the area over which an entire experiment is conducted brings into focus other spatial issues that deserve attention within interaction-strength experiments. The asserted generality of results may be problematic when this type of spatial extent is considered.

In the case of *P. ochraceus* mentioned earlier, this species has been asserted to be a "keystone species" (Power and Mills 1995) with a relatively high interaction strength in intertidal systems "on rocky shores of the Pacific Coast of North America" (Paine 1966). The predation process that imparts to *P. ochraceus* its keystone status is thought to be responsible for a vertical zonation pattern (Paine 1974) of "a remarkably constant association of mussels, barnacles and one starfish" (Paine 1984). These general assertions are based on exclosure experiments conducted by Paine and others on the northern coast of Washington State. Foster (1990, 1991) has questioned the generality of these assertions by pointing out that such experiments have been conducted in only a few regions and that the specified zonation pattern can be observed only in a subset of possible areas on the Pacific Coast of North America. Foster recommends a more systematic placement of exclusion experiments to test the geographic generality of Paine's hypotheses. In terms presented by this chapter, Foster's critique puts a finger on the discrepancy or unbridged gap between the geographic extent of interaction-strength experiments and the geographic extent over which the experimental results have been presumed valid.

We only mention the issue here, but there is also a mismatch between the temporal scale of observed strong interactions in ecological systems and the scales at which the experimental results are presumed valid. Exclosure experiments, like most ecological experiments, generally last only one to several seasons. The duration, or temporal extent, of an exclosure experiment may qualitatively affect the observed interactions among species. The frequency, or temporal resolution, with which data are gathered may also impact analyses of interaction strength.

Summary—Implications for Policy and Research

These considerations of spatiotemporal scale appear critical to the acceptability of ecological generalizations, especially in high profile areas of research (such as interaction strength) that have potentially profound policy implications (Mills et al. 1993). Interaction-strength experimenters often find that a few species considered to be keystone species are observed to have strong interactions, whereas most species, which have been referred to as "wimps" (Power and Mills 1995), appear to have weak interactions. Such observations have led to assertions such as, "Weakly interacting species can be ignored. I believe that most species will eventually fall into that category, leaving the field to a few essential and dominant players" (Paine 1988). Furthermore, the unformalized process of selection and observation of a few species for interaction-strength manipulations, out of many species present in the system, has been asserted to "incorporate all . . . of the ecologically important species" (Paine 1992), even though a similar tendency in early structural food-web work led to inaccurate characterization of food-web properties. Such assertions fail to account for the potential discrepancies between the spatial and temporal scales of observation embodied by the experiment, and the scales of observation at which the experimental results are presumed valid. This is especially important because of the far-reaching consequences of policy makers embracing an ecological theory that holds many or most species to be ecologically unimportant. Labeling large fractions of biodiversity as unimportant could be used to justify policies that would lead to irreversible losses of ecologically "unimportant" biodiversity. These types of assertions should be avoided by ecologists in the absence of critical investigation of other possible interpretations of such experimental results (Schrader-Frechette and McCoy 1993).

An alternative hypothesis to the few-keystone-species/many-wimp-species theory is that the observed rarity of keystone species is a methodological artifact of the limited spatial and temporal scales of experiments. The spatial scale of observation usually includes manipulation areas on the order of 1 to 10 m^2 scattered over a landscape on the order of 100 to 10,000 m^2, and the temporal scale of observation generally extends no longer than 2 years (Menge et al. 1994). These parameters represent a small subset of possible spatial and temporal scales of observation. It is possible that all species are strong interactors at a particular experimental spatiotemporal extent or resolution. The only limit to the number of strongly interacting keystone species identified within a system may be the inability of investigators to choose a full range of extents and resolutions of spatiotemporal scales for experimentation and observation. This possibility is important to investigate in light of prominent hypotheses that most species are unimportant or "redundant" (Walker 1992). Uncritical acceptance of such hypotheses may have the unfortunate and unintentional result of providing an ecological rationale for the attrition or irreversible elimination of objects of ecological study (Martinez 1996).

The effects of varying the extent and resolution of different scales in food-web research need more consideration. This should include, but not be restricted to, space, time, diversity, and productivity. The following hypotheses should be considered: (1) that increasing species richness is correlated with a decrease in average interaction strength, (2) that the constrained connectance pattern breaks down when the scale of observation extends beyond thousands of species, and (3) that varying the temporal and spatial scale of an exclosure experiment greatly alters the observed interaction strengths and the identification of keystone species.

Considering scales other than time and space has been scientifically useful for structural food-web research. Broadening consideration of extent and resolution beyond spatiotemporal scales should be useful to other areas of ecology as well. Increasing the depth of research into issues of spatiotemporal scale will also advance functional food-web studies. Analyses of scale issues have led to formalization of trophic theories into testable hypotheses, empirical evaluation, rejection and revision, and suggestions for further empirical evaluations. Such iteration and refinement of theory and empirical evaluation of hypotheses represent basic scientific progress that aids ecologists to more rigorously assess the generality of their work.

Acknowledgments

Support was provided by National Science Foundation grants BIR-9207426 and DEB-9207588, and by the hospitality of the Rocky Mountain Biological Laboratory.

11

DEFINING ECOLOGICALLY RELEVANT CHANGE IN THE PROCESS OF SCALING UP: IMPLICATIONS FOR MONITORING AT THE "LANDSCAPE" LEVEL

G. A. Bradshaw

Over the past 20 years, a shift in ecological paradigms to a spatially explicit perspective has resulted in a parallel shift in the scale at which we view the environment. The convergence of ecological theory, public land management goals, and data-capture technologies has directed attention toward increasingly larger spatial scales (Levin 1992; Bradshaw and Garman 1994; Michener et al. 1994) (for the present discussion, *larger* scales denote coarser resolutions). Shortages in natural resources and the intense conflict over prioritized use of public lands has precipitated new federal regulations in the United States and elsewhere requiring assessment of ecosystems and indigenous faunal populations within a larger spatial extent. Subsequently, research and land management communities have begun to incorporate spatial and temporal information into their studies. To accommodate this spatial expansion, activity has been directed towards the acquisition and analysis of information at these new scales of interest (e.g., geographic information systems, remote sensing, and landscape analyses) to improve inventory accuracy and to provide a baseline for comparing ecosystem conditions. One task in particular challenges conservation and resource managers: monitoring ecosystems across a range of scales. Monitoring has been employed to mitigate conflict and detect excesses of human-derived disturbances [e.g., land conversion (Munn 1988; Innes, chapter 19)]. The range and extent of monitoring have also expanded to include a greater number of environmental variables and species (USDA Forest Service 1993). From this perspective, monitoring is intended to satisfy both environmental and economic interests by allowing activity on public lands under the watchful eye of federal and other agencies. Monitoring data forms the foundation on which pressing future decisions and policies will be formulated (Rykiel, chapter 21).

Generally speaking, the purpose of monitoring plans is to detect change occurring in a population or ecosystem. In the case of fauna, population parameters and habitat quality are sampled over time. Change is evaluated relative to an established baseline of behavior, abundance, and condition. Typically, changes in population and habitat attributes are observed relative to changes in the environment, such as in air, land, vegetation, and water quality. Experimental and field studies help establish a cause-and-effect functional relationship between a change in physical habitat (as the independent variable) and a population (as the dependent variable). In most cases, this complex relationship is poorly quantified at larger scales. Although there is considerable knowledge of the effects of fine-scale events, their significance within an explicit spatiotemporal framework remains unclear for two reasons. First, insufficient data exist that provide adequate coverage for larger units of interest; most ecological data pertaining to habitat have been gathered at specific locales without the intent of explicitly representing a larger area. Additionally, the successful execution of a robust experiment with adequate controls over the landscape and watershed scales over a long enough period of time (i.e., relative to the frequency and duration of disturbance events) can be very difficult. Large-scale experiments used to ascertain adequate quantification of pattern and process relationships are costly and often logistically intractable (Brown 1994). Second, we do not have a well-developed understanding of the relationship between processes and patterns at the larger scales required to satisfy rigorous monitoring standards. To illustrate, the following is an example from the Pacific Northwest.

Process and Pattern at Large Scales: An Example

Large woody debris has been identified as an important component of anadromous salmon (*Oncorhynchus*) habitat in the Pacific Northwest region of the United States and British Columbia (Canada), and it is one of many stream attributes monitored over time to assess habitat condition (Maser and Sedell 1994; Sedell et al. 1984). Much interest has focused on evaluating the status of salmonid populations relative to habitat quality (e.g., large woody debris distribution) in watersheds within the Columbia River basin. In the framework of landscape ecology, this means identifying the relationship between the faunal population pattern and its habitat pattern. Although a landscape perspective seeks to evaluate salmonid habitat at the basin scale, there is no known target distribution or pattern of large woody debris within a watershed from which a baseline value may be determined. In-stream wood distribution is known to have been affected in areas of intensive timber harvesting, but there are only relative estimates of the natural variation of wood over space and time. Not only is the range of natural patterns not well established, the relationship between human-derived perturbations and their effects on naturally fluctuating environmental conditions are quantified for only limited space–time frames. Typically, studies fall into two categories: (1) intensive (e.g., limited geographical extent over several years) or (2) extensive (e.g., coverage over a broad geographical extent for a single season or year). Additionally, the relationship between habitat quality and population viability is generally not well quantified at any scale and instead is often implicitly modeled as linear and extrapolated to larger physical scales (other habitat/population responses may be modeled as an exponential function to reflect a hypothesized saturation or threshold effect). This type of model can hide the importance of the spatiotemporal dimensions, and it can telescope several relationships into a single relationship, namely the relationship between habitat and population patterns. Explicit temporal dynamics are ignored relative to the spatial context by ignoring the dynamic component of this relationship.

Establishing the relationship between population patterns and habitat patterns for watershed scales actually requires the explicit consideration of interactions between several patterns and processes occurring within the aquatic–riparian landscape. These interactions include effects of land-use activities on wood availability (reflecting patterns of vegetation and geomorphology, and processes of successional dynamics) and wood input (reflecting processes of mass movement and disturbance) and wood distribution (reflecting processes of transport, and precipitation patterns governing flow patterns) through time and within a basin. Failure to include spatial and temporal dimensions when examining the relationship between population and habitat patterns will not lead to successful monitoring protocols. Although components of these patterns and processes have been studied, their integrated function within the landscape context are only beginning to be described with spatially and temporally explicit models (Kareiva 1990).

Despite these knowledge gaps, policy makers demand the same levels of statistical confidence and precision to validate management and legal decisions (Innes, chapter 19; Rykiel, chapter 21). However, before well-informed decisions can be made at the basin scale, scientists and managers must first determine (1) what should be measured to characterize a landscape or watershed unit, and (2) what constitutes significant change at these scales. The term *significant change* requires a multiple-tiered definition. Significant change is defined here as the changes in (spatial or nonspatial) patterns (e.g., wood distribution) that correspond to an appreciable change in ecological functions (e.g., *Oncorhynchus* habitat suitability of the basin). *Appreciable* refers to the level of change regarded as a deviation from "acceptable conditions." *Acceptable conditions* are defined as a set of ecological patterns (e.g., realizations of wood distribution characterizing a basin) that allow sustained support of viable *Oncorhynchus* populations for an indefinite time period and disturbance regime. The design of a satisfactory monitoring plan relies on knowledge of the magnitude and type of significant change of the variable of interest. Once ecologically significant change is quantified, an optimal sampling design will determine the critical states of the variable, detect change, and yield statistical significance in subsequent analyses, and, once these issues are addressed for larger scales, scaling will have been successfully achieved. In short, this discussion may be rephrased as, "How is change altered with scale?"

This chapter argues that despite increased attention on "landscape" and "region," there is insufficient understanding of ecological processes of units at and within these scales. The conceptual difficulties associated with viewing phenomena at different resolutions and extent (i.e., scaling-up) are confounded because of implicit, as opposed to explicit, consideration of spatial and temporal patterns. Landscape indices are calculated to capture changes in landscape pattern, but typically their statistical significance is not well calibrated to commensurate changes in landscape-level functions. The lack of precision relating changes in landscape pattern to landscape function does not allow rigorous assessment of risk for specific management actions. In part, the source of difficulty is derived from the fact that time and space have been oversimplified and their multivariate nature reduced in many applications of scaling-up. It is suggested that efforts be directed at interpreting pattern and change from a functional perspective (e.g., ecosystem processes, spread of wildfire, species' life history patterns) before sampling designs are developed to accommodate spatial and temporal fluctuations.

Determining Landscape Change: Scale of Inference Versus Scale of Sampling

Although not unique to larger scales, changes in environmental quality are often assessed in terms of changes in amount, type, and spatial attributes of resources. For example, the quality of *Strix occidentalis* (spotted owl) and other avian species habitat in a forested landscape depends not only on the amount of suitable forest cover but on factors such as mean stand size and interpatch connectivity (Forsman et al. 1984; Hansen and Urban 1992; Hansen et al. 1992; Spies et al. 1994). Assessment of habitat quality consisting of optimal ranges and combinations of several habitat parameters can require both multivariate and spatial analysis using a combination of classical and spatial statistics (optimal is used here in the sense of low risk of species extinction) (figure 11.1).

In addition to the many available metrics that describe landscape attributes (Milne 1988; Turner et al. 1991; Garcia-Molinar et al. 1993), there is a corresponding range of responses of these metrics to specific landscape processes. The effects of a process on landscape pattern may have differential effects on the direction and magnitude of change as measured by individual landscape metrics. For example, a linear decrease in forest cover caused by wildfire or timber harvest does not necessitate a corresponding decrease in patch size (figure 11.2). A simple one-to-one mapping of pattern (as measured by a single parameter) to process (as represented here by forest removal) is not always possible (Moloney et al. 1991; Bradshaw and Garman 1994). Evaluation of a given landscape metric as a meaningful measure depends on how *ecological significance* is defined and quantified. From the perspective of the wildlife habitat example, which changes in spatial pattern represent a significant functional change in the landscape as habitat? Sampling protocols and methods of measurement are best calibrated to reflect and detect these changes in ecological significance. The scale at which significant change is detectable is a function of the relationship between the scale of inference and the scale of sampling. This relationship will vary with both the selected metric and study objectives (Schneider 1994a; chapter 12).

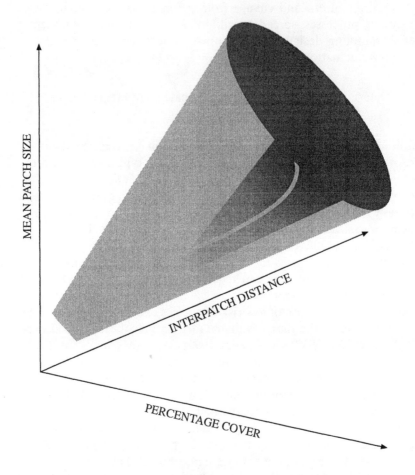

FIGURE 11.1

The range of acceptable habitat may entail a multivariate set of attributes. The set of landscape patterns having characteristics suitable for faunal habitat may be envisioned within this n-space.

FIGURE 11.2

(A) A set of four hypothetical forested landscapes of varying cover and patch sizes. **(B)** Landscape A has a sufficient amount of cover that patch size is not a consideration in terms of habitat suitability. Landscape D has such a low percent cover that patch size does not matter. Landscapes B and C have the same percent cover but vary in their mean patch size.

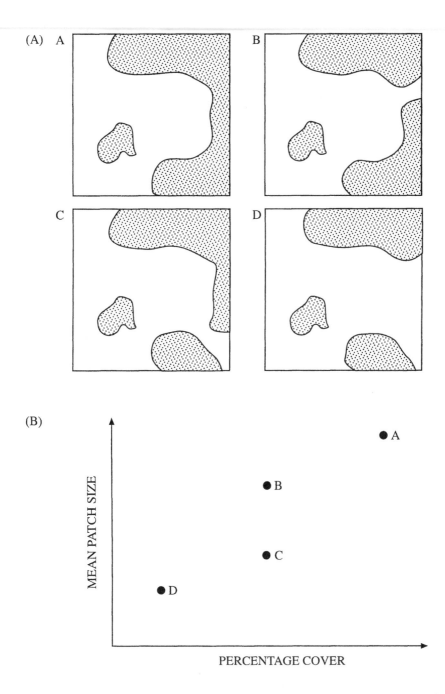

For purposes of illustration, let us assume that habitat quality can be described as a function of two parameters, mean patch size and percent forest cover. Two cases are examined. In the first scenario, habitat quality is conditional both on levels of patch size and on percent cover (figure 11.3A). At the extrema [either very low cover (point D) or very high cover (point A)], patch size does not matter; there is either sufficient cover that patch size is implicitly very large, or sparse cover, implying the existence of very small forest patches. However, to capture landscape change between these extrema, sampling resolution must be calibrated to distinguish patch sizes at a precision that will distinguish the threshold point (i.e., point B from C; figure 11.3A). The sampling scale (i.e., minimum patch size or less) will be finer than the scale of inference (i.e., a defined landscape). In this case, landscape function as habitat is governed by processes that affect patch size. Although it is necessary to detect changes in processes governing patch size, these processes may be governed by patterns at scales other than the patch or landscape (e.g., El Niño southern oscillation). In either case, analysis of population patterns within the landscape should relate to phenomena relating to patch size, and monitoring designs should accommodate sampling at the resolution of changes in patch size. For example, satellite imagery resolution should be sufficiently high to detect and describe patch sizes above and below acceptable levels (Vande Castle, chapter 13).

In the second scenario, if patch size is less important in defining viable habitat, and cover is a sufficient descriptor, a lower resolution is sufficient (figure 11.3B). In this latter case, the scale of inference (e.g., landscape) may match the sample scale (e.g., an image pixel may equal the dimensions of a landscape unit) without loss of information; retention of fine-scale spatial information is not needed. The scale at which significant change is sampled is commensurate with the scale of inference (i.e., landscape). Consequently, population patterns within the landscape should be related to phenomena governing forest cover across the landscape. If sampling is performed at finer scales than the

FIGURE 11.3

(A) The *shaded area* represents a hypothetical range of suitable habitat for the four landscapes in figure 11.4. In this case, sampling resolution would need to distinguish between patch sizes of landscapes B and C. (B) In an alternative scenario, the habitat quality of landscapes B and C are considered viable. Sampling resolution would need to distinguish only percent cover.

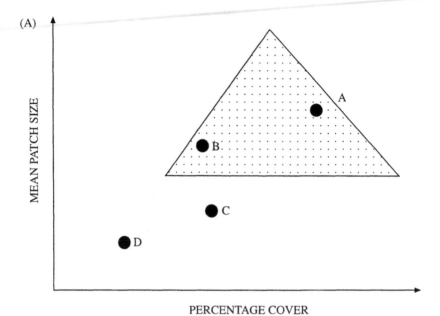

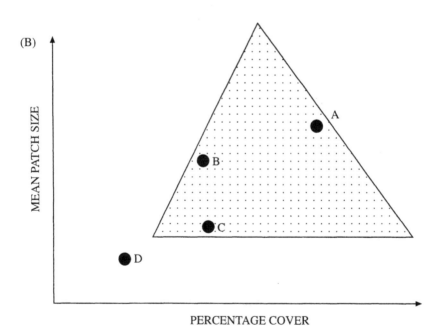

scale of inference, then a test for spatial correlation must made to avoid sampling bias. Estimates of cover made without accounting for the correlated pattern result in a significant bias. If spatial information is not needed, it can be discarded using detrending methods; if it is integral to defining functional habitat, the autocorrelation can be retained as part of the analysis. Alternatively, two scales of measurement may be employed independently and in concert to capture both habitat-dependent variables. Both landscape change analysis and the determination of statistical significance are needed to define ecologically significant change for the mean patch size and percent cover independently. In summary, both scenarios described above are designed to evaluate habitat change for a given landscape. However, the two cases differ in the scale at which significant change occurs and at which monitoring would be successfully executed. The monitoring scale of choice depends on the definition, and hence the scale, of ecological significance.

Interactions Between Spatial Pattern and Resolution

Because the level at which significant change is detected varies with sample resolution, the representation of spatial pattern across scales is very important. Specifically, the manner in which pattern aggregates in the process of scaling-up will contribute to the quality and type of information obtained (Rastetter et al. 1992). For example, error propagation that results from scaling-up forest canopy patterns using remotely sensed imagery can be strongly influenced by spatial pattern at finer scales. To illustrate the effect of spatial pattern on representation quality with a change in sampling resolution, consider a simulation experiment in which conifer forest canopy patterns are emulated as projections in the plane (figure 11.4). All parameters used to generate the individual forest patch sizes and distributions are identical in this example, with the exception of the level of aggregation of patches. The radius of patchiness was gradually relaxed to create a series of four landscapes, ranging from very patchy to increasingly more homogeneous in distribution.

To emulate the process of scaling-up occurring with a decrease in image resolution, four new images were created for each of the four patterns using the wavelet transform (Strang 1984; Mallat 1988). To explore the interactions between scale of resolution and spatial patterns with respect to landscape function, a stochastically governed fire-spread model was executed on each of the 20 images (one unfiltered and four filtered images for each pattern). In this simulation, the measure of ecological function for the landscape (i.e., wildfire spread) is defined a priori as the mean number of cells burned. The source of variation in this behavior depends on both the spatial configuration (patch size and distribution) and total cover. The ability of wildfire to move through the simulated forest within an image is intended to represent a measure of the ecological function of the forest pattern. A change in the amount of landscape burned is used to quantify the degree to which the landscape process changed with spatial pattern and scale of resolution. Each image was burned 30 times to alleviate bias introduced by random wildfire initiation sites; the total number of cells burned for the fixed time period were counted per run and used to calculate the mean number of burned cells per image. The underlying hypothesis is, if no change in landscape function occurs with the scaling process, the mean number of cells burned will remain the same as the original image.

Wildfire model response varies with both pattern and resolution; the forest canopy patterns act differentially at the same scale (figures 11.5 and 11.6). Information governing fire spread is retained only at the original and highest resolutions. The difference in behaviors of the four landscapes at the original scale becomes increasingly pronounced with a decrease in resolution. The behavior of three of the four images remains fairly constant until the coarsest resolution (W4). At the patchiness radius of 500, the behavior of the second pattern departs from the others by the second and third resolutions; pattern and reso-

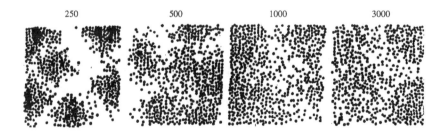

FIGURE 11.4
A set of four simulated forested landscapes shown with mean crown size and crown overlap held constant, while the patchiness constant is gradually relaxed. The number above each image refers to the scales of patchiness.

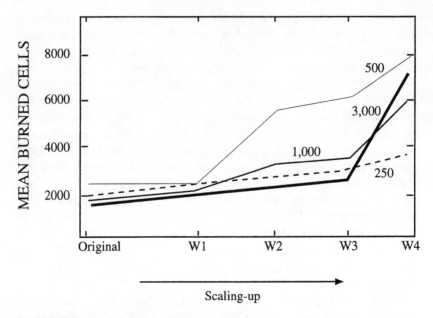

FIGURE 11.5

The mean number of burned cells per image per resolution. The numbers refer to the images in figure 11.4. The symbols W1, W2, W3, and W4 refer to the four scales (resolutions) at which wavelet decomposition was applied.

lution interact most strongly in this case. To examine implications for monitoring, one can envision the four landscapes as a set of possible realizations resulting from four different disturbance scenarios, respectively. Although monitoring directed at scales through the third resolution could be considered for landscapes of radii equal to 250, 1000, and 3000, sampling at coarser resolutions would function poorly for the landscape of radius equal to 500. This landscape pattern has strong elements from both the fine and the coarse scales. To preserve the ability to distinguish differences between the patterns and ecological functions of four landscapes, sampling would best be directed at scales of higher resolution.

Interactions Between Temporal Patterns and Resolution

Implicit to most monitoring plans is the assumption that the sampling design will provide information in a sufficiently timely manner

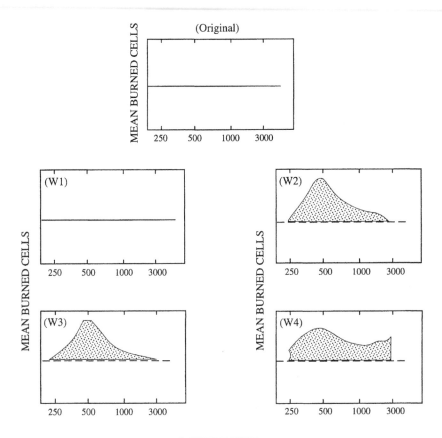

PATTERN TYPE

FIGURE 11.6

The mean number of burned cells from figure 11.5 are displayed. The mean numbers of burned cells are plotted per scale (W1, . . ., W4) as a function of pattern type (250, 500, 1000, 3000). In the original and W1 images, no appreciable differences between the four patterns are apparent; the wildfire model executed at the original or W1 scales does not differ in performance. On the other hand, at the scales of W2, W3, and W4, model performance (as measured by the mean number of burned cells) is a function of spatial pattern (i.e., there is a strong pattern-resolution interaction).

to anticipate change prior to the point of irreversible damage sustained by an ecosystem. There are underlying assumptions regarding the temporal dimension when a landscape is referred to as "suitable" habitat. In the previous examples, time was implicitly held constant; the landscape was considered appropriate habitat in its current state and capable of sustaining a given number of individual animals for an unspecified period of time. This is a static view in which the landscape can exist in perpetuity without change and retain its status as suitable habitat. More realistically, the landscape is but one of many states along the temporal record. Similarly, the faunal population in the landscape follows a trajectory integrating landscape states over time. Viewed in this dynamic way, a landscape may be defined as a set of spatially patterned realizations that may or may not provide suitable habitat for a species for a specified interval. The landscape may fade in and out of the suitability hyperspace over time cycles as it is subjected to a range of conditions [e.g., range of natural variability (Swanson et al. 1993)]. The landscape will qualify as suitable habitat depending on the relationship between the rate and type of change and the integration of population responses with a string of landscape realizations. An appropriate monitoring scheme should detect deviations from this set of landscapes of several attributes varying in space and time.

When individual "member" landscapes are viewed as a temporal array, one needs to consider several factors, including sequencing (timing), frequency, mean, variance, and duration of an event (half-period) (figure 11.7). Commonly used time series methods such as Fourier analysis quantify only one or two attributes of the temporal record and fail to characterize many of these features with sufficient resolution (Ford and Renshaw 1989; Bradshaw 1991). Some of these attributes may be critical to understanding how a population responds to a change in habitat. It is often the case that the timing of an event or current state of the landscape composition and configuration, and not merely its presence, makes the difference between survival and extinction.

For example, field studies have noted that connectivity and areas of aquatic habitat of *Rana cascadae* (Cascade frog) in central Oregon fluctuates significantly with annual variation in precipitation (Olson 1992). The spatially structured habitat may be envisioned as a series of realizations whose particular configuration varies through time. *R. cascadae* population dynamics were simulated using a metapopulation model to investigate the effects of a varying precipitation regime on frog survivability (Bradshaw and Michalik, unpublished manuscript). A set of annual precipitation records were generated using autocorrelation functions with varying parameters (figure 11.8). This allowed evaluation of the effects of changing frequency, timing, and duration of drought (where *drought* is defined as the level below one standard deviation from the mean precipitation). The simulated populations responded differentially to each realization of a given autocorrelation for a given set of parameters (e.g., range and lag identical). Extinction occurred in some realizations, while the population was sustained in others (figure 11.9). In all cases, the autocorrelation, and hence power spectra, were identical; the difference lay in the timing of drought relative to the animal lifespan and relative to the duration of normal rainfall years prior to drought. These two factors varied for each realization generated by the autocorrelation function. A monitoring scheme based solely on information derived from sampling designs used to assess the spectra would not have captured the differences sufficiently to allow prediction of extinction or survivability of a population. In such cases, methods such as wavelet analysis that preserve nonstationarity may be a better measure (Deutchmann et al. 1993; Strang 1994).

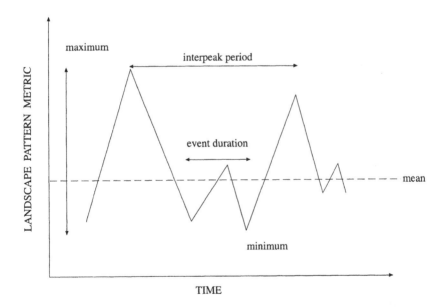

FIGURE 11.7

Two landscapes may experience the same set of patterns over time, as characterized by a particular landscape pattern index (LPI), such as connectivity, mean patch size, etc. However, the order of states (sequence) may be different and produce different trajectories of landscape change. Cumulative effects may differ.

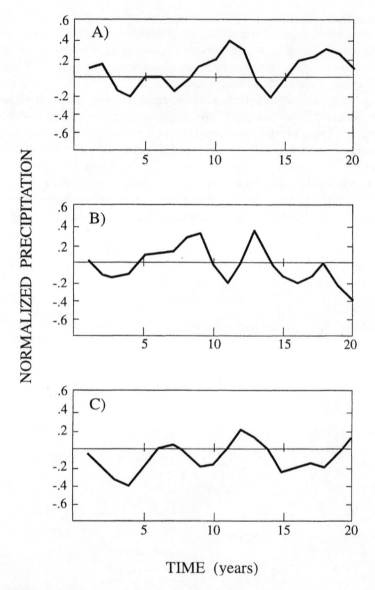

FIGURE 11.8

Annual precipitation records were generated using an autocorrelation function of given parameters. Three realizations of the same autocorrelation function are shown here. Although their forms are similar, the timing and duration of drought years differ. These slight changes have significant effects on modeled populations.

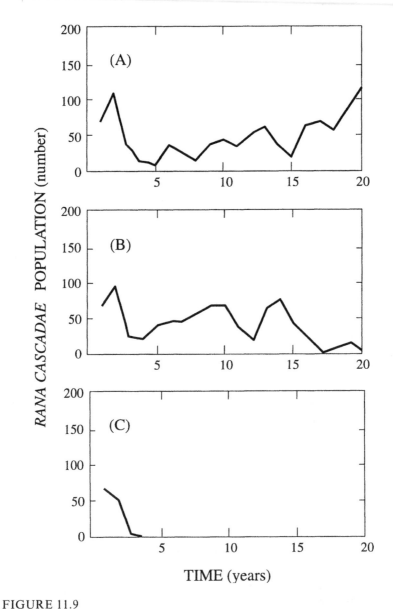

FIGURE 11.9

Corresponding responses of *Rana cascadae* populations for the simulated time series shown in figure 11.8. In the first case, the population is able to survive the drought. In the second and third cases, the differences in drought duration and timing cause the population to crash to extinction.

As a second example of the effects of timing and inter-event intervals, consider the effects of natural resource policy on coniferous forests. The policy of fire suppression has altered the successional trajectory of many stands of ponderosa pine (*Pinus ponderosa*) forests in western North America (Borchers 1995). These forests have become increasingly susceptible to insect depredation and replacement by other, less fire-tolerant, late-successional species. Generally speaking, fire suppression has delayed the occurrence of intermittent wildfire events beyond historic frequencies. As an effort toward ecosystem restoration, it has been suggested that wildfire be allowed to burn and that prescribed fires be used as a resource management tool. However, it is feared that the accumulation of fuels may result in fires more extensive and intensive than the historic range of natural variability, thereby shifting the ecosystem away from its presettlement trajectory. The alteration of fire timing and frequency may be sufficiently large that it will produce a very different set of environmental boundary conditions (Perry et al. 1989).

In addition to the previous discussion regarding the determination of ecologically significant spatial patterns, monitoring designs need to assess the temporal variability of landscapes. This assessment implicitly evaluates a string of states (figure 11.10). If an event is considered ecologically significant, then the range of landscape states and patterns must include the delineation of timing, frequency, and duration of the event, as well as the event intensity. To characterize the state of a spatially configured landscape, the commensurate temporal scales must be included in the analysis. The relevance of one time series versus another is best understood from the perspective of function. Returning to the example of faunal habitat, the suitability of a given landscape configuration may be evaluated in terms of the characteristic time scales of the life history patterns of the animal. The definition of ecologically significant change in a landscape depends on how a given ecosystem process (e.g., faunal life history, wildfire) "reads" this string of states. Change is a spatially and temporally defined deviation from this set.

Analogous to the preceding discussion regarding spatial pattern and resolution, sampling design is also described relative to the temporal scale at which an organism perceives the landscape (Allen, chapter 3; Dutilleul, chapters 17 and 18). For a characteristic time, τ_α, describing a life history attribute (e.g., average lifespan), consider the scenario in which changes in habitat patterns occur rapidly over a period τ_0, where

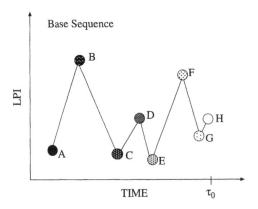

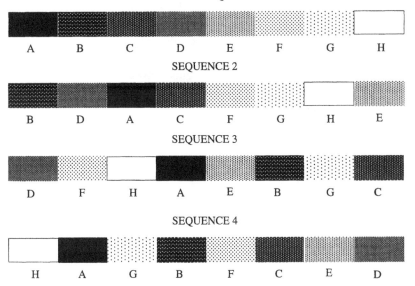

FIGURE 11.10

The sequence of events can be changed to produce a different set of landscape state strings. The way in which a faunal population "reads" these sequences will have an effect on the viability of the population. The relevant question for the land manager is, Which strings are equivalent and under what circumstances?

τ_α is greater than or equal to τ_0 (figure 11.11). If variability within this time period is unimportant, sampling intensity can be directed at the scale τ_α. Even if at one point the landscape condition falls outside the suitability hyperspace (see figure 11.1), and the change to a successive state is sufficiently rapid (therefore the resulting impact on the population is sufficiently small), the habitat may remain qualified as suitable and the population may remain viable. On the other hand, in the case where τ_α is less than τ_0, sampling intensity and level of detection will be most effective at finer temporal scales (figure 11.11). These dynamics may be explored using spatially and temporally explicit population models as a sensitivity analysis to determine an acceptable habitat envelope (Dunning et al. 1995). Such a model can be parameterized based on life history traits of an organism or community, and it can then be subjected to a series of changing landscape patterns following various temporal trajectories. The resulting set of landscapes that define the envelope would form the baseline from which deviations could be measured. Once these spatiotemporal envelopes are established, the susceptibility or predictability to change must be determined to anticipate the point of "irreversible damage." Monitoring plans would best be designed to distinguish this set from other trajectories (figure 11.12).

Summary

There is tremendous pressure to utilize monitoring as a salve for environmental conflict. As such, monitoring is expected to characterize variables expressed at large spatial (i.e., ecosystems, watersheds, faunal populations) and temporal (i.e., population viability) scales. Although there is a substantial body of ecological theory and computer technology available to accommodate these demands, there are a number of critical issues which have yet to be addressed. Consequently, there is a danger of creating sampling designs that will inaccurately describe the state of the environment and ill inform decision-makers. This chapter has attempted to raise a few of these issues; specifically, I suggest that

FIGURE 11.11

By using life history patterns and ecosystem processes as a filter by which landscape patterns may be interpreted over time, an assessment of change may be evaluated statistically and functionally.

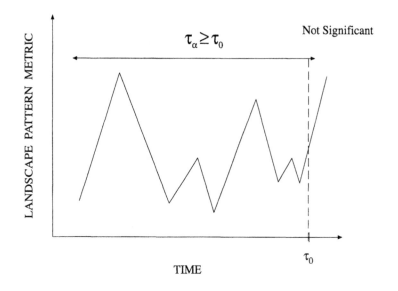

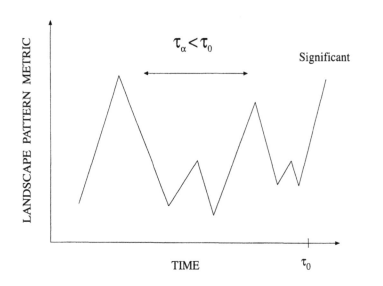

attention shift from the mere characterization of pattern to a rigorous understanding of ecological significance relative to the processes that pattern reflects. This requires a more complex and simultaneous interpretation of pattern in the spatial and temporal domains. Accordingly, the choice of appropriate sampling resolutions will become more transparent with this unraveling of conceptual knots regarding scale and pattern. To this end, there is a recent move toward the development of spatially and temporally explicit models.

From a practical standpoint, decisions will be made much sooner than these models can be implemented and field tested. We will not be able to afford the certainty required for the scale of precision demanded by policy makers (Rykiel, chapter 21). For this reason, the results from the above discussion imply a simpler solution to designing a monitoring plan (i.e., to scale temporal scale to spatial scale). If management and conservation are directed at the watershed and landscape scales, then management decisions and actions must be scaled at commensurate temporal scales that allow the fluctuations and permuta-

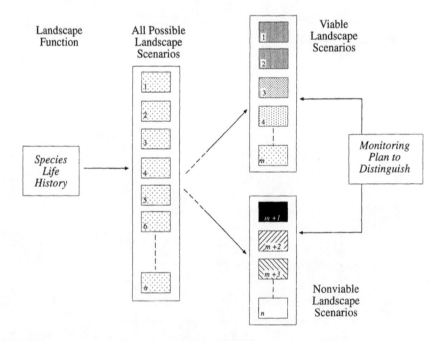

FIGURE 11.12

A schematic depicting how ecosystem processes and faunal life-histories may be used in the development of models and assessment of risk.

tions characteristic of larger areas. This implies that extraction activities and even restoration activities need to be scaled to the level at which certainty is high and risk low relative to our knowledge base. Unfortunately, this implies temporal scales exceeding one to several hundred years in many cases when the spectrum of disturbance regimes is considered; these time periods far exceed the patience of our present society. This will ultimately be a societal decision.

Acknowledgments

I would like to thank two anonymous reviewers for their comments and insight on earlier versions of the manuscript.

PART III

MOVING ACROSS SCALES: ECOLOGICAL
INFERENCE AND APPLICATIONS

Chapter 12

APPLIED SCALING THEORY

David C. Schneider

The increasing use of scale represents an enduring change in the way that ecological research is pursued, rather than another idea in a succession of concepts (often borrowed) that have passed through ecology. Strong interest in the topic is now over a decade old and still burgeoning. Multiscale spatial analysis can be traced back to agricultural trials in the early part of the century (Mercer and Hall 1911), with several notable developments in the 1970s, including recognition of the need for explicit choice of spatial scale in survey design (Wiens 1976; Smith 1978), and the linkage of space and time scales in paleontology (Valentine 1973), terrestrial ecology (Shugart 1978), and aquatic ecology (Haury et al. 1978; Steele 1978a). In the mid 1980s, the rate of publication on spatial scaling expanded rapidly.

The need for multiscale spatial and temporal analysis is now widely recognized. It has become increasingly clear that (1) spatial and temporal patterns depend on the scale of analysis (Platt and Denman 1975; Delcourt et al. 1983), (2) experimental results cannot be extrapolated directly to larger scales (Ricklefs 1987), (3) biological interactions with the environment occur at multiple scales (Harris 1980; O'Neill et al. 1986), (4) population processes do not occur at scales convenient for investigation (Dayton and Tegner 1984), (5) environmental problems arise through propagation of effects across scales, and (6) there is no single or "characteristic" scale for research (Levin 1992).

Verbal and graphical concepts from the 1980s are now evolving into quantitative techniques. Two books (Turner and Gardner 1991a; Schneider 1994a) have appeared, and new techniques continue to be developed (O'Neill and King, chapter 1). There are so many techniques, some highly complex, that no investigator is able to master them all. This chapter suggests a generic approach that is accessible to any ecologist.

This generic approach begins with a careful definition of scaled quantities, which can be rescaled either isometrically or allometrically. The concept of allometric rescaling is extended from mass-based power laws to less familiar length- and time-based power laws. The concept of scope arises naturally from these power laws, and it proves to be widely useful in applying multiscale analysis to practical problems. Examples of application include scope diagrams for natural phenomena, measurement instruments, surveys, and experiments. The principle of homogeneity of scope provides a link between scaling theory and statistical analysis, and it also proves useful in linking experiments (typically at a small scale) to theory (typically at a larger scale).

Scaled Quantities

Ecologists work with definable quantities, not with pure numbers or mathematical abstractions divorced from measurement. The distinction between a quantity and a number might appear to be inconsequential, but it is not. The mathematical rules for working with scaled quantities differ from those for working with numbers. When a number ($N = 78$) replaces a quantity such as world population size—78 *Grus americana* (whooping crane) in 1981—scale is lost.

A well-defined quantity has five parts: (1) a name, (2) a symbol, (3) a procedural statement that prescribes the conditions for measurement, or for calculation from measurement, (4) a set of numbers generated by the procedural statement, and (5) units on one of several types of measurement scales. A measured quantity is conveniently represented as a symbol that equals a set of numbers arranged inside brackets, multiplied by the unit of measurement. An example is the length of *Gadus morhua* (codfish) upon settlement out of the plankton into benthic habitats:

Procedural statement	Name	Symbol	Numbers	•	Units
Length, snout to caudal peduncle	Standard length	sL	$\begin{bmatrix} 55 \\ 83 \\ 48 \end{bmatrix}$	•	mm

Units occur on one of the four types of measurement scales defined by Stevens (1946). On a nominal scale, the units are "yes" or "no" (present or absent, 0/1, etc.). On an ordinal scale, the unit is the rank of an object. On an interval scale, the unit is a count of the number of units that separate one measurement from another. For interval scale measurement, the textbook example is temperature in degrees Celsius: the zero point is arbitrary. On a ratio scale, units are counted from a natural zero point; the textbook example for this scale is temperature in degrees Kelvin. Both ratio scale (e.g., mass) and interval scale quantities (e.g., degrees Celsius) are used in rescaling.

Allometric Rescaling

Any quantity can be rescaled according to its similarity to another quantity. The rescaling is isometric if a direct proportion is used. For example, the volume of an animal scales isometrically with mass because animals are composed largely of water, which is nearly incompressible. Consequently, a rescaling of the volume (say, by a factor of 2) will rescale the mass by the same factor. The idea of isometric rescaling can be expressed as a proportion:

$$\frac{Volume_{big}}{Volume_{small}} = \frac{Mass_{big}}{Mass_{small}} \tag{1}$$

Allometric rescaling occurs if the proportion is other than direct. The volume of an organism will be proportional to the volume and hence to the cube of the length, not to length itself. A rescaling of length (say, by a factor of 2) will change the volume by a factor of 2^3. The idea of allometric rescaling is expressed as a proportion modified by an exponent:

$$\frac{Volume_{big}}{Volume_{small}} = \left(\frac{Length_{big}}{Length_{small}}\right)^3 \tag{2}$$

Another way to look at this is that for isometric rescaling, the scaling exponent is unity, whereas for allometric rescaling, the exponent differs from unity.

Allometry usually refers to a special case of allometric rescaling in which function or form has been scaled to body size. An example is respiration, which scales as $(Mass)^{0.7}$. Allometric rescaling according to body size was developed by D'Arcy Thompson in his 1917 treatise *On Growth and Form* (Thompson 1961). This landmark book in biology illustrates the principle of similitude with examples based largely on body size and geometric similarity. Thompson advocated the more general principle of similitude, which includes geometric, hydrodynamic, thermal, and other forms of similarity. Recent authors (Peters 1983; Schmidt-Nielsen 1984) have tended to expand the term toward its root meaning, which refers to metrics other than body size.

The steps in allometric rescaling are:

1. State the conditions under which two quantities are considered similar.
2. Express similarity as a proportion. The generic expression rescales a quantity Q according to similarity to another quantity Y:

$$\frac{Q_{new}}{Q_{old}} = \left(\frac{Y_{new}}{Y_{old}}\right)^{\beta} \qquad (3)$$

3. Rearrange to permit calculation of the rescaled quantity Q_{new} from Q_{old} according to the allometric rescaling factor $(Y_{new}/Y_{old})^{\beta}$.

$$Q_{new} = Q_{old}\left(\frac{Y_{new}}{Y_{old}}\right)^{\beta} \qquad (4)$$

The rescaling factor is a ratio (Y_{new}/Y_{old}) raised to the power β. This ratio will be called the scope of the rescaling. Y can be any quantity. The most common is mass, for which there is extensive literature (Peters 1983). Allometric rescaling according to length has become increasingly common (Sugihara and May 1990).

Here is an example of allometric rescaling according to body mass. Maximum running speed ($Vmax = m \cdot s^{-1}$) does not increase isometrically with body mass; the scaling exponent as estimated by Bonner (1965) is $\beta = 0.38$, and hence

$$\frac{Vmax_{big}}{Vmax_{small}} = \left(\frac{M_{big}}{M_{small}}\right)^{0.38} \qquad (5)$$

Rearranging this expression results in an equation for calculating the rescaled quantity:

$$Vmax_{big}(M_{small} / M_{big})^{0.38} \quad = \quad Vmax_{small} \quad \text{(6)}$$

$$2\,\text{m\,s}^{-1}\,(1/2)^{0.38} \qquad \Rightarrow \quad 1.5\,\text{m\,s}^{-1}$$

The following calculation is for a halving of body mass:

$$(M_{small}/M_{big} \; = \; 1/2 \; = \; 1\text{kg}/2\text{kg} \; = \; 1\text{g}/2\text{g} \; = \; ...)$$

At half the body mass, speed is rescaled downward by $(2 - 1.5)/2 = 25$ percent, rather than by half.

Here is an example of allometric rescaling according to length. In a mosaic habitat such as a grassland, the area of bare soil does not scale isometrically with total area. Wiens and Milne (1989) estimate the scaling exponent to be 1.8 and hence

$$\frac{Asoil_{big}}{Asoil_{small}} = \left(\frac{L_{big}}{L_{small}}\right)^{1.8} \qquad \text{(7)}$$

This expression says that the area of bare soil ($Asoil = m^2$) scales with length ($L = m$) raised to fractional power (1.8), rather than with total area (L^2). Because the exponent relative to length is not an integer, it is called a fractal dimension. The physical interpretation of this fractal scaling is that barren soil occurs as a rambling network that becomes less evident at larger scales. Thus, doubling the measurement scale L will not, on average, double the area of bare soil.

As before, this expression is rearranged to obtain an equation for calculating the rescaled quantity:

$$Asoil_{small}(L_{big}/L_{small})^{1.8} \quad = \quad Asoil_{big} \qquad \text{(8)}$$

$$2\,\text{m}^2\,(2/1)^{1.8} \qquad \Rightarrow \quad 6.96\,\text{m}^2$$

Three Scaling Relations

The idea of allometric rescaling is often expressed in the form of a power law. For example, allometric rescaling according to body mass is typically written as

$$Q = kM^{\beta} \qquad \text{(9a)}$$

In this abbreviated expression, Q stands for any quantity, and M stands for body mass. The more complete expression is

$$Q(M) = Q(M_o)(M/M_o)^\beta \tag{9b}$$

Translation to the abbreviated version occurs by writing $Q(M)$ as Q and replacing $Q(M_o) \cdot M_o^{-\beta}$ with k. The best known allometric scaling according to body mass is Kleiber's law, which states that energy use at rest (BMR = joules $\cdot$ s^{-1}) scales allometrically with body mass (M = kg), according to an exponent less than one:

$$BMR = k\, M^\beta \tag{10}$$

This expresses, in quantitative terms, the idea that large organisms live less intensely than small ones, which consume more oxygen and food per unit of body mass (Kleiber 1961).

In a similar fashion, allometric rescaling according to Euclidian length L can be written as

$$Q = k\, L^\beta \tag{11a}$$

where Q again stands for any quantity of interest. D (= $1 - \beta$) either represents Euclidian dimensions (D = 1, 2, 3) or it represents fractal dimensions (Mandelbrot 1977), as in the example of soil area. The fractal dimension $1 < D < 2$ represents a convoluted line embedded in a plane. The fractal dimension $2 < D < 3$ represents a convoluted area embedded in a volume. In this expression, k stands for $Q(L_o) \cdot L_o^{-D}$ and the expanded expression for partial allometry is

$$Q(L) = Q(L_o)(L/L_o)^\beta \tag{11b}$$

Many quantities scale allometrically with time. Examples are frequency of measles epidemics (Sugihara and May 1990), daily rainfall (Lovejoy and Shertzer 1986), and the frequency at which animals change their direction of movement (Frontier 1987). The general abbreviated expression for scaling with time is

$$Q = kT^\tau \tag{12a}$$

where Q again represents any quantity of interest. T is time and τ is an exponent expressing the degree of acceleration ($\tau > 1$) or dampening ($\tau < 1$) as the scope of T increases. In this expression, k stands for $Q(T_o) \cdot T_o^{-\tau}$. The full expression is

$$Q(T) = Q(T_o)(T/T_o)^{\tau} \qquad (12b)$$

The examples to date from the ecological literature suggest that τ is often fractal; environmental processes rarely occur according to the ticking of a clock ($\tau = 1$) or as smooth Newtonian acceleration or deceleration ($\tau = 2$). They occur instead as episodic outbursts, with rates somewhere between the regularity of a clock and Newtonian acceleration.

Mass, length, and time are most commonly encountered in allometric rescaling, but in principle any quantity can be used. One could, for example, rescale organism form and function according to energy exchange, for which the units are energy/time (e.g., watt = joule $\cdot$ s^{-1}):

$$Q(E) = Q(E_o)(E/E_o)^{\gamma} \qquad (13)$$

Allometric rescaling can be applied to complex quantities. An example is scaling of spatial heterogeneity with length scale. If we measure spatial heterogeneity (as a variance) of numbers of organisms (N) in contiguous quadrats of 1 cm, then recompute the variance using contiguous quadrats of 10 cm, we can expect the variance to increase if organisms are clumped. In aquatic habitats, we expect heterogeneity to increase in a regular way with increase in quadrat size, reflecting the physical structure of the surrounding medium (Platt and Denman 1975). This regularity can be expressed as an allometric scaling:

$$\text{Var}(N) = \text{Var}(N_o)(L/L_o)^{\beta} \qquad (14)$$

Spatial allometry ($\beta \neq 1$) forces a rethinking of the use of statistical techniques based on constant variances.

Scaling relations could be written for any pair of quantities thought to be similar. The generic expression for allometric rescaling of some quantity Q according to some other quantity Y is:

$$Q(Y) = Q(Y_o)(Y/Y_o)^{\beta} \qquad (15a)$$

This is easily rewritten as a statement of proportion, as in the generic recipe for allometric rescaling listed above. The short form of the generic scaling law is

$$Q = k \, Y^\beta \qquad\qquad (15b)$$

Scope

The ratio Y/Y_o was defined in passing as the scope of the rescaling. This ratio is a dimensionless number: it has neither units (e.g., kg, m, s) nor dimensions (e.g., mass, length, time). This lack of units results from the rules for operations on units; it should not be confused with the practice of omitting units. The concept of scope as a dimensionless ratio can be extended to other applications of scaling theory. Several examples, selected from a larger collection (Schneider 1994a), will be described.

The biological and physical phenomena that ecologists study typically have upper and lower limits in space and time. In common usage, the scale of these phenomena refers either to the upper or lower limit. Equivalent pairs of terms (Sugihara and May 1990) are the *minimum* (or inner) scale and the *maximum* (or outer) scale. Still another pair is *grain* and *extent* (Wiens 1989). The scope of a natural phenomenon is defined as the ratio of the upper to the lower limit. Scope is thus the ratio of the extent to the grain, or of the outer to the inner scale. An example is the frequency of El Niño events. The time between events, on average, is 5 years. The temporal scope is $T/T_o = 8$ years $\div$ 2 years $= 4$.

The spatial and temporal scopes of natural phenomena are often graphed in two dimensions. Figure 12.1 shows an example for El Niño events. Logarithmic axes are used because these show multiplicative changes (such as change in ratio), in contrast to linear axes that show additive changes. The upper and lower limits of the hatched area correspond to the upper and lower frequency of such events. The right and left limits of the hatched area correspond to the spatial range and resolution. The distance between upper and lower limits on this logarithmic plot corresponds to the temporal scope T/T_o. The distance from left to right indicates the spatial scope L/L_o. The larger the hatched area in these diagrams, the greater the scope.

Scope (as a dimensionless ratio) can also be used to express the capability of measurement instruments. The scope of an instrument is defined as the ratio of the maximum measurement to the resolution.

A surveyor's chain, for example, has a scope of approximately 10 m ÷ 0.002 m = 50,000. A satellite positioning system (Vande Castle, chapter 13) has a scope of approximately 20,000 km ÷ 0.1 km = 200,000. This is calculated as the maximum distance between two points on the earth's surface, divided by the resolution. The scope of the satellite system is of course greater, but only by a factor of 4. Intuitively, one would have thought that the satellite positioning system would have hundreds of times the scope of a surveyor's chain.

Still another application is the spatial scope of a survey, defined as the ratio of the area to be surveyed to the area of each measurement. An example is the distribution of *Pinus edulis—Juniperus monosperma* (piñon pine-juniper) woodland in a 100,000-ha reserve (Milne et al. 1992). The spatial scope of an aerial survey, based on a resolution

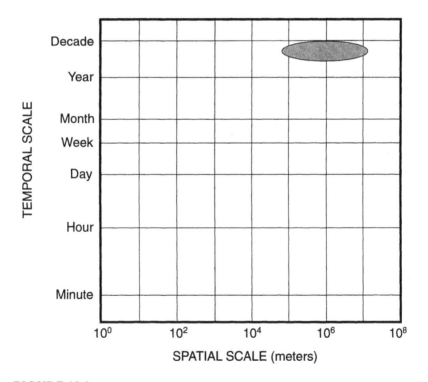

FIGURE 12.1

Scope diagram for El Niño events. Upper limit of *hatched area* is maximum time between events, lower limit is minimum time between events. Left limit is minimum linear extent of affected area. Right limit is maximum linear extent of affected area.

of 2.54 m × 2.54 m pixels in photographs, is 1.6×10^8. The scope of the ground-based survey of 30 randomly selected plots, each 10 m on a side, is 10^5. The scope of the aerial survey is more than 1000 times greater than the 30-plot survey. Computing, then plotting the scope of a survey relative to suspected sources of variation in the quantity of interest, often proves to be informative. Expressed as a graph, scope calculations illustrate the advantages and limitations of a particular instrument for investigating natural phenomena.

The scope of an experiment is a related application. Specifically, the scope of an experiment is the area in which the experiment is set, relative to the minimum area of a sample. Table 12.1 shows a series of scope calculations for an exclosure experiment carried out on a muddy intertidal flat at Punta Mala, Panama (Schneider 1985). To test for effects of avian predation on density of tropical infauna in Panama, invertebrate density was measured at 21 sites distributed over four beaches at the beginning, middle, and end of a 3-month period prior to the migratory departure of shorebirds. At each site in January, cores were taken inside and outside a single roped exclosure measuring 1 m × 1 m. Shorebirds grew accustomed to the exclosures and were observed walking under the ropes. Consequently, square canopies (1 m × 1 m) were placed over the ropes at two sites in March, 1978.

TABLE 12.1

Scope calculations for a single exclosure experiment*

Core = $\pi(^{10}/_2 \text{ cm})^2 = 78.5 \text{ cm}^2$
Sample = 11 cores = 864 cm²
Site = $(3 \text{ m})^2 = 9 \text{ m}^2$
Area = $(20 \text{ m})^2 = 400 \text{ m}^2$
Punta Mala inlet = $(200 \text{ m})^2 = 40,000 \text{ m}^2$
Coast = 500 km × 1 km = 500 km²

Design	Unit	Frame	Scope	Type of inference
A	11 cores	11 cores	1	None
B	11 cores	Site	104	Statistical
C	11 cores	Area	4630	Statistical + judgment
D	Site	Area	44	Informal survey
E	Area	Inlet	100	Informal survey
F	Inlet	Coast	12,500	Informal survey

*The frame is the target of inference (population from which the sample is drawn). The scope is the ratio of the area of the frame to the area of the sampling unit.

This successfully excluded birds. Canopies were placed in areas heavily used by foraging birds. Core samples 10 cm in diameter were taken at all 21 sites in January, March, and April. Cores were taken haphazardly within staked areas and within 1 m of the roped or canopied area.

The scope of the canopy experiment depends on the frame (i.e., the target of inference) and on the sampling unit. The frame and unit correspond (respectively) to the outer and inner scale of the experiment. The inner scale is fixed (a 10-cm diameter core), but there is flexibility (and hence room for judgment) in the choice of outer scale. If the target of inference is the 3 m × 3 m site where cores samples were taken, then the scope is 104 (see table 12.1). This means that a factor of 104 is needed to extrapolate from the sample (11 cores) to the 9-m^2 site from which samples were taken. If the target of inference is the 20 m × 20 m area where the canopy at Punta Mala was placed, then the scope increases to 4630 (see table 12.1). Inference from the sample to this 400-m^2 area is partly from judging that the site was typical of the area. The scope of this judgment is 44, as shown in table 12.1. If the experimental site is judged to be representative of the 200 m × 200 m intertidal area at Punta Mala, then the scope of inference is 100 relative to the experimental area (see table 12.1), and 46,300 relative to the 11-core sample. If the results from this one inlet were judged to be representative of similar intertidal areas stretching 500 km eastward, then the scope or extrapolation factor is 12,500 relative to the inlet (see table 12.1).

Homogeneity of Scope

The idea of scope can be incorporated directly into the statistical machinery used to analyze this experiment, using the principle of homogeneity of scope (Schneider 1994a). This principle says that each term of an equation must have the same scope. Here is an example for the canopy experiment, using the generalized linear model, which includes both normal and nonnormal error structures (Nelder and Wedderburn 1972; McCullagh and Nelder 1989). To evaluate whether the gradient in density across the cage boundary changed during the experiment, the interaction term in a two-way classification of the data must be examined. The generalized linear model relates the response variable (Y = organisms/core) to a random variable ε and a structural model μ composed of explanatory variables.

$$Y = \mu + \varepsilon \tag{16a}$$

$$\mu = \beta_o + \beta_1 X1 + \beta_2 X2 + \beta_{1 \cdot 2} X1 \cdot X_2 \tag{16b}$$

The explanatory variables are X_1 = presence (0) or absence (1) of canopy, X_2 = beginning (0) or end (1) of experiment, and $X_1 \cdot X_2$ = interaction term, or product of X_1 and X_2. If we assume that the random variable ε is normal (i.e., constant variance around any expected value) in estimating the parameters, then the residuals will not be independent of the fitted values (figure 12.2a). Consequently, an F-distribution cannot be used to calculate Type I error.

The response variable is a count, which suggests that a Poisson error structure might be appropriate. A structural model based on proportions, rather than on differences (as in equation 1), is generally used with Poisson counts:

$$Y = k\, e^\mu + \varepsilon \tag{17a}$$

$$\mu = \beta_o + \beta_1 X1 + \beta_2 X2 + \beta_{1 \cdot 2} X1 \cdot X_2 \tag{17b}$$

This is the statistical model for the familiar two-way contingency test using the G-statistic, the test used originally in the analysis of the experiment (Schneider 1985).

The residuals are still correlated with the fitted values (see figure 12.2b). The variance of the counts exceeds the mean, indicating that the organisms are "clumped" at the scale of a core. Consequently, the assumption for a standard two-way contingency test (i.e., a Poisson error structure) is not appropriate for the data, even though the data are counts. The clumping or overdispersion of the response variable suggests that a Gamma error structure may be appropriate. If we assume a Gamma error structure and use the linear rather than the logarithmic model, we obtain residuals that are independent of the fitted values (see figure 12.2c).

Once we have an acceptable model (equations 16a and 16b with a gamma error structure), we can ask if the gradient in density across the cage boundary changed during the experiment. In statistical terms, should the interaction term be included in the model? The overall deviance of the model from the data is $D = 1.828$ (df = 7), using a Gamma error structure. The deviance of the model from the data, if the interaction term is omitted, is $D = 2.091$ (df = 8). The reduction in deviance $\Delta D = 2.091 - 1.828 = 0.263$, a reduction that is not significant at

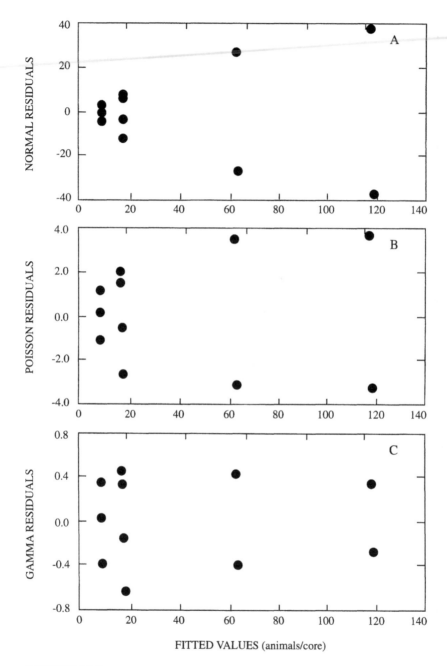

FIGURE 12.2
Residuals from analysis of invertebrate density in core samples, assuming **(A)** normal errors, **(B)** Poisson errors, **(C)** Gamma errors.

$\alpha = 5$ percent, using a Chi-square distribution with a single degree of freedom (McCullagh and Nelder 1989). The null hypothesis, that the radial gradient in density around the canopy boundary did not change during the experiment, cannot be rejected. If we use a negative binomial error structure (also appropriate), the change in deviance is even smaller ($\Delta D = 0.0127$) and hence not significant.

What is the spatial scope of this conclusion? We apply the principle of homogeneity of scope as follows:

$$Y \quad = \quad \beta_{1 \cdot 2} \quad \cdot \quad X_1 \cdot X_2 \quad + \quad \varepsilon \qquad (18)$$

$$\frac{\text{sample}}{\text{core}} = \frac{\text{sample}}{\text{site}} \cdot \frac{\text{site}}{\text{core}} + \frac{\text{sample}}{\text{core}}$$

$$11 \qquad\qquad 0.106 \quad \cdot \quad 104 \qquad\qquad 11$$

The response variable Y has a limited scope of 11. The interaction term has a scope of 104 and hence the parameter $\beta_{1 \cdot 2}$ must have a scope of 0.106 to scale the model down to the data. Another way to look at this is that the parameter scales the data up to the model by a factor of 0.106^{-1}, or about 9. Similar calculations can be made for the temporal scope of $\beta_{1 \cdot 2}$, if this were of interest.

Experimental results in ecology are notoriously variable. One contributing factor is that experiments being reported differ considerably in scope. Calculations of scope, such as that shown previously, are a way of determining the degree of comparability of several different experiments. Experiments with the same scope are completely comparable, and experiments with similar scope are more comparable than experiments with scopes that differ greatly.

Scope calculations suggest what degree and type of inference should be employed. In the case of the single canopy experiment at Punta Mala, statistical inference can be used to scale up from the sample (11 cores) to the scale of the 9-m² site. One might object that this constitutes pseudoreplication (Hurlbert 1984), but the underlying problem of confounding with unmeasured effects can be addressed at this scale by using Monte Carlo methods to compute the probability of obtaining radial gradients at one site by chance alone. Here, radial gradients are defined as an observed mean within a 1 m × 1 m area being less than the mean in the surrounding 9 m² − 1 m² = 8 m² area. Statistical inference cannot be used to scale up from the sample to the scale of the 400-m² area where the canopy was placed, because at this larger scale only one site (experimental unit) was used. Inference to this and larger scales must be based on surveys (Eberhart and Thomas 1992), or judgment, or more experimental units.

The scope calculations shown so far have been isometric, with an exponent of unity. Allometric scalings may well prove to be more appropriate in analyzing the spatial scope of experiments. An allometric relation between density and area would result in a different estimate of the scale-up for the exclosure experiment. If the scaling exponent were 1.8 rather than 2, then the scale-down from model to data is $11^{1.8/2}/104 = 0.083$, rather than $11/104 = 0.106$. The scale-up from data to model is $0.083^{-1} = 12$, rather than 9. This increase in the scaling factor quantifies the intuitive notion that scale-up is less certain in a heterogeneous environment ($D = 1.8$) than in a homogeneous Euclidian environment ($D = 2$).

Summary

This chapter describes a simple and accessible series of concepts and computational methods for applying scaling theory to ecological research. The first key concept uses scaled quantities rather than numbers stripped of units (and hence scale). The second key concept is allometric rescaling of one quantity according to its similarity to another. This concept is readily expressed as a power law. The third key concept is that of scope, defined as the ratio of the range to the resolution of a quantity, instrument, survey, or experiment. The scope of any of these is readily expressed in diagrammatic form, which is already becoming increasingly common in ecology. The principle of homogeneity of scope proves useful in linking scaling theory to statistical analysis.

Rapid expansion of multiscale analysis can be expected in the next decade, as computational machinery (Turner and Gardner 1991b; Rastetter et al. 1992) becomes more familiar. This should lead to a better understanding of environmental problems through improved skill in analyzing physical and biological processes at multiple scales. One area in which progress is likely to be rapid is that of scaling measurements (typically at small space and time scales) up to that of problems of societal interest (typically at larger scales), such as global warming (Innes, chapter 19) or reduced biological diversity. It is now clear that isometric rescaling (the "just multiply" strategy) will not work. Increasing facility in the use of allometric rescaling factors is likely to lead to more accurate computations of the expected effects of human activities on the environment. A second area in which rapid progress seems likely is in the use of fractal spatial scalings. Several interesting applications have already appeared (Lovejoy and Shertzer 1986; Milne et al. 1992), and more are certain to follow.

A third area in which applied scaling theory can potentially increase the effectiveness of ecological practice is in multiscale analysis of the behavior of resource users, and their interaction with the environment (Lee 1993). Humans have always interacted with their environment at multiple spatial scales. The trend in this century has been toward increasing the spatial and energetic scale of environmental alteration (dredging, logging, etc.). Another trend has been increased scale of spatial coordination of resource exploitation. Fish have traditionally been pursued through the uncoordinated activity of individual or recreational fishers. Throughout this century, pursuit has become coordinated at increasingly large scales through technological innovations such as sonar and satellite imagery. The corporations that deploy large fishing fleets are well aware of their ability to locate and harvest every aggregation within extensive areas of the ocean. There has, however, been a lag in the degree to which responsibility for increased resource exploitation has been accepted. Larger-scale institutional arrangements for responsibility have been slow to evolve, leading to overexploitation of renewable resources such as fish and timber. Large-scale responsibility is clearly needed, but large-scale institutional arrangements tend to be rigid and unresponsive (Rykiel, chapter 21). Consequently, environmental science within centrally organized governmental departments may turn out to be irrelevant or wrong. In any case, resource users will consider it irrelevant and wrong (Hobbs, chapter 20). One solution is multiscale or community-based science, in which resource users participate actively in the gathering and assembly of knowledge about those resources. Ecologists involved in conservation-related problems may find that they are more effective working with resource users than they are in relying entirely on technicians. Regardless of the institutional arrangements for ecological research, it is evident that multiscale reasoning is required for understanding the social, legal, and economic settings for ecological problems, as much as for understanding the underlying biological questions.

The concept of scope is based on the principle of similitude, which is widely used in geophysics, oceanography, and meteorology. This principle is used in allometric scaling according to body size, but it has not yet been widely applied in biology at the population or community level. If scaling theory is to become part of normal ecological practice, the first challenge will be the development and mastery of the computational machinery appropriate to applying the principle at the population and community levels of organization. An allied challenge will be integrating this machinery with current statistical methods in ecology. Integration is important because the statistical machinery by itself impedes the use of the principle of similitude. Statistical methods do not distinguish numbers from scaled quantities. This reinforces the widespread (Schneider 1994a) practice of using numbers rather than scaled quantities in both theoretical and experimentally oriented papers in ecology. This practice leads to computational errors because the rules for working with scaled quantities differ from those for working with pure numbers. The practice retards the development of ecological theory by eliminating the mathematical basis for multiscale reasoning. Progress in applying multiscale analysis will accelerate when ecologists become as skilled in applying the principle of similitude as they are now in applying statistical reasoning.

Acknowledgments

Several of the concepts presented here grew out of discussions with Paul Dayton during a sabbatical under the Visiting Researchers program at the Scripps Institution of Oceanography. Funding for subsequent work has been primarily through a continuing research grant from the Natural Sciences and Engineering Research Council (Ottawa). Comments by Donald McKenzie, Claudia Pahl-Wostl, and Mark Simpson improved the manuscript.

13

REMOTE SENSING APPLICATIONS
IN ECOSYSTEM ANALYSIS

John Vande Castle

Remote sensing and associated technologies such as geographic information systems (GIS) have become a crucial part of ecological research. Remote sensing data permit observation of ecological information at different scales of interest and provide the ability to extrapolate information from site-based ecological studies to larger spatial extents. It is perhaps the only tool used to observe ecological phenomena on a global scale. Ecological processes operate at a variety of different scales, and remote sensing data represent a powerful tool for integrating data at these scales. Remote sensing data can also provide the basis for development of theory and applications at large scales. Therefore, remote sensing provides a tool for extraction of large-scale variables not apparent in ground studies. Remote sensing data acquired from the same or similar instruments can be used to compare ecological processes at different places or times. However, using remote sensing data for observations at different scales and for intercomparison is simple in concept but not in practice (Wessman 1992). The limitations involved in extrapolating from small-plot or site-based ecological studies must be recognized. The details involved in scaling ecological information, as described within this volume, are important considerations, as are the technical aspects of integrating ecological information and remote sensing data. The technological aspects of remote sensing systems themselves—their capabilities and their limitations—are also important considerations in ecological applications of remote sensing tools.

Remote sensing data, particularly data from space-based satellites, provide information from observations in a consistent and standard format. These data, when properly used, permit both intercomparisons of regions separated by large distances and extrapolation of information from small to large extents. This chapter provides information related to the scale of observations from remote sensing data as it applies to ecological research. It provides examples of remote sensing image data commonly used in ecological studies, and it explains why the various types of data are important. This discussion considers remote sensing data available from conventional remote sensing tools, with a focus on satellite technology, and looks into information that will be important to ecological research in the future.

Remote Sensing Background

The field of remote sensing research encompasses many scientific disciplines. It covers technology ranging from measurements of biophysical properties to image-based observations, including conventional photography and digital-scanning technologies. General information on remote sensing technologies and their applications can be found in standard textbooks (Jensen 1986; Lillesand and Kiefer 1994). This discussion of scale considerations in remote sensing focuses on examples of image data from space-based satellite technologies, because satellite-based remote sensing usually provides more consistent data in standardized formats with fewer problems inherent in aircraft remote sensing data.

Normally, ecologists consider scale in a very applied sense. That is, small scale is synonymous with small areas (e.g., a 1-m^2 plot) and large scale is synonymous with large areas (e.g., a watershed). For a geographer, this is incorrect because a large scale in a mapping context implies that the ratio of the observation to the actual area is large. For example, a scale of 1:2,000,000 commonly used in landscape analysis and considered large scale by ecologists, is actually a small scale to a geographer. The map format of the data also needs to be considered with respect to strict use of the term *scale.* For remote sensing data, the word *approximately* is often used to describe the spatial grain and extent resulting from the map projection used with the data. Because we are often working with large areas of the earth, the curvature resulting from the round earth becomes important. These data result in some distortion when translated to a flat plane such as a map, or to a data layer within a GIS. Although this is less a problem for small areas, significant spatial errors can accumulate as data from small areas are extrapolated to large ones. These extrapolations need to consider that the earth is not flat, and that the shortest distance between two points on a map is almost always a curved line. The most important factor is the consideration of area (or extent) and resolution (or grain) of the observations (Gardner, chapter 2; Schneider, chapter 12). Although a subtle point, it is important here, because the extent and grain of remote sensing data define the limits to which remote sensing data can be used in ecological research.

Remote sensing images are not just pictures but large blocks of data. An analogy is that a remote sensing image is a computer spreadsheet of data values. The data of remote sensing images are organized into rows and columns similar to a spreadsheet, where the grid cell data of each row and column represent a single measurement for a specific surface area. An example of remote sensing image data can be seen in table 13.1, which shows a subset of a normalized difference vegetation index (NDVI) image, calculated using data from the Thematic Mapper (TM) sensor of a Landsat satellite (Landsat-TM). These same data, when displayed as rows and columns of an image in the color or brightness of a computer display or printed page, form the picture with which we are more familiar. Figure 13.1 displays the same data of table 13.1 in the row-and-column format of a remote sensing image. Each grid cell, or pixel of the image, contains data for a specific area of the ground (or its grain), which is the minimum resolution of the data. The complete spreadsheet of data values represents the area or extent of the data.

The basic image format of remote sensing data is a two-dimensional *x* and *y* surface. Because remote sensing data are based on observations of reflected or emitted electromagnetic energy, the wavelengths observed by a remote sensing system are also critical. Temporal scale adds a third important dimension to applications of the data. This can result in a four-dimensional matrix of information, which must be considered when using remote sensing data. The spatial, temporal, and spectral grain or resolution as well as the spatial, temporal, and spectral range or extent of the data define the boundaries to which remote sensing data can be applied to ecological processes. Combining the needs of fine spatial, temporal, and spectral detail, with large spatial, temporal, and spectral extents requires a trade-off in the type of data used at various spatial scales.

TABLE 13.1
Landsat-TM NDVI data in spreadsheet format*

Row number (latitude)	Column number (longitude)									
	1	2	3	4	5	6	7	8	9	10
1	0.799	0.754	0.682	0.685	0.685	0.690	0.410	0.320	0.304	0.464
2	0.794	0.714	0.727	0.714	0.621	0.600	0.600	0.230	0.142	0.347
3	0.794	0.714	0.727	0.714	0.621	0.486	0.211	0.211	0.365	0.405
4	0.751	0.751	0.762	0.714	0.621	0.486	0.211	0.365	0.365	0.405
5	0.748	0.753	0.688	0.688	0.615	0.598	0.524	-0.000	0.405	0.505
6	0.739	0.753	0.653	0.610	0.610	0.569	0.616	-0.000	0.176	0.434
7	0.697	0.717	0.702	0.633	0.535	0.535	0.616	0.095	0.177	0.347
8	0.697	0.711	0.691	0.692	0.633	0.597	0.082	0.082	0.128	0.045
9	0.720	0.720	0.708	0.658	0.669	0.597	0.118	0.011	0.011	-0.011
10	0.717	0.711	0.711	0.668	0.669	0.616	0.118	-0.037	-0.083	-0.083

*The data, when transformed into color, brightness on a computer screen, or printed page, comprise the remote sensing image. These are data for only the upper left 10 rows and 10 columns. The full data set is approximately 6000 rows by 6000 columns.

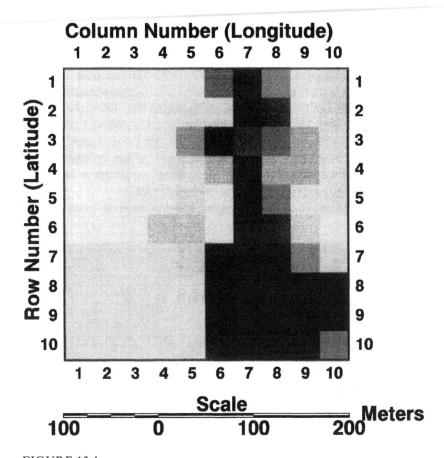

FIGURE 13.1

Image format of remote sensing data displayed as rows and columns of grid cells, or pixels. The data are NDVI data calculated from Landsat-TM data. *Dark pixels* are low data values and represent low vegetation cover; *bright pixels* depict higher vegetation cover.

Spatial, Temporal, and Spectral Scale of Remote Sensing Data

Medium Resolution Data

In practical terms, there is always a trade-off between spatial, temporal, and spectral components of remote sensing data. Landsat-TM is the most common type of image-based remote sensing data used by ecologists. These data are popular because they represent an effective compromise between spatial, temporal, and spectral scales. Landsat-TM data have been available since the launch of Landsat-4 in July 1982, and the generic form of the sensor design is planned for future remote sensing systems. Each TM scene covers an extent of approximately 185 km on a side at a spatial resolution of approximately 30 m. Figure 13.2 shows a sample data set of Landsat-TM. These data cover a region of western Oregon from Portland in the northwest corner, to Eugene in the south, and just beyond the Blue Mountains and Cascade Range to the east. This figure is a composite image in which blue, green, and red wavelengths of light have been combined, resulting in a black-and-white representation of reflected light.

Landsat-TM provides data in seven spectral regions from 0.45 to

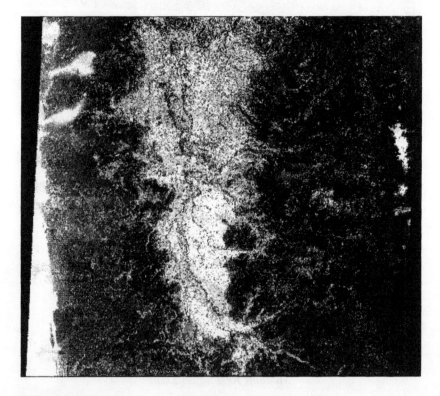

FIGURE 13.2

Landsat-TM data image covering a portion of western Oregon. The figure is a composite image of blue, green, and red wavelengths combined to form a black-and-white image.

12.5 μm. In general terms, this represents data in the green, blue, red, near-infrared, middle-infrared, and thermal channels or bands of the electromagnetic spectrum. An obvious advantage, as with other remote sensing systems, is the ability of the electronic sensors to "see" beyond the range of the human eye, thereby removing a serious bias in ecological research. The spectral resolution of each of the spectral bands is relatively broad, or coarse, and contributes to the lack of specificity to which land-cover classes and other information can be extracted from Landsat-TM data. An example classification for the Landsat-TM data of figure 13.2 is described in Cohen et al. (1995).

The 30 m spatial resolution of Landsat-TM data contains observations of sufficient detail to provide meaningful inferences at the plot or stand level. Landsat-TM also provides sufficient spectral information to separate coarse classes of data (Lillesand and Kiefer 1994). However, the 30 m spatial resolution has disadvantages, because a large amount of detail is averaged in each of the 30 m grid cells. The spectral resolution allows separation of generic classes of data, such as classes of deciduous and coniferous vegetation. The limited spectral resolution or grain is not sufficiently precise, except in special situations, to generate information regarding individual species.

In addition to classification, vegetation indices, such as an NDVI, are often computed for general land cover from Landsat-TM data (Cohen et al. 1995). They provide a simple procedure for assessing vegetation cover on a comparative basis (Vande Castle 1995). For the NDVI form of these data, data values near 1.0 represent large amounts of vegetation, whereas those approaching 0.0 to −1.0 indicate barren land. Figure 13.3 is an example of full-resolution NDVI data from Landsat-TM. The upper left corner of the scene are the data values shown in table 13.1 and figure 13.1; these values are also within the scene of figure 13.2. The headquarters of the H. J. Andrews Experimental Forest (west-central Oregon, 44.2 north latitude, 122.2 west longitude) is located in the center of figure 13.2. Darker areas show lack of or small amounts of vegetation, and lighter areas show more complete vegetation cover. Vegetation indices in figure 13.3 do not provide a large amount of ecological information, but they are one of the few measures that can be used for comparing data collected with different sensors and at different spatial scales (Price 1987).

The Multi-Spectral Scanner (MSS), a component of the Landsat satellites, has supplied important long-term data since July 1972. These MSS data are relatively inexpensive and do not have the license restrictions associated with some of the commercial data systems. However, the data are less popular for ecological research because of their reduced spatial resolution of approximately 80 m. The spectral extent of only four bands between 0.5 and 1.1 μm also reduces their applicability to ecological studies. Because of these limitations, the MSS sensor system has been superseded by more modern systems.

Scale

1000   0   1000   2000   3000   4000   5000   Meters

Landsat-TM Satellite Data
Normalized Difference Vegetation Index
Acquired July 7, 1991

FIGURE 13.3

Full-resolution NDVI image from Landsat-TM data for central Oregon (July 7, 1991). *Dark areas* indicate low vegetation cover; *white areas* indicate more complete vegetation cover.

High Spatial Resolution Data

To detect small objects and to integrate remote sensing information with ecological data from small areas requires systems with high spatial resolution. High spatial resolution data available from aircraft aerial photography are most common, but they are usually less consistent in content than remote sensing data from satellite systems. Data available from newer remote sensing systems will provide data at spatial resolutions approaching those of aerial photography. Data from the SPOT (Système pour l'Observation de la Terre) series of satellites are an example of fine-scale resolution data available from satellite remote sensing platforms. The high resolution visible (HRV) sensor on these satellites acquires data at spatial resolutions of 10 m (panchromatic) and 20 m (multispectral) and an extent of approximately 60 km square. The SPOT-HRV 10 m panchromatic data are shown in figure 13.4, an image of the same region shown in figure 13.3. A comparison of the two figures demonstrates the dramatic difference in data between spatial scales of 30 m and 10 m. The 10 m SPOT-HRV data are clearly able to identify landscape features such as roads, changes in agricultural use, and forest cutting. The detail itself is useful for determining landscape structure and indices (Nellis and Briggs 1988; Milne 1992). A disadvantage of the SPOT 10 m panchromatic data is that they cover a spectral extent from 0.5 to 0.7 μm but contain no distinct spectral information.

SPOT-HRV 20 m multispectral data offer a significant advantage for identifying vegetation cover on the ground, but there is a loss of resolution by a factor of two. SPOT-HRV are also potentially useful to ecological research because of their potential for time-series and change detection. Although the system covers a specific area only once every 26 days, the sensor can be programmed to view off-center from its orbit path. This not only increases the frequency with which an area can be viewed but it has the added capability of providing stereoscopic views, by observing the ground from different positions in space during successive orbits. The parallax created from this spatial displacement allows the data to be ortho-rectified, which corrects for errors caused by landscape relief. The parallax also permits the generation of high-resolution digital-elevation models, which is perhaps one of the primary applications for SPOT-HRV data in ecological research.

High Temporal Resolution Data

The temporal resolution of data is a critical component of remote sensing systems. The temporal resolution and length of the data set (temporal extent) define how well the system can measure change. An important source of time-series data useful for ecological research comes from a system not originally designed for this purpose. The Advanced Very High Resolution Radiometer (AVHRR) on the U.S. National Oceanic and Atmospheric Administration (NOAA) series of meteoro-

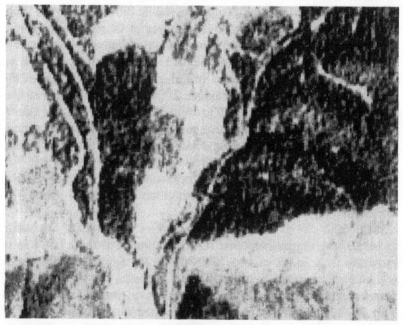

Scale

200   0   200   400   600   800   1000 1200 1400   **Meters**

**10m SPOT-HRV Panchromatic Satellite Data
Acquired July 4, 1991
Image Copyright 1991 CNES**

FIGURE 13.4

SPOT-HRV 10-m panchromatic data image for the region shown in figure 18.2 (July 4, 1991). Note the presence of landscape-scale features, such as roads, agricultural fields, and forest cutting. Image copyright © 1991, CNES.

logical satellites has been providing data since the first launch of the system in 1979. Each of the two operational satellites provides early morning and late afternoon coverage of the entire earth on a daily basis. AVHRR data have a nominal spatial resolution of 1 km with a spatial extent of over 2000 km. Global data at a resolution of 4 km are archived daily on an operational basis.

The AVHRR sensor collects data in the red, near-infrared, middle-infrared, and thermal-infrared (0.58 μm through 12.5 μm). The thermal-infrared wavelengths can provide surface-temperature data within 0.5°C. Similar to the example of Landsat-TM data (figures 13.1 and 13.3), the red and near-infrared wavelengths of AVHRR data can be used for an assessment of vegetation cover. An example of AVHRR-NDVI data is shown in figure 13.5, an April 7, 1991, image of data for western Washington state. Low vegetation-index values indicating a lack of vegetation appear dark in the image, whereas land with high vegetation cover appears bright. These data have provided valuable information regarding vegetation cover on a global scale (Roller and Colwell 1986), and this is the basis for assessing improvements in the technology of global measurements (Running et al. 1994). These new developments will combine the high temporal resolution of AVHRR data with higher spectral and spatial resolutions. These systems, such as the Moderate Resolution Imaging Spectroradiometer (MODIS), will provide products more useful than simple NDVI data on a global scale.

High Spectral Resolution Data

Remote sensing data applications are moving toward development of hyperspectral imaging sensors. A number of airborne systems have already been developed, and satellite systems will provide hyperspectral data in the future. The Airborne Visible/Infrared Imaging Spectrometer (AVIRIS) sensor has been used widely for ecological studies. It provides hyperspectral data of 20 m spatial resolution for relatively small areas. The 224 spectral channels provide the fine spectral resolution needed to more clearly identify land-cover information and vegetation parameters related to productivity. These hyperspectral data also provide information for atmospheric correction that is lacking in other broad-band remote sensing systems.

New satellites are taking advantage of changing technology and are reducing the costs associated with design, integration, launch, and operation of the systems. An example is the proposed Lewis hyperspectral imager (HSI) with 30 m resolution and 384 bands or channels covering a spectral range from 400 to 2500 nm. An example of simulated HSI data is shown in figure 13.6, which depicts an area within the Hanford Reservation (Washington). The data shown represent a single near-infrared channel of data at 910 nm. Similar to the AVIRIS data, each band covers a spectral bandwidth of less than 10 nm. Figure 13.7 shows a plot of the spectral information these data provide. The data are relative reflectance extracted from the 384 spectral channels

	Scale				
100	**0**	**100**	**200**	**300**	**400** **Kilometers**

NOAA-AVHRR Satellite Data
Normalized Difference Vegetation Index
Acquired July 7, 1991

FIGURE 13.5
AVHRR-NDVI data image for western Washington (July 7, 1991). *Dark areas* indicate low vegetation cover, and *bright areas* indicate high vegetation cover.

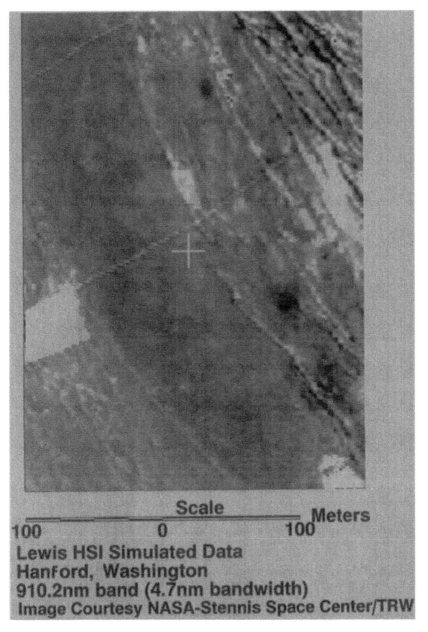

FIGURE 13.6

Simulated Lewis data image for the Hanford Reservation area of Washington. Data used to represent the image show a single near-infrared channel of data at 910 nm. Image courtesy of NASA-Jet Propulsion Lab/TRW.

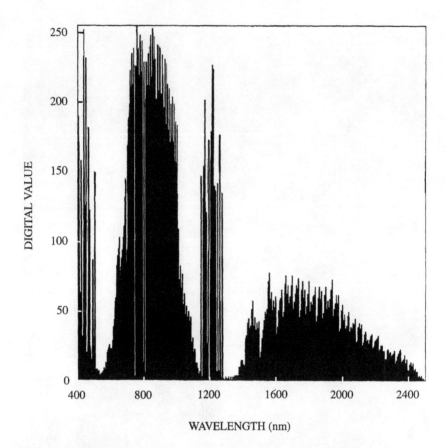

FIGURE 13.7

Plot of spectral information for data in figure 13.5. Data are relative reflectance from 384 spectral channels for one 30 m area of the image in figure 13.5.

or data layers for one 30 m area of the image (indicated by the "+" in figure 13.6). It is this kind of information that makes hyperspectral data unique. More than 15 channels of these data fit within the bandwidth of the near infrared channel of Landsat-TM data. This high degree of spectral resolution is necessary to improve the information available from the coarse vegetation indices and simple land-cover classifications available from conventional broad-band remote sensing systems.

Special Purpose Remote Sensing Data—Radar

Remote sensing information from relatively new radar-imaging systems offer important information for ecological research that other systems cannot provide. An example of these data is shown in figure 13.8. This is a Shuttle SIRC-C/X-SAR radar image of Mount Rainier

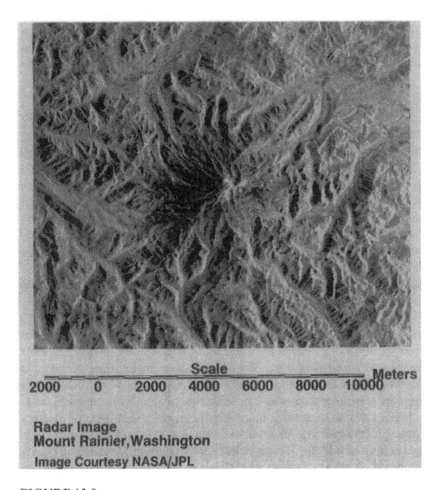

FIGURE 13.8
Shuttle SIRC-C/X-SAR radar data image of Mount Rainier, Washington. The image, which is a composite of three microwave wavelengths, provides information on landscape roughness, vegetation cover, and vegetation roughness. Image courtesy of NASA-Jet Propulsion Lab.

(Washington). The image is a black-and-white composite of 3 cm, 6 cm, and 24 cm microwave wavelengths known as X-, C-, and L-bands. These data provide information based on structural detail, and the different wavelengths and the method by which they are processed provide various information on landscape roughness, vegetation cover, and vegetation structure. Clouds and haze, which severely limit optical systems, are transparent to radar data. This is particularly important in tropical ecosystems and mountains, where information from optical systems is often unavailable. Perhaps the greatest limitation of these data is the inability of ecologists to utilize them, because less is understood about the data characteristics when compared to more conventional optical systems.

Summary—Future Applications of Remote Sensing

An important consideration for ecological research with respect to remote sensing is the increasing amount and rate of data being amassed. The Earth Observing System (EOS) of the U.S. National Aeronautics and Space Administration (NASA) has promised a number of new space-based remote sensing instruments that will be useful for ecological research. A warning—for lack of a better word—was that these new instruments will generate a data stream greater than anything remote sensing scientists, much less ecologists, will be able to use. For EOS data, an elaborate infrastructure was developed to ingest, process, distribute, and archive data from the EOS, but the data crush from NASA-EOS is not the only problem ecologists face.

The large amount of data from all available remote sensing systems, as well as the historical record from past systems, presents a serious limitation for ecologists interested in time series and historical trends using remote sensing data. We already have an inherent problem in tapping the real power of remote sensing because of our inability to process very large data sets. The daily flow of data from currently operational satellite systems is managed by an elaborate but poorly integrated infrastructure. Most of the data are acquired for specific purposes such as weather forecasts, but a large amount is unused or poorly archived. A good example of this is the global 1 km data generated more than once an hour by the global network of geostationary satellites. Many of these data are archived, but retrieving time-series data from the archives can be a difficult task.

Other data, including Landsat-TM and even NOAA-AVHRR, are difficult to manipulate for large areas because of the massive volume of data. A single Landsat-TM scene requires from 300 to more than 600 Mb for data storage, depending on the map projection and the specific sampling resolution used. Current computer technology has allowed ecologists to use these data for many applications, but data volumes for large areas or long time periods become problematic for storage as well as computational speed. With rapid changes in technology, the capability of using hyperspectral data will shift interest to these new data sources and require a change in our perspectives of what comprises a large data set. High performance computing facilities exist to alleviate this problem, but costs and expertise for using them are beyond the reach of most ecologists. As a result, ecologists are barely scratching the surface of ecologically significant information available from remote sensing data. Data available from future systems promise to increase this complexity.

The NDVI data used as examples in this chapter are only a starting point for extrapolation between various scales. We need to pay careful attention to calibration and validation of products from remote sensing data. Integration of ground-based measurements of more conventional ecological measurements, such as productivity and leaf-area index, into operational remote sensing systems requires that these data be validated. Extrapolation of single-point measurements to valid information at regional and global scales is perhaps the largest task facing the remote sensing community. Ground-based measurements and validation systems are in place to accomplish this task. For instance, the wide-bandwidth data of conventional Landsat-TM, SPOT-HRV, and AVHRR data can be corrected with attention to instrument calibration and ancillary ground-based measurements such as that from sun photometers and other instruments (Hall et al. 1991b; Vande Castle and Vermote 1996), but these data are not usually collected on a routine basis. The added spectral resolution of hyperspectral systems such as HSI and MODIS may reduce the need for such ancillary measurements. However, the products generated from these systems must first be verified with ground-based measurements. The validated products from these data promise to offer standardized products for comparative ecological studies (Running et al. 1994). This is where the true strength of space-based hyperspectral instruments such as HSI and MODIS prevail.

14

FIELD STUDIES OF LARGE MOBILE ORGANISMS: SCALE, MOVEMENT, AND HABITAT UTILIZATION

S. Jonathan Stern

Mobility is common to all taxa. Vegetation moves, albeit at rates requiring many years and generations of individuals to detect. At the other end of the mobility scale, large mobile organisms (LMOs) such as *Balaenoptera acutorostrata* (minke whale) (figure 14.1) can move over large areas in a short time. LMOs are unique because rapid large-scale movement can occur at all levels of organization—individuals, groups, and populations. Typically, LMOs are large terrestrial and marine mammals, although not all large, mobile species exhibit wide-ranging behavior. Some species of birds, fishes, and marine reptiles can be considered LMOs because they are mobile at similar spatial and temporal scales as mammalian LMOs. Large mammalian body size has physiological and ecological implications, which makes combining or comparing mammals with other taxa a problem for some applications (Peters 1983; Calder 1984; Schmidt-Nielsen 1984).

Can LMOs be viewed essentially as small animals that are able to rapidly move on a large scale? The same types of questions about small animals apply to large animals. However, some properties relating to the ecology of an organism are a function of body size (Peters 1983; Calder 1984; Schmidt-Nielsen 1984). For example, large body size means that large animals require a large quantity of food to fulfill energetic demands, and that large animals sample the environment at a coarser grain than smaller animals. Thus, they need areas of large absolute size. The ability to sample large areas relatively rapidly allows

LMOs to compensate for these scale-dependent properties (Holling 1992).

Despite their large ranges, individual LMOs repeatedly use particular areas and locations as part of their home range. Thus, patchiness is an essential component of home range use but can introduce logistical difficulties in measuring home range. For example, large animals are relatively less abundant than smaller ones and require larger home ranges (Peters 1983; Calder 1984). Therefore, unless field studies are done at sufficiently large scales, patches and interpatch areas will be difficult to detect, and because areas are used repeatedly, long-term studies are required to determine temporal utilization patterns. This becomes essential when assessing how important an area is to a particular population.

Useful temporal scales are often divided into annual, seasonal, and within-seasonal scales. The annual scale encompasses the annual cycle of a species, population, and individual. Years can be compared and analyzed as a time series to determine cycles and variability. Within a year, the next temporal scale differentiates productive and nonproduc-

FIGURE 14.1

Balaenoptera acutorostrata (minke whale), an example of an LMO that can move over large areas in a short time. Photo by S. J. Stern.

tive seasons. Much of the life histories of LMOs, and of animals in general, are linked to these seasonal cycles and what occurs in a particular season. The other major time scale is that of the particular season. In some species, the ability to live off stored energy reserves is critical, and the ability to do so increases with increasing body size. The consequence is that smaller animals must eat more often as energy stores are more rapidly depleted. Therefore, it is more difficult for a small animal to survive long periods of poor resource levels during less productive seasons. The options for small animals are to store food or to reduce metabolic rate. Large size offers the individual the option to move great distances, an option generally not available to smaller animals (Peters 1983). Thus, LMOs may reside in an area all year, living off stored fat with reduced foraging and hibernation, or they could seasonally migrate long distances to suitable areas.

It has been argued that, because of their lower abundance and their scale-dependent physiological processes, larger animals are less ecologically significant within a community than smaller organisms (Peters 1983). LMOs can be considered rare events in a particular study plot (e.g., a 1-ha plot) because they operate at such large spatial and temporal scales. However, at local scales, LMOs contribute to ecosystem structure and function through feeding and through biogeochemical cycling in the form of excretion (Botkin et al. 1981; Northcote 1988; Pastor et al. 1988; Steinauer and Collins 1995), decomposition, and litter (Northcote 1988; Pastor et al. 1988) that affects all trophic levels (Naiman 1988). LMOs are able to store material for long periods and potentially move it great distances (Kitchell et al. 1979; Northcote 1988; Zijlstra 1988; Deegan 1993). Through these processes, as well as physical disturbance, they structure the environment in a way that climate and geology alone could not reproduce (Naiman 1988), and repeated use serves to maintain local heterogeneity. The scale-dependent impact of LMOs on an ecosystem is largely dictated by their distribution and movements.

Field studies addressing distribution and movements use one of two general sampling strategies. A lagrangian study consists of measurements of individuals, and in the present context, it focuses on the location and movement of individuals. Lagrangian data consist of a series of x-y coordinates over time for a given individual. A eulerian study focuses on the distribution of individuals in an area. Data consist of a pair of y-y coordinates for the locations of individuals in a surveyed area. The lagrangian sampling unit is the individual, while the eulerian sampling unit is the area. Characteristics of both eulerian and lagran-

TABLE 14.1
Techniques of eulerian and lagrangian methods of study

Technique	Eulerian	Lagrangian
Unit of study	Area	Individual
Type of study	Transect, survey	Mark-and-recapture, telemetry
Movement models	Diffusion models, partial differential equations, cellular automata	Random walk
Distribution	Plotted locations of sighted animals, density estimates	Locations and movements of individuals
Model	Analytic/simulation, spatially explicit population models	Analytic/simulation, spatially explicit individual-based models
Grain	Observer- or protocol-defined (e.g., km²)	Observer-, animal-, or protocol-defined (e.g., patches)
Extent	Observer- or protocol-defined (e.g., hectare, 100 km²)	Observer-, animal- or protocol-defined (e.g., home range)

gian studies as they relate to studies of LMOs are summarized in table 14.1. The choice of approach depends on the objectives of the study, and in some instances, both techniques are used. For example, a lagrangian study may be limited to a specific area where sampling occurs. Studies of movement patterns are most effective if focusing on the individual, whereas distribution can focus at the level of either the individual or the area. However, lagrangian distribution studies are limited by the number of tagged individuals.

This chapter focuses on movement by large animals and the implications of their movement in assessing three aspects of habitat utilization—distribution, abundance estimates, and foraging dynamics—measures that are commonly addressed in field studies of LMOs.

The Effect of Mobility on Estimates of Distribution Pattern

One common feature of most field studies of LMOs is the great deal of time spent simply looking for them, due to mobile observers looking for mobile animals over large areas (Koopman 1980). Even when a particular area is known to be used at a particular time, the area is often so large that finding individuals or groups is difficult. Further, low abundance, and behavioral traits such as observer avoidance, can add to the difficulty in finding animals. On the other hand, some animals are prone to being captured; although this makes it easier to find them, it violates the assumptions of some population estimators, and it may obscure naturally occurring movements and distribution patterns. The ability to find animals is also partially dependent on social behavior, because it is easier to find large groups of organisms than a single individual.

Home range defines areal extent of an individual or population and is a naturally defined study area. Home range combines aspects of habitat productivity, energy requirements (Harested and Bunnell 1979), resource distribution in space and time, foraging dynamics (Holling 1992), and both interspecific and intraspecific interactions. Species-specific response to variability in these factors results in different strategies, as mentioned earlier, and ultimately determines home-range size.

Although home range scales to body size (McNab 1963; Peters 1983; Calder 1984; Schmidt-Nielsen 1984; Lindstedt et al. 1986), it is difficult to predict absolute dimensions of a home range. The difficulty is exemplified in Bertram's (1979) study of home ranges in *Acinonyx jubatas* (cheetah) and *Panthera leo* (lion) of the Serengeti Plains of Kenya and Tanzania, East Africa. Based on the relationship between body size and home range, *P. leo* (100 to 200 kg) would be expected to have a larger home range than *A. jubatas* (40 to 60 kg). However, the observed home range of *P. leo* is 20 to 150 km², whereas the home range of *A. jubatas* is about 1300 km². Clearly, factors other than body size determine home-range size. *P. leo* lives in prides of about nine (range, 4 to 15) individuals, in which several females cooperate in killing prey. *P. leo* prides stay in limited range during the year, feeding on *Conochaetes taurinus* (wildebeest), *Equis burchelli* (zebra), and *Gazella thomsonii* (Thompson's gazelle) that migrate through the area seasonally, and on resident *Syncerus caffer* (water buffalo), *Phacochoerus aethiopicus* (warthog), and smaller mammals. *A. jubatas* are generally solitary hunters, following herds of migrating *G. thompsonii,* but they will also take *G. grantii* (Grant's gazelle) and *Aepyceros malampus* (impala). Social behavior and large body size allow lions to feed on a wider size-range of prey than *A. jubatas.* These attributes also reduce the chances of losing food to other species, while increasing the chances of successfully stealing food from other species. *A. jubatas* loses 10 to 12 percent of its kills to other predators, such as *P. leo* and hyenas (Hyaenidae), whereas *P. leo* rarely lose prey to other species. Further, scavenging does not play a role in the diet of *A. jubatas,* whereas 10 to 15 percent of the *P. leo* diet consists of scavenged food, allowing *P. leo* to utilize more of the production available from the area than *A. jubatas*. *A. jubatas* responds to a change in prey distribution, as patchiness in the spatial domain, by following herds as they migrate, resulting in a large home range. *P. leo* prides exploit changes in prey within an area, as patchiness in the temporal domain, by feeding on whatever is locally abundant at a given time. Differences in exploitation of patchiness in either space or time allow for increased species diversity and abundance in an area.

The relationship of a species or population to its environment is reflected in distribution and abundance in both space and time. Therefore, patterns and processes of spatial variation in abundance within a species are key factors in large-scale spatial and temporal distribution patterns (Brown 1995). However, in very large-scale studies, coarse sampling grain makes the underlying dynamics of a change difficult to determine. A new distribution can be the result of a redistribution of the same individuals from one sampling period to the next, or the result of individuals immigrating into and emigrating out of the area. Further, the distribution pattern can be the same, with only the individuals changing. This problem is intractable in eulerian studies.

Mark-and-recapture studies can be used to estimate home range. These are explicitly lagrangian because individuals are tagged, but sampling is often carried out within arbitrarily defined areas. The underlying assumption is that animals sample the environment at relevant scales, and the location and time of recapture yield information about movements. The assumption of homogeneous sampling required by such studies (Swihart and Slade 1985a,b, 1986; Swihart et al. 1988) is violated when an animal's habitat use is variable because of environmental heterogeneity at a variety of spatial and temporal scales (Minta 1992). This violation is a way of life for LMOs. Statistically dependent points may be used to estimate home range, provided they are biologically independent (Lair 1987; Aebischer et al. 1993). An appropriate sampling lag can be calculated, given information on the movement patterns of individuals (Swihart et al. 1988).

Another lagrangian technique, telemetry instrumentation, has expanded the range and duration of measurements of movement patterns of animals. The resolution of patterns depends on the frequency the signal are received. Radiotelemetry is best suited to determine small-scale movement patterns because continuous tracking is possible, although, because of limited range of reception, the individual must be followed as it moves, or repeatedly approached to within reception range. Satellite tracking allows for monitoring on a much larger scale (Vande Castle, chapter 13), and proximity to the animal need not be maintained to receive the signal. Resolution for these signals depends on the frequency of satellite passes overhead of the tagged animal. In many cases, the fine-scale measurements possible with radio tags cannot be made with satellite tags. The choice of method depends on the objectives and logistical considerations. If following an animal to maintain contact is too difficult or results in an adverse behavioral reaction, then resolution of small-scale movement patterns is sacrificed in favor of information on larger-scale movement patterns. A global positioning system in a retrievable data-recording package combines the best features of both systems, resulting in continuous large-scale coverage without having to follow the animal. A significant logistical problem is deployment and retrieval of the package. In addition, such long-term, continuous data recording results in a large volume of data that requires subsampling at appropriate intervals to minimize the effects of autocorrelation and aliasing. For example, location data may be autocorrelated if collected too close together in space or time (Kareiva and Shigesada 1983; Johnson et al. 1992); therefore, analysis requires an appropriate spatial and temporal lag.

The spatial distribution of individuals or groups of animals can be viewed as nested within hierarchical scales of patchiness (Kolitar and Wiens 1990; Wu and Loucks 1995; Stern et al., unpublished manuscript) (figure 14.2). Animals search for patches, defined as areas of abundant resources such as food, within which they search for randomly distributed prey (Goodwin and Fahrig 1997, chapter 9). Search behavior can be broken down into between-patch and within-patch search (Bobisud and Voxman 1979). Movement patterns can provide clues to how LMOs search for resources at these two scales. The assumptions are that (1) the general distribution of patches is known to the individual and (2) resources are randomly distributed within the patch. The first assumption is reasonable for long-lived individuals that repeatedly sample the same areas year after year. Movement can be quite directional between patches, but can appear random within patches. At scales within a patch, once a target has been localized, movement can be highly directional (Johnson et al. 1992). This "scale of detection" indicates a change from search to active pursuit, and if the resource is relatively immobile, the directions are in a straight line. If the resource is mobile, the movements may be highly nondirectional. In this case, it may be impossible to differentiate active pursuit from search movements within a patch. In practice, measurements recorded after a target has been detected should not be included in the analysis, because they are not related to search. At large scales, there can be areas of patches; at even larger scales, there can be regions of areas. Animals may use different search cues and distribution and movement patterns for each of these levels. However, to a foraging animal, the most relevant dimensions are those within and between patches. At larger scales, there may be nested scales of patchiness (Kotliar and Wiens 1990; Wu and Louks 1995).

Analysis of movement patterns may reveal differences in within-patch and between-patch behavior. The *r*-test measures uniformity or randomness in directions of travel (Batschelet 1980), and the data are collected at spatially or temporally discrete intervals:

$$r = \frac{1}{n}\left[\left(\Sigma \cos \alpha\right)^2 + \left(\Sigma \sin \alpha\right)^2\right]$$

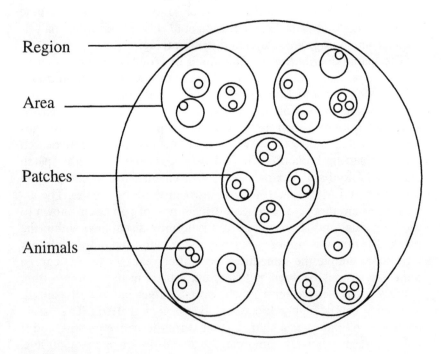

FIGURE 14.2

A hierarchical view of the environment. This is how an LMO might view the world in terms of itself and resource distribution.

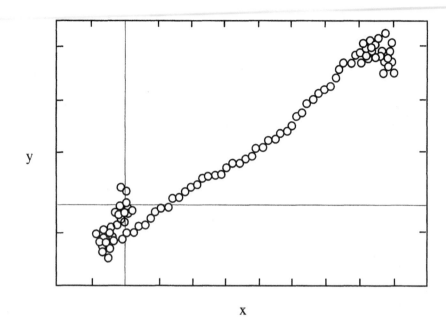

y

x

FIGURE 14.3

A track of a *B. acutorostrata* (minke whale).

where α is the angle of the direction of travel, and *n* is the sample size. The calculated values range from 0 to 1, indicating random to uniform direction of travel, respectively. An example is given for *B. acutorostrata* traveling between feeding areas. Figure 14.3 shows the track of the whale, and figure 14.4 shows the corresponding *r*-values of the track. Random directions, those within a patch, are indicated by decreasing *r*-values, points 0 to 25 and 75 to 100. The relatively uniform directions of travel between patches are indicated by increasing *r*-values in points 26 to 74. From these *r*-values, spatial information on patch size and distance between patches relevant to the sampling by individuals can be estimated.

Similar analyses can be done using an index of linearity, which is the ratio of net-to-cumulative distance. Net distance is the distance traveled from the beginning point to the ending point of the track. The cumulative distance is the sum of the distance from one sampling point to the next over the length of the track. The calculated values range from 0 to 1, indicating random and straight line, respectively (Batschelet 1981). Using the same data set as for the *r*-test, the overall index of linearity is 0. 4. However, if the first and last 25 points, representing travel within the patch, are analyzed separately, the calculated indices are 0.23 and 0.04, respectively. This suggests random movements within the patch. The index for the middle 50 points (0.71), representing travel between the patches, is fairly linear. The *r*-test and index of linearity provide similar information, the difference being in the type of data collected, and how the data are manipulated. However, either measure can be calculated from *x-y* coordinate data.

Fractal analysis (Hastings and Sugihara 1993; Dutilleul, chapter 18; Gardner, chapter 2) provides another technique for analyzing habitat utilization. In terms of animal movement, the fractal dimension describes movement as the proportion of a line, surface, or volume, representing one, two, or three dimensions, respectively. Marine mammals, birds, fish, and arboreal and mountain-dwelling mammals and birds sample the environment in three dimensions, because there is a significant vertical component to their movement patterns. The realized production in two versus three dimensions is possibly an important distinction that has yet to be investigated. The fractal dimension is calculated by plotting the log of the cumulative distance, over the log of some reference or step length, for a number of tracks of the

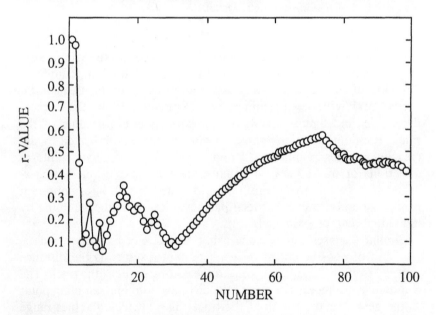

FIGURE 14.4

The r-values of a *B. acutorostrata* (minke whale) track.

same length. The slope of this line is the fractal dimension indicating how an organism uses the environment in one, two, or three dimensions. Additional information on fractals is contained in Morse et al. (1985), Milne (1991), Holling (1992), and Johnson et al. (1992).

The Effect of Movement on Abundance Estimation

Habitat use is partially a function of the number of individuals using the habitat. For this reason, abundance estimates are an important component of field studies. However, mobility can have a confounding effect on abundance estimates. Effort-related measures of abundance that use density can be biased as a result of mobility. For mobile animals that congregate in patchy distributions, a decline in abundance is generally accompanied by a reduction of range, with little or no change in density within the range (Murphy 1966; Smith 1978). Abundance estimates can be seriously biased if the range is reduced while density remains constant, and if this range reduction is not taken into account. This may result in counts being extrapolated over an artificially large scale and therefore overestimated. This is in contrast to less mobile populations in which declines are accompanied by a reduction in the density within the range, rather than a reduction in range itself (Schneider 1994a).

Transect sampling assumes that no movement of individuals occurs between transect lines. Bias occurs if animals move faster than the observer, who may unintentionally count the same individuals more than once (figure 14.5). In contrast, mobility coupled with intermittent sightability can lead to missed sightings of animals that are actually present. Correction factors may be used but may lead to uncertainty in population estimates. For LMOs, larger areas need to be surveyed to achieve an adequate sample size (Buckland et al. 1993). One problem is that as the size of the area increases, the time required to sample the area increases as well. Simply "getting a faster boat" will not necessarily solve the problem. For example, the probability of sighting an intermittently visible animal decreases as observer speed increases (Stern 1992). There is a trade-off between quickly covering as much area as possible and maximizing the probability of sighting animals.

Mark-and-recapture methodology assumes that all individuals in a population have the same probability of being captured, with random mixing of the population between successive samples. However, if not enough time elapses between samplings, recently marked individuals have a higher probability of recapture, resulting in population esti-

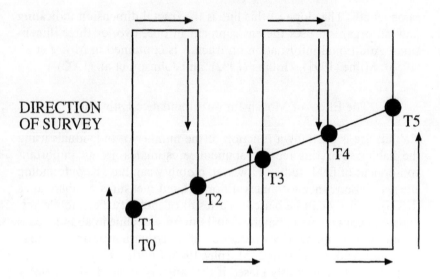

FIGURE 14.5

The survey starts at point T0, and an animal is sighted at T1. The animal could be recounted at T2, T3, T4, and T5 positions. In general, animal movement can confound efforts to accurately estimate population size in eulerian studies. Adjustments in transect design and correction factors can be used to minimize this effect.

mates being biased downward. Therefore, care must be taken to correct for temporal autocorrelation that occurs when there is insufficient time for random mixing of individuals between samplings. Once again, the time lag necessary between sequential observations may be difficult to determine for most species (Swihart and Slade 1986; Swihart et al. 1988), but it is essential for analysis.

The Effect of Movement on Foraging Studies

The objective of many field studies is to determine how consumers are distributed relative to their resources. This link between distribution and resource requirements is based on the assumption that through evolution and adaptation, a dependence exists between ecological and physical/chemical processes (Steele 1989a,b). Although LMOs are linked to physical processes through the distribution of resources, mobility allows them to escape the influence of a particular physical regime (Murphy et al. 1988) and exploit other areas.

Eulerian studies directed at correlating LMO distribution to physical processes are problematic because there may be few individuals present during a survey through an area where they "should" occur. One analytical method is to lump animals into conveniently sized sampling grids, resulting in an increased sample size per grid with a corresponding loss of resolution. For example, sea birds are located by scanning an area around a research vessel while their prey are located by acoustic methods, sampling directly below the research vessel (Schneider and Piatt 1986; Schneider 1994b). Although both bird and prey samples are simultaneously collected in transects through an area, the areas being effectively searched are different, and sightings of birds and prey are lumped into blocks of arbitrary size, thus defining the level of resolution.

Is there actually no correlation, or is the particular resource yet to be found? Correlation of LMOs and their resources is often scale dependent (Schneider and Piatt 1986; Rose and Leggett 1990; Veit et al. 1993; Schneider 1994a). For example, Rose and Leggett (1990) found that *Gadus morhua* (cod) and *Mallotus villosus* (capelin) were positively correlated at large scales (4 to 10 km) and negatively correlated at small scales (2 to 5 km). However, when *G. morhua* were actively feeding on *M. villosus,* there was a positive correlation at very small scales (3.5 m). Large-scale correlation relates to distribution of both predators and prey in patches, whereas negative correlations at small scales is caused by the random distribution of prey and search behavior of predators within patches. The fine-scale positive correlation represents predators actively feeding on prey that are within the scale of detection. There is no reason to expect complete coherence of predators with prey; search is a temporal process. Thus, correlation is scale dependent in time as well as space.

A lagrangian approach is to track individuals as they move about, each one moving within and between patches, and into and out of areas. This technique requires that many individuals be tracked, and it may be difficult and expensive. Simple correlations can be made by plotting locations of animals on resource distribution maps using a geographic information system (GIS). Although these correlations provide a description of animal distribution relative to resources, other factors are important when estimating resource utilization within a habitat.

Feeding rate is a function of both the spatiotemporal distribution of prey and the ability of a consumer to find it. Therefore, not only resource abundance, but also resource availability in space and time, dictates exploitation patterns of those resources (Levin 1978; Briand

1983; Wiens 1984; Getty and Pulliam 1991). Feeding rates can be calculated as an encounter rate that is a function of prey density (Mangel and Clark 1988), often averaged over large scales. Prey densities of resource-rich patches and prey densities of resource-poor areas between patches are often averaged together. Scale-dependent aspects of search for patchily distributed resources are missed. Patches represent energy sources, providing a net energy gain, whereas areas between patches represent an energy sink, providing either no net energy gain or a net energy loss. Information on the effects of the spatial and temporal pattern of resource availability on intake and storage for later use is essential to understanding the dynamics of foraging ecology of wide-ranging individuals, populations, and species. Given the patchiness of the environment, variable feeding rate within and between patches influences both short-term and long-term success, including growth, reproduction, and survival.

Large-scale mobility is an effective foraging strategy in environments that are heterogeneous at multiple scales. However, in some cases, the frequency or amplitude of environmental variability increases to the point that the area can no longer support a population. A potential scenario resulting from a decline in forage available for a grazer would be as follows. As resources decline, so does density of resources within a patch, resulting in less energy available per patch. Although the overall range of a resource does not necessarily decline, the density of patches decreases, resulting in a greater distance to travel between patches. The result is an increase in time spent in an energy-sink environment. Animals increasingly resample patches, resulting in an increased patch revisitation rate that does not allow for sufficient patch regeneration. Resources within patches are reduced, and patches are eventually suppressed or destroyed; dispersal is then the only option for an LMO. However, the ability of an individual to disperse may not be limited only by available resources but by social, behavioral, physiological, or other factors. In the case of one population of *Loxodonta africana* (African elephant), when local resources declined, food that would have normally been spared for later use had already been consumed (Botkin et al. 1981; Laws 1981). Food that elephants were able to find and consume was deprived of essential nutrients. Ranges of *L. africana* became reduced and fragmented, resulting in abnormally high densities, until dispersal occurred. Individuals that had a limited dispersal ability (e.g., those needing to care for calves) died. Individual variability in survival ability through dispersal or tolerance of severe conditions is critical for the survival of populations.

Interpreting Mobility at Large Spatial Scales

Habitat utilization by a species is the result of many factors. Distribution, abundance estimates, and foraging are not the only aspects relevant to habitat utilization, but they illustrate the impact of movement on design and implementation of LMO studies. Determining relevant sampling areas and scales that are biologically relevant to the organisms and study objectives is a critical first step. At large spatial and temporal scales, metapopulation-level questions, such as movement of individuals between populations, are relevant.

The study of mobility is the study of individuals. Variation between individuals influences survival success, including finding hosts, foraging, aggression, and reproductive success, and it occurs on individual, population, and evolutionary time scales. However, the success of species and populations depends on the variability of individuals within the population (Taylor 1992). An individual-based approach to conceptual and simulation models is a relatively new, and potentially powerful, approach to addressing dynamics at these scales. LMO populations are relatively small, and, because it is possible to collect data on individual behavior, movement, distribution, and foraging can be modeled for an entire population. Not all individuals must be explicitly modeled. For example, *Rangifer tarandus* (caribou) could be modeled as a herd rather than as a number of individuals, with several herds interacting on a larger scale. These models allow for the study of patterns at a variety of levels of organization and use data that are commonly collected in field studies (Huston et al. 1988; DeAngelis and Gross 1992; Judson 1994). Spatially explicit models integrating individual movement in relation to resources can be used to explore a number of processes, such as population response to changing environmental conditions (Folse et al. 1989, 1990; Coulson et al. 1991; Hyman et al. 1991; Roese et al. 1991; Dunning et al. 1995; Holt et al. 1995).

Density dependence is an explicitly spatial process, and the use of space through mobility should be included in studies of foraging, competition, host–parasite interactions, reproduction, and the effect of harvesting. Through movement, the impacts of density-dependent processes on a community of LMOs are reduced, allowing for a greater species diversity and abundance than would otherwise be possible (Taylor 1992). For example, succession of herbivore species in the Serengeti Plains occurs as species enter and leave an area. The different species feed on the same plants, but each species specializes in a particular part of the plant. For example, *Conochaetes taurinus* and its predominant forage species are coadapted in such a way that grazing stimulates plant growth, allowing for increased biomass available for species that subsequently migrate into the area (McNaughton 1979). Temporal patchiness is an aspect of mobility that introduces complexity to the question of habitat utilization by populations and communities.

Caution must be exercised when making statements about animal movement in relation to resources based on field observations or simulation experiments. Models must have rigorous validation and sensitivity analysis before any confidence can be placed in their predictive abilities (Kareiva and Wennergren 1995). Care must be taken when attempting to interpret the result of changes in distribution patterns. In addition to a change of individual distribution versus a change in individuals, difficulties in interpretations arise when trying to distinguish between dispersal, death (Zeng and Brown 1987), capture shyness, and chance. Development of models to test for processes such as dispersal (Porter and Dooley 1993) are critical for differentiating these processes from one another. Modeling studies are important components of this analysis, although there is a temptation to assign an underlying process to an observed or generated pattern. Unless the process is explicitly sampled in the field or incorporated into the model, no underlying processes can be assumed. These observations and models are valuable for generating hypotheses about underlying processes that can then be sampled and explicitly modeled.

Null models make it possible to test whether the observed pattern of distribution or movement would be expected by chance (Caswell 1976; Gotelli and Graves 1996). Two such models could compare observed distributions and movement patterns with randomly simulated distribution and movement patterns. A third model would attempt to reproduce the observed large-scale distribution pattern, given a random movement pattern. A correlated random walk is one such model because it represents organisms that tend to travel in a particular direction (Okubo 1980; Kareiva and Shigesada 1983). A spatially explicit model, such as an object-oriented simulation model or an intelligent GIS model (Folse et al. 1989, 1990; Coulson 1992) could be constructed to include resource distribution in the simulations and to test whether the correlated random walks result in movement and distribution patterns similar to movements and distribution observed in the field. Linking distribution and movement of LMOs to their resources allows for testing the effect of changes in resource distribution on distribution and movement patterns. Generally, there are no underlying assumptions about the use of cues or memory in movement or distribution patterns, because only the patterns, not the underlying processes, are modeled. However, the results of simulations can be used to generate further hypotheses to test the effects of changes in resource distribution on LMO distribution and movement patterns.

Summary — Management and Environmental Implications

Mobility of LMOs has management-related implications other than logistical considerations of census methods. The use of an area by individuals over time must be resolved for effective long-term management. As mentioned earlier, if mixing of a population is mistakenly assumed to be occurring, a portion of the population in an area will be subject to overexploitation. In these instances, the scale of the hunt should be set to minimize the chance that an isolated pocket of individuals is being exploited in a manner prescribed for a larger population, that is, one that is assumed to move through an area where exploitation occurs.

Species may be managed on the basis of competitive interactions, either between a species and humans for resources, or between species where one species is being outcompeted by another species. A species being outcompeted, partially through reduced abundance from human exploitation, is often a commercially important species. Interactions between species is local, occurring between individuals, although the extent of these interactions may be of a larger scale. The proposed solution for this problem is often to cull the competitor species, that is, the species competing with humans or the commercially valuable species. The rationale is that the resources that are freed up are then available for the more favorable species. However, food webs usually have more than two species feeding on the same resource (Martinez and Dunne 1997, chapter 10), and there is no way to ensure that the freed resources will be used by the intended species. This management strategy was proposed for the endangered *Balaenoptera musculus* (blue whale) in the Southern Hemisphere. Because minke whales are numerous in these waters and because both species feed on *Euphausia superba* (Antarctic krill), the assumption is that a reduction in the number of *B. acutorostrata* would result in more food available for *B. musculus.* However, many other species feed on *E. superba,* most of which are smaller and more abundant than *B. acutorostrata.* The only effective way to implement this strategy would be to follow a *B. musculus* and kill any

B. acutorostrata feeding in the immediate area. Otherwise, given the mobility and abundance of other predators, any resources not taken by *B. musculus* are more likely to be taken by the other species.

Many species of LMOs have been subject to hunting by humans, resulting in subsequent population declines. As these populations recover, separating range extensions of increasing populations from a shift in distribution in response to environmental change may be difficult because mobility allows for relatively rapid dispersion. Simultaneous measurements of abundance, foraging, and distribution may allow for distinction between response to environmental change, population increase, and dispersal.

A macroecological approach to study environmental change (Brown and Maurer 1989; McDonald and Brown 1992; Lawton et al. 1994; Brown 1995) uses information on large spatial- and temporal-scale species distribution to estimate the potential impact and likelihood of extinction based on predicted range reduction in response to climatic change. Because LMOs move large distances in a short amount of time as part of an annual cycle, it may be difficult to distinguish distribution shifts in response to short-term environmental fluctuations from longer-term climatic shifts. Short-term studies will miss these dynamics. Studies of animal movement do not reveal information about the spatial scales required for populations to maintain themselves over long time scales (May 1994). However, important dynamics are occurring below the metapopulation scale, which influences the rate of dispersal and survival of individuals, population, and species. Integrating both approaches, as a combination of short-term "experiments" within long-term distribution studies, may be the best approach to predict both the outcomes and underlying dynamics. Focusing on annual, seasonal, and within-seasonal cycles and on within- and between-patch dynamics is likely the best approach. Variability in foraging occurs within and between patches, and when this variability becomes extreme, large-scale redistribution is the only option for survival.

Studies of the interaction of LMOs with the environment must include the mobile aspect of the animals at relevant scales. Data can be collected on movement patterns using telemetry instrumentation; by using null models, hypotheses of how animals use the environment can be constructed and tested. Wide-ranging distribution and movement patterns illustrate that LMOs use and influence the environment on a variety of spatial and temporal scales. Predicting and measuring the response of LMOs to environmental change at a number of scales are issues that require further study.

Acknowledgments

I would like to thank Tim Hollibaugh, Allisa Arp, and the staff of the Romberg Tiburon Center for Environmental Studies, San Francisco State University, for support during preparation of this chapter. An anonymous reviewer provided valuable comments and suggestions on an earlier draft of the manuscript, and Edward J. Rykiel Jr. and Stuart McKay provided insights into scale issues relevant to my work. This is Contribution No. 50 of the Marine Mammal Research Program, Texas A&M University.

15

SCALING AND INTEGRATION IN TREES

T. M. Hinckley, D. G. Sprugel, J. R. Brooks, K. J. Brown,
T. A. Martin, D. A. Roberts, W. Schaap, and D. Wang

One of the most difficult obstacles in predicting the consequences of climate change for vegetation is the fact that the questions are generally posed at a regional or a continental level, whereas the information available to answer them is often based on studies at the individual organ or organelle. One would like to be able to easily collect data for large blocks of land—ideally by remote sensing—and use it to make predictions about how the vegetation in those areas will respond to increased temperatures, decreased precipitation, carbon dioxide (CO_2), etc. (i.e., a top-down approach). Top-down approaches have the appeal of parsimony but the disadvantage of uncertainty as to whether the critical controlling factor has been identified (or misidentified as a result of statistical artifacts), and if the critical controlling factor has been identified, whether this factor will be important in new environments or situations. The alternative to a top-down approach is a bottom-up approach, which involves making detailed measurements on small units (e.g., leaves or shoots) and then somehow scaling up to larger units [e.g., ecosystems or landscapes (Lertzman and Fall, chapter 16)]. Bottom-up approaches have the elegance of starting with first principles and clear mechanistic responses, but they may rapidly turn into sampling and scaling nightmares as one extrapolates, for example, chloroplast responses to the ecosystem level.

Variation in processes exists at all levels of spatial and temporal resolution and must be considered when conducting scaling research. In some cases, we may choose to ignore this variation. For instance, stomatal conductance has been shown to vary considerably within individual leaves (Beyschlag et al. 1994), yet most investigators choose to work under the assumption that a porometric measurement of stomatal conductance taken from a portion of the leaf or a population of needles represents the average conductance of that leaf or group of needles. This average may provide an excellent predictor of some larger scale response, such as foliar water loss, but it may fail when plant or environmental circumstances not covered during data collection are encountered. Documenting the spatial and temporal heterogeneity or variation associated with the mean may give biological and, therefore, experimental insight. Understanding the nature of (and mechanisms associated with) variation is critical for interpreting physiological phenomena, such as the influence of patchy stomatal behavior on A versus c_i curves (net photosynthesis versus internal carbon dioxide concentration). Therefore, efforts in scaling should include both a consideration of the mean as well as the variance around the mean.

For almost a decade, we have been conducting studies in which scaling or integration have been either a central focus or an important component. In the early 1980s, our work was largely oriented to problems best addressed using observations taken at smaller scales of biological organization. We occasionally extrapolated (Teskey et al. 1984) or linked these leaf level data to larger scales (Hinckley et al. 1984). Later, there was increasing emphasis on using leaf-level information to explain observations at stand and ecosystem levels. The first example presented here discusses the frustrating experience of not being able to make such a clear link in our studies on the influence of the eruption of Mt. St. Helens (Cascade Range, Washington State) on forest lands northeast of the volcano. Then studies are presented in which scaling was included in the research; these are in chronological order, beginning with efforts to scale up measurements made at the leaf level and ending with our more recent studies involving remote sensing. Practical examples of both top-down and bottom-up approaches to scaling are used, because the bottom-up approach allows a linkage to the vast body of observations that have been made by plant physiologists, and the top-down approach illustrates how closure to our efforts to scale or integrate might be achieved.

Background: Eruption of Mt. St. Helens

The May 1980 eruption of Mt. St. Helens deposited airfall tephra over a large area of forested land northeast of the volcano. Forest stands representing a large range of ages and species compositions were directly and indirectly affected by this disturbance (Zobel and Antos 1982; Seymour et al. 1983; Hinckley et al. 1984). In the decade after the eruption, we conducted a series of investigations at the needle, tree, stand, and regional scales. This section outlines the results we found at various levels of organization and describes how our observations could have been better focused by choosing appropriate scales of observation for additional study. In retrospect, including studies at intermediate levels of biological organization (branch scale) would have greatly improved our ability to make connections between the initial impacts observed in the needle- and shoot-scale studies and the large-scale, heterogeneous patterns of tree decline observed later at the stand and regional scales.

Immediately after the eruption, we initiated studies on the impact of airfall tephra on foliage and trees. By combining field observations with experimentation and laboratory analyses, we were able to isolate the cause of needle mortality, abscission, and morbidity on trees covered with airfall tephra (Seymour et al. 1983; Hinckley et al. 1984; Schulte et al. 1985). Our observations appeared to be valid for all of the major conifer species located northeast of Mt. St. Helens. From these observations, we were able to speculate what impacts might exist at the tree and stand levels, and measurements of individual-tree and whole-stand growth losses initially confirmed our predictions. Linking these disparate scales assured us that we had effectively documented the relationship between airfall tephra, foliar interception, and subsequent decreases in tree and stand growth in 1980.

Two very different patterns of tree responses emerged in subsequent growing seasons. For mature trees of *Abies amabilis* (Pacific silver fir), no clear picture emerged until late 1985 and 1986, when increased decline (i.e., loss of crown vigor) and mortality were observed. In contrast, complete recovery of height and diameter growth of sapling-sized trees was evident during the second growing season after the eruption (Hinckley, unpublished data). We were able to document dramatic, compensatory increases in new foliage production in sapling-sized trees, and the magnitude of the increases was related to the increase in total tree growth (height, diameter, branch length). Again, we appeared to have a strong link between foliage level responses and responses of the whole tree.

Explaining the delayed response of mature *A. amabilis* was difficult. Studies implemented in 1987 addressed the following questions: what features of the area northeast of Mt. St. Helens, what features of a stand, and what attributes of a tree were most responsible for the observed decline and mortality of *A. amabilis* (Segura et al. 1994, 1995a,b)? Although a number of regional and stand features were found to be important [such as depth of the fine ash layer (A_3), elevation of the stand, and percentage of stand basal area in species other than *A. amabilis*], we were unable to identify features at the tree level or within the canopy that could explain the wide range of stand and individual tree responses. A detailed examination of the tree-ring-width chronologies, when studied either as a stand level composite or at the individual tree level, provided insight about the timing and pattern of decline, but it did not help us understand the mechanisms linking foliage function to loss of growth and tree vigor.

Although the absence of either observational or experimental data from lower scales of biological organization hindered our ability to interpret data on the decline of mature *A. amabilis,* we explored another conceptual approach. If we assume that the initial impact of tephra occurred at the needle level, then we can ask the following question: How did this impact propagate its way to the branch and whole-tree level? Observations at the branch level might have offered an appropriate intermediate scale in making the jump from the foliar to the whole-tree level (figure 15.1). Using this intermediate scale approach, a tree would be considered a population of branches, and these branches would respond in a minimum of three ways.

The response of the branch would be a function of its history in terms of (1) the amount of tephra it received and retained, (2) the position of the tree in the canopy, and (3) the condition of the tree itself (i.e., the presence of root or stem diseases and insects). The model BRANCH would be used to keep track of each branch's response as well as the contribution of the branch population to the whole tree (Ford et al. 1990), and of the whole tree to the stand (Sorrenson-Cothern et al. 1993). To utilize this model, we would have needed details about the physiology and morphology of branches from mature trees, and experimental data on the carbon economy of tephra-covered versus uncovered branches. Such an approach would have provided the necessary scaling linkage to allow us to connect the large amount of foliar-level information to observed individual tree responses. Although the opportunity is now lost to exploit these intermediate scales, our experiences, difficulties, and assessments have guided subsequent attempts to link small-scale physiological function with larger-scale phenomena.

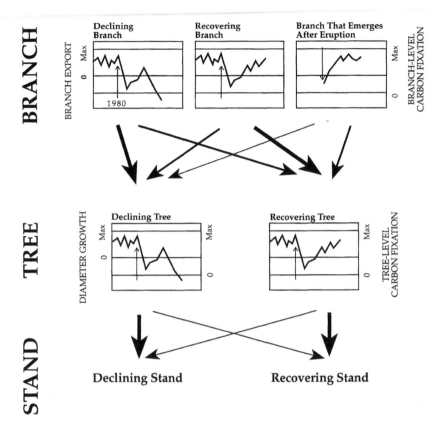

FIGURE 15.1

Simple diagram illustrating how branches, trees, and stands might be linked. Three broadly different types of branches might be found on a tree (many more options exist) (*top panel*). The first two branches are branches that were on the tree prior to the eruption; one recovers and the other does not. The last branch (*farthest to the right*) emerged the growing season after the eruption. Carbon fixation (*right side*) and export abilities (*left side*) are illustrated for each of these branch types. The combination of branches determines tree behavior and other characteristics.

Scaling Examples

Foliage Respiration

Since the advent of the infrared gas analyzer and field studies of carbon exchange in the early 1950s (Pisek and Winkler 1958), physiological ecologists have been measuring photosynthesis and respiration of foliage in the field in diverse species and under a wide range of environmental conditions. These studies have led to a detailed knowledge of how foliage functions under different environments, but they do not explain how these processes integrate to the whole tree or to the stand. In our efforts at scaling these processes, we first measured dark respiration at the leaf level, and then estimated responses at higher levels of organization—branches, trees, and stand. We used a bottom-up approach in which we tried to understand the mechanisms for spatial and temporal variation in the process, and then we used this information to estimate the process at higher levels of organization. This approach was very labor intensive, yet it was simple and straightforward in that each step was clearly linked to the next.

Brooks et al. (1991) documented how foliage respiration varies within *A. amabilis* canopies. As expected, foliage respiration increased with leaf temperature, but the nature of this relationship changed with the age of the foliage and position in the canopy. Young, newly developing foliage had the highest rates of growth and maintenance respiration. Older cohorts of foliage lacked growth respiration but varied considerably in the rates of maintenance respiration. Variation in respiration with foliage age was similar to variation in photosynthetic rates (Teskey et al. 1984); 1-year-old foliage had the highest rates of both respiration and photosynthesis, and these rates were observed to decrease with age. For a given age of foliage, the highest rates of respiration were observed in the top of the canopy, and rates decreased with lower positions within the canopy. Photosynthesis followed a similar pattern, as rates decreased with light attenuation (Brooks et al. 1996). It appeared that foliage maintenance-respiration rates were closely coupled to the photosynthetic capacity of the foliage. Photosynthetic capacity decreased with both foliage age and light environment (Brooks et al. 1996). The spatial distribution of foliage respiration can be explained by knowing the spatial distribution of light and foliage age within the canopy.

To estimate respiration rates for branches, trees, and ultimately the stand, we needed to know the amount of foliage in each age class and its relative position within the canopy. Foliage position in the canopy provided us with an approximation of its light environment. We returned to the branch as the unit for extrapolation in order to obtain biomass information. We harvested 150 branches from 14 trees covering a large range of branch ages and canopy positions. For each branch, we separated the foliage into five age classes (current, 1-year-old, 2- to 4-year-old, and 5-year-old and older) and obtained the dry weight. We developed biomass equations using branch diameter, length, and position within the canopy. From these equations, we predicted the total foliage biomass for a branch and the percentage of biomass in each age class. Using respiration–temperature equations from Brooks et al. (1991) and our biomass estimates, we estimated branch respiration for branches throughout the canopy. At a given temperature, upper canopy branches with the same foliage biomass respired 6 times the rate of branches in the understory (Brooks 1987). This respiration difference was the result of an interaction between age and light. Branches in the lower canopy were not only at lower light levels, but they contained mostly older foliage compared with branches at the top of the canopy. Because upper canopy branches were often at higher temperatures, their respiration was more than 7 times that of understory branches with similar biomass (Brooks 1987, Brooks et al. 1991).

We calculated whole-tree respiration from the assemblage of branches in each of 14 sampled trees, ranging from dominant to suppressed crown classes. Foliage biomass and respiration were calculated for each branch on a tree, and summed to represent whole-tree foliage respiration. Dominant trees had greater respiration rates per gram of foliage because they had more young foliage higher in the canopy than other trees (figure 15.2). Dominant trees also had more foliage biomass, thus respiring at more than 10 times the rate of median trees within the stand.

For scaling to the stand, two approaches were used: a categorical blocking approach and a nonlinear modeling approach. In the blocking approach, we divided all the trees in the stand into 14 categories based on their diameters. Each category was represented by one of the 14 sample trees in which whole-tree respiration was estimated. The respiration rate of a category was estimated by calculating respiration per unit basal area for the representative trees, then multiplying that by the basal area of the category. Summing the categories together gave an estimate of stand respiration. In the nonlinear modeling approach, we developed a model for whole-tree foliage biomass and respiration based on tree diameter and the length of the live crown (Brooks et al. 1991). Using these equations, we calculated respiration and foliage biomass for all the trees in the stand and summed them for an estimate of stand respiration. The two techniques gave very similar estimates (4.88 versus 4.56 mmol m^{-2} s^{-1} at 15°C), which adds confidence in the approach. Scaling the estimates temporally required temperature information. We saw little change in respiration rates over the season that was not explained by temperature (Brooks et al. 1991), thus daily and monthly temperatures were used to estimate stand respiration over a day (0.5 mol m^{-2} d^{-1}) and a growing season (40 mol m^{-2} y^{-1}, or 480 gC m^{-2} y^{-1} in grams of carbon).

Scaling from measurements made on the shoot in μmol m^{-2} s^{-1} (leaf area) to mol m^{-2} y^{-1} (ground area) required detailed spatial information on biomass and respiration. It also required temporal information on the respiration response to temperature and temperature variation over the season. Although we have these data for the stand of *A. amabilis,* none of this information can be generalized for other ecosystems, and thus the approach will work only if such spatial and tem-

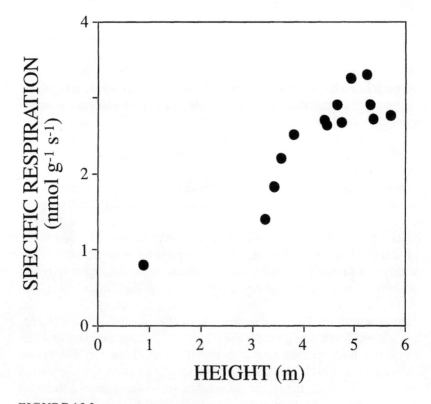

FIGURE 15.2

Relationship between specific respiration of the foliage, and tree height in young (37-year-old) *Abies amabilis* trees.

From data cited in Brooks et al. (1991).

poral information exist. Without any top-down measurements such as eddy covariance fluxes, we have no closure to our system and are unsure how much error was propagated up in the scaling exercise. Despite these limitations, our estimates are similar to other estimates of conifer foliage respiration. Ryan (1991) used a top-down model to estimate foliage respiration for this stand and estimated 460 gC m^{-2} y^{-1}. Cropper and Gholz (1990) estimated foliage respiration for a *Pinus elliottii* (slash pine) stand at 308 gC m^{-2} y^{-1}. Yokota et al. (1994) estimated 556 gC m^{-2} y^{-1} for *Chamaecyparis obtusa* (hinoki cypress) stands in Japan. Both of these estimates are of the same magnitude as ours, which also lends confidence to the estimates. The bottom-up scaling approach that we used also allowed us to examine the influences of branch position within the canopy and of crown class on foliage respiration. These components cannot be examined using a top-down approach; ideally, the two approaches should be used together. This effort to scale shoot-level information to the stand, where the branch was used as an intermediate unit of integration, suggests a critical need for a more rigorous examination of branch function.

Branch Autonomy

Branch function and branch autonomy have been topics of interest to physiologists and silviculturists for several decades (Sprugel et al. 1991). Currently, there is interest in branches as experimental units for research on pollutant exposure (Houpis et al. 1991; Teskey et al. 1991; Barton et al. 1993; Dufrene et al. 1993), global climate change (Liu and Teskey 1995; Teskey 1995), and water relations (Meinzer et al. 1993; Hinckley et al. 1994). The use of branches as an experimental unit is neither a top-down nor a bottom-up approach; rather, it constitutes a "middle-out" approach using an experimental unit that is intermediate in size between a single needle or shoot and an entire tree or stand. Using branches as the study unit offers experimental advantages that must be weighed against some potentially significant limitations. Compared with whole trees or stands, branches can be more easily treated with pollutants or other stresses, sample sizes can be larger, statistically powerful paired-branch designs are possible (i.e., a treated branch can be compared with a genetically identical control branch on the same tree), and physiological and morphological changes can be more easily monitored to advance our understanding of tree response to abiotic and biotic factors (Brooks et al. 1996). Compared with individual needles or shoots, branches are a more integrated unit, and one can look at cumulative effects on wood production as well as specific effects on gas exchange or other processes. However, results from these studies can be extrapolated to whole trees only if the response of a study branch to a stress is not affected by the level of stress to which the rest of the tree is exposed (Sprugel et al. 1991). For example, if the results from branch exposure systems are to be scaled to higher levels of organization with confidence, the response of a pollutant-exposed branch on a "clean" tree must be the same as the response of a pollutant-exposed branch on a pollutant-stressed tree.

In 1988, we initiated an integrated research program to address branch autonomy, or branch functional independence (Schaap 1992). The study used controlled ozone exposures to *Pseudotsuga menziesii* (Douglas-fir) foliage to learn (1) how reductions in photosynthate production, as induced by a foliar level ozone stress, would be translated into branch and tree level responses, and (2) whether the condition of the whole tree would affect the response of a branch to a treatment different from that received by the whole tree. Branches and trees were independently exposed to ozone (figure 15.3), and physiological function and growth were examined. The study used six 9-year old, field-grown *P. menziesii* enclosed in three-tiered open-top chambers. From August 1989 to October 1990, half the trees were exposed to charcoal-filtered air and the other half to 200 to 300 parts per billion on a volume basis (ppbv) ozone. Simultaneously, individual branches in the upper and lower crown on these trees were enclosed in branch chambers of our own design and exposed to either charcoal-filtered air or 200 to 300 ppbv ozone, so that all combinations of branch and tree exposures were studied.

Photosynthesis, respiration, chlorophyll fluorescence, stomatal conductance, transpiration, needle and shoot growth, needle chlorophyll, and nutrient status were measured during both growing seasons. After the first 6 weeks of ozone treatment (approximately 30 ppmv · hr ozone exposure), reductions in the chlorophyll fluorescence parameter, F_v/F_m, and net photosynthesis were noted within the second half of the daily exposure period. Complete photosynthetic recovery was observed by the following day. After 3 months of ozone treatment (approximately 50 ppmv·hr of ozone exposure), there was incomplete recovery. In addition, clear impacts on photochemistry, carboxylation efficiency (based on A versus c_i curves), and stomatal conductance were observed. A residual effect on net photosynthesis, resulting from ozone exposure, was observed at the beginning of the second exposure season (1990). Even more pronounced decreases in net photosynthesis and stomatal conductance were noted the second year. The ozone treatment clearly impacted the photosynthetic productivity of the foliage; the impact was significant over a wide range of foliar age classes and canopy positions, and it was severe to the point that accelerated foliar senescence and abscission were observed. Was this a sufficient stress to result in reductions in branch and whole-tree growth?

After the end of the second growing season, all the study trees were harvested. At the time of harvest, trees were 12 years old and they averaged 5.1 m in height and 6.9 cm in diameter (at breast height, 1.4 m). All of the upper treated branches were 5 years old at the time of harvest, and lower treated branches were between 7 and 9 years old. Treated branches, the main stem, and six to ten major roots were collected and sectioned, and internode length and annual ring width were measured. For upper branches, there was no significant difference between treatments for either main axes or total annual elongation increments. Thus, primary (extension) growth was not affected by this exposure to ozone. There were no significant differences in either ring width or area growth in 1988 and 1989, but ring width in 1990 was significantly less in upper branches treated with ozone rather than charcoal-filtered air. Secondary growth was impacted, but only slightly and only in the second year of treatment. In contrast, lower branches exposed

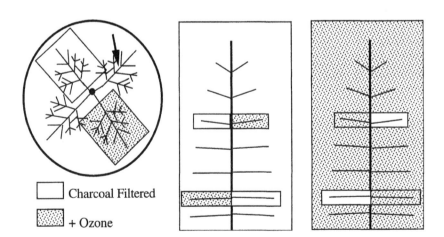

FIGURE 15.3

Study design showing top view of a *Pseudotsuga menziesii* tree in an open-top chamber (*upper left*) and side views of a charcoal-filtered (*middle*) and ozone-fumigated tree (*right*).

to ozone had less elongation of main axes in 1990, had significantly less total shoot elongation (210 versus 551 cm), and had large reductions in ring width and ring area. For lower branches, both primary and secondary growth was affected. For the entire tree, average height increment of ozone-treated trees was less than trees treated with charcoal-filtered air in 1990, although the difference was not statistically significant. Ozone significantly reduced ring width in 1990 at all stem positions from the top to the bottom of the tree. For the main stem, secondary growth appeared most sensitive to ozone treatment. Finally, ozone significantly reduced secondary growth of primary roots. Ozone caused a foliar-level stress that was differentially propagated within the tree. Growth processes of lower priority, such as root and cambial growth, were affected more than primary growth. Primary growth in the upper crown was not affected, yet there was some evidence that primary growth in the lower crown was significantly impacted.

Was the response of a branch to treatment affected by the relative condition of the tree to which it was attached? In other words, do stressed branches on a healthy tree respond in a different manner from that of stressed branches on a stressed tree (and vice versa)? Many different processes and parameters were measured during the study, and each responded differently to ozone treatment at both the branch and the tree level. However, the following patterns were observed:

1. *Branch effects are the most important.* Every parameter or process measured in this study (photosynthesis rates, A versus c_i relationships, water relations, nitrogen content, and branch growth) was controlled primarily by the atmosphere to which the branch was exposed. Thus, at least at the first level of analysis, branches are a good experimental unit for studies of pollutant effects on trees.

2. *Tree effects are common but not universal.* In a number of cases, (e.g., photosynthesis, A versus c_i relationships, growth rates, and nitrogen concentrations) branches exposed to a given environment on ozone-treated trees performed better (or had higher nutrient concentrations) than branches exposed to the same environment on charcoal-filtered trees.

3. *Significant branch–tree interactions are rare.* In only a few isolated cases did ozone treatment affect branches on ozone-treated trees differently from the way it affected branches on charcoal-filtered trees.

Because ozone generally had negative effects on tree growth, why did branches on ozone-treated trees have consistently higher photosynthetic and growth rates than branches in the same atmosphere on charcoal-filtered trees? The answer may lie in the contrast between the conditions experienced by the treated branch and those experienced by the rest of the tree. A charcoal-filtered branch on an ozone-treated tree is in a better environment than the rest of the branches on that tree, which apparently allows it to have to disproportionately high photosynthesis, growth, and nitrogen concentrations compared with a charcoal-filtered branch on a charcoal-filtered tree. Conversely, an ozone-treated branch on a charcoal-filtered tree is in a poorer environment than the other branches on the tree, which apparently leads to disproportionately poor performance compared with ozone-treated branches on ozone-treated trees.

There are several mechanisms that might have produced these results. Differences in access to root-derived resources may be part of the answer. Ozone reduced stomatal conductance [either directly or indirectly (Schaap 1992)] in a major part of the crown of ozone-treated trees. Therefore, a branch receiving charcoal-filtered air on an ozone-exposed tree would have access to a greater pool of root-derived resources compared with other branches on the same tree, whereas an ozone-treated branch on a charcoal-filtered tree would have reduced access to root-derived resources. Alternatively, source–sink relationships might be involved. The strength of assimilate demand (sink strength) in crop plants and in tree canopies can affect the rate of photosynthesis in source leaves (Sprugel et al. 1991). Treating trees with ozone presumably reduces source strength (i.e., net photosynthesis throughout the tree) without altering sink strength, much like a reduction in effective leaf area. Perhaps the charcoal-filtered branches on ozone-treated trees respond to this reduction in source strength from surrounding branches with increased photosynthesis, whereas ozone-treated branches on charcoal-filtered trees respond to the comparatively high surrounding source strength with a down-regulation of photosynthesis. Regardless of the mechanism, the result can be summarized by the phrase, "The rich get richer and the poor get poorer."

For scaling purposes, however, the absence of branch × tree interactions is probably the most important result. The absence of significant branch × tree interactions means that whereas the *absolute* photosynthesis rate of ozone-treated current foliage on ozone-treated trees was greater than the rate for ozone-treated foliage on charcoal-filtered trees, the *reduction* caused by ozone treatment (in which experimentalists are usually interested) was the same regardless of the tree treatment. Thus, the results of this study tend to support the use of branch chambers to study pollution effects, at least for short- to medium-term studies of the parameters measured here.

Although branch responses are not absolutely independent of the atmosphere to which the tree is exposed, the results of this study indicate that, for the most part, branches may serve as a sampling tool by which physiological, morphological, and growth measures may be scaled to a whole-tree level. By remembering possible shortcomings (i.e., position, duration of exposure, and long-term feedbacks between the parent tree and study branches), and by placing one's findings in a source–sink context, studies at the branch scale may be very useful in understanding the response of large trees to stress.

Shoot, Branch, and Tree Water Loss

Many efforts have been made to integrate measurements of stomatal conductance at the leaf scale upward to estimate transpiration at the whole-tree crown or canopy scales. A common approach to this problem is to treat the canopy as a single "big leaf" with the average physiological properties of the canopy. The Penman-Monteith combination equation (Monteith 1965) is the most common expression of the relationship among environmental variables, the physiological properties, and transpiration of the big leaf. The Penman-Monteith equation for a leaf with stomata only on the lower surface can be written as

$$E_{crown} = \frac{sR_n + \rho_a C_p D g_a}{\lambda \left[s + \gamma \left(1 + \dfrac{g_a}{g_{crown}} \right) \right]} \qquad (1)$$

where E_{crown} is transpiration in kg m^{-2} s^{-1}. The environmental drivers for transpiration, usually measured at a reference point above the stand, are represented by D (vapor pressure deficit, Pa and R_n (net radiation, W m^{-2}). The plant regulators of transpiration, crown conductance and boundary layer conductance, are represented by g_{crown}(m s^{-1}) and g_a(m s^{-1}), respectively. The constants and temperature-dependent variables are s (the slope of the saturation vapor pressure versus temperature curve at air temperature, Pa K^{-1}), ρ_a (the density of dry air, kg m^{-3}), C_p (the specific heat capacity of air, J kg^{-1} K^{-1}), λ (the latent heat of evaporation of water, J kg^{-1}), and γ (the psychrometer constant, Pa K^{-1}).

The biggest challenge in using equation 1 for big-leaf integration is defining g_{crown}, the crown conductance. Usually, g_{crown} is estimated by averaging many porometric measurements of stomatal conductance taken within the tree crown. However, factors such as current and historical leaf microenvironment, leaf age, leaf damage, and random chance conspire to make stomatal conductance an exceedingly variable property. As a result, even with an appropriately stratified sampling scheme, it may require more than 100 measurements of stomatal conductance to adequately characterize canopy conductance for one moment in time (Leverenz et al. 1982). In addition, recent discussions have highlighted the difficulties of determining an appropriate weighting scheme to obtain values of g_{crown} that are consistent across scales or for scalars other than water vapor (Raupach 1991; Lhomme 1992; McNaughton 1994; Raupach 1995).

Shoot-to-Crown Integration   Our attempts to integrate shoot-scale stomatal conductance measurements upward to crown-scale estimates of whole-tree transpiration provide an example of the problems encountered with porometric estimation of g_{crown}. We measured stomatal conductance of needles at six different positions within an *A. amabilis* canopy, using a null-balance diffusion porometer. At each position, we sampled four age classes of needles (current year, and 1, 3, and 6 years old). By continuously measuring stomatal conductance throughout the day, we were able to make four to five cycles through a complete set of 24 sample points. The stomatal conductance measurements were divided into sets of current and older-than-current leaf measurements, and each set was averaged for each cycle. These measurements were then scaled to the level of the tree crown (crown conductance) by weighting the averages by the amount of current and older-than-current leaf area on that tree. Because we did not take stomatal conductance measurements constantly from sunrise until dark, we had to make assumptions about patterns of g_{crown} for the periods before and after the measurements. Crown conductance was assumed to increase linearly from sunrise until the first cycle average; it followed the average throughout our sampling period and then decreased linearly from the final-cycle average until sunset (figure 15.4a).

The other plant control on transpiration, g_a, or boundary layer conductance, was estimated from a relationship between wind speed and boundary layer conductance measured on branch models in a wind tunnel (T. A. Martin, unpublished data). These conductances, along with net radiation and vapor pressure deficit measured from a weather station on a tower above the stand and averaged over 15-minute intervals, were substituted in equation 1 to estimate whole-tree transpiration (figure 15.4b). Although the shoot-to-crown integration matched the pattern of actual whole-tree water loss measured by stem heat-balance sap-flow gauges (with an allowance for time lags), the magnitudes of the rates were considerably different. On the 5 days we sampled stomatal conductance, estimates of crown transpiration exceeded independent measurements of tree water loss by a factor of 2 or more. There was clearly a problem with our integration.

There are two possible explanations for these problems. First, it is possible that the estimate of leaf area for the tree was incorrect. If we overestimated tree leaf area, or if we incorrectly estimated the proportion of current and older leaves within the canopy, the estimate of whole-tree transpiration would have been inaccurate. This seems unlikely, considering that projected leaf-area index (LAI) for the stand estimated from our measurements (9.4) was reasonable. In addition, the Beer's law extinction coefficient calculated from our vertical leaf-area distribution and the stand light-extinction profile was 0.47, which agrees with other values for coniferous forest in the literature (Jarvis et al. 1976). Second, we may have overestimated g_{crown} with the porometer measurements. This could have happened if the sample weighted undamaged, well-lit, or younger leaves (all more likely to have high sto-

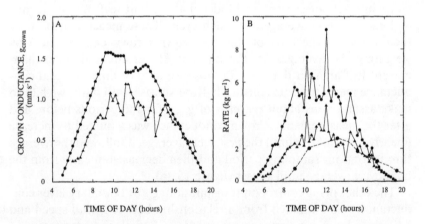

FIGURE 15.4

(A) Crown conductance for an *Abies amabilis* tree on August 19, 1993, calculated from shoot (●) and branch (▲) integration schemes as described in the text. (B) Whole-tree water loss on August 19, 1993, measured with stem heat balance gauges located at breast height (1.37 m) on the tree stem (*thin dashed line*), and estimated by shoot-to-crown (●) and branch-to-crown (▲) Penman-Monteith big leaf integrations. Note the time lag between the big leaf integrations and stem sap flow, presumably a result of capacitance.

matal conductance) more heavily than they were represented within the crown, resulting in erroneously high values for g_{crown}.

Branch-to-Crown Integration   A more successful approach to the estimation of g_{crown} and whole-tree transpiration involved the use of branch sap-flow gauges. These gauges, using a heat-balance technology similar to that utilized to measure whole-tree sap flow, enabled us to measure the water loss (over 15-minute intervals) of branches at six different positions in the canopy. The branches had leaf areas ranging from 0.5 to 2.0 m². Five of them were located in positions within the canopy where they were exposed to light for most of the day, and the sixth was located deep within the crown where it was shaded throughout the day. The branch sap-flow gauges were used as an alternative to porometers for sampling conductance on a subcrown scale. Because transpiration for each branch was a known quantity, we could solve equation 1 to determine stomatal conductance at the branch scale (g_{branch}, m s^{-1}):

$$\frac{1}{g_{branch}} = \left[\frac{s}{\gamma} \left(\frac{R_n - \lambda E_{branch}}{\lambda E_{branch}} \right) - 1 \right] \frac{1}{g_a} + \frac{\rho_a}{E_{branch}} \frac{C_p}{\lambda \gamma} D \qquad (2)$$

where all variables are defined as in equation 1, except that the terms are evaluated at the branch rather than the crown scale. For the sunlit branches, the values of R_n, D, and wind speed (for the determination of g_a) were assumed to be equivalent to those measured at the weather station. For the shaded branches, R_n was calculated from the weather-station values, modified to reflect the radiation attenuation with depth in the canopy measured in this stand. The attenuation of wind speed with depth into the canopy was estimated from within-canopy wind-speed measurements taken in a younger stand of *A. amabilis* with a similar LAI (T. A. Martin, unpublished data). We could then define, for each 15-minute period, a shaded and a sunlit branch conductance using equation 2. We translated these conductances to the crown scale using the proportions of sunlit and shaded leaf area in the crown. To determine the amount of sunlit leaf area in the crown, we calculated the solar equivalent leaf area (Cermák 1989) for the tree. Briefly, solar equivalent leaf area (LA_{solar}, m²) was calculated as

$$LA_{solar} = \sum_{i=1}^{n} (LA_{layer} R_{layer})_i \qquad (3)$$

where LA_{layer} is leaf area (m²), and R_{layer} is radiation (fraction of above-canopy radiation) at each of *n* layers within the crown. Crown conductance (g_{crown}) was then calculated as a weighted average of the sun and shade branch conductances (g_{sun} and g_{shaded}):

$$g_{crown} = \left\{ g_{sun}\left(\frac{LA_{solar}}{LA_{total}}\right) + g_{shaded}\left(1 - \frac{LA_{solar}}{LA_{total}}\right)\right\} \qquad (4)$$

where LA_{total} is the total leaf area (m²) for the tree. It should be noted that although McNaughton (1994) and others have warned against arithmetic averaging of small-scale conductances, it has been shown in modeled canopies that the differences between arithmetic averaging and other weighting schemes is actually very small, especially in canopies with higher LAI (Raupach 1995).

Figure 15.4 shows a typical diurnal pattern of g_{crown} and E_{crown} calculated by the branch-integration method. There are several important differences between the results of the shoot- and branch-integration schemes. First, the branch-integration scheme produced smaller values for crown conductance. Second, the branch-integration scheme has much finer time resolution with a greater time coverage than does the shoot-integration scheme. Finally, and most important, the branch-integration values for tree transpiration match independent sap-flow measurements both in magnitude and in pattern (allowing for capacitance) much more closely than do the integrated shoot results.

These results demonstrate several advantages of the branch-integration approach over our shoot integration for estimation of g_{crown} and tree transpiration. First, the branch sap-flow gauges overcame the temporal limitations inherent with porometric sampling. Many porometric measurements must be taken to characterize g_{crown} for any point in time, limiting the resolution of changes in g_{crown} throughout the day (note the plateaus in the shoot-integrated g_{crown} in fig. 15.4a). In contrast, branch sap-flow measurements are recorded automatically, so conductance can be monitored throughout the day on a relatively fine time-scale resolution. Another advantage of the branch-integration approach is that conductance can easily be estimated for days when there are no research personnel on site. This clearly cannot be accomplished with porometer sampling. We were able to calculate whole-tree transpiration for 27 different days throughout the summer using branch integration (figure 15.5), compared to the 5 days we made porometer measurements. The branch integration approach also overcame sampling shortcomings inherent to the shoot-integration method. Sap-flow

gauges monitored over 6.0 m² of leaf area, whereas all of the porometric measurements together sampled only about 0.048 m². As a result,
the branch sap-flow measurements were much more likely to monitor
a representative sample of leaves than was the porometric sampling,
which, although stratified by age and light environment, was too small
to fully cover the range of variation within the crown.

This study demonstrated the difficulty of characterizing variation
on a smaller scale and then successfully translating that variation to a
larger scale. Although porometer sample size was small, it encapsu-
lated the temporal and spatial limitations inherent in this type of sampling. Because of the limits of human movement, only a small number

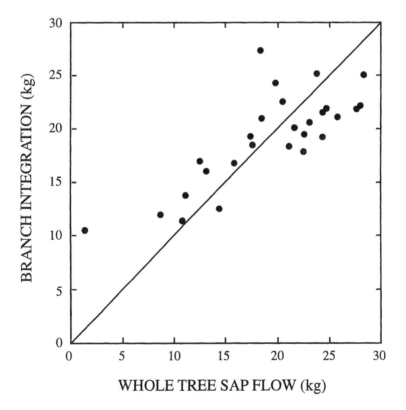

FIGURE 15.5

Comparison of whole-tree water loss as estimated by branch-to-crown Penman-
Monteith big leaf integration and measured by stem heat balance sap flow gauges.

of porometer samples can be taken in 1 day. However, because of the multitude of microenvironments, leaf ages, and conditions, variation of stomatal conductance within a tree crown is large. In addition, actual measurements of stomatal conductance can be scaled only to the crown level for days on which measurements are taken. We addressed these issues by using branch sap-flow gauges as large porometers. This allowed us (1) to sample much more frequently, which overcame the temporal limitations of porometry, and (2) to sample a much larger amount of leaf area, which helped us overcome the sample-size and stratification problems.

An alternative approach to the temporal sampling problems is to model stomatal conductance as a function of microclimatic variables, then simulate variations in conductance using weather station data for model forcing. This approach is fairly common (Dye and Olbrich 1993; J. Roberts et al. 1993b; Granier and Loustau 1994) and often success- ful, but it still requires large numbers of porometer measurements if stomatal conductance at the shoot scale is modeled. There is also the danger of extrapolating beyond the "data space": encountering envi- ronmental or physiological situations that are not accounted for in the model-building process.

Finally, this exercise emphasizes the importance of having indepen- dent measurements of the variable of interest at a larger scale against which to compare the integration from the smaller scale. Although this is not always easily accomplished, the increasing frequency and reliability of ecosystem-level gas exchange measurements (Diawara et al. 1991; Kelliher et al. 1992; Baldocchi 1994; Hollinger et al. 1994; Grace et al. 1995) and organism-level water use measurements (Cer- mák et al. 1982; Hinckley et al. 1994; Cienciala and Lindroth 1995; Granier et al. 1996; Martin et al. 1997) are very helpful in providing independent checks on modeling and integration efforts. As Jarvis (1995) pointed out, "Because upscaling is the only way of analyzing ecosystem function and predicting global environmental change at the larger spatial and longer temporal scales, it is important that we 'get it right.' However, since the process is essentially open-ended, there is a major danger that we may 'get it wrong.' Thus it is essential that we steadily improve the models that we use and constrain them by measurements of activity by the system at each traverse of scale.

Remote Sensing

A common problem in physiological ecology involves scaling up physiological measurements that may be labor intensive, spatially limited, and extremely variable in time and space. The utility of remote sensing is that the technology accurately measures reflected energy from large areas of the earth's surface, successfully making integrated measurements across large spatial scales (Vande Castle, chapter 13). The challenge is to interpret the causality between spectral data and the nature of the surface vegetation from which the energy is reflected. This contrasts with the bottom-up physiological approach, in which many point measurements are directly made of vegetation, but many errors can become propagated when attempts are made to extrapolate the data spatially (figure 15.6).

Remotely sensed data can also be used to provide top-down information about vegetation over large spatial scales, or as a check on scaled-up landscape estimates of variables such as stand-level nitrogen and radiation intercepted by a forest canopy. Vegetation indices, such as the normalized difference vegetation index (NDVI), a ratio of near-infrared-to-red reflectance, have been most commonly used to estimate vegetation cover, LAI, and the amount of light intercepted by vegetation (Tucker 1979; Sellers 1985, 1987; Sellers et al. 1992). The physical structure of forests may be inferred by applying textural analyses to remotely sensed data, such as variances, variograms, and semivariograms (Woodcock and Strahler 1987; Woodcock et al. 1988a,b; Cohen et al. 1990b; Dutilleul, chapter 18).

An example of a bottom-up approach to scaling remotely sensed data is the utilization of leaf-level spectra to interpret and classify hyperspectral images of landscapes. This type of analysis, called spectral mixture analysis (SMA), explicitly addresses the issue of scale. SMA utilizes a library of reference spectra of the materials that make up an image, such as needles, leaves, wood, and soils. SMA is used to model a canopy as mixtures of green leaves, nonphotosynthetic vegetation (NPV: branches, stems, etc.), and architecturally derived shade and shadows, each weighted by the aerial proportion of the component within the field of view (Adams et al. 1993; D. Roberts et al. 1993a; Ustin et al. 1993).

In this section, we illustrate some important scaling concepts with examples from an investigation using the Airborne Visible/Infrared Imaging Spectrometer (AVIRIS). AVIRIS is valuable because this sensor provides a continuous 224-band spectrum comparable to some of the best laboratory instruments, at a spatial resolution (17.4 m) that cannot be obtained in the laboratory or by using field instruments. Specific objectives of our study were to evaluate how reflected light, measured on single leaves, scales upwards to the branch and whole stand, and to evaluate the role of architecture in modifying that signal. A related objective was to use spectral and textural information derived from the image data to estimate canopy roughness, and to compare the image data between architecturally distinct forests.

Examples are derived from three sites, two in central Washington (Cedar River Watershed) and one in southeastern Washington (Wal-

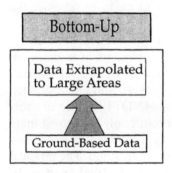   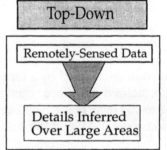

CONSIDERATIONS:

Analytical Methods
  • Kriging
  • Regression
  • Conditional Simulation

Sampling Issues
  • Adequate Design
  • Experimental Error
  • Sampling Error
  • Identify Appropriate
    Indices for Scaling up

CONSIDERATIONS:

Instrumental Interpretation

Identifying the Causality of the Signal:
  • Atmospheric Correction
  • Topographic Effects on Data
  • Pixel Mixing
  • Spectral Similarities Between
    Different Land Cover Types
  • Classification Uncertainties

FIGURE 15.6

Two approaches, bottom-up and top-down, to the following question: What are the alternatives and trade-offs for providing large-scale measurements of ecologically relevant variables (such as leaf nitrogen or LAI)?

lula). The Cedar River sites are comprised of stands of old-growth and second-growth *A. amabilis,* and the Wallula site consists of a plantation of several architecturally distinct clones of *Populus* (cottonwood). These include a *P. trichocarpa* × *P. deltoides* (TXD) hybrid that has relatively horizontally oriented leaves, and a *P. deltoides* × *P. nigra* (DXNa) hybrid with more vertically oriented leaves. The AVIRIS data used in this study were collected on September 22, 1994. At the time of the overflight, branch and whole-tree sap-flow rates and microclimatological data were being measured at the old-growth *A. amabilis* site from a 44-meter canopy-access tower. Architecture at the subpixel scale was assessed using SMA to model canopies as real mixtures of green leaves, NPV, exposed soil, and architecturally derived shade and shadows. Larger-scale canopy roughness was assessed by calculating interpixel variance for shade and green leaves for each stand.

The high stand heterogeneity and dead wood content in old-growth forest are readily illustrated by textural measures and spectral fractions (figure 15.7). Old-growth *A. amabilis* has higher interpixel variance than second growth (figure 15.7A), and higher subpixel shadowing and higher wood content as expressed by NPV (figure 15.7B). A comparison between *Populus* and *A. amabilis* also illustrates how coniferous and broad-leaved deciduous vegetation differ. At the subpixel scale, conifers are associated with higher shadowing (figure 15.7B) than *Populus;* this concurs with Williams (1991) who compared spectra from several conifers with that from *Acer saccharum* (sugar maple) at leaf, branch, and canopy scales. However, in terms of texture, second growth *A. amabilis* shows no significant difference from 3-year-old stands of clonal *Populus* (figure 15.7A). Both could be characterized as even-aged with relatively uniform canopy heights and few gaps.

The role of leaf-angle distribution in modifying canopy reflectance is illustrated by comparing mixture models for the TXD (horizontal leaf display) and DXNa (vertical leaf display) clones (figure 15.7B). At this scale (17 m) the more horizontal leaves of TXD produce a significantly higher green leaf fraction and lower shade fraction than DXNa. A comparison of AVIRIS spectra illustrates how architecture influences the entire spectrum (figure 15.8). Canopy reflectance is high in *Populus* relative to *A. amabilis* at all wavelengths. Within the *Populus* clones, TXD has higher reflectance than DXNa, whereas old-growth *A. amabilis* is significantly darker than second growth.

The manner in which light reflecting off individual leaves in a crown is modified by canopy-scale light-scattering processes can be illustrated

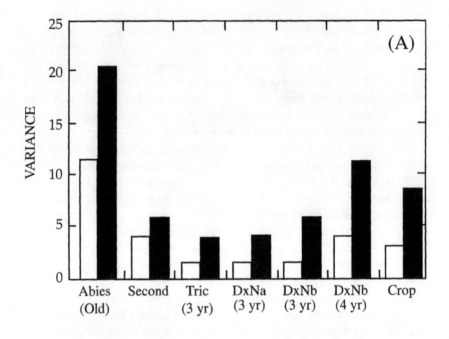

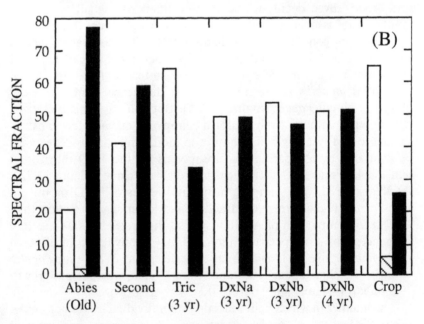

by tracing reflected light upwards from leaf to AVIRIS scales. This exercise illustrates the extent to which measurements made at one scale are modified by processes operating at another scale (Williams 1991). Furthermore, this exercise illustrates how new technologies, such as imaging spectrometry, provide information that cannot be obtained using less advanced remote sensing technologies. Spectra at leaf and branch scales in the clonal *Populus* stands were collected in September 1995, when phenology was comparable to the 1994 AVIRIS data. Leaves were harvested from high, medium, and low positions in the crown for TXD and DXNa, and they were measured in the laboratory. Branch-scale spectra were collected at heights of 3 meters or more above each stand, providing a field of view of at least 1 meter.

A comparison of spectra derived at the leaf scale illustrates how canopy-scale processes modify spectra (figure 15.9). At the leaf scale, reflectance was not significantly different between the two clones (figure 15.9A), whereas at larger scales the two clones are readily distinguished (figure 15.9B). Subpixel shadowing at branch and stand scales results in a significant decrease in canopy reflectance in the visible (400 to 700 nm) and short-wave infrared (1900 to 2500 nm) (figure 15.8), spectral regions where absorption processes dominate. In spectral regions where absorption is low, such as the photographic infrared (700 to 1000 nm), reflectance at the larger scales is comparable to laboratory measurements, and multiple scattering is the dominant process (Roberts et al. 1990). As a result, any absorption that is present (such as the 960- and 1200-nm liquid water absorptions) is accentuated. A comparison of leaf spectra to AVIRIS spectra (figure 15.8) illustrates this concept. At 960 nm, multiple scattering greatly accentuates the depth of the liquid water band. The greater strength of the 1200-nm water band produces a significant decrease in reflectance, greatly altering the shape of canopy spectra. Field measurements between 400 and 1000 nm illustrate that most of these processes are functioning at a 1-meter scale.

FIGURE 15.7

Interpixel and subpixel textural measures. (A) Variance for both the green vegetation (*left*) and shade fractions (*right*). (B) Subpixel fractions for green vegetation (*left*), nonphotosynthetic vegetation (*central*), and shade (*right*). Seven different vegetation types are interpreted: old-growth *Abies amabilis*, second-growth *A. amabilis*, 3-year-old plantation of hybrid of *Populus trichocarpa* and *P. deltoides*, three different 3- to 4-year-old plantations of hybrids of *P. deltoides* and *P. nigra* (Na and Nb refer to two different clones), and agricultural crops.

Stand heterogeneity, as expressed by a gradient from smooth canopies (*Populus*) to rough canopies (old-growth *A. amabilis*) is clearly expressed as an increase in interpixel variance and an increase in shade content. Architectural differences between second-growth and old-growth *A. amabilis,* such as an increase in exposed-wood content (e.g., snags) are also expressed as an increase in subpixel NPV fractions. A comparison of reflectance spectra, measured at scales ranging from individual leaves to AVIRIS pixels, illustrates how processes operating at different scales modify the signal measured using remote sensing (Williams 1991). At the leaf scale, a comparison between TXD and DXNa spectra shows few differences because of the occurrence of similar physical processes of internal light scattering and light absorption

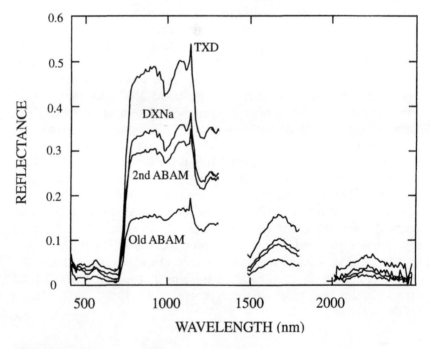

FIGURE 15.8

AVIRIS spectra for the *Populus* clones TXD (*P. trichocarpa* × *P. deltoides*) and DXNa (hybrid clone a of *P. deltoides* × *P. nigra*), and old-growth and second-growth *Abies amabilis* (ABAM).

FIGURE 15.9

Leaf **(A)** and branch **(B)** level spectra of *Populus* hybrid clones, TXD and DXNa.

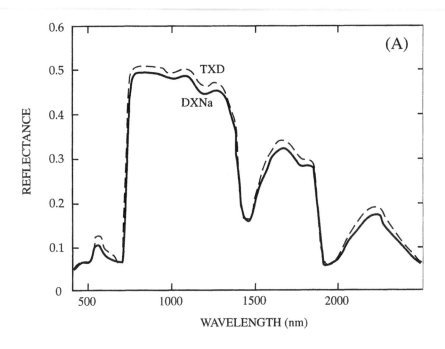

(A)

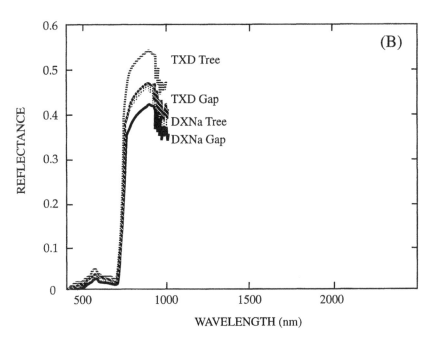

(B)

by liquid water. At the branch scale, multiple scattering of near-infrared (NIR) light and architectural differences between the two clones combine to increase the contrast between visible and NIR reflectance, accentuate absorption features, and better distinguish the high reflectance TXD (horizontal) from the darker DXNa clone (more vertical leaf display). At the stand scale, architecture and multiple scattering play an even greater role, resulting in a TXD spectrum that is similar to the 1-meter scale but a DXNa spectrum that shows an even greater darkening because of shadows.

Scale clearly influences remotely sensed data and how remote sensing can provide useful measures at different scales, but many challenges remain. For example, variation in canopy roughness is expressed at the subpixel and interpixel scale, yet it is unclear how this information can be used to directly estimate parameters of interest such as canopy coupling and roughness length. Multiple scattering of NIR at the canopy scale may provide greater leverage in estimating parameters such as LAI through greater expression of liquid water absorptions at the canopy scale (Roberts et al. 1997), but this relationship has not been rigorously tested.

Summary

Scaling and integration are conceptually and technically difficult, although there are a variety of tools and ideas that can be used to accomplish these links (Valentini et al. 1995). Our experience with different physiological measurements and methods suggests that there are several promising approaches for understanding scaling in trees.

It is essential that patterns of variation in data be considered when using either the top-down or the bottom-up scaling approach. The tendency, when faced with large-scale questions, to average responses and to fold the variation into the average may obscure important biological and evolutionary information. We must first understand the nature of this variation before we can be conceptually or experimentally comfortable with a top-down approach (Hunter and Price 1992; Wessman 1992; Dawson and Bliss 1993; Dawson and Ehleringer 1993).

Branches, larger and more integrated than a single leaf or shoot, but smaller and more experimentally manageable than an entire tree, offer promise as a practical spatial scale for study of mechanistic physiology. Although we have found that there are caveats for the use of branch studies, this unit of biological organization offers a tractable means of understanding water and carbon exchange in large trees.

In addition, remote sensing is clearly a promising tool. The opportunity to use large clonal blocks of trees of different architecture provides an important means of testing the power and the information associated with this technique. Remote sensing demonstrates how we can use a large-scale technique or tool to pose small-scale questions. As these questions are answered (or not), we then address the larger scale again. This is an iterative, interactive approach to scaling (Root and Schneider 1995). Remote sensing also provides a larger-scale, independent measurement that may serve as a check or constraint on integrations from lower scales.

Acknowledgments

We thank R. Ceulemans, J. Cermák, S. Herold, J. Kucera, R. Meinzer, M. Obersteiner, M. Redlin, J. Rombold, and G. Segura for technical and intellectual input into much of the material found in this chapter. The research reported herein has been supported by the U.S. Department of Ecology (Biofuels, WESTGEC), the U.S. Environmental Protection Agency, The National Science Foundation, the U.S. Department of Agriculture, and the U.S. Forest Service.

16

FROM FOREST STANDS TO LANDSCAPES: SPATIAL SCALES AND THE ROLES OF DISTURBANCES

Ken Lertzman and Joseph Fall

Forests are characterized by patterns and processes distributed over a broad range of scales. We can find explanations for ecological patterns in forests at spatial scales varying from centimeters (Hartgerink and Bazzaz 1984; McConnaughay and Bazzaz 1987; Perry et al. 1987) to many thousands of hectares (Pastor and Broschart 1990; Hall et al. 1991a; Spies et al. 1994). Similarly, meaningful ecological relationships can be described that operate at time scales from the seasonal phenology of litterfall and seedling population dynamics, through changes in the geographic distribution of species in relation to long-term fluctuations in climate and disturbance regime (Davis 1986; Hopkins et al. 1993). Whereas the roles of fine-scale processes creating structure within forest stands have been the focus of a substantial research effort, larger-scale patterns and processes are poorly understood. Further, the degree to which conclusions about ecological processes at one scale may be generalized across other scales remains largely a subject of speculation. Applied research and management prescriptions have also focused at smaller scales: larger-scale patterns of land-use and management have generally been the unplanned consequences of smaller-scale decisions and actions. This chapter discusses the role of disturbances, "natural" and anthropogenic, at scales from forest stands to landscapes.

Forest Stands and Landscapes

The terms *stand* and *landscape* imply to the forest ecologist or manager a particular range of scales. Stands are generally defined as areas of forest that are relatively homogeneous with respect to the variables of interest, usually disturbance history, species mix, and edaphic factors. In practice, stands typically range from a few hectares to a few hundred hectares in size, depending on the grain of the soil mosaic, the geomorphic history, and the kind of disturbance regime in the forest. Whereas stands are usually delimited on the basis of internal homogeneity, forest landscapes are necessarily heterogeneous: a forest landscape is a mosaic dominated by different stands. Minimally, forest landscapes occur at the smallest scale at which more than one stand can be delimited. These may be stands of the same type but of different age classes, or they may be a more heterogeneous mix of stands of varying ages and types. Stand age is generally defined as a function of the time since the last major disturbance, so the idea of disturbance regimes is fundamental to the definition of the observation units at the scales of forest stands and landscapes.

The presence of stands of different types is generally a consequence of spatial variation in soils, hydrology, and geomorphic history. Patches of nonforest vegetation (e.g., grassland) or nonvegetated surfaces (e.g., rock outcrops or water bodies) also make up varying proportions of forest landscape mosaics. The scale of a very fine grained forest landscape is potentially only a few hectares in size, but forest landscapes are generally conceived of and analyzed at scales ranging from 10^3 to 10^6 ha (e.g., Mladenoff et al. 1993, 1994; Spies et al. 1994).

When ecologists consider a forest landscape as a unit of analysis, they are usually concerned with processes linking different elements of the landscape (e.g., flows of water through a drainage network, movements of animals, or spread of disturbances) (Swanson et al. 1988). No formal maximal limit has been set on the scale of forest landscapes. However, the maximal spatial scale at which it may be useful to think about the ecology of forest landscapes will likely be determined by the size of area that can be linked by such ecological processes, by the scale at which forest-dominated mosaics grade into other vegetation types in a given region, and by gradients in climate and physiography at regional to subcontinental scales.

Most research in forest ecology and management has focused on processes at the stand scale. This has resulted in a reasonable understanding of the stand-level structure and dynamics of many forest types (Franklin and Spies 1991). Much of this research, however, has focused on particular subsets of the forest types within any given landscape. Characteristically, much more is known about the common types of forest patches than the uncommon, about productive stand types than unproductive ones, and about economically significant stand types than those with little economic value.

One reason for the failure of some past attempts to scale stand-level conclusions up to the landscape level is that the data on which those conclusions were based represent a biased selection of landscape elements (Lertzman et al. 1996). Considering larger scales of analysis in forests necessarily means living with substantially more variability among the objects being studied than traditional studies have generally allowed. It also means considering a broader range of forest types than traditional research and management objectives have led ecologists to examine.

Disturbance Regimes

All forests are subject to disturbances that kill or influence the growth of the trees and other organisms in them. Some forests are subject primarily to large, intense, stand-initiating disturbances that kill trees in big patches, creating a coarse-grained pattern of relatively young, even-aged forest patches. Many boreal forests follow this type of pattern (e.g., Mladenoff et al. 1993; DeLong and Tanner 1996). Other forests are subject to smaller or less intense disturbances that kill trees in patches of only a few trees, resulting in a finer-grained mosaic of older, often uneven-aged forest patches. The wetter types of temperate rainforests often follow this second pattern (Lertzman et al. 1996, 1997). The frequency and type of disturbances a forest experiences have many consequences for forest structure, the kinds of habitats the forest provides, and the suite of animals living in them (Bunnell 1995; Carey and Johnson 1995; Hansen et al. 1995; Huff 1995).

The natural disturbance regime of a forest can be defined as the long-term pattern of the frequency, intensity, spatial extent, and internal heterogeneity of disturbances (Pickett and White 1985). The regime of disturbances a forest has experienced is thus a phenomenon of larger scales of space and time. Because it is difficult to study such large-scale phenomena, conclusions about disturbance regimes are generally based on inferences about the history of disturbances at smaller scales. It is relatively straightforward, at least in concept, to describe the history of disturbances in an individual stand (Parker and

Pickett, chapter 8), either by dating the origin of a stand after a major disturbance or reconstructing the pattern of small-scale, stand-maintaining disturbances, such as low-intensity fires or other gap-forming disturbances (Kilgore and Taylor 1979; Morrison and Swanson 1990; Frelich and Reich 1995; Huff 1995).

Scaling up from these stand-level reconstructions to a landscape-level description can be problematic. For instance, different parts of a landscape are often subject to different types of disturbance regimes (Romme and Knight 1981; Hemstrom and Franklin 1982), so averages over larger areas are unlikely to be ecologically meaningful. The description of dynamics in even a relatively large subsection of a landscape may provide a poor estimate of the dynamics of the landscape as a whole (Baker 1989).

Most forests in the world currently experience disturbance regimes that result from the interaction between natural disturbance processes and human activities. In some cases, human disturbances overwhelm the ability of forest ecosystems to persist, and a land-use conversion occurs. In other cases, forest ecosystems persist, but with substantially different characteristics from those they would have had in the absence of human intervention. Modern industrial forestry is a major kind of disturbance to forest stands and landscapes in many parts of the world, and it generally results in forests very different from those produced by natural disturbance processes (Hansen et al. 1991; Swanson and Franklin 1992).

Understanding how natural disturbance regimes shape forest ecosystems has become a key element of new approaches to forest management. Many current problems in reconciling conservation and timber objectives in forest management relate directly to the differences between modern industrial management regimes and the natural disturbance regimes that dominated the forests before the arrival of Europeans. A variety of recent initiatives in forest policy in both the United States and Canada have emphasized natural disturbance processes and their structural consequences as models for management objectives (Forest Ecosystem Management Assessment Teams 1993; Swanson et al. 1993; Scientific Panel for Sustainable Forest Practices in Clayoquot Sound 1995; Lertzman et al. 1996, 1997). However, implementing this approach is often limited by our understanding of natural disturbance regimes, by our ability to make direct management recommendations from what we know, and by constraints imposed by the social context in which forest management operates.

In the remainder of this chapter, we discuss seven significant ideas that have emerged from the study of disturbances in forests at the scales of stands and landscapes. We highlight key problems and challenges for both the research and resource management communities.

Seven Lessons from Stands and Landscapes

1. Paying Attention to Scale Matters

> No single mechanism explains pattern on all scales.
>
> (Levin 1992)

Ecologists have long recognized that the scale of an investigation is an important determinant of the ecological patterns that are observed (Greig-Smith 1952; Rosenzweig 1995). Both the grain and extent of a study limit the range of scales, in space and time, at which processes responsible for creating ecological patterns may be detected and interpreted. In addition, methods of sampling and analysis impose filters that further scale observations of the system under study (Allen and Hoekstra 1991). A mismatch between the scale of an investigation and the system's scale of heterogeneity may lead to erroneous conclusions based on observed patterns that are simply artifacts of scale (Wiens 1989; Allen and Hoekstra 1991).

Despite its importance, scale has rarely been dealt with explicitly in traditional studies of forestry and forest ecology. These studies tend to focus on a single scale, often chosen arbitrarily based on methodological constraints and previous studies. Because conclusions drawn at one scale may not be valid at other scales, results from scientific inquiry of this sort are often limited in their application. For example, studies of small gap-forming disturbances in forests have typically focused on particular types and ages of stands at a fairly small scale of analysis, yet substantial variation in patterns of tree mortality occur over the range of sites and forest age classes in a landscape (Lertzman et al. 1996). Similarly, ecological studies conducted over short periods of time often lead to conclusions different from those found in similar studies of the same system over longer periods (Wiens 1977; Davis 1986).

The critical factors influencing the behavior and effects of a natural disturbance change with the scale of analysis. For example, at smaller scales, topographic position and fuel load may be the key determinants of fire initiation, intensity, and propagation (Romme and Knight 1981;

Romme and Despain 1989). However, over larger spatial and temporal scales, the frequency, intensity, and extent of fires may be explained better by longer-term trends in climate and weather (Clark 1988; Johnson and Larsen 1991; Swetnam 1993; Bessie and Johnson 1995). Similarly, the apparent nature of a disturbance regime may vary with the scale being considered. For instance, in forests with very long intervals between large stand-initiating disturbances, the disturbance regime appears to be dominated by small stand-maintaining disturbances.

Forest landscapes often reflect a complex spatial and temporal interaction between both large- and small-scale disturbances (Lorimer 1989; Spies and Franklin 1989). In the mosaic of forest age classes in Mt. Rainier National Park (Washington State), pockets of very old forest are juxtaposed with large patches of younger forest that burned repeatedly over the past millennium (Hemstrom and Franklin 1982). At the stand scale, individual trees of long-lived early seral species, such as *Pseudotsuga menziesii* (Douglas-fir) and *Picea sitchensis* (Sitka spruce), can influence the structure and composition of stands for more than 1000 years after a stand-initiating disturbance. Some very long lived species, such as *Thuja plicata* (western redcedar) and *Chamaecyparis nootkatensis* (yellow-cedar), can exhibit highly episodic recruitment in response to changing environmental conditions and patterns of disturbances (Daniels et al. 1995). This can lead to very complex stand dynamics as regeneration tracks current climatic conditions in a matrix of legacies from very different past conditions (Lertzman 1995). Resolving the relative importance of large- and small-scale processes requires a multiscaled understanding of the role of disturbance in forest ecosystems.

2. Multiscale Study and Analysis Is Critical for Understanding the Roles of Disturbances

Many ecological processes act on multiple scales or have consequences over a range of scales. Local processes can have larger-scale consequences, such as the downstream effects of a small landslide that deposits sediment in a stream. Some local processes have consequences that are observable only over larger scales of analysis. The cumulative effects of stand-level management practices are expressed over a landscape scale of analysis. For example, small patch clearcuts may individually remove a trivial amount of habitat for late successional species, but cumulatively, populations may be threatened in managed watersheds through conversion and fragmentation of the uncut forest.

Ecologists are accustomed to studying some processes at small scales and others at larger scales, but they rarely examine how a given process acts across a range of scales. A substantial amount of attention is currently being paid to examining larger scales in ecology —much of it under the rubric of "landscape ecology"—but forest landscapes can be defined at multiple scales and encompass processes acting at multiple scales. We suggest that multiscaled research is likely to provide more useful results than merely larger-scaled research.

Research conducted at one scale, or over a narrow range of scales, will not likely identify the extent to which its conclusions are scale dependent. Conclusions about parts of forest landscapes are often not applicable to the broader landscape (Baker 1989; Bergeron 1991; Lertzman et al. 1996), but it would be difficult to know this from having studied only the smaller areas. In the following discussions on top-down and bottom-up processes, we describe a variety of scale-dependent interactions that act together to create pattern in forest landscapes.

The methodological constraints of research at different scales provide both problems and opportunities for multiscale research. Research at smaller spatial scales is more amenable to field studies using traditional approaches to experimental design and manipulation. However, the time scales of response in forests often preclude these approaches for any but the fastest of processes. This has led to extensive use of chronosequence studies in a "space-for-time substitution" (e.g., Huff 1995). Research at larger spatial scales allows the use of newer technologies such as digital satellite data (Vande Castle, chapter 13), but the problems of interpretation and error analysis of such data are potentially critical (Hall et al. 1991a).

Perhaps more challenging is the variation in attention ecologists have paid to forest composition, forest structure, and ecological function at different scales. We can describe forest composition well at a range of scales, but as the size of the area increases, the classification becomes increasingly coarse. For instance, many classifications of forests over large areas are binary—forested/nonforested or old forest/ young forest (e.g., Spies et al. 1994). A substantial amount of research has focused on forest structure at both small and large scales, but the structural attributes examined are different at each scale. At small scales, structural descriptions focus on variables such as tree size and age, characteristics of coarse woody debris, gaps, and tree-spacing (see papers in Ruggiero et al. 1991). At larger scales, structural analyses generally focus on characteristics such as patch shape, size, and spatial pattern (Pastor and Broschart 1990; Mladenoff et al. 1993; Spies et al. 1994). Few studies have integrated analyses of these different aspects of structure, although some methodological approaches appear promising (e.g., Cohen et al. 1990b).

Ecological function in forests has been considered in a variety of ways at different scales. Some aspects of ecological function, such as production and biomass dynamics, have been studied over a wide range of scales, but generally not in the same study. A number of modeling efforts have recently attempted to build from smaller-scale data bases to very large scales (Kasischke et al. 1995; Kurz and Apps 1996). Many other aspects of ecological function, especially at large scales, remain largely speculative. It is common to hear references to "landscape function" in conversations, and modeling approaches have begun to build a framework for considering function in landscapes (Turner et al. 1993, 1994), but the disparate aspects of ecological function at large scales have not been integrated. This is an important management concern: the putative attributes of "functional landscapes" are often discussed by managers with little substantive input from formal analysis. A substantial body of research attempts to examine how landscape-level attributes influence the ways in which landscapes function as habitat for various animals, but there is little agreement among studies, and these studies are poorly integrated with studies of landscape dynamics.

3. Should We Expect Equilibrium at Any Scale?

Disturbance and succession work in concert across a landscape to create stands in a variety of successional states over time. A minimum dynamic area (MDA) is the smallest area that is large enough to always contain patches in every successional state over time (Pickett and Thompson 1978). The MDA indicates the scale of a natural disturbance regime, so it is an important theoretical concept. In principle, planning for MDAs should be a critical aspect of the design of nature reserves to ensure that they continue to provide the range of habitats with which species have evolved (Wright 1974; Sullivan and Shaffer 1975; Pickett and Thompson 1978). However, the search for an MDA will not be useful unless the landscape is at equilibrium at some scale. If the landscape exhibits instability at all scales, then it will not be possible to identify a single "natural" combination of scale and condition from which to derive management objectives.

At the scale of individual stands, forests are dynamic and rarely in equilibrium (Sprugel 1991; Turner et al. 1993). Although the time-scales of stand development and disturbance vary greatly among eco-systems, virtually all stands are subject to catastrophic disturbance at some time. However, some ecologists have suggested that over substan-tially larger areas there is a balance between the rate of disturbance and recovery, which maintains a mosaic of patches in an equilibrium pattern. Empirical evidence suggests that some landscapes may exhibit this kind of stable mosaic (Sprugel 1976), but that others do not (Romme 1982; Baker 1989).

The "shifting-mosaic steady-state" hypothesis (Bormann and Lik-ens 1994) is an example of how fluctuating stand-level dynamics may produce stable landscape-level attributes. The shifting-mosaic steady state requires that forests exhibit two properties, one at the stand scale, one at the landscape scale:

1. After a disturbance, individual stands should follow a more-or-less deterministic trajectory of development in the variables of interest (e.g., biomass, species composition, etc.).

2. The frequency and scale of disturbances must be relatively constant in time and not episodic, so that a stable age distribution of forest patches on the landscape is achieved (figure 16.1).

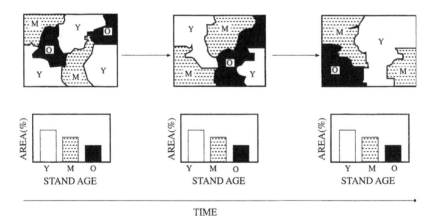

FIGURE 16.1

The "shifting-mosaic steady-state" forest maintains the same area in each age class (Y = young, M = mature, O = old) over time through shifting of patch types on the landscape. Such shifts are mediated by, in turn, disturbance and succession.

Both of these assumptions have problematic elements. There may be typical patterns of biomass accumulation in ecosystems, but the idea of deterministic successional pathways is unlikely to be generally applicable (Drury and Nisbet 1973; McCune and Allen 1985; Fastie 1995), and it cannot be easily assumed for any system. Even if successional pathways were deterministic on similar sites, forest landscapes exhibit variation in successional pathways among site types, nonrandomness in the spatial distribution of sites of different types, and differences in recovery time on sites of varying productivity. On real, heterogeneous landscapes, forest patches cannot freely "shift," making the first assumption difficult to meet. The second assumption requires constant environmental conditions and disturbance agents, a condition rarely borne out by empirical test (Davis 1986; Sprugel 1991). Disturbance frequencies vary in time (Clark 1988; Swetnam 1993) and are often episodic (Franklin et al. 1985; Romme and Despain 1989; Boose et al. 1994).

Studies that have attempted to determine the MDA for real forest landscapes have generally been unsuccessful. For instance, Baker (1989) attempted to determine the minimal area for perpetuating the "historic" landscape-age class structure in a fire-prone forest landscape. Baker used a combination of empirical data (Heinselman 1973) and simulation models to reconstruct forest-age distributions over space and time in the 404,000-ha Boundary Waters Canoe Area (Minnesota). He compared the reconstructed age distributions at a series of nested spatial scales with the steady-state distributions expected from his models. Baker found no evidence of a steady-state landscape at any scale. He concluded that "significant spatial variation in the fire-patch regime and/or environment" meant that the "shifting-mosaic simply did not shift into 29.5% of the study area." He suggested that his study area was "a mosaic of different non-steady-state mosaics."

The shifting-mosaic steady-state model provides a useful heuristic for thinking about how the attributes of a forest landscape can persist through time by shifting location in space. This model is a useful point of departure for considering the ecologically significant reasons why forest ecosystems exhibit non-steady-state dynamics (Turner et al. 1993). A multiscaled view of forest dynamics leads us to a new ecological paradigm that focuses on the pervasiveness of change rather than on a search for constancy (Swetnam 1993).

4. Small-Scale Processes Interact to Create Bottom-Up Controls of Landscape Pattern and Processes

The effects of local conditions of soils and microclimate on vegetation are ubiquitous and represent one of the better-studied aspects of plant ecology. The ways in which these local conditions cumulatively influence broader-scale patterns are more poorly understood.

The few studies that have assessed the relative roles of local and broader-scale processes in influencing landscape-level patterns or dynamics in forests have found a complex interaction between controls exerted by small-scale, local processes and larger-scale regional processes (figure 16.2). For example, Pastor and Broschart (1990) found that local edaphic factors were largely responsible for the distribution of the most common stand type in a northern hemlock–hardwood landscape. *Tsuga canadensis* (eastern hemlock) stands were the most common and most continuous stand type and formed the background matrix of the landscape. Edaphic factors also explained the pattern of a less common mixed hemlock–hardwood stand type associated with

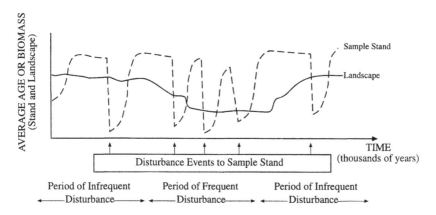

FIGURE 16.2
Hypothetical trajectory for forest dynamics on a non-steady-state shifting-mosaic landscape. At the stand scale (*dotted line*), fluctuations in biomass or age class are caused by disturbance and succession. At the landscape scale (*solid line*), fluctuations in biomass arise from the cumulative stand level dynamics. Conversely, regional changes in climate exert top-down influence in driving the disturbance regimes experienced by individual stands, as well as the rate and pathways of succession.

better-drained, sandier soils. In contrast, a coarser pattern of nearly pure hardwood patches resulted from disturbances to this matrix.

Large and small disturbances play different roles in forest dynamics and are important during different periods of stand development and succession. Larger disturbances generally synchronize ecological patterns and processes across scales, resulting in larger, more homogeneous patches of forest dominated by early seral species. Over time, fine-scaled processes create heterogeneity within this coarser pattern and often act to shift species dominance to later seral species (Glitzenstein et al. 1986; Lertzman 1992).

Frelich and Reich (1995) described the dynamics of spatial pattern in southern boreal forest in Minnesota at scales from 0.01 to 16 ha. They proposed a suite of related changes in species composition, age–class structure, and spatial pattern with time after a catastrophic fire. In this conceptual model, initial postfire communities are coarsely patterned and even-aged, with a relatively uniform canopy dominated by early-successional species. Periodic windstorms are the mechanism for shifts to a finer-scale spatial pattern and to dominance by later seral species. These windstorms create small canopy openings in the even-aged forest (10 to 30 m across on average), which break up the coarse-grained mosaic and enhance regeneration and recruitment of late-seral species. Eventually the mosaic has a mean patch area of 35 m^2, with the largest patch measuring about 0.1 ha. The timing of such shifts is thus determined by the temporal distribution of windstorms of various magnitudes. With a recent reduction in fire frequency, the mosaic as a whole is shifting from dominance by young, even-aged, coarse-grained *Pinus banksiana* (jack pine) stands to older, uneven-aged, fine-grained, mixed-species stands.

Some types of forests experience large-scale disturbances so infrequently that the patterns that appear are primarily the consequence of accumulated small-scale disturbance and recovery processes. In the wet, temperate rainforests of the outer coast of British Columbia (Canada), undeveloped watersheds are characteristically a fine-grained matrix of old, uneven-aged forest with small openings. The intervals between large-scale fires may be more than 1000 years or more (Gavin et al. 1996) and these uneven-aged forests are shaped largely by small gap-forming disturbances (Lertzman et al. 1996). Openings caused by the mortality of two to ten trees are common and make up about 15 percent of forest area. A further 15 percent of forest area is in small openings associated with edaphic factors (e.g., rock outcrops, stream courses).

Although many forest ecosystems are characterized by the influence of fine-scale processes, in others the influence of larger scale disturbances is so overwhelming that small-scale processes have received relatively little attention. Until recently, research in the Douglas-fir-dominated forests of the American Pacific Northwest focused on recovery from major stand-initiating disturbances, primarily large fires (Spies and Franklin 1989). Relatively little attention was paid to the dynamics within late-successional forests until these forests became the preeminent management concern of the past decades. More recent work suggests that the historical dynamics of these forests resulted from a complex interaction between disturbances of varying size and intensity (Spies and Franklin 1989; Morrison and Swanson 1990).

5. Larger-Scale Processes Exert Top-Down Control, Forming a Context for Local- and Stand-Level Dynamics

Over a range of temporal scales, forest dynamics are punctuated by disturbance events that appear extreme against the background of smaller-scale or less intense disturbances. For instance, fires in Yellowstone National Park (Wyoming) from the early 1700s to 1988 were generally small, and most fires between 1972 and 1988 went out without intervention after burning less than a hectare (Romme and Despain 1989; Turner et al. 1994a). Such a fire regime could contribute to a steady-state mosaic in an area the size of Yellowstone. The Yellowstone landscape mosaic, however, was not in steady state during this period. The 1988 fires, and those around 1700, burned a substantial proportion of the park area, setting up a disequilibrium in species mix and age structure capable of persisting for hundreds of years. Not only was more area burned, but fire behavior during the most intense days of the fire was different from that during periods when less was burned (Turner et al. 1994a). Large, catastrophic fires are generally less constrained by local fuel conditions, fire breaks (natural or anthropogenic), and topographic barriers (Johnson and Larsen 1991; Turner and Romme 1994), and they are often more heterogeneous (Eberhard and Woodard 1987; Turner et al. 1994b; DeLong and Tanner 1996) than smaller fires. Infrequent, extreme events can play a significant role in determining ecosystem dynamics over long periods of time, even in fire-dominated forests.

A number of authors have argued that fluctuations in regional- to subcontinental-scale climate is the key variable in determining the possibility of such extreme events (Johnson and Larsen 1991; Swetnam

1993; Bessie and Johnson 1995). They propose that large-scale physical geography (e.g., orientation of mountain ranges in relation to continental patterns of air circulation) combines with warming trends at the decadal-to-centennial time scale to create weather that, given an ignition source, will result in widespread regional fires irrespective of local conditions. It is well established that fire frequency is sensitive to climatic fluctuations and that such fluctuations can play a significant role in driving forest dynamics over long periods of time (Davis 1986; Cwynar 1987; Clark 1988; Campbell and Andrews 1990).

The topography of a forest landscape can act at an intermediate scale to isolate patches of forest from major disturbances, resulting in spatial subdivision of the disturbance regime. This can be true even in cases where large-scale weather patterns appear to drive fire behavior. For instance, subdrainages oriented perpendicular to prevailing winds may not burn even during extreme fires (Johnson and Larsen 1991). Bergeron (1991) found that boreal forest on islands in a lake experienced different disturbance regimes and exhibited a different age-class mosaic structure than similar forest on nearby lakeshores. The isolation of the islands affected both the frequency and types of fires experienced by the forest. We still have a relatively poor understanding of the ways in which fire weather at large scales, topographic structure of the landscape at intermediate scales, and local forest conditions interact to determine forest dynamics.

6. Human Activities Change Disturbance Regimes in Important Ways

Humans have impacts on the disturbance regime of a forest by attempting to suppress some agents of disturbance (e.g., fire and insect outbreaks), and by replacing them with a different set of disturbances (e.g., logging and grazing). We have become very effective at this, and the resulting anthropogenic disturbance regimes differ from the historical "natural" disturbance regimes in a variety of ways at various spatial and temporal scales (Swanson et al. 1993; DeLong and Tanner 1996; Lertzman et al. 1997) (figure 16.3). We can characterize these differences between disturbance regimes by comparing their size, intensity, frequency, and internal heterogeneity, and the variability of each of these factors in space and time. Logging is a ubiquitous anthropogenic disturbance in forests around the world. We use it here as an example of how alterations to the historical disturbance regime change and rescale patterns in forests.

Industrial disturbances and natural disturbances differ substantially in their variability. Many kinds of natural disturbances create patches of forest distributed over a broad range of sizes. For example, the effects of fire or insects may be limited to a single tree or may spread over thousands of hectares. Patch sizes in a given landscape may vary over more than an order of magnitude (DeLong and Tanner 1996). On the other hand, the spatial extent of areas cut in industrial logging tends to fall within a range limited by policy and engineering (e.g., Forest Practices Code of British Columbia, Government of British Columbia 1995). This often results in anthropogenic landscapes with a regular pattern and a uniform grain. The grain may be coarser than the prelogging landscape, as when large clearcuts are introduced into

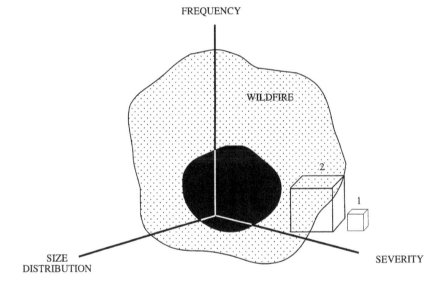

FIGURE 16.3

Hypothesized differences between a natural disturbance regime, a traditional management regime, and an example of a management regime based on the natural disturbance regime (from Swanson et al. 1993). The *large irregular cloud* represents a probability distribution of natural wildfires as interpreted from dendrochronology. *Box 1* represents a management regime of dispersed clearcuts with broadcast burning. *Box 2* represents the disturbance regime that results from the silvicultural disturbances of Box 1 interacting with ongoing natural disturbance processes and other factors. The *dark core* at the center represents the range of conditions that might be allowed under an "ecosystem management" scenario. Under that scenario, the range of historical variability is intended to bound the possible states created by management; some natural, but extreme, conditions are avoided for a variety of reasons, including the social context of management.

a forest dominated by small gap-forming disturbances, or it may be finer, as when small-patch clearcuts are introduced into a mosaic previously dominated by larger fires. This rescaling is exacerbated when the range of sizes of management units falls outside the range of natural variability in disturbances (Swanson et al. 1993; Lertzman et al. 1996, 1997).

Natural disturbances tend to be internally heterogeneous. For instance, wildfires often miss individual trees or clumps of trees within burned areas, resulting in regenerating stands with complex shapes and a substantial structural diversity. This postdisturbance structural diversity plays an important role in the ability of those stands to provide habitat for a variety of organisms (Carey and Johnson 1995; Hansen et al. 1995; Hutto 1995; Turner et al. 1997). Individual disturbances often vary in intensity (Morrison and Swanson 1990) (figure 16.4), and this internal variation in intensity varies among disturbance events.

Franklin (1990, 1993c) described as "legacies" those elements of the predisturbance forest that persist through a disturbance and contribute to the structural, functional, and compositional diversity of the developing stand. In figure 16.5, structural diversity remains above the origin immediately after a disturbance because of the presence of legacies. At the landscape level, entire patches that are skipped by disturbances can be considered legacies. Landscape-level legacies create complex patterns of more- and less-disturbed forest in the postdisturbance landscape (figures 16.4 and 16.5A). Standard even-aged systems of forest management, such as clearcut logging, exhibit a relatively high and uniform disturbance intensity: an area of prescribed size and generally simple shape is cleared with uniform 100 percent mortality of canopy and subcanopy trees. In many areas of western North America, large-scale, even-aged management has been applied uniformly across the landscape, resulting in both stands and landscapes with low internal heterogeneity. Under standard approaches to timber-oriented management, the maturing stand is harvested before it has recovered to its predisturbance state, and, unless it is specifically planned, subsequent rotations will not benefit from the structural legacies potentially left from the original stand (figure 16.5) or landscape (Hansen et al. 1991). A substantial literature has developed that proposes a variety of solutions to these problems at scales ranging from individual habitat elements within stands to large landscapes (Swanson and Franklin 1992; McComb et al. 1993; Franklin 1993b,c).

Natural disturbances vary in time as well as in space. The time between disturbance events can change substantially over periods of de-

■ High-Mortality Patch
☐ Medium-Mortality Patch
▥ Low-Mortality Patch

▲
N

Scale: 1 km

FIGURE 16.4

Reconstruction of fire mortality in patches of forest burned between 1800 and 1900 at a site in the central western Cascades of Oregon. High mortality was considered 70 to 100 percent, moderate mortality 30 to 70 percent, and low severity, <30 percent mortality. In this location, more area burned with moderate to low mortality (31 and 42 percent of the area, respectively) than with high mortality (27 percent).

Reprinted from Morrison and Swanson (1990).

cades to centuries (Swetnam 1993; Cwynar 1987; Clark 1988). In contrast, industrial logging is intended to operate on a strict schedule of fixed-length rotations designed to produce a sustained flow of timber output. In fact, a forest landscape managed on this model is intended to approximate a shifting-mosaic steady state. Over long periods, the silvicultural disturbance regime reduces the number of patch types on the landscape by reducing the duration of early successional stages and truncating stand development before the old-growth stage is reached

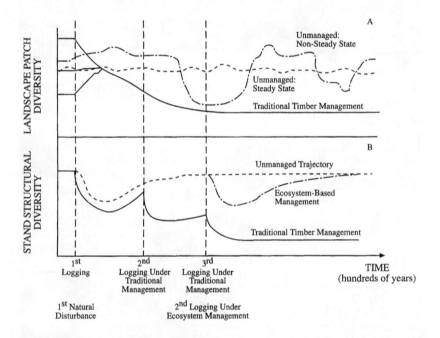

FIGURE 16.5

Hypothetical trajectory for structural diversity under different disturbance regimes. The immediate impact of forest management on patch diversity at the landscape level will vary with initial conditions. Timber management simplifies landscapes by increasing the disturbance frequency and truncating the patch age distribution (A). Timber management is generally intended to create and maintain a steady-state forest (a shifting mosaic, in fact), but with lower patch diversity and less temporal variability than with natural disturbance regimes. Under any management regime, logging has immediate impacts at the stand level (B), but the magnitude of these impacts varies, depending on management objectives. In this example, ecosystem-based approaches to timber management maintain conditions more like those in the natural disturbance regime by lengthening the rotation and making a less intrusive entry during harvest.

(Franklin et al. 1986; Hansen et al. 1991; Swanson and Franklin 1992). In many forests, this results in both a shift downward in the mean patch age and a reduction in the variance about that mean.

Management regimes intended to "protect" forest landscapes from natural disturbances have also resulted in changes that rescale the pattern of the forest mosaic. Intensive fire suppression, carried out over the latter half of this century in many parks (and elsewhere), results in a more connected landscape free of patches of younger forest. Such landscapes are more susceptible to large, uncontrollable, catastrophic disturbances (Romme and Despain 1989; Romme and Knight 1981; Turner et al. 1989). This is especially true if susceptibility to disturbance increases with forest age. Similar kinds of effects occur over a broad range of forest types in the interior of British Columbia and the inland western United States, where large areas of *Pinus contorta* (lodgepole pine) forest are aging and becoming susceptible to *Dendroctonus ponderosae* (mountain pine beetle). These forests would have experienced relatively frequent fires in the past, which probably maintained a mosaic less likely to propagate insect outbreaks and with less forest in a susceptible state at any given time.

In parks and other protected areas, many of these problems are related to conflicting management goals and the difficulties associated with defining and managing for "naturalness" in non-steady-state systems (Romme and Despain 1989; Sprugel 1991; Turner et al. 1993). An ecosystem-based management regime, whether for timber or parks, requires substantial site- to subregional-scale information on forest structure and disturbance history, and on landscape dynamics (Franklin 1993c; Swanson et al. 1993; Lertzman et al. 1997). Unfortunately, such information is frequently lacking in sufficient detail to provide clear guidance for the management of specific landscapes.

7. Tools for Multiscaled, Spatial Analysis are Available but Are Challenging to Integrate and Use Effectively

There are many new tools available to ecologists for doing multiscaled, spatially explicit analysis over the range of scales addressed in this chapter (figure 16.6). Global positioning systems (GPS) and geographic information systems (GIS) provide tools for collecting, storing, and manipulating spatial data. Remotely sensed data are available at a variety of resolutions and can be used to complement more traditional field-based data (Vande Castle, chapter 13). Spatial statistics and indices of landscape pattern (Bradshaw, chapter 11; Gardner, chapter 2;

Dutilleul, chapter 18) offer methods for analyzing spatial data and determining the scale of patterns. Finally, the availability of high-speed computers in conjunction with large storage media makes spatially explicit modeling more feasible. If used together, these tools allow for cross-scale integration, but they have generally not been used to their full potential.

Remote sensing (satellite images and air photos) has traditionally been applied to forests as a stand classification tool. This has produced quite detailed maps of forest cover, with varying degrees of (in)accuracy. Better methods and models for error representation and propaga-

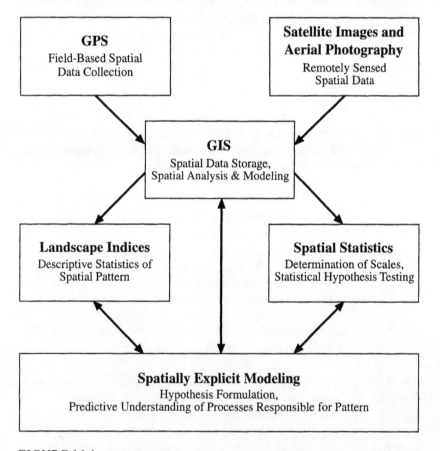

FIGURE 16.6

Many new tools are available to ecologists for doing multiscaled, spatially explicit analysis. Used in combination, these tools provide powerful methods for elucidating cross-scale relationships.

tion are required to make these data more useful for testing ecological hypotheses. More recently, ecologists have used satellite images to produce interesting results at larger scales (e.g., Hall et al. 1991a; Spies et al. 1994). In addition, temporal sequences of remotely sensed images are becoming long enough to provide useful descriptions of landscape change (Spies et al. 1994). Further exploration of these techniques and integration with field-based, finer-scaled studies holds the promise of revealing cross-scale relationships (Breininger et al. 1995; Knick and Rotenberry 1995; Robinson et al. 1995b).

Geographic information systems have been used predominantly as spatial databases for storing large, spatially referenced datasets and making maps from them (Krumell et al. 1987). Many GIS come equipped with sophisticated modeling tools, and these are becoming more accessible to ecologists. However, for GIS to become a useful tool in ecology, data manipulation and transfer, and evaluation of error propagation need to be easier, more explicit, and more commonplace. Also, most GIS are not well integrated with statistical packages for performing the kinds of pattern analysis required by landscape ecology. If GIS is to fulfill its potential for ecologists, it must be taken beyond its use as a database and mapping tool to become an effective and efficient environment for spatial modeling and analysis (Berry 1993; Sample 1994).

Global positioning systems use signals from a network of orbiting satellites to allow a hand-held receiver to locate itself accurately on the surface of the earth. The actual accuracy obtained varies with the price and quality of the hand-held unit, local vegetation and topography, and timing of observation. In practice, 2- to 4-m accuracy is a reasonable expectation in forested environments. GPS is important in this set of tools because it allows field-based studies to be georeferenced and linked with relative ease to digital, remotely sensed data via GIS.

A profusion of indices for describing spatial pattern have been developed or borrowed from other fields to describe the patterns of landscapes (Bradshaw, chapter 11; Dutilleul, chapter 18; Gardner, chapter 2; Schneider, chapter 12). These landscape indices measure the geometry and composition of individual patches, and the configuration of patches in the landscape (O'Neill et al. 1988; Turner 1989; Rogers 1993; McGarigal and Marks 1994). Although a few popular indices are becoming more widely applied, most of them have poorly understood ecological significance (Riitters et al. 1995). Perhaps for this reason, most applications have focused on relatively simple nonspatial indices such as core area, perimeter-to-area ratio, and percentage area in focal patch type. However, many indices are quite scale dependent and behave counterintuitively in response to known inputs (Rogers 1993). For these approaches to become more useful tools, a robust and parsimonious set of indicators must be identified; these indicators must have known behavior and incorporate scale and spatial relationships.

Spatial statistics and geostatistics have their historical roots in other fields. However, these techniques can be applied to ecological problems with some utility (Legendre and Fortin 1989; Rossi et al. 1992). The semivariogram, spatial correlogram, and variance staircase incorporate space and scale as an explicit, independent variable. These methods generally involve plotting some measure of ecological variability at different scales of observation, and they may elucidate important scaled relationships. The Mantel and Partial Mantel tests (Legendre and Fortin 1989) provide a method of doing spatial cross-correlations, or a spatial regression. Some of these methods are amenable to formal statistical inference about spatial patterns and spatial cross-correlations (Leduc et al. 1992) but are typically applied more for exploratory data analysis (e.g., Cohen et al. 1990b). These tools, along with others [e.g. fractal analysis (Milne et al. 1996)], can allow ecologists to identify the scale of a pattern or process, and they will be important elements of any toolkit for cross-scale analysis.

Some new simulation tools integrate GIS databases with the analytical models more familiar to ecologists. These tools allow ecologists to perform complex, spatially explicit simulations of landscape dynamics (e.g., Hargrove 1994; Fall and Fall 1996; Finney 1996; McKenzie et al. 1996). Models of this sort will be useful for developing hypotheses and characterizing the spatial patterns that emerge from different disturbance regimes. We should now be able to combine these techniques to integrate data and models across scales of analysis (see figure 16.6). Unfortunately, many of the tools themselves are not integrated or easy to use. The assumptions for applying some methods are not well understood, and confidence estimates are often difficult to obtain. Actual integration of multiscaled data and models remains challenging.

Where To from Here? Archetypes of Landscape Type and Behavior

In the ecological literature, forest landscapes have been characterized by a limited range of types of mosaic structure and a few kinds of behaviors in response to disturbance regimes. For instance, the shifting-mosaic steady state discussed above represents one kind of landscape behavior expressed on a mosaic made up, in concept at least, solely of different age classes or successional stages (table 16.1). We can refer to each such combination of landscape type and behavior as an archetype.

Archetypes are important because they tend to define the range of models, implicit or explicit, we consider for the possible states of nature. As explicit models, archetypes can be useful tools against which to evaluate empirical data (Lertzman and Gass 1983). However, because archetypes influence our conception of the possible states of nature, adopting the wrong archetype can limit our ability to resolve real ecological dynamics. For instance, the assumption that, historically, forests were in carbon steady state at larger scales has led to significant miscalculation of continental-to-global scale carbon budgets. Temporal fluctuations in disturbance regime have led to non-steady-state carbon dynamics even at relatively large scales of space and time (Lugo and Brown 1986; Kurz and Apps 1996).

Some progress has been made in developing a set of landscape archetypes that represent the range of possible behaviors and types of landscapes. For instance, Turner et al. (1993, 1994b) identify a range of possible landscape behaviors based on two ratios: disturbance interval to recovery interval, and disturbance extent to landscape extent (figure 16.7) (Turner et al. 1994b). Six general types of landscape behavior occur in the space defined by these two axes, ranging from a

TABLE 16.1

A typology of archetypes of landscape behaviors and mosaic structures*

	Type of mosaic structure			
Type of landscape behavior	*Age-class mosaic*	*Site mosaic*	*Age by site mosaic*	*Age by site by disturbance regime mosaic*
Stable landscape	◄─────── Possible at short time scales only ───────►			
Shifting steady-state dynamics	Possible			
Non-steady-state dynamics	Possible	Possible	Possible	
Spatially heterogeneous dynamics			Possible	Possible

*See text for a discussion of the meaning of the term *stable* in the landscape behavior column.

steady state to an unstable bifurcating system. All of these behavioral patterns arise from a model that represents the forest landscape solely as a mosaic of patches of different age-class.

Landscape archetypes should represent variation in both the nature of the mosaic structure and the range of behaviors that arise from that structure. Table 16.1 illustrates such a typology as a matrix of landscape behaviors by mosaic structure. This typology is not intended to be exhaustive or definitive, but rather to illustrate the usefulness of such an approach. The four types of mosaic structures are distinguished by the nature of the differences among the elements of which

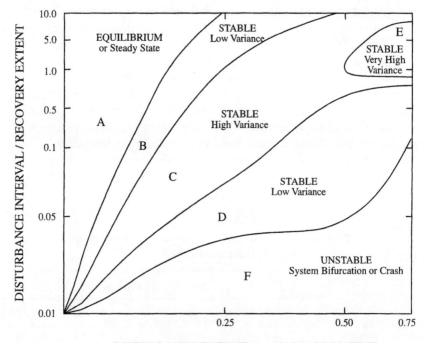

DISTURBANCE EXTENT / LANDSCAPE EXTENT

FIGURE 16.7

A state–space diagram illustrating the set of qualitatively different landscape behaviors generated by a simple, spatially explicit simulation model of a forest landscape. The x-axis is a spatial axis representing the ratio of area disturbed to the total landscape. The y-axis is a temporal axis representing the ratio of the time between disturbances to the time required for the system to recover from each disturbance.

they are composed. Landscape behaviors are distinguished by their temporal and spatial patterns of variability.

The first mosaic category is a simple age-class mosaic overlying an otherwise uniform landscape. We also use age-classes here as a proxy for successional stages. This is the type of landscape modeled by Turner et al. (1993, 1994b). Both steady-state and a variety of non-steady-state behaviors are possible (see figure 16.7), and this is the only kind of landscape that can truly exhibit a shifting-mosaic steady-state: all elements of the mosaic are equally capable of existing in all possible states of the system. This is the landscape type that was assumed by Baker (1989) as a null hypothesis in his search for a steady-state spatial scale in the Boundary Waters Canoe Area, and by Romme (1982) in his reconstruction of forest dynamics of landscape structure in Yellowstone National Park. Baker had to abandon this model because spatial subdivision in disturbance regimes meant that some patches "simply did not shift." Romme had to abandon this model because very large fires occurring on roughly 300-year intervals led to a landscape that is always recovering from the last major disturbance event. The recovery period of age structure and spatial pattern on forest landscapes is likely to be long, even if directed by human intervention (Wallin et al. 1994).

The second mosaic category is a mosaic of site types that vary in vegetation because of differences in soils or microclimate. Although this underlying mosaic of site conditions is a major contributor to forest pattern (Pastor and Broschart 1990; Leduc et al. 1992), it is difficult to imagine such a forest landscape without age-class structure superimposed on it as well, which leads to the third category: the age by site mosaic. This is the first category that the field ecologist is likely to feel bears any resemblance to his or her experience.

These simple mosaics of age- or site-classes are conceptually important and can generate complex temporal and spatial dynamics. Neutral models of landscape pattern (Gardner et al. 1987; Turner et al. 1989) to date have examined binary landscapes of patches that are susceptible to a disturbance or are not. These can be considered to represent a site mosaic with the nonsusceptible patches being sites of low flammability or fuel continuity. These models produce important conclusions about the permeability of landscapes to disturbances and how landscape dynamics are influenced by landscape structure. More recently, this approach has been implemented as the age-class mosaic used to generate figure 16.7 (Turner et al. 1993, 1994b). These classes of mosaics map fairly well onto the conceptual models of forest landscape dynamics that dominated the thinking of ecologists for much of this century (Sprugel 1991). These dynamics can be either steady state or nonsteady state, depending solely on how disturbances are scaled in space and time relative to recovery processes and landscape extent.

The final mosaic category is one in which the age by site mosaic is spatially divided, with different disturbance regimes dominating different parts of the landscape. Individual submosaics within such a divided landscape may exhibit any of the behaviors in table 16.1. Although Baker's (1989) analysis started out considering an age-class mosaic, he concluded that the shifting-mosaic steady state was not achieved because his study area was a "mosaic of different non-steady-state mosaics." We suspect that any consideration of disturbance regimes and mosaic dynamics in forest landscapes of any significant size (where size is scaled relative to the largest disturbance event) will find this "mosaic of different mosaics."

We further expect that most cases of spatial subdivision of disturbance regimes can be attributed to underlying intermediate-scale topographic structure in the landscape. Bergeron's (1991) demonstration of different disturbance regimes on islands and nearby lakeshores provides a good example of this. However, examples also exist in which physiographic barriers appear incapable of inducing spatial subdivision in disturbance parameters (Johnson and Larsen 1991). The ways in which combinations of small-scale variables (e.g., fuels), middle-scale variables (e.g., topography), and larger-scale variables [e.g., regional weather patterns (Johnson and Wowchuk 1993; Bessie and Johnson 1995)] interact to spatially structure disturbance on landscapes remain a subject for future research.

There are two meanings for the term *stability* in the context of landscape behavior. The first is merely the relative constancy of age-class or patch composition over time. All categories of mosaic type are capable of apparent stability of this type over short periods of time; a forest may appear constant in species composition, age-class, and spatial pattern over a 20-year period and still be in a state of substantial disequilibrium (Lertzman 1995). It is important to recognize that the appearance of a steady state over more than the short term is unlikely in both small and large landscapes, but for different reasons. Small landscapes have a greater likelihood of being overwhelmed by a single, infrequent disturbance event (Romme 1982). Larger landscapes are likely to be a mosaic of sublandscapes with different disturbance parameters and different dynamics (Baker 1989).

The second way of defining stability is the tendency of the system to return to its previous state or trajectory after perturbation. These are homeostatic and homeorhetic stability, respectively (Turner et al. 1993). Under a temporally unvarying disturbance regime, age-class mosaics should exhibit homeostatic stability in summary statistics, such as age-class distribution, but not in the spatial configuration of age classes within the mosaic. Site mosaics should exhibit homeostatic stability that includes spatial relationships if disturbances do not change the actual structure of the site mosaic. To the extent that the spatial subdivisions within age by site by disturbance regime mosaics are a function of the middle-scale structure of the landscape, they should impose some degree of homeorhetic stability on the mosaic, given no change in disturbance regime.

We are far from being able to make firm statements about the dynamics to be expected from any but the most simply structured mosaics. Modeling efforts have generally examined the upper left portion of the matrix in table 16.1, whereas we suspect that most forest landscapes will fall towards the lower right of the matrix. Consideration of this typology of landscapes poses challenges for landscape modeling that relate directly to problems of conservation and management under a changing climate. What kind of dynamics can we expect from spatially subdivided, non-steady-state mosaics under a changing disturbance regime?

Summary

Managed Forests

Modern anthropogenic disturbance regimes differ from historical "natural" disturbance regimes in a variety of ways at various spatial and temporal scales. This is both a problem of the scale of management practices and the scales on which planning processes have focused. Small-scale planning (e.g., at the stand level) without adequate landscape planning has resulted in bottom-up processes driving managed landscape dynamics. Landscape structure in the managed forest is driven by a "tyranny of small decisions."

Forest management has traditionally been planned, carried out, and assessed at a limited range of scales, with the most detailed planning being carried out at the stand scale. Cumulative effects of stand-level planning have resulted in unintended and unanticipated landscape-level problems. In many jurisdictions, the idea of formal landscape planning is very new. We require much better integration of quantitative tools for multiscale analysis with practical decision-making frameworks for management. We must be able to integrate spatially explicit data and models to allow long-term assessment of a variety of management options over a broad range of spatial scales.

Conservation

Franklin's (1993b,c) notion of integration between the intensively managed matrix and the reserves embedded within it is essentially a statement of the multiscale requirements of a successful regional conservation strategy. Stand-level prescriptions for wildlife habitat and large-scale regional reserve systems are both unlikely to meet their objectives unless they are part of a coordinated, multiscale conservation strategy. Patterns in forests represent the consequences of both bottom-up and top-down controls: conservation strategies for forest ecosystems or forest biota must recognize this range of processes. In particular, the lack of steady-state behavior in forest landscapes poses serious challenges for both the definition of management objectives for "natural areas" and the development of policies to achieve those objectives.

The relative benefits of investing in conservation at various scales in any given situation remain unclear. Assessing these trade-offs is at the stage of hypothesis development, rather than hypothesis testing. There are substantial opportunities for experimental management regimes that address alternative strategies for combining tactics at large, medium, and small scales. Successful design of these experiments requires a more systematic approach to the design of conservation options at multiple scales than has generally been considered to date.

Natural Disturbance Regimes and Management

Using the "natural" states or historical dynamics of an ecosystem or landscape as a basis for determining appropriate management regimes is the central focus of a variety of recent policy initiatives. The apparent ubiquity of non-steady-state dynamics in many forest ecosystems challenges any simple concept of naturalness. It also makes it difficult to identify a meaningful time frame for identifying the relevant historical dynamics of a system. However, we continue to believe this general approach is useful. For instance, it has great potential as a way to set boundary conditions for management objectives. The historical range of variability in a forest landscape is the only set of conditions known to have a high probability of maintaining the elements of biodiversity that have in fact been maintained in that landscape. Research needs to focus on establishing a clearer linkage between patterns in ecosystem dynamics at various scales and the particular aspects of ecological function and biodiversity that management strategies are intended to maintain.

We expect that anticipated future changes in climate will result in temporally variable and novel disturbance regimes. This will contribute substantially to the non-steady-state character of forest landscapes. Disturbance regimes are likely to be very sensitive to changing climate and to drive a variety of other ecological processes, such as those contributing to carbon dynamics. Understanding the ecological consequences of such dynamics is likely to be one of the more challenging problems facing forest ecologists and resource managers over the next few decades.

Acknowledgments

We thank the following people for helpful comments on earlier versions of this chapter: Brigitte Dorner, Eva Riccius, Andrew Fall, and Holger Sandmann. The research from which our ideas on scale arose has been supported by the Natural Sciences and Engineering Research Council, Forest Renewal British Columbia, the Science Council of British Columbia, and Forestry Canada.

17

INCORPORATING SCALE IN ECOLOGICAL EXPERIMENTS: STUDY DESIGN

Pierre Dutilleul

Only recently has scale been recognized as an intrinsic component of ecological heterogeneity. The heterogeneity of an ecological system is scale dependent or scale specific (Allen et al. 1987; Milne 1988; Levin 1992; Schneider 1994a). The sampling scale also affects our perception, as observers, of the heterogeneity of a system. Using a single, ill-adapted sampling scale can lead to incorrect conclusions about ecological heterogeneity (Allen and Hoekstra 1991; Milne 1991a; He et al. 1994). This chapter focuses on the adjustment of a study design to the heterogeneity of an ecosystem at the scale considered. Chapter 18 addresses the interrelationships between sampling scale and measured heterogeneity. Organizational aspects of functional heterogeneity are discussed in other chapters (Kolasa and Waltho, chapter 4; Pahl-Wostl, chapter 7).

The emphasis here is on temporal and spatial heterogeneities. It is also on the surface-pattern approach, where heterogeneity refers to the variability of a quantitative or categorical response, rather than on the point-pattern approach, where heterogeneity concerns the distribution of events over time or the dispersion of ecological entities through space (Dutilleul and Legendre 1993). In fact, the latter is essentially based on the intensity function of the generating-point process, which is quantitative and continuously defined in time as well as in space. When the intensity function is constant, the corresponding point process is the classical, homogeneous Poisson process. Heterogeneity in the surface-pattern approach may concern the mean, the variance, or the autocorrelation function of a quantitative process (Dutilleul and Legendre 1993).

Time and space are two main axes of investigation of heterogeneity. The definition of scale hereafter applies to either axis. This does not mean that the temporal and spatial heterogeneities are equivalent. First, the time axis is ordered and one-dimensional, whereas the spatial axis is not ordered and may be one-, two-, or three-dimensional. A consequence of the absence of order in space is that the value observed in a given location may be influenced by all nearest neighbors, whether left-right, east-west, north-south, or top-bottom, but future observations will never influence past observations in time. Accordingly, the definition of the theoretical autocorrelation function may be restricted to positive lags in time, with only the cross-correlation function being evaluated at negative lags (e.g., Dutilleul 1995). Second, differences in temporal heterogeneity between locations will induce spatial heterogeneity, but consistent spatial heterogeneity between locations over time does not necessarily imply temporal heterogeneity in each site (Kolasa and Rollo 1991).

Schneider (1994a) writes: "One . . . definition is that scale denotes the resolution within the range of a measured quantity." David Schneider certainly did not want to give us the impression that measurement was a prerequisite to the definition of scale and, of course, it is not. The scale may be defined independently of any measurement (i.e., data sampling) and refers then to the level of resolution of parts of an ecosystem (e.g., cell, tree ring, tree, forest) and the length of natural periods of time (day, season, year, lunar tide, life cycle). Scale is also a relative concept. For example, a large scale for most insects will look small to most birds, in space as well as in time. Associating scale and measurement units (Schneider 1994a) removes any relativity from the definition. An alternative way to deal with that relativity is to define the scale of a study as the interaction of grain and extent, where the grain of observation is the finest distinction that can be made between two sampling units and, in contrast, the extent is the span of all measurements in a given study (Allen and Starr 1982; Allen et al. 1987; Allen and Hoekstra 1991). In earth sciences, the equivalent of grain is called support; supports commonly used in geostatistics are the point and the block, a prescribed local area (Isaaks and Srivastava 1989). The length of sampling intervals and size of sampling quadrats follow from the compromise between fine and coarse resolution, and between narrow and wide extent, given methodological arguments and practical constraints (Gardner, chapter 2; Schneider, chapter 12).

This chapter discusses how the design of an experiment can be accommodated to the heterogeneity of a given type at a given scale, with or without treatment application, and whether heterogeneity is of interest or not. A range of experimental situations involving time or space, or both, is covered. In the time case, large-scale heterogeneity of the mean, small-scale autocorrelation and rhythms are treated; the rhythm part is illustrated by motor activity data for the Wistar rat. The space case is based on Dutilleul (1993b) with some new examples, most of them using simulated data and one regarding the impact of a bird population on the vegetation cover of an island. The chapter is concluded with the most general, spatiotemporal case and the spatiotemporal repeated-measures design.

Preamble

A general principle of experimental design (Hurlbert 1984; Mead 1988; Dutilleul 1993b) is that a valid design is based on reliable a priori information regarding the experimental material and field, which must be as complete as possible, especially in the absence of natural scaling. Presampling can provide such a priori information if it is conducted under conditions similar to those of the experiment. The design of the experiment also depends on the statistical methods of analysis to which the experimenter intends to submit the data. Good knowledge of those statistical methods not only in terms of test power and sample size, but also in terms of ability or inability of the method to provide information at a particular scale, is a prerequisite.

The Time Case

General Aspects

Some temporal scales are imposed by nature to the observer as well as to the object of the observation, including the day–night circadian rhythm in chronobiology, the effects of the season on plant growth and animal behavior, the yearly tree-ring growth in dendrochronology, and planetary behaviors, such as the 18.6-year lunar nodal tide. Some other temporal scales beside the "natural" ones result from the way human beings commonly divide time: seconds and minutes, the 24 hours of the day, 7 days of the week (of religious origin), 12 months of the year (of unequal length, but about the 29 days, 12 hours, 44 minutes and two seconds of the lunar month), centuries and millenaries. Although matter may be decomposed into cells and light into photons at the microscale, the time axis is continuous so that one could argue that any continuous-time process has no scale until the observer makes the time axis discrete in place of nature (for instance, by integrating the response over given time intervals) (Priestley 1981; Dutilleul 1995). In the following example of the Wistar rat, a given motor activity is recorded in the animal, providing a binary response: the rat is, or is not, in the activity considered at a given time. Integration over regular time intervals of a given length (e.g., 15 minutes or 2 hours) provides a time-series realization of a discrete-time process varying with the interval length used in integration (e.g., the time spent by the rat in the activity every 15 minutes or 2 hours over the experiment). In the same example, integration transforms the scale of observation from binary data and state transitions in continuous time to continuously quantitative data bounded by a true zero point and the length of the integration interval in discrete time.

Sources of variation usually studied in time-series analysis are (1) large-scale heterogeneity of the mean, called trend, (2) small-scale autocorrelation, which is often associated with heterogeneity of variance or heteroscedasticity at larger scale, and (3) rhythms and heterogeneity of the mean of periodic type. The heterogeneities of the mean and variance are sources of nonstationarity that prevent the data analyst from performing autocorrelogram analysis on raw data (Box et al. 1994). It is crucial to take both types of heterogeneity into account by designing the experiment adequately, whether or not heterogeneity is of interest per se.

Trend in Time or Temporal Heterogeneity of the Mean at Large Scale

It may seem that any sampling scale will allow the detection of large-scale heterogeneity of trend type, and this is true as far as the extent covers the growing season or lifespan of the biological entity reasonably well. If the extent is restricted to the beginning or end of the growth curve, a plateau may be observed (figure 17.1). Sometimes trend is supposedly detected, but the observed increase or decrease is only temporary and actually results from the use of a narrow extent that prevents the observer from detecting a reverse pattern outside the time interval considered. On the other hand, the degree of resolution of the growth curve (i.e., if it can be fitted by linear, curvilinear, or nonlinear regression) depends on the fineness of the grain for a suitable extent.

The recommended design to compare growth curves observed under different experimental conditions is a repeated-measures design. Repeated-measures data arise whenever measurements are made of the same characteristic on the same observational unit but on more than one occasion (Crowder and Hand 1990). Time can then be incorporated into the design as a repetition factor with as many levels as sampling time points. The time factor is usually crossed with the other (treatment or classification) factors. It is also generally considered fixed, so that large-scale heterogeneity is taken into account by a mean function, and specific comparisons between time levels are performed by using contrasts. Furthermore, time is ordered and sampling time points are usually equally spaced, so the experimenter commonly wants to perform specific comparisons between time levels by using contrasts, which would not be permitted if the time factor was random. Repeated-measures designs are thus not split-plot designs (Yates 1982), where subtreatments are randomly assigned to subplots constituting the main plots. Following Winer et al. (1991), the repeated-measures designs discussed here will not include the cross-over designs, where more than one treatment is applied to each individual over time (Jones and Kenward 1988).

In the example of growth curves (see figure 17.1), the annual plant or the tree represents the experimental unit followed over time, with a

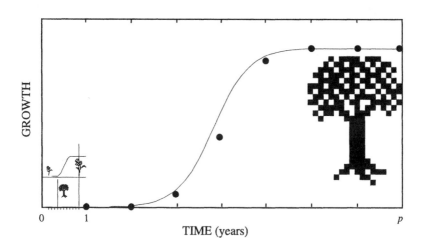

FIGURE 17.1

Schematic representation of plant growth curves, within a year (annual plant) and over years (tree), as illustrations of temporal heterogeneity of the mean at large scale for different extents.

given plant at a given time being the sampling unit. It is convenient to consider a repeated-measures data table with n rows for the individual units and p columns for the levels of the repetition factor. A minimum sample size required by the multivariate methods of data analysis is $n - g > p$, where g is the number of classification groups (Timm 1980). Repeated-measures data of 15 years ($p = 15$) of radial growth from 40 individual trees ($n = 40$), 20 slow-grown and 20 fast-grown ($g = 2$), are analyzed in chapter 18.

Temporal Autocorrelation at Small Scale

The finer the grain, the more likely the sample data are to be autocorrelated at small scale; that is, the correlation between successive observations tends to be higher if the time interval separating them is shorter. Temporal autocorrelation can be defined as the statistical property of a random variable that takes values at two sampling time points, which are more similar (positive autocorrelation) or less similar (negative autocorrelation) than expected for pairs of randomly associated values (Dutilleul 1995). Autocorrelation violates the postulate of independence of the classical methods of statistical inference (Sokal and Rohlf 1995).

The random errors superimposed on the mean function in the analysis of growth curves (see figure 17.1) can be autocorrelated. Therefore, the observer will tend to space the collection of data over time. The counterpart of that grain increase is a loss of resolution in curve fitting. A better solution is provided by a repeated-measures design with a sufficient number of individuals (i.e., $n > p + 1$ in the case of one group), so that the sample covariance matrix computed over the profile vectors of repeated measures can be inverted and used in a weighted least-squares regression (Searle 1971) to reduce the variance of the slope estimates. It must be noted that in the analysis of variance (ANOVA), the independence of observations and the homogeneity of the variance (homoscedasticity) are not the most general necessary and sufficient condition ensuring valid unmodified F-testing (see chapter 18). It must also be noted that for a limited extent, it is sometimes difficult to distinguish between strong positive autocorrelation and linear trend in practice. Widening the extent may help solve the problem because the tendency should finish by being reversed in the former case.

Rhythms and Temporal Heterogeneity of the Mean of Periodic Type

Two main types of periodicity in time series are (1) when the mean function can be modeled by sine and cosine waves at given frequencies, on which random errors are superimposed (harmonic model), and (2) when the observation at time *t* is directly but partly dependent on the observation at time *t − S*, where *S* is the period (seasonal autoregressive model) (Dutilleul 1995). The emphasis here is on the former type that can be defined as temporal heterogeneity in the mean at medium scale. In fact, when the objective is the detection of periodicities in time series, it is crucial to define a sufficiently wide extent so that the periodic signal is reproduced several times over the series. A minimum number of cycles (i.e., signal reproductions) usually reported in the literature is five in the analysis of discrete spectra and 10 or more in the analysis of continuous spectra characterized by a spectral density function (Brillinger 1981; Priestley 1981); the detection of periodicities falls in the analysis of discrete or mixed spectra. The definition of the extent depends on the length of the target periodicities, and its adequacy relies on the length of hidden periodicities, with or without target.

If the extent should be wide enough to reproduce the signal several times, what about the grain? In dendroclimatology, where periodicities in climate are related to periodicities in ring-width chronologies (Fritts 1976), the grain is imposed: it is the year, because of the yearly tree-ring growth, although weather data involved in the tree response function may come from specific months. In chronobiology, targets are the lunar (about 29 days) and semilunar rhythms on the one hand, and the circadian (24 hours) rhythm and the discrepancies from it under given experimental conditions on the other hand. When the lunar or semilunar rhythms are of interest [in fish; see Leatherland et al. (1992)], the grain should be finer than a week; a grain of 1 to 4 days is appropriate. When maintained in constant darkness during several days (DD conditions), the rhythm of an animal tends to free-run and lengthen. This lengthening of the rhythm can be up to 1 or 2 hours; it can also be manipulated by short light pulses temporally interrupting the DD conditions (Morgan and Minors 1995). A suitable grain should then fall below the hour; classically, it is 15 minutes.

A main difference between the examples in dendroclimatology and chronobiology is that climatic periodicities are empirical (i.e., they must be detected and are estimated first), whereas the lunar-related rhythms and the discrepancies from the circadian rhythm are expected (i.e., there is a theoretical basis for them). On the other hand, the relation between climate and tree-ring growth is well established (Fritts 1976), but periodicities in climate are not "universal" as the circadian rhythm and planetary tides are. Figure 17.2 presents time-series data of motor activity for the Wistar rat, as provided by sampling scales of 15 minutes to 2 hours. The series plotted are mean series computed from the individual series, after these are standardized to a zero mean and a unitary variance so that each individual rat contributes uniformly to the mean; the individual series follow from integration of the original binary data over successive time intervals. Whatever the sampling scale (figure 17.2), a signal appears to be reproduced about five times over the 5 days of observation; these data are analyzed in chapter 18. In tree-ring series analysis, the individual chronologies are "indexed" prior to computation of the master chronology (Fritts 1976).

The collection of replicated time series in an experimental design (Brillinger 1973; Shumway 1983; Dutilleul 1990) has the advantage of increasing the signal-to-noise ratio (i.e., the squared amplitude of the signal fitted to the mean series, divided by the unexplained variance attributable to the noise). This is a result of the basic statistical property by which the sample mean (the mean series) approaches the population mean (the signal) when the sample size (number of replicated time series) increases. Furthermore, that particular type of repeated-measures design permits an ANOVA in the frequency domain (see chapter 18). In rhythm analysis, the choice of the grain and extent can be discussed in terms of the signal-to-noise ratio. When the opportunity is offered to the experimenter to collect more data while the signal-to-noise ratio is low, and while reliable a priori information is available about the target periodicity, it is not recommended to refine the grain but instead to widen the extent in order to reproduce the signal further (Dutilleul 1990, 1995) (see chapter 18 for numerical results).

FIGURE 17.2

Time-series realization for motor activity in the Wistar rat, using a sampling scale of **(A)** 15 minutes, **(B)** 30 minutes, **(C)** 1 hour, and **(D)** 2 hours, over 5 days in darkness. The *large black circles* in panel **(D)** are the overall daily means of standardized data. Data source: Unité de Psychobiologie, Université catholique de Louvain, Louvain-la-Neuve, Belgium.

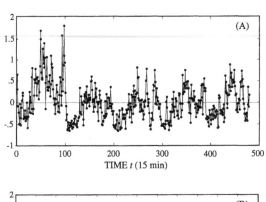

(A)

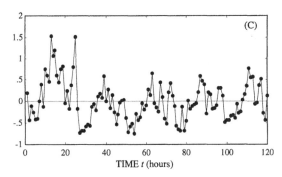

(B)

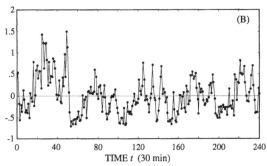

(C)

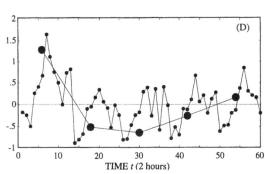

(D)

MEAN STANDARDIZED MOTOR ACTIVITY DURATION x_t

The Space Case

General Aspects

It may seem that space provides fewer "natural" functional and observational scales than does time. This is only partly true. In fact, plants or animals must grow or live beyond a minimum distance from one another to survive competition. In one-dimensional (1D) space, vertical sampling scales may be provided by the leaves along a plant stem, the branches of a tree, or the thermal strata of the water column in a lake. In two-dimensional (2D) space, controlled experiments in forestry and agronomy may direct or impose the working scale by fixing the thinning intensity in a planted stand or the row spacing in a cultivated field. When the data collection is conducted around the trees or among the rows, half the average distance between trees or between rows defines an upper boundary for the grain (e.g., when soil samples are collected); there is no lower boundary in principle, but in practice this depends on technical aspects of field conditions. Three-dimensional (3D) spatial scales arise, for example, when plant architecture is studied vertically (top–bottom) and horizontally (perpendicular or parallel to light or wind gradients), or when the thermal strata of a lake are followed horizontally (north–south and east–west). Outside the controlled experiments and when no "natural" scale is imposed, the experimenter needs to design a sampling procedure.

Scale and Sampling: An Outline

In any sampling procedure, the main parameters to be fixed (not the statistical population parameters to be estimated) for given boundaries are (1) the number of sampling units, (2) their spatial location, and (3) their size in the case of quadrats or volumes. The number of sampling units must be large enough to ensure sufficient precision in estimating the spatial mean, and their location must be defined so that all distance classes can be evenly represented in the autocorrelogram or variogram analysis (see chapter 18). The number of pairs of points falling in each distance class should be nearly the same from class to class, which provides even levels of accuracy and precision among classes in estimating the spatial autocorrelation; *point* refers here to a sampling point or the centroid of a sampling quadrat or volume. To ensure sufficient representativeness at small distance classes, it is recommended that the grain be reduced around some nodes of the main sampling grid or lattice, whether regular or irregular, and to collect data on a (e.g., 3 × 3) subgrid around each of these nodes.

Three sampling methods are outlined here: uniform random sampling, systematic sampling, and stratified random sampling (Green 1967; Krishnaiah and Rao 1988). Random spatial sampling consists of randomizing the coordinates of the units to be sampled. Although randomization is recommended for assigning treatments in experimental design (Winer et al. 1991; Dutilleul 1993b; Potvin 1993; Sokal and Rohlf 1995), uniform random sampling (i.e., unrestricted random sampling within the limits of the field) is not recommended in space (or time). The only exception may be in the "design-based approach" (de Gruijter and ter Braak 1990), which postulates that the underlying spatial process is a deterministic surface; there is then no variance other than that of the locations randomly sampled, because the surface, once sampled, is known without any error. In terms of field cover, uniform random sampling may overcover some parts of the field and under-cover some others, so that it is difficult to know what the grain actually is.

On the other hand, the grain may be fixed in systematic sampling. In fact, when sampling points are located on a lattice of regular geometrical form, the grain is given by the distance between nearest neighbors, when constant. It must be noted that the geometrical form may be a square lattice with smaller distance between nodes along one axis than along the other, so that the resolution (i.e., the grain) is finer in one dimension than in the other; this is called geometric anisotropy (Ripley 1981). In contrast, when the lattice is rectangular with constant distance between nodes along both axes, the grain is unique but the extent differs between axes.

In stratified random sampling, the field is divided into plots, called strata, so as to maximize heterogeneity among strata and homogeneity within them (Dutilleul 1993b). For a given number of sampling units per stratum (that number may be proportional to the size of the stratum), their location is randomized within each stratum. For a given field size, increasing the number of strata provides smaller strata and supposedly higher homogeneity within strata, resulting in smaller error variance (Ripley 1981). When the number of strata is increased and the number of sampling units is one per stratum, stratified random sampling tends toward centric systematic sampling where sampling points and strata centroids coincide.

Except under the postulate of the "design-based approach" (i.e., the surface sampled is deterministic), any random spatial sampling is actually doubly random (1) in the location of the units and (2) in the values of the process. The error variance is thus largest for uniform random sampling. In general, systematic sampling is preferred to stratified random sampling; the reverse is true in the case of periodicity (Ripley 1981) or strong positive autocorrelation at small to medium scale (Haining 1990). Finally, none of the procedures of random spatial sampling can really ensure that the sample data are spatially independent, outside the postulate of an underlying deterministic surface.

Spatial Heterogeneity and Study Design

In the following experimental designs, the sampling units are not only points: they can be continuous intervals, subareas (plots, subplots, quadrats), or volumes, depending on the spatial dimension (1D, 2D, or 3D, respectively). The emphasis is on the 2D space.

For an ecological hypothesis to be rigorously tested, the experiment must be thoughtfully designed, especially when conducted in the field (Hurlbert 1984; Dutilleul 1993b; Ver Hoef and Cressie 1993). Ecological field experiments present some specificities with respect to the "controlled" experiments conducted indoors in the lab or the greenhouse, or outdoors in planted stands or lake enclosures. To some degree, however, similar experimental designs may be applied. The main problem in the field is in the arrangement of blocks (1D, 2D, or 3D) so they match the spatial heterogeneity, on the basis of a priori information to be collected without perturbing the ecosystem. As a follow-up to Dutilleul (1993b), the objective is to examine how the design of an ecological field experiment can accommodate small-scale and patchy heterogeneity, as well as environmental gradients, either as a nuisance or as a characteristic of interest, with or without treatment application.

Spatial heterogeneity can be taken into account in experimental design by defining random ANOVA factors whose variance components allow a quantification of heterogeneity in the linear model used for data analysis. Plot splitting and blocking, simple or multiple, complement this idea by allowing heterogeneity analysis at smaller and larger scales, respectively. In fact, these techniques provide random ANOVA factors whose levels (i.e., plots and subplots, simple and multiple blocks) are defined so as to maximize variation among subplots or blocks, and to minimize variation among plots or within blocks, depending on the type and scale (i.e., small or large) of spatial heterogeneity. Replication is hardly applicable at the macroscale (Carpenter

1990), whereas pseudoreplication (a fault committed each time replicates are collected at a scale finer than the one for which the conclusions of the study are intended to be drawn) is likely to arise at the microscale (Hurlbert 1984).

The examples discussed here are illustrated in figure 17.3. The corresponding types of spatial heterogeneity are (1) small-scale heterogeneity with large-scale homogeneity (figure 17.3A), (2) large-scale heterogeneity of gradient type, only in the mean (figure 17.3B) and in both the mean and variance (figure 17.3C), and (3) patchy heterogeneity at medium scale (figure 17.3D).

The ecological example of small-scale spatial heterogeneity in Dutilleul (1993b) was provided by high variability of dissolved organic nitrogen among surface water samples less than 2 meters apart from above a limestone reef system. That heterogeneity can be captured and the homogeneity at large scale maintained, by defining 4 × 4 m sampling stations (plots) randomly defined over the reef and divided into 1 × 1 m quadrats (subplots) used as replicates. When small-scale heterogeneity is combined with large-scale homogeneity, the recommended designs in the presence of treatment assignment are (1) when spatial heterogeneity is of interest, a completely randomized design with plot splitting, because it takes spatial variability at small scale into account, and (2) when only the treatment effects are of interest, a classical (i.e., without plot splitting) completely randomized design, because there is no heterogeneity to remove at the scale of plots.

In some other cases (soil nutrients or biomass in a watershed), the field may clearly be heterogeneous at large scale. This heterogeneity of gradient type may then be of two types: only in the mean (figure 17.3B) or both in the mean and in the variance (figure 17.3C) (Dutilleul and Legendre 1993). In the absence of treatment assignment, it is recommended that blocks be arranged perpendicularly to the gradient in the former case (see the arrow in figure 17.3B), with each block being divided into plots to minimize variation within blocks; in the latter case, multiple blocking with row and column blocks is recommended. In both cases, blocking aims to capture the spatial heterogeneity at large scale. Blocking is not restricted to space but applies to time as well, with similar aims; in either case, gradients need not be monotonic, increasing or decreasing. In the presence of treatment assignment, when spatial heterogeneity is of interest, recommended designs are (1) a randomized complete block design with more than one observation per treatment per block in figure 17.3B, because it takes spatial heterogeneity at large scale into account and allows the quantification

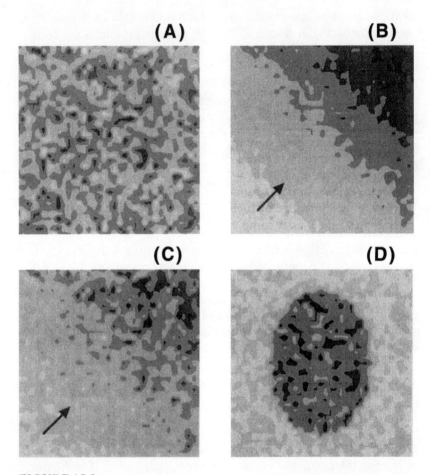

FIGURE 17.3

Contour plots for four surface patterns of spatial heterogeneity. **(A)** Heterogeneity at small scale, with large-scale homogeneity. **(B)** Gradient in the mean parameter of the spatially continuous variable plotted; the mean increases linearly from the bottom left to the upper right corner of the figure (*arrow*). **(C)** Gradient in both the mean and the variance parameters, which increase linearly from the bottom left to the upper right corner of the figure (*arrow*). **(D)** Single patch; the example in text is provided by different types of vegetation cover in the center of an island compared to its periphery. The gray tones range from the low value (*very light gray*) of a continuous quantitative variable (e.g., biomass) or given state of a discrete categorical variable (e.g., vegetation cover), to the high value (*black*) or different state.

of the spatial variation resulting from inconstancy of treatment effects over blocks; and for similar reasons, (2) multiple Latin squares (Mead 1988) laid out in accordance with the spatial patterns in the mean and in the variance in figure 17.3C (e.g., two Latin squares on the bottom, left and right, and two others on the top). When only treatment effects are of interest and spatial heterogeneity is considered as a nuisance to be removed, a randomized complete block design with one observation per treatment per block is recommended (figure 17.3B), because it separates the spatial heterogeneity at large scale from the effective experimental error; similarly, a single Latin square is then appropriate in figure 17.3C.

The last example is provided by a study of the impact of a bird population on the vegetation cover of an island (N. Bays, unpublished data). Substrate (sandy or rocky) and the associated vegetation define a patch of elliptical form in the center of the island, as opposed to its periphery (figure 17.3D). These two distinct parts of the island clearly define two blocks naturally laid out perpendicularly to the periphery-to-center gradient. The shape of these blocks is "circular," and not longitudinal as usual. (This circularity and the circularity condition in the repeated-measures ANOVA are entirely unrelated.) A certain number of exclosures (microplots to which birds have no access) were randomly laid out within each block, with perimeters (quadrats to which birds have free access) nearby. Monitoring the vegetation composition in the exclosures and the companion perimeters at a given time provides paired observations (exclosure–perimeter) in a randomized complete block design with two blocks (center versus periphery of the island) and more than one pair of observations as small-scale replicates per block. When there is more than one patch, the blocks have to match the patches as much as possible, to capture the medium-scale patchy heterogeneity by maximizing the among-block spatial variability. Patches may be provided by low and high values of a continuous quantitative variable, or by various states of a discrete categorical variable as in the example (figure 17.3D).

Spatial Repeated Measures

Repeated-measures data are not confined to time. They can also arise in space, and even outside the spatiotemporal framework (Dutilleul and Potvin 1995). In space, repeated measurements are not restricted to a given dimension (1D, 2D, or 3D) or a given axis or plane. At small scale, they may be vertical, whereas plot splitting or blocking takes into account horizontal small-scale or large-scale heterogeneity, respectively. For example, nitrate content of a water table can be measured at different depths, in locations with identical or different soil composition; zooplankton biomass can be sampled in the thermal strata of a lake, in-shore versus off-shore, or only at a given distance from the shore. The spatial repetition factor (depth, stratum) is crossed with the spatial block or plot factor, that is, nitrate content is measured at the same depths in all locations, and the same strata are sampled over the lake for zooplankton. This crossing may be seen as a constraint, but it allows the data analyst to use the full capacity of the repeated-measures ANOVA technique (see chapter 18). Spatial repeated measures may also be horizontal. When they are, they can be sequential, as when soil nutrients are sampled at fixed distances from given trees in parts of a forest with controlled tree density (block factor). Spatial repeated measures arise in 3D when, for instance, bird or insect damage on a given tree is studied as a function of elevation (vertical axis) combined with geographical orientation (horizontal axis). Repeated measures are rarely collected only in space and not in time; they arise more often in both space and time.

The Spatiotemporal Case

General Aspects

Although a multiscale approach to heterogeneity analysis has been advocated by some (e.g., Milne 1991a), until now this potential approach tended to be limited either to space or to time. This limitation must be removed. In fact, in the same way that the pattern of a process may change drastically from scale to scale, the distribution of a variable may change depending on the space–time point of observation. Combining space and time in the same approach is not easy, however, because of their distinct nature; in the 3D space already, heterogeneity along the vertical axis is different from that in the horizontal plane. As a result, it is not possible to define and calculate a relevant scalar distance between space–time points.

It does not mean that there is no way to adopt a spatiotemporal approach to heterogeneity and undertake its scaling in that framework. The sampling procedures and experimental designs presented above in the purely temporal and purely spatial cases apply to the spatiotemporal case, with some restrictions. For instance, when the same 4 × 4 lattice over a limestone reef system is resampled for dissolved organic nitrogen sequentially over time, the problem is twofold: (1) most statistical methods of spatiotemporal data analysis require the stationarity of the spatial mean and spatial autocovariance over time, and (2) without replication in space, stationarity may be postulated, but it cannot be tested.

Spatiotemporal Repeated Measures

Replication is difficult to achieve in 2D space. When is node (i, j) of a sampling grid in actual correspondence with node (i, j) of another grid of same size? The location of nodes on grids may be the same, but while two grids should provide replicates of the spatial variability (i.e., variance) of dissolved organic nitrogen among subplots, is spatial autocorrelation among nodes the same from grid to grid and from time to time? Replication is more difficult to achieve when time series are longer and lattices are larger. A compromise solution exists for short time series and small-size lattices: the spatiotemporal repeated-measures design, where time and space are considered as two repetition ANOVA factors with given numbers of levels. When the spatial and temporal repetition factors are crossed (i.e., all individuals are observed on the same space–time points), it is then possible to test for their interaction. This design relaxes the condition of stationarity on the mean and the autocovariance, both in space and time, whenever the number of individuals minus the number of groups, $n - g$, is greater than the number of spatiotemporal repeated measures, p. This condition is required, without further assumption concerning the spatiotemporal autocovariance structure, by the multivariate procedure of the repeated-measures ANOVA. This design may seem restrictive, but it is not so for small-scale autocorrelation, including the spatiotemporal autocorrelation between space–time points. Through replication, large lattices and long time series are not required. Even a short extent allows precise estimation of autocorrelation over a given range of spatial distances and temporal lags. Classically (i.e., without replication), such a range does not exceed half the maximum distance between sampling units.

Another example of spatiotemporal repeated measures is given by the fractal dimension of a plant canopy, when computed week after week for two different sides (i.e., two spatial levels) of the same plant during the growing season. In such a case, the fractal dimension (Gardner, chapter 2) is a complexity measure of space occupation.

(This example is developed in chapter 18.) Chlorophyll fluorescence measured on three positions (ear − 1, ear, and ear + 1 leaf) of individual corn plants (Zhou et al. 1997), and nitrate concentration recorded at three depths of a water table (Serem et al. 1997), both on a weekly basis, are examples with three space levels.

Summary

In all cases, thoughtful experimental design is required prior to analysis of heterogeneity and scale. In the absence of universal optimal design, the exercise must be repeated on a case-by-case basis. Plot splitting and blocking, simple or multiple, are classical ways to take into account small-scale and large-scale spatial heterogeneity.

A multiscale approach to any ecological study must be defined in both time and space. From this respect, the spatiotemporal repeated-measures design provides a framework that allows an efficient and complete study in terms of scale and pattern of heterogeneity (i.e., large-scale gradients, small-scale autocorrelation), at the cost of few assumptions. Gradients in the mean, less than other patterns of heterogeneity, are affected by change in sampling scale.

Acknowledgments

Research was supported by operating grants from The Natural Sciences and Engineering Research Council of Canada, and Le Fonds pour la Formation de Chercheurs et l'Aide à la Recherche du Québec. The graduate course on temporal and spatial statistics that I taught in winter 1995, combined with fruitful discussions with my colleague Martin J. Lechowicz about ecological aspects, helped to significantly improve the content of the chapter. Many graduate students and post-doctoral fellows contributed through discussions, especially Wafa Aboul Hosn, Magda Burgess, Nathalie Bays, Wambua Kaluli, Vincent Serem, Jason Stockwell, and Xiaomin Zhou. Special thanks go to Isabelle Schelstraete and Marc Weyers for permission to use part of their Wistar rat data set for illustration. I am grateful to Pierre Legendre and two anonymous reviewers for their helpful comments on the manuscript.

18

INCORPORATING SCALE IN ECOLOGICAL
EXPERIMENTS: DATA ANALYSIS

Pierre Dutilleul

This chapter focuses on the interrelationships between sampling scale and measured heterogeneity, with emphasis on temporal, spatial, and spatiotemporal heterogeneities in the surface-pattern approach (Dutilleul and Legendre 1993). The major points include (1) models used for data analysis, their definition and how they can incorporate temporal and spatial scales, and (2) the importance of sampling scale and data transformation, such as smoothing, integration, and definition of distance classes, on measured heterogeneity. The following questions are addressed: How can sampling scale affect the estimation of a periodicity in time and the investigation of a pattern in space? and How can preliminary data transformation and classification alter our perception of a pattern in scale and heterogeneity analyses?

Statistical methods of data analysis involved in the discussion are (1) univariate procedures of analysis of variance (ANOVA), (2) multivariate procedures of ANOVA, (3) repeated-measures ANOVA, (4) autocorrelogram analysis, (5) periodogram analysis in the time case, (6) semivariogram analysis, and (7) fractal dimension estimation in the space case. The focus is on specific methods that are not widespread in the literature, such as the multifrequential periodogram analysis, the ANOVA of the finite Fourier transform, and the spatiotemporal repeated-measures ANOVA. Reference is made to chapter 17 for study design, and some examples from that chapter (e.g., the Wistar rat example in the time case) are discussed further. Zooplankton biomass data from Lake Erie (Ontario, Canada) are analyzed in the space case, and changes in the fractal dimension of a plant canopy are studied in the spatiotemporal approach. Numerical results are completed by using simulated data. Guidelines on how to perform the statistical methods of data analysis are given in the appendix to this chapter; references to statistical packages are made when relevant.

The Time Case

This section discusses temporal heterogeneity of the mean (1) at large scale, when combined with autocorrelation and heteroscedasticity at small to large scales, and (2) at medium scale, of periodic type. The relevant methods of data analysis are, respectively, (1) the repeated-measures ANOVA and (2) the autocorrelogram and periodogram analyses, and the ANOVA of the finite Fourier transform.

The Repeated-Measures ANOVA

The statistical method is illustrated on a set of *Picea abies* (Norway spruce) radial growth data. The experimental stand is located in the Belgian Ardennes near Rendeux. It was planted in 1949, and thinning started in 1969. Circumference at breast height (1.5 m) was measured on all trees of the stand at the end of the growing season in 1969 and in the spring of 1987. From the difference recorded between the two dates, an average annual increment of circumference growth was computed for each individual tree. Three categories of growth (slow, intermediate, fast) were defined. Twenty trees were randomly sampled from the slow-growth category and 20 others from the fast-growth category. The objective here is to assess if the differences observed among trees in radial growth from 1969 to 1987 were preconditioned, that is, if differences existed before thinning started. Data analysis covers the 1955 to 1969 period.

The data table is thus 40×15 (i.e., $20 + 20$ trees $= 40$ rows, and 15 years $= 15$ columns). The repetition factor (fixed) is the year, and the classification factor (fixed) is the growth category. The factors are crossed, and the individual trees within each category provide the experimental units or subjects (random). The resulting mixed ANOVA model may be written as:

$$X_{ijt} = \mu + a_i + B_{j|i} + c_t + (ac)_{it} + \varepsilon_{ijt} \tag{1}$$

$$(i = 1, 2; j = 1, \ldots n_i = 20 \text{ for } i = 1, 2; t = 1, \ldots p = 15)$$

where X_{ijt} represents the radial growth of tree j from category i up to year t; μ is the overall population mean or intercept; a_i is the fixed effect of growth category i; $B_{j|i}$ is the random effect of tree j nested within growth category i; c_t is the fixed effect of year t; $(ac)_{it}$ is the fixed effect of interaction between growth category i and year t; ε_{ijt} is a random error. The fixed effects are assumed to satisfy the estimability condition (Searle 1971), and the random effects to be normally distributed with a zero mean (Sokal and Rohlf 1995). Equation (model) 1 follows Winer (1971) and von Ende (1993), except that it combines the subject-by-year interaction with the true experimental error in the random error ε_{ijt} (the two effects are indistinguishable). Differences in temporal trajectory of tree radial growth are tested here between categories, not among individual trees, which are considered random. When subjects are considered fixed, the subject-by-time interaction can be tested using polynomial time contrasts (Yates 1982).

There are two approaches to the repeated-measures ANOVA: (1) a modified univariate, based on equation 1 and the adjustment of the probabilities of significance of all effects involving the repetition factor, in order to take into account the lack of independence and homoscedasticity of the repeated measures, and (2) a multivariate, based on equation (model) 2 (see later), where the basic object of the analysis is the profile vector of p repeated measures, and the usual sums of squares are replaced by $p \times p$ matrices of sums of squares and cross-products (Crowder and Hand 1990; Potvin et al. 1990; von Ende 1993). The multivariate ANOVA (i.e., MANOVA) model corresponding to univariate equation 1 may be derived as follows. First, the terms in equation 1 with no subscript, μ, or depending only on t, c_t, are grouped, providing $\mu + c_t$; those depending only on i, a_i, or on i and t, $(ac)_{it}$, are grouped apart, providing $a_i + (ac)_{it}$, so that the sum of the remaining terms is $B_{j|i} + \varepsilon_{ijt}$. Let $\underline{x}_{ij} = (X_{ij1}, \ldots, X_{ijp})'$ denote the profile vector of p repeated measures on tree j from growth category i; $\underline{m} = (\mu + c_1, \ldots, \mu + c_p)'$, the multivariate intercept; $\underline{a}_i = [a_i + (ac)_{i1}, \ldots, a_i + (ac)_{ip}]'$, the multivariate growth category main effects; and $\underline{e}_{ij} = (B_{j|i} + \varepsilon_{ij1}, \ldots, B_{j|i} + \varepsilon_{ijp})'$, the multivariate random errors. The MANOVA model may then be written as:

$$\underline{x}_{ij} = \underline{m} + \underline{a}_i + \underline{e}_{ij} \tag{2}$$

where subscripts i and j have the meanings they had in equation 1.

The procedure followed above to derive the MANOVA model from the ANOVA model is quite general [the SAS procedure GLM requires the MANOVA model (see the appendix to this chapter)]. Derivation of equation 2 from equation 1 shows that (1) the multivariate intercept contains the univariate main effects of year, (2) the multivariate main effects of growth category combine the univariate growth category main effects, whose testing will not be affected by any lack of independence or homoscedasticity from the repeated measures, and the interaction between growth category and year, and (3) the interaction between individual tree and year is combined with the true experimental error and the univariate random effects of individual tree in the multivariate error term.

From a scale perspective, (1) the terms $\underline{m}$ and $\underline{a}_i$ in equation 2 take into account the temporal heterogeneity of the mean at large scale across growth categories and the departures from it for each growth category, respectively; (2) equations 1 and 2 apply to the individual series of the Wistar rat example as well, and point (1) holds, except that temporal heterogeneity of the mean is then at medium scale; (3) the covariance matrix of the multivariate random errors $\underline{e}_{ij}$ characterizes the temporal autocovariance structure of the profile vectors $\underline{x}_{ij}$ at small to large scales, within the extent considered; and (4) as a complement to (3), the diagonal entries of the covariance matrix (e.g., the within-year error variances) define the magnitude and scale of heteroscedasticity, as do the off-diagonal entries (e.g., the between-year error covariances), once standardized, for the autocorrelation.

In the *P. abies* example, the within-year error variance increases by a ratio of 5 from year 1 to year 15. This heteroscedasticity is at large scale only, the error variances being similar from one year to the next. Autocorrelation is strong and positive up to lag 14 (maximum lag), as it slowly decreases from 0.93 (lag 1) to 0.36 (lag 14). Multivariate homoscedasticity between growth categories (i.e., homogeneity of within-group covariance matrices) can be assessed in the SAS procedure DISCRIM (SAS Institute Inc. 1989), by using Bartlett's modification of the likelihood ratio test (Morrison 1990). The probability of significance for multivariate heteroscedasticity was 0.0158 in the example. The impact of such multivariate heteroscedasticity among groups on modified F testing in the repeated-measures ANOVA is limited (Crowder and Hand 1990).

Concerning the MANOVA, all test statistics (Wilks' lambda, Pillai's trace, Hotelling-Lawley trace, Roy's greatest root) provide the same F-equivalent statistic in the *P. abies* example. When they do not, the F statistic for Roy's greatest root is an upper bound, and the F statistic for Wilks' lambda is exact (Timm 1980). In the modified univariate

ANOVA, the number of degrees of freedom associated with the mean square of any effect (main effects or interaction) involving the repetition factor is reduced by the Greenhouse and Geisser (1959) or Huynh and Feldt (1976) estimate of Box's epsilon. These coefficients measure the discrepancies of the autocovariance structure of p-1 orthonormal contrasts of the repetition factor from sphericity (i.e., autocorrelation at no scale, and homoscedasticity at all scales). When the autocovariance structure of the contrasts is spherical, that of the raw data is said to be circular; the smaller the Box epsilon estimates, the larger the discrepancies from circularity (Potvin et al. 1990; Dutilleul and Potvin 1995). The values observed for the Box epsilon estimates are very low (table 18.1), attesting to the heteroscedasticity and positive autocorrelation discussed above.

The year main effects are highly significant whether the univariate or multivariate procedure is used; the first three polynomial contrasts

TABLE 18.1

Analysis of variance of temporal repeated measures: results of the multivariate and univariate procedures in terms of probabilities of significance $P > F$ for the *Picea abies* example

Variable	Effect	MANOVA*	ANOVA[†]	Modified ANOVA[‡]	
				G-G	H-F
Cumulative	GC[§]	0.8060	0.2357		
tree-ring width	Year	0.0001	0.0001	0.0001	0.0001
	GC-by-year	0.7397	0.0002	0.0806	0.0782
			Epsilon =	0.0936	0.0982

Year linear contrast across GC	0.0001
Year linear contrast by GC	0.0767
Year quadratic contrast across GC	0.0001
Year quadratic contrast by GC	0.2459
Year cubic contrast across GC	0.0001
Year cubic contrast by GC	0.5520

*MANOVA = multivariate analysis of variance. For each multivariate test statistic, there is an equivalent F-statistic (see text).

[†]ANOVA = univariate analysis of variance

[‡]In the modified ANOVA procedure, probabilities of significance $P > F$ are adjusted by using the Greenhouse-Geisser (G-G) and Huynh-Feldt (H-F) estimates of Box's epsilon correction factor (see text). This correction is applied to the number of degrees of freedom of the F-statistic, in order to take the among-year heteroscedasticity and autocorrelation into account.

[§]GC = growth category

[‖]GC-by-year = growth category-by-year interaction

of the year factor are also highly significant (table 18.1). The most striking result is provided by the nonsignificance of the interaction between growth category and year at the 0.05 level after adjustment, although this effect was highly significant before adjustment. After taking into account the temporal heteroscedasticity at large scale and autocorrelation at small to large scales, the two categories of trees do not differ in the temporal heterogeneity of the mean radial growth at large scale. The univariate main effects of growth category are not significant. In conclusion, there is no evidence in these results for a preconditioning of the radial growth of trees from either category.

The Autocorrelogram and Periodogram Analyses

The following statistical methods do not require the data to be collected under a temporal repeated-measures design, so they apply to unreplicated time series as well as to individual series and mean series when replicated time series are available. Autocorrelogram analysis is performed in the time domain where the data are collected; periodogram analysis is carried out after transformation into the frequency domain, for example, by application of the finite Fourier transform (Brillinger 1981; Priestley 1981; Diggle 1990; Box et al. 1994).

The theoretical autocorrelation function extends the simultaneous correlation coefficient between variables X and Y; it is defined as the correlation between variable X at time t and variable X at time t plus some lag k (e.g., Legendre and Dutilleul 1992). Its use was promoted in a forecasting perspective (Box and Jenkins 1976). The sample autocorrelation estimate used in the Wistar rat example (figure 18.1) is based on the sum of cross-products of the increments divided by the length of the series, and not by the length of the series minus the lag (Dutilleul 1995); this is also the way it is calculated in the SAS/ETS procedure ARIMA (SAS Institute Inc. 1988). The autocorrelogram results from the plot of the sample autocorrelation estimate against lag k.

The principle underlying the periodogram analysis is the decomposition of the time series into its frequency components, that is, into cosine and sine waves at different frequencies corresponding to the periodicities present in the series. With Schuster's (1898) periodogram, the time series, as a function of time, is submitted to a Fourier series expansion. As a result, the only frequencies considered, called Fourier frequencies, correspond to integer numbers of cycles over the time se-

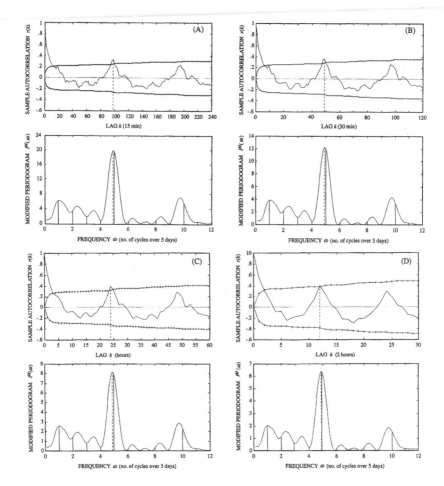

FIGURE 18.1

Sample autocorrelogram and periodogram computed from time-series realizations of motor activity in the Wistar rat, for sampling scales of **(A)** 15 minutes, **(B)** 30 minutes, **(C)** 1 hour, and **(D)** 2 hours, over 5 days in darkness. Time-series realizations are plotted in figure 17.2 of chapter 17, where a description of the data is given in text. Data source: Unité de Psychobiologie, Université catholique de Louvain, Louvain-la-Neuve, Belgium.

ries, and the classical periodogram statistic is defined, up to a multiplicative constant, as the squared amplitude of the cosine wave fitted to the series at each Fourier frequency. This approach has many limitations from a periodicity analysis perspective. It is restricted to the Fourier frequencies, although the series length is unlikely to correspond to an integer number of cycles of the periodic components because most periodicities are unknown in practice. The classical periodogram cannot detect or separate frequency components whose periods are close to one another, because it is unifrequential and will thus present one spurious inflated peak at an intermediate frequency instead of the two corresponding to the actual periodicities (Dutilleul 1990; Legendre and Dutilleul 1992; Girling 1995).

Some of the tools designed for the analysis of continuous spectra, including the maximum entropy spectral estimate (Brillinger 1981; Priestley 1981), can be useful in the analysis of discrete spectra. Nevertheless, the maximum entropy spectral estimate requires an autoregressive model of order 156 to split two frequencies of 1.9 and 2 cycles per day over a 5-day period (Girling 1995). A new periodogram statistic is thus required, modified in its computation with respect to the classical and, above all, multifrequential (Dutilleul 1988, 1990). The modified, multifrequential periodogram statistic benefits from the following properties: (1) it represents exactly the sum of squares of the trigonometric model fitted by least squares to the series at the frequencies considered, (2) under the normality assumption for the series, it follows a chi-square distribution, in small as well as in large samples, and (3) when the number of frequencies involved in its computation are equal to the number of periodicities of the series, the frequencies maximizing the periodogram statistic are unbiased estimates of the "real" frequencies. A stepwise procedure was designed to estimate the number of "real" periodicities (see the appendix to this chapter). Applications of the statistical method are reported in Dutilleul and Till (1992), Schelstraete et al. (1995), Aboul-Hosn et al. (1997), and Tardif et al. (1998).

From a scale perspective, the experimenter can increase the number of sampling time points in periodicity analysis by following two scenarios: (1) by reducing the grain and keeping the extent fixed, or (2) by keeping the same grain and increasing the extent. Which scenario should be preferred and why? Two situations are considered here (table 18.2) for a series length of 25, both with two periodicities, either similar to one another with periods 12.5 and 15, or separate with periods 5 and 12.5. Differences between the two scenarios in terms of empirical bias and variance are substantial and in favor of scenario 1 when the two periodicities are close to one another, especially when the signal-to-noise ratio is low, whereas differences are sensible and in favor of one or the other of the scenarios when the periodicities are

TABLE 18.2

Periodicity analysis: simulation results in terms of empirical bias, empirical variance, and asymptotic variance of frequency estimates in the bi-frequential case

Series length	Signal-to-noise ratio	Period	Frequency parameter	Empirical bias	Empirical variance	Asymptotic variance	Number of simulations
25	25	12.5	2	-15.10^{-3}	20.10^{-3}	22.10^{-3}	4000
		15	5/3	43.10^{-3}	29.10^{-3}	16.10^{-3}	
50	25	12.5	4	-20.10^{-4}	19.10^{-4}	19.10^{-4}	4000
		15	10/3	-6.10^{-4}	21.10^{-4}	21.10^{-4}	
50	25	25	2	-10.10^{-3}	10.10^{-3}	11.10^{-3}	4000
		30	5/3	-25.10^{-3}	14.10^{-3}	9.10^{-3}	
100	25	12.5	8	4.10^{-4}	26.10^{-5}	27.10^{-5}	4000
		15	20/3	-8.10^{-4}	28.10^{-5}	29.10^{-5}	
100	25	50	2	-8.10^{-3}	54.10^{-4}	57.10^{-4}	4000
		60	5/3	10.10^{-3}	58.10^{-4}	46.10^{-4}	
200	25	12.5	16	-2.10^{-4}	12.10^{-5}	13.10^{-5}	2000
		15	40/3	-2.10^{-4}	14.10^{-5}	13.10^{-5}	
200	25	100	2	-45.10^{-4}	26.10^{-4}	28.10^{-4}	2000
		120	5/3	63.10^{-4}	27.10^{-4}	24.10^{-4}	
25	1	12.5	2	-37.10^{-2}	18.10^{-1}	56.10^{-2}	4000
		15	5/3	-30.10^{-2}	13.10^{-1}	41.10^{-2}	
50	1	12.5	4	-55.10^{-3}	65.10^{-3}	46.10^{-3}	4000
		15	10/3	-31.10^{-3}	40.10^{-3}	52.10^{-3}	
50	1	25	2	-20.10^{-2}	51.10^{-2}	28.10^{-2}	4000
		30	5/3	24.10^{-2}	19.10^{-2}	22.10^{-2}	
100	1	12.5	8	40.10^{-4}	71.10^{-4}	68.10^{-4}	2000
		15	20/3	-45.10^{-4}	79.10^{-4}	73.10^{-4}	
100	1	50	2	-12.10^{-2}	20.10^{-2}	14.10^{-2}	2000
		60	5/3	17.10^{-2}	13.10^{-2}	12.10^{-2}	
200	1	12.5	16	-13.10^{-4}	32.10^{-4}	32.10^{-4}	2000
		15	40/3	-11.10^{-4}	31.10^{-4}	33.10^{-4}	
200	1	100	2	-57.10^{-3}	55.10^{-3}	71.10^{-3}	2000
		120	5/3	12.10^{-2}	81.10^{-3}	59.10^{-3}	
25	25	5	5	-9.10^{-4}	10.10^{-4}	11.10^{-4}	4000
		12.5	2	6.10^{-4}	11.10^{-4}	10.10^{-4}	
50	25	5	10	-9.10^{-4}	52.10^{-5}	50.10^{-5}	4000
		12.5	4	10.10^{-4}	48.10^{-5}	49.10^{-5}	
50	25	10	5	-13.10^{-5}	52.10^{-5}	52.10^{-5}	4000
		25	2	4.10^{-4}	50.10^{-5}	52.10^{-5}	
100	25	5	20	-5.10^{-4}	24.10^{-5}	24.10^{-5}	4000
		12.5	8	2.10^{-4}	26.10^{-5}	24.10^{-5}	
100	25	20	5	-6.10^{-4}	25.10^{-5}	26.10^{-5}	4000
		50	2	3.10^{-4}	26.10^{-4}	26.10^{-5}	
200	25	5	40	-9.10^{-4}	14.10^{-5}	12.10^{-5}	4000
		12.5	16	-4.10^{-4}	14.10^{-5}	12.10^{-5}	

TABLE 18.2
Continued

Series length	Signal-to-noise ratio	Period	Frequency parameter	Empirical bias	Empirical variance	Asymptotic variance	Number of simulations
200	25	20	5	3.10^{-4}	13.10^{-5}	13.10^{-5}	4000
		50	2	2.10^{-4}	13.10^{-5}	13.10^{-5}	
25	1	5	5	-24.10^{-3}	86.10^{-3}	27.10^{-3}	4000
		12.5	2	4.10^{-4}	48.10^{-3}	26.10^{-3}	
50	1	5	10	-54.10^{-4}	16.10^{-3}	12.10^{-3}	4000
		12.5	4	-6.10^{-4}	15.10^{-3}	12.10^{-3}	
50	1	10	5	-38.10^{-4}	23.10^{-3}	13.10^{-3}	4000
		25	2	47.10^{-4}	18.10^{-3}	13.10^{-3}	
100	1	5	20	$<1.10^{-5}$	64.10^{-4}	61.10^{-4}	4000
		12.5	8	-23.10^{-4}	65.10^{-4}	61.10^{-4}	
100	1	20	5	-16.10^{-4}	67.10^{-4}	64.10^{-4}	4000
		50	2	8.10^{-4}	71.10^{-4}	65.10^{-4}	
200	1	5	40	20.10^{-4}	32.10^{-4}	30.10^{-4}	4000
		12.5	16	-9.10^{-4}	32.10^{-4}	30.10^{-4}	
200	1	20	5	14.10^{-4}	32.10^{-4}	32.10^{-4}	4000
		50	2	-10.10^{-4}	32.10^{-4}	32.10^{-4}	

From Dutilleul (1990).

separate. The empirical bias and variance decrease when the series length increases, especially when the two periodicities are close to one another; this feature holds, to a lesser degree, when the periodicities are separate, because the accuracy and precision are then high from the beginning (i.e., $p = 25$), whether the signal-to-noise ratio is high or low. Except when the signal-to-noise ratio is low (i.e., 1.0) and the series is short (i.e., $p = 25$), scenario 1 provides frequency estimates with empirical bias and variance of the order of 1%. When comparing the empirical variance with the asymptotic variance (Gallant 1987), good concordance between variances is observed for both scenarios when the periodicities are separate. When the periodicities are close to one another, this holds more for scenario 1 than for scenario 2, especially when the signal-to-noise ratio is low.

In the Wistar rat example, the number of time intervals over which the motor activity of the young is accumulated decreases from 480 to 60, as the grain increases from 15 minutes to 2 hours (see figure 17.2). The data submitted to autocorrelogram and periodogram analyses are from days 23 to 27, during which the experimental young previously submitted to an atypical *Zeitgeber* were maintained in constant darkness (DD conditions) (see chapter 17; Schelstraete et al. 1992, 1993); a *Zeitgeber* designates a forcing oscillation that entrains a biological rhythm (e.g., light–dark, temperature cycle). The objective here is to detect the free-running rhythm with a period supposedly greater than the 24 hours of the circadian rhythm. A five-cycle extent and a 15-minute grain were recommended in chapter 17.

The autocorrelogram and periodogram analyses perform well when the grain is 15 and 30 minutes; both methods then allow the detection of a periodicity of about 24.5 hours (see figure 18.1A,B). The sample autocorrelation estimate is significant and locally maximized at lags 98 and 49 when the grain is 15 and 30 minutes, respectively; the modified periodogram statistic is significant and globally maximized around frequency 4.9 in both cases. The periodogram plot has been restricted to the 0 to 12 frequency band because of the flatness of the statistic outside this band; spectrum lines are for the Fourier frequencies. The sample autocorrelation estimate at lag 48 (24-hour period) is almost equal to the one at lag 49 in figure 18.1B, but in figure 18.1C,D, the sample autocorrelation is significant and locally maximized at lags 24 and 12, and not at lags 25 and 13. The sample autocorrelation estimate is constrained to be evaluated at multiples of the grain, so the observation of a peak was merely probable at lag 25 (25-hour period) in figure 18.1C, and unlikely at lag 13 (26-hour period) in figure 18.1D, if the "real" rhythm was of 24.5 hours. The striking result here is that whatever the grain in figure 18.1, the modified periodogram is significant and globally maximized around frequency 4.9 (24.5-hour period), and it continues to be, up to the very coarse grain of 8 hours (results not reported in figure 18.1). Except for coarse grains, the classical Schuster periodogram performed as well as the unifrequential version of the modified periodogram in the Wistar rat example (results not reported), essentially because there was only one rhythm. A centering of the time series to a zero mean is required by the periodogram analysis (standardization provides such centering in the Wistar rat example) because the intercept represents a cosine wave at zero frequency and acts as a noncentrality parameter for the periodogram statistic when evaluated at non-Fourier frequencies. In the case of free-running rhythms, the day loses its character of natural time unit, in favor of another evolving over time. Contrary to the 24.5-hour rhythm, the detection of positive autocorrelation at small lags does not seem to be affected by the sampling scale.

In summary, the use of an appropriate statistical tool is critical, whatever the sampling scale. The frequency-domain approach based on the periodogram analysis allows a fine resolution, whatever the grain for a given extent, by screening a time series for hidden periodicities from low to high frequencies continuously. The time-domain approach based on the autocorrelogram analysis relies heavily on a suitable sampling scale to be effective. Accumulating or integrating a continuous-time process over successive time intervals is the only data transformation discussed here.

The ANOVA of the Finite Fourier Transform

When replicated time series are collected in accordance with an experimental design, the ANOVA of the finite Fourier transform (Brillinger 1973, 1980; Dutilleul 1990) (1) allows the assessment of differences in periodicity among the levels of the classification factors and (2) provides information on the degree to which replicated time series collected under different experimental conditions share the same periodicities. This particular type of ANOVA complements the periodogram analysis in a multiscale approach and is available when the time series are long and the number of replicated time series are small; in this case, the condition $n - g > p$ is not satisfied, the multivariate procedure of repeated-measures ANOVA in the time domain is not available, and the estimation of Box's epsilon in the modified univariate procedure suffers from lack of precision. The use of the finite Fourier transform is justified in the ANOVA framework because the cosine and sine functions evaluated at the Fourier frequencies represent orthogonal contrasts of the repetition factor and allow the orthogonal decomposition of all sums of squares involving time, as in a Fourier series expansion. Multiple testing in this spectral context can be taken into account by using the Bonferroni correction (Dutilleul and Till 1992). Applications are reported in Deswysen et al. (1989, 1993), Dutilleul and Till (1992), Schelstraete et al. (1992), Aboul-Hosn et al. (1997), and Tardif et al. (1998).

Let ω_k denote the angular Fourier frequency of k cycles over the time series and $e^{i.\omega_k.t} = \cos(\omega_k.t) + i.\sin(\omega_k.t)$ be the complex exponential function, the finite Fourier transform of the time series $x = (X_1, \ldots, X_t, \ldots, X_p)'$ at frequency ω_k is then defined by

$$\underline{X}^{\mathrm{FFT}}(\omega_k) = \frac{1}{\sqrt{2\pi p}} \sum_{t=1}^{p} X_t e^{i.\omega_k.t} \quad \text{(e.g., Priestley 1981, p. 418).} \quad (3)$$

The model for the ANOVA of the finite Fourier transform at frequency ω_k that corresponds to equation 1 follows, after all terms not depending on t have vanished:

$$\underline{X}_{ij}^{FFT}(\omega_k) = \underline{C}^{FFT}(\omega_k) + \underline{A}_i^{FFT}(\omega_k) + \underline{E}_{ij}^{FFT}(\omega_k) \qquad (4)$$

where subscripts i and j are as in equations 1 and 2; $\underline{C}^{FFT}(\omega_k)$, $\underline{A}_i^{FFT}(\omega_k)$, and $\underline{E}_{ij}^{FFT}(\omega_k)$ are the finite Fourier transform at frequency ω_k of $\{c_t, (ac)_{it}, \varepsilon_{ijt}; t = 1, \ldots, p\}$, respectively; $\underline{C}^{FFT}(\omega_k)$ is the intercept in equation 4; $A_i^{FFT}(\omega_k)$ represents the fixed effect of treatment or classification level i; and $\underline{E}_{ij}^{FFT}(\omega_k)$ is the random error. When k is greater than 0, the model is bivariate, the two components of the random vector $\underline{X}_{ij}^{FFT}(\omega_k)$ are the real and imaginary parts of the finite Fourier transform, and the MANOVA test statistics (Wilks' lambda, Pillai's trace, Hotelling-Lawley trace, Roy's greatest root) can be used to test the main effects $\underline{A}_i^{FFT}(\omega_k)$ against the errors $\underline{E}_{ij}^{FFT}(\omega_k)$. The finite Fourier transform benefits from the statistical properties of asymptotic normality and independence (Brillinger 1983). The multiplicative constant before summation in equation 3 may differ depending on the author (Schuster 1898; Brillinger 1983) or the statistical package (SAS/ETS; S-PLUS), with no repercussion in the ANOVA.

The Space Case

General Aspects

The ANOVA models corresponding to experimental designs recommended in companion chapter 17 for given types of spatial heterogeneity are available in Dutilleul (1993b) and will not be discussed here. As an alternative to the ANOVA approach based on random factors, space may be incorporated into data analysis in the form of explanatory variables of spatial location (e.g., geographical coordinates) in a regression approach called trend-surface analysis (e.g., "Student" 1914; Ripley 1981; Legendre 1993), where a polynomial, for instance, is fitted to the spatial mean function. Space levels are then quantitative and continuous; a stepwise or forward procedure may help solve the problem of selecting the polynomial order. Between the ANOVA and regression approaches lies the analysis of covariance (ANCOVA), where the random factor may be defined by large-scale spatial units (e.g., blocks), and the location of smaller-scale units (e.g., points on a subgrid) may provide the spatial explanatory variables. Each spatial term refers to a given scale in the ANOVA and ANCOVA models, whereas trend-surface analysis tends to be restricted to large-scale heterogeneity. The relationship between ecological variables may also vary with scale, but correlation analysis between presampled and actual experimental data must be performed with allowance for the autocorrelation of spatial (and temporal) data (Dutilleul 1993a).

Concerning the problem of spatial autocorrelation in the ANOVA, different types of solutions have been proposed outside the repeated-measures ANOVA. Some are model-based and take spatial autocorrelation into account by correcting the error variance estimate; examples include the nearest-neighbor analysis (Bartlett 1978; Wilkinson et al. 1983) and the GLS-variogram method (Ver Hoef and Cressie 1993). Others are model-free and adopt a nonparametric approach by incorporating autocorrelation in a permutational procedure of testing (Legendre et al. 1990).

The spatial version of spectral analysis exists (Ripley 1981), but it is not popular in spatial statistics because of the lack of periodicities or regularities in space, patches excepted. Concerning spatial repeated measures, the 1D case can be treated in a manner similar to that used for the temporal one; for the 2D spatial repeated measures, I refer to the spatiotemporal case, where time is to be replaced by one of the two spatial repetition factors.

This section poses three questions. First, how does fractal dimension, as a measure of complexity in space occupation, change with spatial pattern and sampling? This point will be illustrated by simulated data. Second, how does spatial autocorrelation vary anisotropically (i.e., unequally), horizontally versus vertically? Zooplankton biomass sampled along transects in Lake Erie will provide a case study as illustration. Third, how does the sample support affect the perception of a spatial pattern (e.g., patches), visually in mapping and numerically in the analysis of Geary's c correlogram and in the computation of the fractal dimension as a measure of complexity in surface pattern? The incorporation of scale into the analysis of ecological multivariate data is also discussed.

Fractal Dimension and Space Occupation

Fractal dimension is considered here less as a tool to assess the "fractality" of an object (i.e., its ability to reproduce similar structures at different scales) (Mandelbrot 1977, 1983), than as a measure of complexity in space occupation or in surface pattern (Frontier 1987; Milne

1988, 1991b). As a measure of complexity in space occupation, the fractal dimension may be computed by using the box counting method. In the 2D space, the area containing the shape of interest is covered by "boxes" (i.e., 2D squares) of a given size. For a given box size r, the number of boxes intersecting the shape, $N(r)$, is counted. Counting is repeated for various box sizes, generally geometrically decreasing, and the resulting counts are plotted in a $\log[N(r)]$ versus $\log(r)$ graph. The fractal dimension estimate is given by the slope of the straight line fitted. The procedure is started with two box sizes, and it is stopped when no change is observed in the fractal dimension estimate by considering one more box size.

Various spatial patterns of vegetation cover were simulated, and the fractal dimension was estimated at different sampling scales. Quantitative continuous data were first simulated on a 40×40 grid (see figure 17.3, chapter 17). Binary data were then provided, with 0 and 1 whenever the quantitative data were falling below and above the 50th percentile of the sample data distribution, respectively. Finally, two types of subsampling were performed: one by retaining only the upper left corner of the 400 2×2 subgrids nested within the 40×40, which resulted in a 20×20 grid; the other by retaining similarly the upper left corner of the 100 4×4 subgrids nested within the main one, with a 10×10 grid as result. The area covered remains almost the same, and so does the extent, whereas the grain increases as the size of the grid (i.e., number of nodes) decreases. The interpolation parameters used for the mapping in MacGridzo (RockWare, Inc. 1991) were kept fixed whatever the sampling scale, in order to avoid any confounding between interpolation and sampling scale effects.

The increase in sampling scale (figure 18.2) mostly affects our perception of complexity in space occupation for small-scale heterogeneity, which appears at the finest sampling scale but tends toward patches as the sampling scale increases. To some degree, the linear gradient in the mean and variance and the single patch of ellipsoid shape follow a similar pattern of apparent decreasing complexity in space occupation with increasing sampling scale. The linear gradient in the mean is the least affected.

The features observed above are confirmed numerically (table 18.3); the interpretation rule is that the closer the fractal dimension is to 2.0 (1.0), the more (less) complex the space occupation. The highest fractal dimension (1.89) is estimated at the 40×40 sampling scale for small-scale heterogeneity combined with large-scale homogeneity. For this type of spatial heterogeneity, the fractal dimension estimate drastically

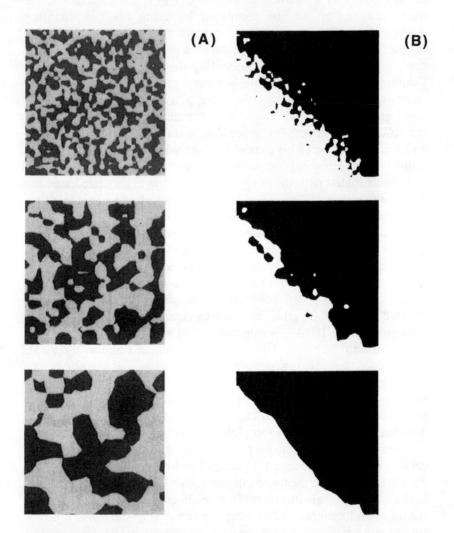

FIGURE 18.2
Simulated patterns of vegetation cover, as a function of sampling scale. Data were initially simulated as in figure 17.3A–D of chapter 17 and then discretized (*light gray* = not covered; *black* = covered), below and above the 50th percentile of the sample data distribution. For the same area, sampling grids are (1) 40 × 40, (2) 20 × 20, and (3) 10 × 10. Fractal dimension, as a measure of complexity in space occupation, is reported for these patterns and others in table 18.3.

decreases with sampling scale to reach a level (1.47) comparable to an average type of patchy heterogeneity. When large-scale heterogeneity is of gradient type, the fractal dimension estimate changes slightly with sampling scale (from 1.87 to 1.85) for the linear gradient in the mean, and substantially (from 1.74 to 1.52) for the linear gradient in the mean and variance. In the case of patchy heterogeneity, contiguity and separation of patches, as well as their number, size, and shape, variously affect the impact of sampling scale on perceived complexity in space occupation. In particular, complexity increases with decreasing sam-

TABLE 18.3
Fractal dimension* estimate in relation with spatial pattern and sampling scale

Spatial pattern	Scale 1 (40 × 40)	Scale 2 (20 × 20)	Scale 3 (10 × 10)
Small-scale heterogeneity Large-scale homogeneity (see figure 18.2A)	1.89	1.77	1.47
Large-scale heterogeneity Linear trend in the mean (see figure 18.2B)	1.87	1.85	1.85
Quadratic trend in the mean	1.73	1.78	1.74
Linear trend in the mean and variance (see figure 18.2C)	1.48	1.52	1.74
Patchy heterogeneity One central ellipsoid patch (see figure 18.2D)	1.56	1.51	1.49
One central square patch	1.47	1.38	1.34
Patch size	(20 × 20)	(10 × 10)	(5 × 5)
Two square, contiguous patches on opposite corners of the field (i.e., bottom left, top right)	1.74	1.74	1.67
Patch size	(20 × 20)	(10 × 10)	(5 × 5)
Four square patches			
Separate, located on a 2 × 2 lattice[†]	1.49	1.35	1.50
On each corner of the field	1.50	1.48	1.59
Contiguous, along the diagonal	1.39	1.50	1.56
Separate, randomly distributed	1.50	1.47	1.45
Patch size	(10 × 10)	(5 × 5)	(2.5 × 2.5)

*The fractal dimension is computed here as a measure of complexity in space occupation (e.g., by vegetation cover). Simulated data were used; see text for the simulation procedure.
†The nodes of the 2 × 2 lattice, where the four patches are centered, are nodes (10, 10), (10, 30), (30, 10) and (30, 30), of the 40 × 40 sampling grid (Scale 1).

pling scale in the case of four contiguous square patches located sequentially (the fractal dimension estimate then increases from 1.39 to 1.56) and, to a lesser degree, when four square patches are located on a 2×2 lattice on each corner of the field (from 1.50 to 1.59). A reverse pattern of decreasing fractal dimension estimate (from 1.50 to 1.35) is observed when four square patches are separately located on a 2×2 grid; the spatial pattern then tends to be repeated four times over the field, resulting in a loss of complexity in the overall pattern rather than in the contours of the shape as in figure 18.2D.

Spatial Autocorrelation Analysis: A Case Study

Spatial autocorrelation can be defined as the statistical property of a random variable to take values at two sampling locations, which are more similar (positive autocorrelation) or less similar (negative autocorrelation) than expected for pairs of randomly associated values (e.g., Cliff and Ord 1973, 1981; Griffith 1987; Legendre 1993). It can be investigated by the (auto)correlogram analysis based on Moran's I or Geary's c statistic, and the (semi-)variogram analysis. Moran's I statistic essentially is a sample autocorrelation coefficient, so its interpretation is similar to Pearson's r linear correlation coefficient; Moran's I, however, refers to the correlation between a variable at a given location and the same variable some distance apart. The idea underlying Geary's c and semivariance statistics is somewhat different because it uses the property that the variance of the difference between two random variables depends on their covariance; in a temporal or spatial context, the two variables are the same variable observed on distinct sampling points, and the covariance is to be replaced by the theoretical autocovariance function. The R package (Legendre and Vaudor 1991) has been used for the computation of spatial correlograms (here and in the next subsection).

The definition of distance classes is required prior to computing any correlogram or variogram statistic, the observations of the pairs of points falling in each distance class being used in the computation of the corresponding correlogram or variogram ordinate. To ensure even precision in the estimation and even power in testing across distance classes (i.e., across scales), equal-frequency distance classes should be used instead of equally spaced distance classes (Dutilleul and Legendre 1993). The total number of distance classes must also be fixed so that the smallest distance class matches the grain, or two times the grain, as much as possible.

The purpose here is to show that the same variable may show different patterns of spatial autocorrelation: (1) in a given experimental field depending on the horizontal axis of investigation and (2) within a given horizontal axis depending on the vertical range. This will be illustrated by the spatial distribution of zooplankton biomass in Lake Erie. Data are from two transects (5654 and 3942) sampled in September 1994, which run from south (nearshore) to north (midlake) in the central basin and from midlake south (deepest part of the lake) to the south shore in the east basin, respectively (Stockwell and Sprules 1995; Stockwell 1996). The epilimnion stratum covers the entire water column in transect 5654, whereas transect 3942 has been split into four parts: (1) the entire water column from 3 m down to sampling depth of the hydrodynamic V-fin, (2) the epilimnion (greater than or equal to 19.25°C), (3) the upper half of the epilimnion (near the surface), and (4) the lower half of the epilimnion. Offshore data cover 40 km in transect 5654 and 30 km in transect 3942 (figure 18.3). Most data are equally spaced in transect 5654, with 93.67 percent of the observations less than 0.5 km apart; the sampling distance between successive observations is more dispersed in transect 3942, with 40.82 percent of the sampling distances ranging between 0.3 and 0.4 km, and about 55 percent between 0.5 and 1.0 km. The grain thus seems to be well defined in transect 5654; this is true to a lesser degree for transect 3942. Fifteen equal-frequency distance classes were retained, after trying 10 and 20, because 15 provided an upper bound of 1.0 km for the smallest distance class, which corresponds to twice the grain in transect 5654 and approximates the grain in transect 3942. Fifteen distance classes also provide a good compromise in terms of resolution and statistical efficiency. The number of pairs of observations involved in the compu-

FIGURE 18.3

Spatial autocorrelation analysis for zooplankton biomass in Lake Erie (Ontario, Canada). Data plot and corresponding Moran's *I* correlogram for **(A)** transect 5654, entire water column, (A2) without and (A3) with stationarity of the mean, following trend removal (see *dashed line* in A1), **(B)** transect 3942, entire water column, (B2) without and (B3) with stationarity of the mean, following trend removal (see *dashed line* in B1), **(C)** transect 3942, epilimnion stratum, **(D)** transect 3942, upper half of the epilimnion (near surface), **(E)** transect 3942, lower half of the epilimnion. *Black squares* in correlograms indicate the ordinates significant at the 0.05 level (without Bonferroni correction). Only offshore data were used for illustration and analysis. See text for detail; see also Stockwell (1996). Data source: Department of Zoology, University of Toronto, Erindale College, Mississauga.

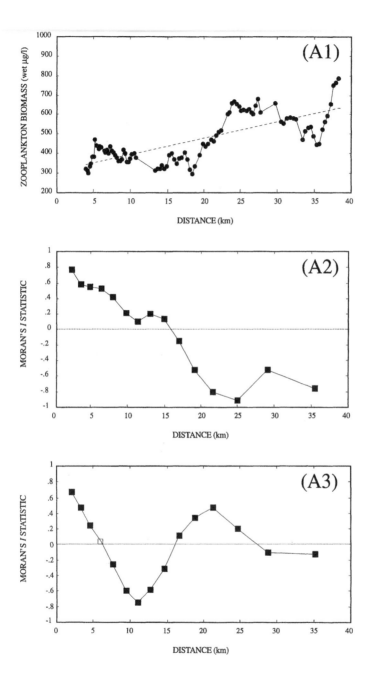

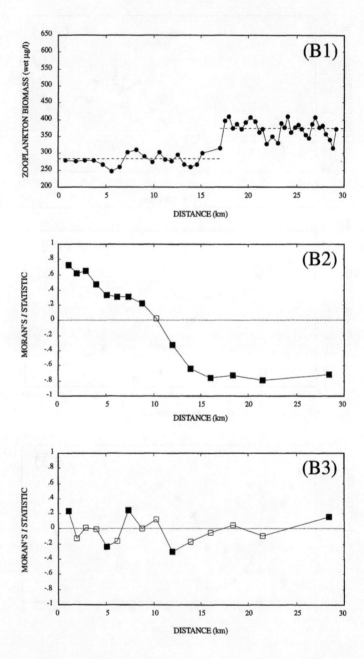

tation of the correlogram ordinates was 297 to 298 for transect 5654, and 82 for transect 3942; it would have been 446 or 447 and 223 or 224 for transect 5654, and 121 to 123 and 61 or 62 for transect 394, for 10 and 20 distance classes, respectively.

Changes in spatial autocorrelation may be observed between transects (figure 18.3A–B) and within transect 3942 (figure 18.3B–E). The nonstationarity in the mean of zooplankton biomass in the entire water column from both transects (of gradient type with a cyclic component in transect 5654 and of two-level type in transect 3942) is reflected by the sustained decrease of Moran's I statistic from 1.0 at small distances to -1.0 at large distances; it is also responsible for spurious significant autocorrelation [figure 18.3(A2,B2)]. After removal of that nonstationarity in the mean [by fitting a simple linear regression based on spatial location along the transect in figure 18.3(A1) and by cen-

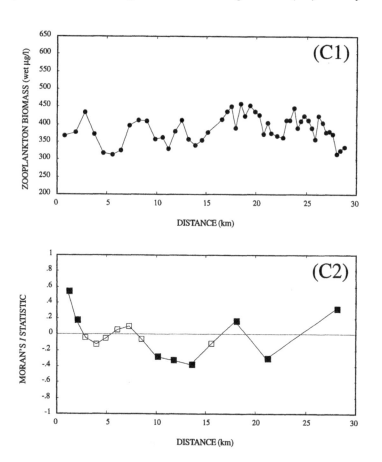

tering the mean of both portions of the transect to zero in figure 18.3(B1)], pseudocycling every 18.5 km combined with strong positive small-scale autocorrelation appears for transect 5654 [figure 18.3(A3)]; only slight, positive small-scale autocorrelation remains for transect 3942 [figure 18.3(B3)]. The other correlograms translate the regularities in the spatial distribution of zooplankton biomass (i.e., sequences of peaks and troughs more or less equally spaced) and their change, depending on the part of the epilimnion stratum considered. After identifying small-scale autocorrelation once the nonstationarity in the mean has been removed, the simple linear regression and centering of the mean should be performed again, using the generalized least-squares estimation method (Searle 1971). Regional analysis can be performed at the macro-scale over the whole lake, but the two approaches provide totally different pictures (Stockwell 1996).

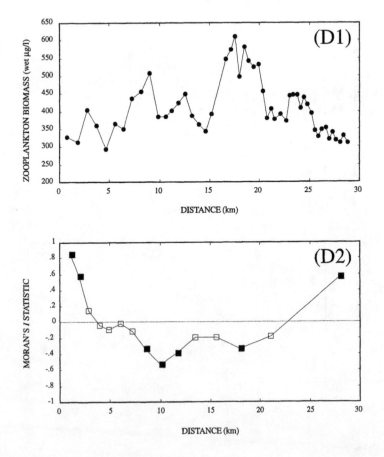

Data Transformation and Fractal Dimension of Surface Pattern

Normality is a classical prerequisite for most parametric statistical methods, and the transformations aiming to improve the normal distribution of data (Box and Cox 1964) will necessarily affect the measurement scale. Prewhitening procedures (Cliff and Ord 1981) and variance-stabilizing techniques (Box and Cox 1964; Dutilleul and Potvin 1995) are other examples. Moving averages have varied effects because they can remove autocorrelation when present or introduce it when absent. In general, a moving average fills in the troughs and polishes the peaks, thereby reducing variation in the data. In geostatistics, "block averaging" is another smoothing method with similar effects, in which the values of a variable are averaged within a prescribed local

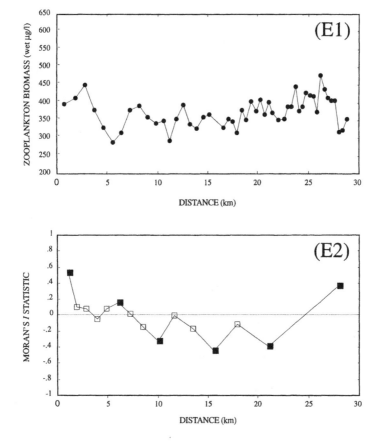

area (Isaaks and Srivastava 1989). When block kriging is preferred to ordinary point kriging, the change of sample support (i.e., size of sampling unit), and thus of scale, will affect the results of data analysis. Kriging is a spatial interpolation method that minimizes the prediction error variance; it requires the fitting of a model to the variogram computed from the data (Isaaks and Srivastava 1989; Cressie 1993).

Spatial averaging, moving or not, can affect the fractal dimension as a measure of complexity in surface pattern. I simulated a 25×25 grid that can be seen as 25 5×5 subgrids. Within each 5×5 subgrid, a 2×2 patch, whose upper left corner was randomly located at node (2, 2), (2, 3), (3, 2), or (3, 3), was defined. A spatial moving averaging was then performed within each subgrid, from quarter to quarter (each quarter is of size 3×3, and two contiguous quarters have three nodes in common), and similarly across subgrids; this resulted in a 10×10 grid. Another (nonmoving) spatial averaging was carried out per subgrid, resulting in a 5×5 grid. The fractal dimension was computed for each map, following Burrough (1981), except that Geary's c statistic was used instead of the semivariance, because the two statistics differ by a multiplicative factor (i.e., the process variance), and the fractal dimension estimation uses a log-log plot. The statistical significance of the ordinates can be assessed in Geary's c correlogram, whereas the nugget effect or small-scale sampling variance and the sill or process variance are parameters of interest in the variogram; both share the property of showing the range or minimum distance after which data are spatially independent (Legendre and Fortin 1989; Dutilleul and Legendre 1993).

The closer to 3.0 (2.0) the fractal dimension, the more complex (smoother) is the surface pattern. The results are a fractal dimension estimate of 2.14 for the original data, and values of 2.25 and 2.03 for the moving and nonmoving spatial averages, respectively. At the finest scale, the regular (i.e., patchy) pattern of heterogeneity is clearly apparent for the original data, both in the mapping and in the correlogram [figure 18.4(A1,2)]. The lower fractal dimension estimated for the original data, compared to the moving spatial averages, follows from the sampling scale that matches the scale of the phenomenon in the former case (Burrough 1981). The moving spatial averaging provides spatial means computed at a medium scale where patchiness is distorted; there is residual patchiness in some parts of the field, whereas no peak remains or troughs are created in some others [figure 18.4(B1,2)]; as a result, the complexity in surface pattern is increased. The initial

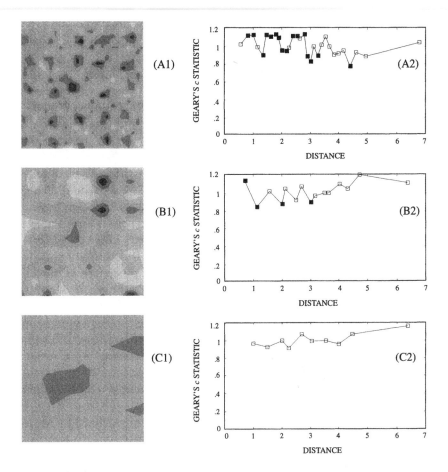

FIGURE 18.4

Patchy heterogeneity in the mean parameter (25 patches of size 2 × 2 located in 25 quadrats of size 5 × 5, one patch per quadrat) and the corresponding Geary's c correlogram, as a function of the sample support. **(A)** Original data. **(B)** Spatial means computed per quarter of quadrat. **(C)** Spatial means computed per quadrat. *Gray tones* used in mapping are the same for all panels; the absence of certain tones in some maps indicates the absence of data for the corresponding contour levels. *Black squares* in correlograms indicate the ordinates significant at the 0.05 level (without Bonferroni correction). See text for detail.

patchiness has totally disappeared from figure 18.4(C1,2), where the grain is twice the patch size and the surface smoothness is high.

Scale and Multivariate Data Analysis: An Outline

Important tools in multivariate data analysis are (1) canonical ordination methods such as redundancy analysis (RDA) and canonical correspondence analysis (CCA) (e.g., Jongman et al. 1987), and (2) the Mantel test (Mantel 1967). Concerning the RDA and CCA, Borcard et al. (1992) showed how to incorporate space into the analysis of multivariate ecological data. Only large-scale spatial heterogeneity in the mean parameter is taken into account by a trend–surface polynomial in the original method of variation partitioning. Legendre and Borcard (1994) recommend the use of matrices with binary entries 1–0 (1 = sampling locations i and j are nearest neighbors; 0 = they are not), to locally fit small-scale spatial heterogeneity. The procedure is model-free, that is, it does not require a unique model for the whole data set, although the isotropic spatial first-order autoregressive model (e.g., Cliff and Ord 1981) underlies the idea. Its drawback is that a regression tool designed for analysis of the spatial mean function is used to incorporate what is likely to be small-scale autocorrelation. As a result, there may be inefficiency in the fitting of that small-scale component of spatial heterogeneity, and a maximum likelihood procedure might be preferable (Ord 1975).

Concerning the Mantel test, the statistic originally defined by Mantel (1967) is a generalized regression coefficient between two matrices of distances (spatial and temporal, respectively), whose significance is assessed by permutation. Its use is not restricted to spatial and temporal distances but applies to any ecological proximities (see Legendre and Legendre 1997 for a review). The use of Mantel's statistic has been extended to the computation of a multivariate correlogram and the hypothesis testing of spatial patterns (see Legendre 1993 for a review). The statistic is then computed between a distance matrix (or more generally, a proximity matrix) and a binary matrix corresponding to a given distance class or a given spatial pattern to be tested.

When a distance in the Mantel correlogram is spatial, it may seem that the interpretation rule is similar to that of the correlograms displayed in figures 18.3 and 18.4; actually it is not. A clear distinction must be made between the "raw data world" in which the autocorrelogram analysis and the ANOVA are performed, and the "distance world" in which the Mantel test is carried out. Accordingly, the hypotheses to be tested are different. In contrast with the absence of autocorrelation at a given spatial distance in the autocorrelogram analysis and the absence of differences among the mean values of the variable(s) in the (M)ANOVA, the hypothesis in the Mantel correlogram is that there is no difference between the average monospatial distance within a given group and the average monospatial distance across the other groups, where the group is a spatial distance class. This restricts the comparison among the methods with respect to their performance. Because the conclusions may change with the distances used, special caution must be devoted to the choice of the distances in the Mantel test. When more than one variable is involved, standardization of the variables (to a zero mean and a unitary variance) prior to computation of the distances is recommended. Such a measurement rescaling is performed in multivariate data analysis (RDA, CCA, principal components analysis) whenever the material is heterogeneous in nature.

The Spatiotemporal Case

General Aspects

In the absence of replication, there are spatiotemporal versions of the autocorrelogram and semivariogram analyses (Cliff and Ord 1981; Rouhani and Wackernagel 1990). Despite their sophistication, these techniques are demanding in terms of application conditions (e.g., stationarity). In the presence of replication, the spatiotemporal repeated-measures ANOVA (see later) is available if $n - g > p;$ when the condition is not met, a procedure of maximum likelihood estimation under the doubly multivariate model of the matrix normal distribution may be used (Dutilleul 1994). The latter procedure relaxes the stationarity conditions on the mean and autocovariance, and it allows modified ANOVA and modified correlation analysis with spatiotemporal data (Dutilleul and Pinel-Alloul 1996), by using Box's correction and Dutilleul's (1993a) effective sample size for the number of degrees of freedom, respectively. The use of multivariate data analysis methods such as RDA and CCA, combined with the variation partitioning of Borcard et al. (1992), is valid as long as space and time are considered as ANOVA factors in the spatiotemporal approach. In the absence of reliable spatiotemporal distance, the Mantel test and its derived forms are not applicable.

The Spatiotemporal Repeated-Measures ANOVA

The experimental design is discussed in companion chapter 17. The example here is provided by the fractal dimension of a plant canopy, computed four times on a weekly basis for two sides of the same plants in a randomized complete block design with three blocks and four treatments. Let X_{ijst} be the fractal dimension on week $t = 1, \ldots, 4$, for side $s = 1, 2$ of a plant in the plot of block j receiving treatment i. The univariate ANOVA model is counting 16 terms: (1) the three usual terms for a randomized complete block design (i.e., the intercept, and the treatment and block main effects), (2) three terms for the main effects of time and its interactions with treatment and block, (3) three similar terms for space, (4) three more for the time–space combinations, and (5) four error terms. For convenience, only the MANOVA model is written. Its derivation from the ANOVA model is similar to that of equation 2 from equation 1. Let $\underline{x}_{ij}$ denote the profile vector of $2 \times 4 = 8$ spatiotemporal repeated measures for treatment i in block j, i.e., $\underline{x}_{ij} = (X_{ij11}, X_{ij12}, \ldots, X_{ij41}, X_{ij42})'$. The MANOVA model may then be written as

$$\underline{x}_{ij} = \underline{m} + \underline{a}_i + \underline{B}_j + \underline{e}_{ij} \tag{5}$$

$$(i = 1, \ldots, 4; j = 1, \ldots, 3)$$

where **m** is the multivariate intercept including the univariate intercept plus the purely temporal, purely spatial and spatiotemporal main effects; $\underline{a}_i$ is the multivariate treatment main effects (fixed), including the univariate main effects of treatment and its interactions with time, space, and time–space; $\underline{B}_j$ is the multivariate block main effects (random), including similar univariate main effects and interactions; and $\underline{e}_{ij}$ is the multivariate errors (random) grouping the four error terms of the ANOVA model.

Spatiotemporal means of fractal dimension for the plant canopy of two cultivated species (i.e., corn and soybean) are plotted in figure 18.5; numerical results are reported for soybean in table 18.4. Differences in fractal dimension between the two sides of the plant are not constant over time, with the largest differences at the beginning of the growing season for corn and a crossing between sides after week 1 for soybean. Spatiotemporal main effects are highly significant for soybean, even after modified testing; the same holds true for the three-factor interaction Week-by-Side-by-Trt, and for Week and Side alone,

TABLE 18.4
Analysis of variance of spatiotemporal repeated measures: results of the multivariate and univariate procedures in terms of probabilities of significance $P > F$

Variable	Effect	MANOVA*	ANOVA	Modified ANOVA G-G	Modified ANOVA H-F
Fractal	Treatment[†]		0.0001		
dimension of	Block[†]		0.3416		
soybean					
plant canopy					
	Time				
	Week	0.0005	0.0001	0.0001	0.0001
	Week-by-Trt	0.0001	0.0001	0.0001	0.0001
	Week-by-Block	0.1796	0.0609	0.0785	0.0609
			Epsilon =	0.8181	1.0000[‡]
	Space[§]				
	Side	0.0001	0.0001		
	Side-by-Trt	0.0005	0.0005		
	Side-by-Block	0.5350	0.5350		
	Time-by-Space				
	Week-by-Side	0.0067	0.0001	0.0002	0.0001
	Week-by-Side-by-Trt	0.0014	0.0001	0.0004	0.0001
	Week-by-Side-by-Block	0.1924	0.4998	0.4800	0.4998
			Epsilon =	0.6553	1.0000[‡]

From Foroutan-pour et al. (unpublished manuscript).

NOTE: Same notations as in table 18.1. The experimental design is a randomized complete block design with four treatments related to plant density and cropping system (soybean monocropping versus corn–soybean intercropping) and three blocks. The spatial and temporal repetition factors are provided by two sides of the plant and four weeks of observation, respectively. See text for detail and figure 18.5B for illustration of the spatiotemporal effects week-by-side.

*The four MANOVA test statistics did not provide the same probability of significance for Time and Time-by-Space. The probabilities reported are for Wilks' lambda.

[†]No multivariate test was available for Treatment and Block.

[‡]The actual value of the H-F estimate of Box's epsilon was greater than 1.0. It has been rounded to 1.0, which is the upper bound of the theoretical value.

[§]The multivariate and univariate procedures provided identical probabilities of significance for the purely spatial effects because of the number (two) of spatial repetition levels. For the same reason, the covariance matrix of the spatial profile vectors is circular, and Box's epsilon is equal to 1.0 (Huynh and Feldt 1970; Rouanet and Lépine 1970).

and their respective interactions with Trt. The decrease in fractal dimension observed for corn from week 3 to week 4 [figure 18.5(A)] follows from lengthening of the plant along the vertical axis, resulting in a more rectilinear shape and decreased complexity in space occupation.

Summary

A multiscale approach, preferably in space and in time, is highly recommended in ecological experiments. This includes data analysis, proceeding from the plot or mapping of the data to numerical analysis based on correlograms or variograms. Fractal dimension, as a measure of complexity in space occupation or surface pattern, cannot be circumvented in scale analysis (e.g., Burrough 1981; Milne 1988, 1991b; He et al. 1994); it must be promoted as a summary statistic of spatial and temporal patterns. Whenever applicable, the spatiotemporal repeated-measures ANOVA permits an efficient and complete data analysis in terms of scale and pattern of heterogeneity to be tested.

More specifically, spectral analysis, including the multifrequential periodogram and the ANOVA of the finite Fourier transform, has the advantage of allowing fine resolution from low to high frequency components of time series for almost any sampling scale; it is a multiscale analysis. The autocorrelogram analysis provides a suitable statistical tool if the sampling scale is well adapted and autocorrelation at small to medium scale is of interest. Multivariate geostatistical methods (Wackernagel et al. 1988), once promoted and introduced appropriately in ecological analysis, should help incorporate scale in ecological multivariate data analysis.

Acknowledgments

Research was supported by operating grants from The Natural Sciences and Engineering Research Council of Canada and Le Fonds pour la Formation de Chercheurs et l'Aide à la Recherche du Québec. For the data analyses, I benefited from the computing facilities of McGill University. The graduate course on temporal and spatial statistics that I taught in winter 1995 helped to improve significantly the presentation of the chapter. Special thanks go to Kayhan Foroutan for the fractal dimension estimation, and to my collaborators for permission to use part of their data for illustration: Marc Herman and

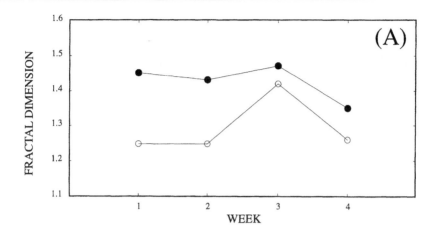

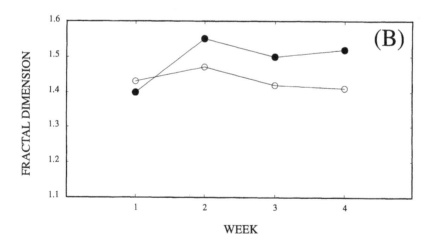

FIGURE 18.5

Spatiotemporal changes of fractal dimension for the plant canopy of **(A)** corn and **(B)** soybean. *Black* and *white circles* are for two different sides of the plant with respect to the plane of developed leaves for corn and to the field rows for soybean (from Foroutan-pour et al., unpublished data).

Thomas Avella for the *P. abies* data, Isabelle Schelstraete and Marc Weyers for the Wistar rat data, and Jason Stockwell and Gary Sprules for the zooplankton biomass data from Lake Erie. I am grateful to Pierre Legendre and two anonymous reviewers for their helpful comments on the manuscript.

Appendix

Repeated-Measures ANOVA in SAS

Three SAS procedures perform the repeated-measures ANOVA: (1) ANOVA for balanced designs, (2) GLM for balanced or unbalanced designs, both procedures using the least-squares estimation method to fit the model (SAS Institute Inc. 1989), and (3) MIXED for balanced or unbalanced data, using maximum likelihood estimation methods (SAS Institute Inc. 1994). In particular, the ANOVA and GLM procedures require the MANOVA model (e.g., equations 2 or 5), whereas the MIXED procedure uses the univariate ANOVA model (e.g., equation 1). The data set must be prepared accordingly. For the GLM procedure used below, the profile vectors of observations are to be read as SAS variables X1-Xp, where p denotes the total number of levels of the repetition factor(s) (i.e., the total number of repeated measures). This rule applies whether the repeated measures are temporal, spatial, or spatiotemporal. Notations in italic are specific to the examples. There are five types of contrasts among the levels of the repetition factor(s) that can be run under the REPEATED statement; they are discussed in some detail by Potvin et al. (1990) and von Ende (1993).

Temporal repeated measures and completely randomized assignment of treatments among subjects (output summarized in table 18.1):

```
DATA tree;
INPUT gc ind X1-X15;
CARDS;
. . .
;
PROC GLM DATA = tree;
CLASS gc;
MODEL X1-X15 = gc/NOUNI;
MANOVA H = gc;
REPEATED year 15 POLYNOMIAL/SUMMARY;
```

Temporal repeated measures and randomized complete blocking in treatment assignment, with one replication of each treatment per block:
    PROC GLM;
    CLASS *trt block*;
    MODEL X1-X*p* = *trt block*/NOUNI;
    MANOVA H = *trt*;
    MANOVA H = *block*;
    REPEATED *time p* HELMERT/SUMMARY;

Temporal repeated measures and randomized complete blocking in treatment assignment, with more than one replication of each treatment per block:
    PROC GLM;
    CLASS *trt block*;
    MODEL X1-X*p* = *trt block trt*block*/NOUNI;
    MANOVA H = *trt* E=*trt*block*;
    MANOVA H = *block*;
    MANOVA H = *trt*block*;
    REPEATED *time p* MEAN/SUMMARY;

Temporal repeated measures and Latin square blocking in treatment assignment, with one replication of each treatment per cell:
    PROC GLM;
    CLASS *trt rowblock colblock*;
    MODEL X1-X*p* = *trt rowblock colblock*/NOUNI;
    MANOVA H = *trt*;
    MANOVA H = *rowblock*;
    MANOVA H = *colblock*;
    REPEATED *time p* CONTRAST(1)/SUMMARY;

Temporal repeated measures and Latin square blocking in treatment assignment, with more than one replication of each treatment per cell:
    PROC GLM;
    CLASS *trt rowblock colblock*;
    MODEL X1-X*p* = *trt rowblock colblock trt*rowblock*colblock*/NOUNI;
    MANOVA H = *trt* E = *trt*rowblock*colblock*;
    MANOVA H = *rowblock*;
    MANOVA H = *colblock*;
    MANOVA H = *trt*rowblock*colblock*;
    REPEATED *time p* PROFILE/SUMMARY;

In the SAS codes above, the *block, rowblock,* and *colblock* factors may or may not be of a spatial nature; they can be provided by genetic lines or animal litters when nonspatial, or by groups of plots blocked perpendicularly to some environmental gradient(s) when spatial. If the blocks, when spatial, are not confounded with the spatial repetition levels (e.g., position on a plant, depth in a water column), the SAS codes for temporal repeated measures can be used for 1D spatial repeated measures. The analysis of 2D spatial repeated measures is similar to that of spatiotemporal repeated measures below.

Spatiotemporal repeated measures and randomized complete blocking in treatment assignment, with one replication of each treatment per block (output summarized in table 18.4):
  Before considering the SAS code, it is worth noting the following:

1. The arrangement of the SAS data set *soybean.* Because of the statement "REPEATED *time 4* . . ., *side 2*," variables X1–X8 must be arranged as follows:

Variable	X1	X2	X3	X4	X5	X6	X7	X8
time	1	1	2	2	3	3	4	4
side	1	2	1	2	1	2	1	2

In other words, the repetition factor appearing first in the REPEATED statement is the one of the two factors whose levels change the least frequently over the profile vector X1-Xp (here *time*).
  2. The number of plant sides (these define the spatial repetition factor in the example). This number is two, so that the circularity condition (Huynh and Feldt 1970; Rouanet and Lépine 1970) is satisfied, and the univariate ANOVA requires no modification for the analysis of the purely spatial repeated measures.
  3. The ANOVA of spatiotemporal repeated measures in SAS. The implication of the statement "REPEATED *time 4* . . ., *side 2*" for data analysis is that the ANOVA of the spatiotemporal repeated-measures data is carried out after the temporal means computed across space levels and the spatial means computed across time levels are submitted to the ANOVA, providing a total of three repeated-measures ANOVA outputs.
  4. The use of contrasts in the spatiotemporal repeated-measures ANOVA. Different types of contrasts may be used for the two repetition factors. The direct or Kronecker product (Graybill 1983) of two contrasts (one for each repetition factor) is then tested as a one-degree-of-freedom

contrast of the interaction between the two factors. In the example, polynomial contrasts are chosen for *time,* whereas no specific choice is made
for *side* (the default is CONTRAST) because the number of levels of the
spatial repetition factor is two (i.e., any other type of contrast is equivalent).

```
DATA soybean;
INPUT trt block X1-X8;
CARDS;
    . . .

;
PROC GLM DATA = soybean;
CLASS trt block;
MODEL X1-X8 = trt block/NOUNI;
MANOVA H = trt;
MANOVA H = block;
REPEATED time 4 POLYNOMIAL, side 2 /SUMMARY;
```

ANOVA of the Finite Fourier Transform in SAS

An example from Dutilleul and Till (1992).

By definition (see equation 3) and the way the finite Fourier transform is implemented in SAS/ETS procedure SPECTRA (SAS Institute
Inc. 1988), the data set *cedar* below must be arranged differently from
soybean above; the replicated time series data for the different levels of
the classification factor *region* are listed in a single variable X, whereas
variable *year* identifies the levels of the time factor. Statement "*fre* =
INT(*120*/(8*ATAN(1.0))*FREQ)" in the data set *Fourier2* transforms
the angular frequency FREQ from the output of SPECTRA, expressed in radians, into frequency *fre* expressed in number of cycles
over the series; the ring-width chronologies cover the 1850 to 1969 period ($p = 120$ years). Notations *RFFT* and *IFFT* are for the real and
imaginary parts of the finite Fourier transform, respectively. The SAS
code below may be adapted with slight modification for different experimental situations; see Deswysen et al. (1989, 1993) for applications
under a cross-over design.

```
DATA cedar;
INPUT region ind year X;
CARDS;
    . . .
```

```
PROC SORT DATA = cedar;
BY region ind year;
PROC SPECTRA DATA = cedar COEF OUT = Fourier;
VAR X;
BY region ind;
DATA Fourier2;
SET Fourier;
fre = INT(120/(8*ATAN(1))*FREQ);
RFFT = COS_01;
IFFT = SIN_01;
DROP FREQ COS_01 SIN_01;
PROC SORT DATA = Fourier2;
BY fre region ind;
PROC GLM DATA = Fourier2;
CLASS region;
MODEL RFFT IFFT = region/NOUNI;
MANOVA H = region;
BY fre;
```

Periodogram Analysis

If $\{y_t; t = 1, \ldots, p\}$ represents the time series data and ω denotes any nonangular frequency (number of cycles over the series) between zero and half-length of the series, the value of Schuster's (1898) classical periodogram statistic at frequency ω is given by the following equation:

$$I^c(\omega) = \frac{2}{p}\left\{\left[\sum_{t=1}^{p} y_t \cdot \cos(\omega \cdot t)\right]^2 + \left[\sum_{t=1}^{p} y_t \cdot \sin(\omega \cdot t)\right]^2\right\} \qquad (6)$$

In matrix notation, if $\underline{y}$ denotes the vector of time series data, $\underline{\omega}$ is a vector of frequencies and $X(\underline{\omega})$ is the matrix of cosine and sine terms at frequencies $\underline{\omega}$, the value of Dutilleul's (1990) multifrequential periodogram statistic is defined by the sum of squares of the trigonometric model $\underline{y} = X(\underline{\omega})\underline{b} + \underline{e}$ (where $\underline{b}$ is the vector of slopes and $\underline{e}$ is the random vector of errors) fitted to the series by least squares, namely

$$I^M(\underline{\omega}) = \underline{y}'X(\underline{\omega})(X(\underline{\omega})'X(\underline{\omega}))^{-1}X(\underline{\omega})'\underline{y} \qquad (7)$$

where $'$ and $^{-1}$ denote the transpose and inverse operators, respectively (Graybill 1983).

The stepwise procedure involving the analysis of the multifrequential periodogram (Dutilleul 1990) can be described as follows, where K is the number of frequencies considered at each step and $\hat{K}_0$ denotes the final estimate of the number of periodicities. The procedure is based on the increase at each step in the sum of squares of the trigonometric model $\underline{y} = X(\underline{\omega}) \underline{b} + \underline{e}$ [i.e., in the value of $I^M(\underline{\omega})$]. At step K, the periodogram statistic evaluated for K frequencies is maximized iteratively; the same operation is carried out for $K + 1$ frequencies. Let $\underline{\hat{\omega}}_K$ and $\underline{\hat{\omega}}_{K+1}$ denote, respectively, the vectors of K and $K + 1$ frequencies maximizing the periodogram at steps K and $K + 1$. Pseudo-likelihood ratio testing (Gallant 1987) is then performed, based on the difference $I^M(\underline{\hat{\omega}}_{K+1}) - I^M(\underline{\hat{\omega}}_K)$ divided by two times the error mean square of the trigonometric model fitted with the $K + 1$ frequencies $\underline{\hat{\omega}}_{K+1}$. The procedure is stopped when the addition of one more frequency does not increase the periodogram statistic significantly, and the $\hat{K}_0$ estimate is then provided by the current K. This procedure and the involved multifrequential periodogram analysis were shown to work in an extensive simulation study (Dutilleul 1989, 1990; see table 18.2). The statistical method is available in the PerioMod software version 1.2; copies of PerioMod (Macintosh and PowerPC versions) are available, with technical notes, from the author upon request.

PART IV

INCORPORATING SCALE CONCEPTS IN
ECOLOGICAL APPLICATIONS

19

MEASURING ENVIRONMENTAL CHANGE

John L. Innes

One must recognize that the description of the system will vary with the choice of scales; that each species, including the human species, will sample and experience the environment on a unique range of scales; and that, rather than trying to determine the correct scale, we must understand how the system description changes across scales.

(Levin 1992)

Environmental change occurs over a wide range of temporal and spatial scales. Which of these scales is relevant depends on the perspective of the observer (May 1989; Peters 1991; Levin 1992). The normal tendency is to look at environmental change from a human perspective, with recent emphasis being placed on global change over the past (and next) 50 years. Within this context, issues such as human population growth, changes in atmospheric chemistry, changes in land use, and climatic changes are prominent. Of these, human population growth seems to be the driving force (Vitousek 1994). However, other organisms have very different temporal and spatial perspectives, and in ecology, these must be considered if concepts such as species assemblages (Kelt et al. 1994; Niemelä and Spence 1994) and metapopulation dynamics (Hanski 1994a) are to be addressed adequately. Other organisms may be much more affected by short-term, small-scale changes in their environment, such as the loss of a host species from a particular site or even a rapid change in temperature over the period of a few hours. Consequently, when looking at those aspects of environmental change that affect ecosystems, it is essential to use the appropriate temporal and spatial scales.

In all cases, different processes operate at different scales, and the choice of working-scale depends on the process being studied. Similarly, patterns can vary according to scale, an important factor when interpreting environmental monitoring data. Failure to reconcile the different scales can lead to significant problems, particularly in relation to resource management (Knopf and Samson 1994; Ruggiero et al. 1994). However, it is not possible to state that any particular perspective is best. There is an optimal scale over which any particular type of environmental change is best measured; this scale varies not only with the nature of the environmental change, but also with the type of organism experiencing the change. This chapter examines how scale affects the measurement of environmental change and the interpretation and application of the subsequent results.

What Is Environmental Change?

Environmental change can occur in two dimensions: space and time. The most commonly viewed change is that which occurs through time at a single point in space. The spatial dimensions of this temporal change may vary from the micrometer to the global scale, but the changes are always characterized by being directional, in that time moves only forward. It is also possible to consider environmental change as the variation that occurs across space at a point in time. Most ecosystems are characterized by spatiotemporal mosaics—they show variation in both space and time (Aubreville 1932; Watt 1947; Whittaker and Levin 1977; Bormann and Likens 1994)—partly in response to environmental variations and other factors such as stage of development.

Spatial and temporal variations have sometimes been combined through the use of chronosequences, whereby space is used as a proxy for time [e.g., plant succession on glacier forelands, Matthews (1992); soil development, Jenny (1941)]. For natural ecosystems, a variety of factors can superimpose an unpredictable element in the relationship between time and space. The results obtained from studies of chronosequences can be severely misleading because of differences in the initial conditions and differences in environmental factors between sites. However, the investigations undertaken on glacier forelands [particularly those from Storbreen in west Norway (Matthews 1978; Whittaker 1985, 1991) and Glacier Bay, Alaska (Cooper 1923; Reiners et al. 1971)] reveal the potential of the method.

The assessment of change depends on the identification of the inherent variability of the phenomenon under study: if the degree of variation has been established, then it will be possible to determine changes over and above this variability. This is in itself a scale-dependent process, and the difference between a variation and an environmental change depends on the perspective being taken. Stochastic variation is normally of little concern in studies of environmental change (although it may be particularly interesting for some aspects of disturbance ecology). However, the point at which stochastic variation becomes environmental change is not fixed and depends on the object in question. As Levin (1992) points out, the main method of scientific inquiry is to reduce this stochastic variation in an attempt to identify regularities about which generalizations can be made. In the current context, these generalizations relate to trends in time or space that reflect a true shift in the mean rather than fluctuations about the mean over time.

The extent of medium- and long-term variation in environmental phenomena, particularly through time, is of great interest and of direct relevance within the context of environmental change. An early but good discussion of equilibrium ideas can be found in Chorley and Kennedy (1971). A number of different types of equilibrium can be identified, all strictly dependent on scale. These have been summarized by Heal (1991) as constant, cyclical, directional, episodic, and catastrophic. Constant time series can contain stochastic variation, but there is no trend in the medium- or long-term means. Cyclical changes show major short- and medium-term variation, but they have a constant mean when a sufficient time scale is taken into account. Directional trends are those in which the direction of the change (e.g., towards larger or smaller values) is continuous. Episodic events generally involve a significant departure from short-term patterns of variability, which may or may not result in a change in the long-term mean. Catastrophic trends, on the other hand, involve a discontinuous change in the long-term mean. Naturally, these different types of trend may be superimposed on one another.

Taking the example of a river (Pahl-Wostl, chapter 7), it is possible to provide an average rate of discharge on a daily, seasonal, or even annual basis. However, long-term records (Federer et al. 1990; Décamps and Fortuné 1991) suggest that it may be difficult to identify the variability around a mean over a series of years, because progressive changes may occur in the mean (i.e., the series is nonstationary). Consequently, different approaches are required depending on the operational scale of the investigation. At some scales, it is possible to specify the variability about a mean. At others, it is necessary to specify variability around a moving average. The latter is more complex, particularly if there is a relationship between the variability and the mean value. Various techniques have been derived for dealing with such data (Brockwell and Davis 1991), and a full discussion is outside the scope of this chapter.

The occurrence of gradual changes over time means that temporal scales must be extended beyond those commonly envisaged in the majority of research projects (Strayer et al. 1986). At their broadest, environmental changes encompass "global change," namely changes in those processes that occur at a global scale (figure 19.1) and which are primarily related to time scales of years to decades. These processes, together with all others, are bounded both spatially and temporally (figure 19.2). At their narrowest, environmental changes involve changes over distances measured in micrometers. It is important that the different scales not be confused.

It would be impossible to cover all aspects of environmental change in a single chapter. This chapter concentrates mainly on measurements of environmental change at the ecosystem scale (10^{-1} to 10^5 hectares; years to decades) (see figure 19.2). In recent years, considerable attention has been focused on this level of organization (Bell et al. 1991; Hansson et al. 1995), and it is particularly appropriate for both scientific and management applications.

Environmental Changes and Ecosystems

At the scale of ecosystems, environmental change involves atmospheric, oceanic, and terrestrial processes, which operate in different spatial and temporal domains. This is shown schematically in figure 19.3 (derived from May 1989), where atmospheric, marine, and terrestrial processes have been separated. As May explains, physical and biological processes operating in these three domains have different relationships between time and space. In the case of oceans, a classic study by Stommel (1963) provides the basis for looking at the scale controls

FIGURE 19.1

Conceptual model of the earth system operating over time scales from decades to centuries. Redrawn from information in NASA (1988) and Roberts (1994).

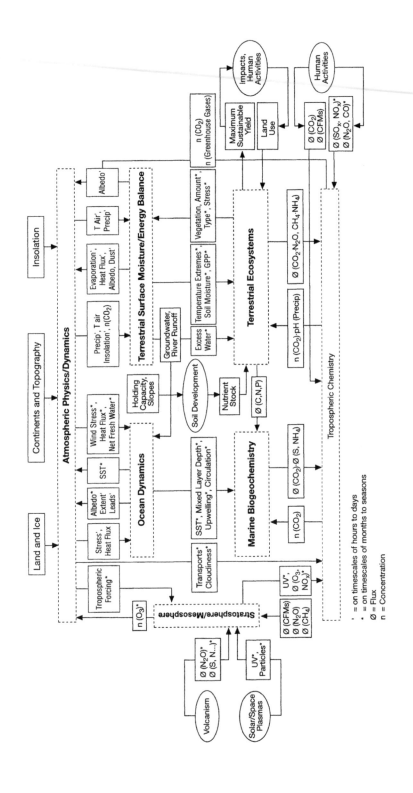

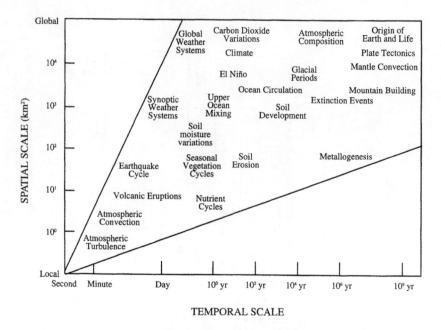

FIGURE 19.2

Spatial and temporal scales characteristic of different earth system processes.
From NASA (1988).

operating over physical processes. In oceans, the relationships between space and time for biological and physical processes are similar. This is not the case for terrestrial environments, where physical (e.g., atmospheric) and biological processes have very different space–time relations. Phenomena such as El Niño involve interactions between atmosphere and oceans, whereas issues such as the effects of desertification in the Sahel (Africa) on changes in global carbon cycling—essentially problems related to terrestrial ecology—occur at time–space scales other than those normally associated with terrestrial ecosystems. May (1989) suggests that this may represent a change in the space–time relationships between biological and physical processes that has been induced by human activities.

The processes affecting ecosystems occur at different points within the time–space boundaries shown in figures 19.2 and 19.3, and not all can be directly related to the ecosystem scale. Consequently, to reach the scale of the ecosystem, it is often necessary to scale assessments either up or down (Addicott et al. 1987; Gyalistras et al. 1994). Some of the process scales shown in figure 19.2 could be questioned. For example, speciation is given as occurring over time scales of more than 1 million years and over spatial scales of 1000 to 10,000 km. However, speciation is probably much more rapid (Grant 1986; Grant and Grant 1993), and it may even be accelerating as a result of environmental change.

Differences in processes operating over different scales can be illustrated with an example. The temporal and spatial nature of environmental change at the ecosystem scale is intrinsic to the concept of vegetation patch dynamics. A particular forest ecosystem may appear stable, but it actually reflects a series of environmental changes oc-

Organization Levels

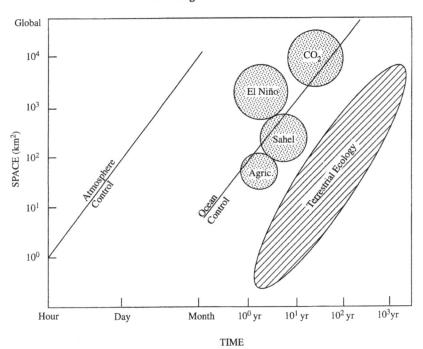

FIGURE 19.3

Schematic representation of characteristic spatial and temporal scales for physical and biological processes on land and in the sea. Events with large spatial scales are generally associated with long time scales: physical processes in the atmosphere exert control over time scales of weeks or less; physical processes in the ocean exert control over years or more.

From *Climate Change*, vol. 7, pp. 5–27, by W. C. Clark. Copyright © 1985 by Kluwer Academic Publishers. Reprinted with kind permission of the publisher.

curring over a variety of spatial and temporal scales (Bycroft et al. 1993). At any particular point within the forest, the environment may be changing but, at the scale of the forest, the entire forest ecosystem may be relatively (to individual points) unchanging (Whitmore 1988). The patch dynamics model is more difficult to apply over longer time scales. For example, it does not readily fit the environmental changes observed in East African savanna ecosystems. A variety of models have been proposed to account for the changes between grassland and woodland (Belsky 1995), where a number of factors interact to change the vegetation at a point from one type to another (see later).

Direct Versus Indirect Measurements of Environmental Change

Environmental change can occur over time scales that are short to human perception (hours, days) and that can be assessed relatively easily; long-term commitments (years, decades) are required when the system under study is expected to respond to changes taking years or decades or when the responses themselves take long periods of time. According to Strayer et al. (1986), there are four classes of ecological phenomena that require long-term commitments. These are (1) processes that operate very slowly, such as succession, (2) rare events, (3) subtle processes, involving large amounts of interannual variance, and (4) complex phenomena. Both direct and indirect measurements of environmental change can be made. The former consist of measurements of environmental parameters such as temperature, ozone concentrations, or soil acidity. The second approach uses physical and biological indicators as a measure of environmental change: a change in an indicator is assumed to reflect a change in one or more environmental variables that would otherwise be impractical or difficult to measure. An example is the Penman model for the calculation of potential evapotranspiration, which is based on meteorological variables that are more easily acquired than evapotranspiration itself. In both cases, there is a need to place environmental monitoring within a theoretical framework based on the notion of temporal and spatial scales (Bradshaw, chapter 11). This requires the identification of a number of steps in the planning of the measurements.

Detailed information on the development of ecological monitoring programs is given in Spellerberg (1991). The steps involved in the development of an environmental monitoring plan can be best illustrated with some examples. In figures 19.4 to 19.6, the environmental deter-

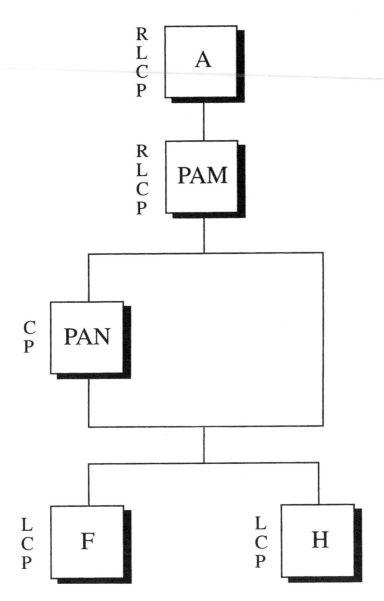

FIGURE 19.4

Suggested hierarchy of savanna determinants. A = catastrophic anthropogenic factors (historical events); F = fire, both natural and human-induced; H = herbivory, both natural and human pastoral systems; PAM = plant available moisture; PAN = plant available nutrients; P = the patch level; C = the catena level; L = the landscape level; R = the regional level.

From *Biology International*, Special Issue No. 24, pp. 1–47, O. T. Solbrig, ed. Copyright © 1991 by International Union of Biological Sciences. Reprinted with permission of the editor.

minants of savanna landscapes are given (Solbrig 1991a; Stott 1994). As an intermediate stage between thorn woodland and moister forest (Coughenour and Ellis 1993), these ecosystems are much more dynamic than previously recognized. Traditionally, five main factors have been considered to determine the form of savanna: plant available moisture (PAM), plant available nutrients (PAN), fire, herbivory, and major anthropogenic events (Stott 1994). PAM and PAN have tended to be viewed as the two most important determinants (Tinley 1982; Scholes 1991), resulting in the use of a plane with axes defined by PAM and PAN, and fire and herbivory have generally been assumed to be of

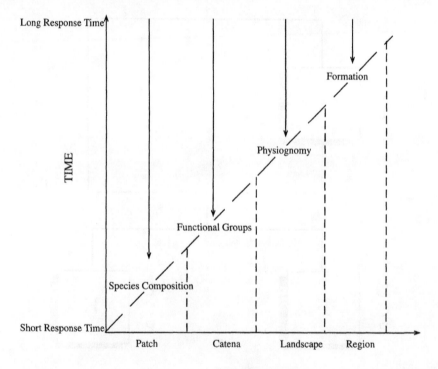

RELATIVE SPACE

FIGURE 19.5

A diagrammatic conceptual framework for evaluating savanna hierarchy determinants at various spatial and temporal scales. The horizontal axis represents the relative spatial scales; the vertical axis, temporal response scales. The appropriate level of biological analysis is indicated on the diagonal.

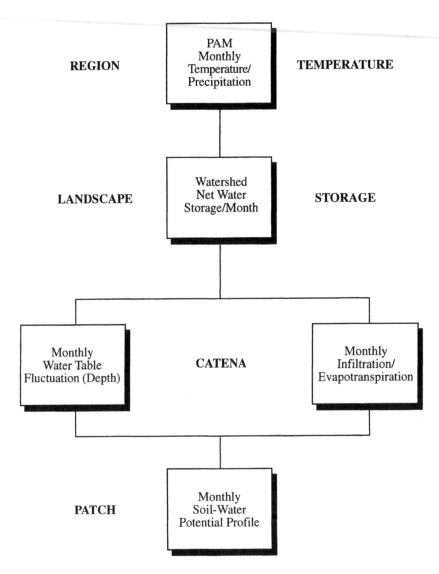

FIGURE 19.6

Suggested hierarchy of measurable variables of PAM in savannas, at different spatial scales.

From *Biology International*, Special Issue No. 24, pp. 1–47, O. T. Solbrig, ed. Copyright © 1991 by International Union of Biological Sciences. Reprinted with permission of the editor.

lesser importance. Moving from the regional scale to the plot scale, it is possible to measure those processes that are appropriate to the spatial scale chosen. For example, catastrophic anthropogenic factors can be assessed at the regional scale, whereas the influence of fire and herbivory is best assessed at the patch scale. Much the same applies to temporal scales, with the response time varying according to the nature of the process.

In figure 19.4, fire and herbivory are given equal status and constrained by the determinants higher in the hierarchy. Historical anthropogenic events are placed at the top of the hierarchy. This is one of a number of alternative hierarchies of the determinants for savanna ecosystems (Stott 1994). Within the hierarchical system, a number of different subsystems (termed holons) are identified for each level, with these subsystems interacting strongly with each other. Thus, the interaction between fire and herbivory is greater than, for example, the interaction between fire and PAM.

Stott (1994) has further elaborated the system. Each subsystem is constrained by those above it in the hierarchy, as can be seen in figure 19.5, where the scale constraints are added and the relevant point of study (e.g., species composition) is included. The spatial scale has been divided into the four units that have commonly been assessed, namely patches, catenas, landscapes, and regions; the subsystems relevant to these are also indicated in figure 19.4.

Measurements of change in the determinants within a subsystem can actually be undertaken at several scales, as shown in figure 19.6 for PAM. The most important aspect of this example is to recognize that global environmental change will affect savanna ecosystems by cascading downwards through the hierarchy shown in figure 19.4. Conversely, savanna ecosystems themselves may affect environmental change, with processes and effects operating upwards through the hierarchy (Stott 1994). Therefore, any study of the relationships between global environmental change and savanna ecosystems must therefore use methods and objects that are appropriate to the relevant scales. The construction of similar hierarchical systems for other environments provides an important and basic way of approaching the myriad of possible environmental measurements.

The scaling approach has been summarized by Caldwell et al. (1993) as follows:

1. Assessing the scale of the phenomenon of interest.
2. Identifying the boundary conditions and constraints.

3. Searching for consistencies at different scales.
4. Stream-lining bottom-up models to incorporate only the salient features.
5. Incorporating feedbacks (both positive and negative) that may operate on some scales but not necessarily on others.
6. Testing the results on different scales with different estimates.
7. Exploring the possibility and usefulness of aggregating species into functional groups.

These seven themes are all relevant to scaling environmental measurements. However, in practice, the measurements are often made without a great deal of thought as to their relevance. For example, meteorological stations are routinely established with a standard set of attributes, some of which are then hardly ever used.

The value of functional groups is debatable, depending on the degree of aggregation used. For example the use of the "big leaf model" as a single parameter for the photosynthetic characteristics of ecosystems (Hinckley et al., chapter 15) is used in general circulation models (GCMs), but it is questioned by ecologists who are reluctant to aggregate all photosynthesizing species in an ecosystem into a single unit (Bazzaz 1993). A variety of other problems exist with the scheme proposed by Caldwell et al. (1993), although it remains conceptually attractive. For example, the steps are best taken when good baseline information is available for an ecosystem. Unfortunately, such information is rarely available and reliable, and accurate data on feedback loops is often absent.

It is important to recognize appropriate relaxation times for processes operating over different scales (Caldwell et al. 1993). A flood may have a major impact at a particular point within a stream channel, but its effects within the catchment as a whole are usually relatively small. It is often difficult to place monitoring data in the correct context because the processes involved are imperfectly understood. This is one of the most important justifications for long-term monitoring (Likens 1989; Magnuson et al. 1991). Without such long-term perspectives, it is virtually impossible to interpret short-term data. Likens (1983) argues the case for such long-term studies particularly well, emphasizing that the experience needed to develop meaningful hypotheses about the environment comes from routine qualitative and quantitative observations undertaken over long time periods. This argument can be extended to management and policy: informed decision making is best done with the support of the experience gained from long-term observation (Peterson et al. 1992; Bormann et al. 1994; Covington et al. 1994; Marcot et al. 1994; Rykiel 1997, chapter 21).

Direct Measurements

Virtually all direct measurements of environmental change involve scaling [the use of information from one spatial or temporal scale to infer characteristics of another scale (Norman 1993)] when applied to ecology. Such scaling requires a mechanistic knowledge of the processes involved (Levin 1993; Norman 1993). The scaling may occur either upwards or downwards, although there seems to be a bias in ecology towards upwards, with the measurements being made at a smaller spatial or shorter temporal scale and applied to larger or longer scales. A well-known example of a bottom-up approach is the use of measurements from meteorological stations, in which it is normally assumed that the measurements made at a particular point in space can be applied to a wider area.

The success of this extrapolation depends on a variety of factors, particularly the homogeneity of the environment and the length of the extrapolation (i.e., how many orders of magnitude it crosses). In the majority of cases, studies of boundary-layer meteorology suggest that such extrapolations need to be made with considerable care. The acceptance of any extrapolation needs to be based on previously defined levels of precision and accuracy, and calibration is essential. Many examples of such calibrations in the literature deal with dendroclimatology (Dutilleul, chapter 17). Normally, a climatic response function is developed for a period of time in which both ring widths and climatic data are available. The calculation of the response function uses only a subset of the period of common measurement, with the remainder of the series being used to test the predictive capability of the response function (Fritts 1976).

Extrapolation presents a predominant problem within the whole area of scaling, namely that most processes and patterns are scale-dependent (Greig-Smith 1979; Wiens 1989). This means that the ecological processes active at one scale may not be active at another, with the implication that extrapolations from one scale to another cannot be made. However, where there is a continuous transition in process with changing scale, then some form of extrapolation may be possible (Wiens 1995). If the change is discontinuous, then it may be possible to extend the results only within specific process domains. The borders of these domains remain poorly known (Wiens 1995).

Developing the meteorological example, wind speed and direction differ according to the scale being assessed. Meteorological measurements made within the vegetation of a forest in an alpine valley cannot be related to synoptic wind speed and direction because of the different processes that are involved. Even if the measurements were made above the forest canopy, they would still be affected by topographic roughness and perhaps also by the presence of anabatic-katabatic (upvalley-downvalley) wind systems, making the use of the measurements for tracking air masses virtually impossible. The difficulties involved in interpreting such data at a point in time are multiplied when trying to assess change over time, because the factors controlling the changes in the processes are also scale dependent. Despite these difficulties, meteorological measurements are quite amenable to scaling, because many of the processes (e.g., heat exchange and momentum) are based on fluid dynamics and can be rigorously defined (Norman 1993).

A variety of techniques have been developed to reduce the impact of problems associated with extrapolations. In particular, there have been a number of recent developments in geostatistics that are especially useful. For example, regionalized variable theory (Matheron 1971), popularly known as geostatistics, has been used to model the spatial variation of physical processes. A variety of sophisticated analytical techniques, such as spectral plots and nested analyses of variance, are also becoming increasingly available (Burrough 1986; Haining 1990; Webster and Oliver 1990; Turner et al. 1991; Rossi et al. 1992; De Cola 1994). These have considerable potential for applications in relation to environmental change.

There have also been many recent developments in the theory of the spatial structure of ecosystems. For example, the fusion of patch dynamics with metapopulation models (Hanski 1991) has been particularly useful. In some cases, these developments have included temporal variation (Fahrig 1991). Another set of theory with scaling problems is that associated with hierarchies (Allen and Starr 1982; O'Neill et al. 1986; Allen, chapter 3; Kolasa and Waltho, chapter 4; O'Neill and King, chapter 1), and hierarchy theory may be useful in establishing the spatiotemporal sequence of factors that affect a particular ecosystem, provided that these factors are linked to the appropriate scale of the problem. Hierarchy theory may also help identify suitable indicators of change (Noss 1990); once these indicators are established, appropriate measurements can be chosen.

Many measures of environmental change are important as basic information for driving ecological models. These models may require information as specific as hourly temperature values. Because of the heterogeneity of meteorological conditions, such measurements need to be tied to a specific site. At a global scale, great reliance is placed on GCMs, which are used to predict what might happen under various greenhouse scenarios. These are by definition global, and the generalized results may not be applicable to specific sites. This is because the models are not designed to answer questions related to individual sites: their resolution is designed to provide information for large regions. To be useful as inputs for ecological models, it is generally necessary to introduce down-scaling (top-down) methods (Gyalistras et al. 1994; Matyasovszky and Bogardi 1994). If these methods are not used, then considerable errors can occur when models with the spatial resolution of a GCM are applied to specific sites (figure 19.7; Vitousek 1993). Down-scaling provides an empirically based method for applying the results from global modeling exercises to regional or local scales, and it is likely to become increasingly important.

Use of Indicators

The use of indicators as proxy measures of environmental change has a long history. For example, pollen assemblages at specific depths in lake sediments have been used to reconstruct the vegetation communities in the vicinity of the lake at a specific point in time, and these in turn have been used to infer climatic conditions over broad regions (Schoonmaker, chapter 5). Sequential analyses from different sediment depths facilitate an environmental reconstruction. The paleoecological record provides much additional information and, for example, it has been possible to reconstruct pollution histories from lake sediments (Battarbee et al. 1990). Similarly, ice cores from the Arctic, Antarctic, and other areas have provided long-term records not only of climate change (Meese et al. 1994; Thompson et al. 1995) but also of various pollutants and other atmospheric gases (Barnola et al. 1987; Chappellaz et al. 1993), although some problems have been encountered when attempts have been made to replicate the results from the very long records (Grootes et al. 1993).

The above examples involve the use of indicators to reconstruct past climates because no climatic measurements are available, but indicators also have other roles. Most experimental studies on the effects of environmental change concentrate (of necessity) on a few variables. Thus, experiments may be conducted on the effects of nitrate in rainfall on plant metabolism and growth, perhaps adding ozone as a second variable and controlling water supply as a third. However, if such experiments are to be adequately replicated, then the number of required replications rapidly increases with the number of factors being assessed. In our current environment, everything is mixed together and the traditional experimental approach may therefore run into serious

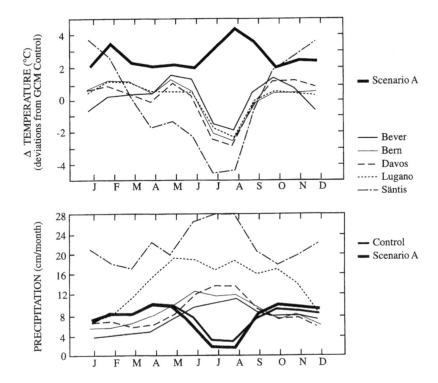

FIGURE 19.7

Comparison of the climate simulated by the ECHAM1/LSG global circulation model in the vicinity of the Alps (averages from three gridpoints) to observations. **Upper graph:** Monthly mean temperatures; *thick line* denotes differences between the means simulated in the years 2075 to 2084 of the GCM "IPCC Scenario A" experiment and the means of the "control" experiment. The following mean annual temperatures were subtracted prior to calculation of the deviations: Bever, 1.5°C; Bern, 8.4°C; Davos, 3.0°C; Lugano, 11.7°C; Säntis, −2.1°C. **Lower graph:** Monthly precipitation totals; *thicker line* denotes simulated annual cycles; *thinner line* denotes observed annual cycles.

Reprinted from *Climate Research*, vol. 4, pp. 167–189, by D. Gyalistras et al. Copyright © 1994 by Inter-Research. Reprinted with permission of the publisher.

difficulties, not only because of the problem of up-scaling from the experiment to the ecosystem, but also because the range of variables that is changing in the environment today is so great that it would be impossible to determine the effects of all possible combinations.

This is where indicators become particularly useful. Different approaches are possible when indicators sensitive to a particular environmental variable are monitored. For example, egg and tissue analyses of *Falco peregrinus* (peregrine falcon) were used to document changes in environmental concentrations of dieldrin and other organochlorines (Furness 1993). Once a relationship has been established, perhaps by supplementary experimental methods (Taylor 1989), monitoring can provide an indication that a particular environmental variable or combination of variables is changing. The precise mechanisms involved are of course useful to know, but this is not always essential. This is the case for such diverse phenomena as the decline of the *F. peregrinus* and the development of stratospheric ozone holes. Another example is provided by paleolimnology, in which environmental changes have been inferred from specific measurements of trace metals, sulfur, and polycyclic aromatic hydrocarbons (Rippey 1990) and inferences based on diatom and chrysophyte assemblages (Charles 1990). In all cases, the measurement of the indicator should be easier than direct measurement of the environmental variable itself.

Evaluating the extent of changes in indicators provides a basis on which experimental work can be focused. A change in some aspect of a population's dynamics normally indicates that an aspect of its environment has changed. For example, the loss of *Lutra lutra* (otter) in Switzerland is attributable to a combination of losses in habitat combined with high levels of polychlorinated biphenyls (PCBs) (which prevented *L. lutra* from successfully utilizing the remaining areas of suitable habitat). PCBs have also been strongly implicated in the breeding failure of marine mammals in the North Sea (Reijnders 1986). In these cases, an effect was observed and then an explanation sought, and it is this approach that, although often considered scientifically weak, may provide the best means of assessing some forms of environmental change. However, there is always the danger with this approach that a correlation is assumed to reflect a cause–effect relationship. The studies of PCBs described above were not confirmed by experimental studies; a cause–effect relationship was assumed because of the high levels of PCBs observed. In at least some of the experiments that have been undertaken under field conditions (Harris and Osborn 1981), no significant effect has been observed, although responses may be species specific.

A major criticism of indicators arises from the difficulties surrounding their selection. By choosing a specific indicator, the possibilities of recognizing environmental changes are immediately narrowed to those that affect that indicator. In some circumstances, this can be beneficial (e.g., if it results in a study becoming more focused). However, the value of the indicator depends on how well it operates (i.e., if changes in its properties reflect changes in the environmental property being considered). For example, the Commission of the European Communities issued a Regulation (No. 1696/87 of 10 June 1987) laying down rules on the implementation of an earlier Regulation on the protection of the Community's forests against atmospheric pollution. Crown transparency was adopted as the indicator that revealed damage to forests, caused in particular by air pollution. Because crown transparency is influenced by a wide range of intrinsic and extrinsic factors, its choice as an air pollution indicator was inappropriate. Further research has attempted to provide better indicators of air pollution stress in trees. For example, Ekeberg et al. (1995) examined whether phytol could be used as an indicator of ozone stress in *Picea abies* (Norway spruce). The study concluded that phytol could be used as an ozone indicator, despite phytol concentrations being related to a number of different stresses, including high and low temperatures, drought, and frost. Clearly, when such ambiguities exist, the value of the indicator is severely compromised, and thus none of the proposed biochemical and physiological indicators have been widely adopted in studies of pollution impacts on forest ecosystems.

In another example from forestry, the calcium-to-aluminum ratio of the soil solution has been suggested as an indicator of whether a forest is under stress from aluminum toxicity and nutrient imbalances. A ratio of 1 has been used in critical loads modeling (de Vries et al. 1992), whereby ratios lower than this indicate that the forest is under stress. However, in many studies, no response was detected at this value, and a molar ratio of less than 1 was required before any effects are observed. Conversely, some experiments indicate that effects could occur when the ratio is greater than 1. Consequently, Cronan and Grigal (1995) proposed that the indicator be associated with risk assessments, such that a ratio of 1.0 is associated with a 50 percent risk of adverse effects on tree growth, a ratio of 0.5 is associated with a 75 percent risk, and a ratio of 0.2 is associated with a 100 percent risk. The uncertainty associated with these figures is estimated at ±50 percent. The use of such quantitative risk assessments represents a major development in the use of potentially ambiguous indicators.

Because concentration on a single indicator can lead to difficulties, the use of suites of indicators often has more potential, particularly if these include different types of indicators (Landres et al. 1988). There may also be considerable merit in combining indicators designed for different scales (Turner 1995). Vegetation assemblages have frequently been used as indicators of site condition (Lahti 1995), and similar techniques have been used in paleolimnology (Pienitz and Smol 1993; Pienitz et al. 1995) and other, related sciences. Specialized statistical techniques have been developed to analyze such data sets (e.g., Indicator Species Analysis, TWINSPAN). However, it is dangerous to include measurements of an indicator simply because the technology exists to do so, rather than because it actually represents something. Many chemical analyses fall into this category, as increasingly sophisticated technology enables a range of different elements to be analyzed quickly and cheaply. A further problem is that scale dependency may exist (Weaver 1995). For example, the species–area relationship is well known: the number of species increases with increasing sampling area. The number of species also may vary over time; for example, fish populations of some aquatic ecosystems are particularly unstable.

These examples have been more concerned with physical or biological properties than with the use of species as indicators. The use of indicators presents a number of problems, yet some indicators have been used for more than 100 years and are still quite widely used (Patrick and Palavage 1994), particularly for looking at spatial environmental variation. As with other types of indicators, the greatest problem is ensuring that the presence, absence, or abundance of a particular species actually reflects the environmental variable in question. As an example, the presence of *Strix occidentalis* subsp. *caurina* (northern spotted owl) in the Pacific Northwest region of North America is not necessarily a reliable indicator of the presence of old-growth forests, although the owls are generally much more common in old-growth than in secondary forests (Forsman et al. 1977; Carey 1985). Many criticisms of the use of indicator species have arisen through the use of the dynamics of one species to predict the dynamics in others, particularly for vertebrates (Landres et al. 1988; Morrison et al. 1992). However, this is not the problem being addressed here, and most of the difficulties associated with the use of species as environmental indicators can be resolved by careful selection of indicator species.

Most of the criteria that Pearson (1994) identified for the selection of indicator species in biodiversity studies are applicable to the selection of environmental indicator species:

1. The taxa should have a stable and well-known taxonomy.
2. They should have a well-known natural history.
3. They should be relatively easy to survey and manipulate.
4. The higher taxa should be broadly distributed over a range of different habitat types.
5. The lower taxa should be specialized and sensitive to habitat changes.
6. Patterns of biodiversity should be reflected in other related and unrelated taxa.
7. The taxa should be of potential economic significance.

An additional criterion, that the indicator is independent of sample size (Noss 1990), should be added to the list. In relation to environmental criteria, criterion 5 above would best be restated as the recommendation that lower taxa have well-defined environmental thresholds. The width of the tolerance is perhaps less important as long as the threshold value of the environmental control is similar to that observed in the system. For example, there is little point in looking for the ingress of tropical species in the boreal zone, but an increase in the occurrence of species characteristic of the temperate zone could suggest the presence of an environmental change. Francour et al. (1994) used this approach to conclude that an increase in the occurrence and frequency of thermophilic species in the Mediterranean Sea reflected an increase in the mean temperature of shallow waters.

Whatever method is used to select indicators, it is important that they be relevant to the scale of the problem. Each problem has an optimal scale at which it can be approached, and only a restricted set of indicators is appropriate to this scale. When the problem transgresses more than one scale, then indicators from the different scales involved may be necessary. In terms of acceptability, indicators that reflect well-defined environmental phenomena are more likely to be successful than those that are used for more conceptual phenomena. This is illustrated by difficulties surrounding the selection of biodiversity indicators and is even more apparent in the debate over indicators for issues such as sustainable forest management.

Remote Sensing

Many of the scale problems associated with the measurement of environmental change can be overcome through the use of remote sensing (Vande Castle, chapter 13). The majority of techniques involving remote sensing also use indicators, although in this case the indicators are often more directly related to specific attributes, such as numbers of individual trees. However, most of the sensors used in satellite imagery measure reflectance, which is an integrated measure of a number of different phenomena.

The use of satellite imagery to monitor environmental change has a number of advantages. Satellite imagery provides global-scale coverage with reasonably frequent repetition and with an increasingly fine spatial resolution. Features occurring over a variety of different scales can also be addressed (figure 19.8). On the negative side, coverage of certain reflectance bands is often impeded by cloud cover, the true costs of the imagery are enormous, and the ground-truthing of the imagery is often rather limited.

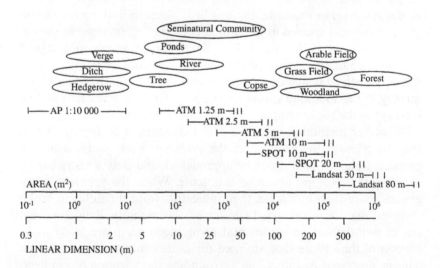

FIGURE 19.8

Variation in the minimum classifiable area for a range of sensors.

Despite the problems, remote sensing is playing an increasingly important role in the measurement of global change. It is included in all three of the global observation systems (climate, ocean, terrestrial) and is widely used and applied in a variety of different contexts (Ustin et al. 1991; Haines-Young 1994). In addition, the digital nature of much of the data make it ideal for incorporation into geographic information systems, greatly speeding up potential analyses (Haines-Young et al. 1993).

Scale Considerations and Monitoring

Environmental monitoring is extremely complex and needs to be undertaken at a variety of scales. Can these general points be brought together in some firm recommendations for monitoring specialists, managers, and policy makers? With many national monitoring programs currently in trouble or even being completely redesigned [e.g., the Environmental Monitoring and Assessment Program (EMAP) in the United States], it is appropriate to examine some of the potential errors and how they can be best avoided.

A recurring problem in monitoring is the apparent conflict between obtaining sufficient observations to generate useful information and the perceived requirement for statements that are statistically reliable and representative. The best and most interesting long-term ecological studies have involved detailed investigations undertaken at individual sites. Examples include Hubbard Brook (Bormann and Likens 1994; Likens and Bormann 1995) and Walker Branch (Johnson and Van Hook 1989) watershed studies in the United States, and Solling in Germany (Ellenberg 1971; Ellenberg et al. 1986). No attempt has been made in these studies to claim that the results are, for example, in the case of Hubbard Brook, representative of northern hardwoods ecosystems. Yet, this has by no means detracted from the value of the results. Investigators with an inventory background have been slow to acknowledge the role of case studies in furthering ecological knowledge.

Most ecologists, on the other hand, would question many of the assumptions built into statistically designed inventories, particularly in relation to "representative" results. What do the results actually represent? This is a question that involves, among other things, scale, pattern of sampling, intensity of sampling, and sampling design (Fortin et al. 1989). For example, the Co-operative Program for Monitoring and Evaluation of the Long-range Transmission of Air Pollutants in Europe (EMEP) grid for pollution estimates in Europe is based on measurements taken using a 150 km × 150 km grid. These provide representative data (for Europe), but they are certainly not representative for a country such as Liechtenstein (total area, 160 km²). Some of these problems are related to the fractal nature of many natural phenomena, a characteristic that means that estimates do not necessarily become more reliable with increasing sample size (Dutilleul, chapter 18; Gardner, chapter 2; Schneider, chapter 12). Consequently, monitoring programs that look at phenomena on the basis of large numbers of representative samples need to carefully examine at which scales the data are representative.

Problems associated with representativeness of results may cause the scientific credibility of many observational studies to be questioned. Is replication necessary? Large-scale, long-term studies are impossible to replicate, as are many smaller studies, simply because every ecosystem is unique. The result is that observational studies need to be supported by experimental studies and by modeling (Tilman 1989; Gutiérrez 1994). Carefully replicated inventories can provide valuable information that sometimes helps to determine whether the results from single sites can be applied over wider areas. However, such inventories rarely provide information on processes, and the cause–effect inferences that can be made from observed correlations are generally weaker than those that come from long-term, intensive observations and those that are based on a knowledge and understanding of the processes involved.

Temporal scales are a major problem in monitoring. At what point does a series of observations become a trend, and how can any such trend be put into the context of long-term dynamics? A good example is provided by the European, and to a lesser extent the North American, studies of forest-crown condition. There are now 10 years of European data. If the data are reliable [a questionable assumption (Innes 1993)], then it appears that a marked deterioration in forest health is occurring. However, long-term studies of growth using tree rings indicate that a 10-year period is of little value when looking at the long-term dynamics of forests. Such a conclusion is obvious when the longevity of trees is considered, yet policy makers in some European countries instigated massive pollution-control measures, despite not only the absence of a reliable trend but the absence of any satisfactory cause–effect mechanism between crown condition and trans-boundary air pollution.

The extrapolation of short-term results to longer periods of time and the establishment of cause–effect relationships represent major limitations of environmental monitoring. The problems have generally been overcome by using simulation models. Modeling, using both accumulated knowledge and site-specific empirical data, can provide an indication of the likely continuation of any trend and enables temporal coverage of the monitoring program to be expanded. In forest monitoring, models have been used to extrapolate short-term stand dynamics to successional predictions over time spans of centuries. Such extrapolations are often made under "business as usual" scenarios or under strictly defined changes (e.g., a specific temperature change by a specific date). However, historical precedence suggests that any environmental prediction that extends to decades is likely to bear little resemblance to reality. Despite this problem, the careful combination of monitoring and modeling can greatly reduce the time period between the start of data collection and the derivation of policy-relevant results (Rykiel, chapter 21). When this is combined with a precautionary approach to policy making, then policy actions may occur relatively quickly. A good example of this is the identification of an "ozone hole" in the Antarctic, followed by atmospheric chemistry modeling and the establishment of the Montreal Protocol on the release of chlorofluorocarbons into the atmosphere, despite incomplete knowledge of the processes involved.

Another dilemma facing the design of monitoring programs is the choice of environmental variables (or their indicators) to be monitored. Inevitably, this depends on the questions being asked. Within the context of global change, these questions are often poorly formulated, and monitoring programs have suffered as a result. This is partly because of the problems described above: results are expected at one spatial or temporal scale, but data may be collected at another. It may be particularly true in ecosystem studies, because the individual components of the ecosystems operate over a range of different spatial and temporal scales. This means that the design phase of monitoring programs needs to be undertaken with great care (Spellerberg 1991; Innes 1994, 1995), with specific detail on the expected use of any measurements. Although this may seem obvious, several monitoring programs have failed to specify in sufficient detail why measurements are being undertaken. The intensive forest monitoring plots of the Commission of the European Community fall into this category, primarily because insufficient time was allocated for the design of the system and selection of appropriate criteria for the choice of the plots. In every case of long-term monitoring, there should never be a concern such as, "We have the data. Now what can we use them for?" Instead, the data should be collected with a specific purpose in mind. This does not preclude subsequent use of the data for other purposes unforeseen at the time of the program design.

When combined with the scientific difficulties surrounding the choice of temporal and spatial scale, the high costs and uncertain results can raise questions about the validity of monitoring. However, most scientists implicitly acknowledge the value of environmental monitoring. Why is this? The answer is relatively simple. Environmental monitoring has had a huge number of success stories, ranging from the quantification of changes in the earth's climate to the determination that sulfur-control policies have resulted in an improvement in the chemical and biological quality of acidified lakes. For engineers and policy makers, environmental monitoring provides the scientific basis for risk assessment (Rykiel, chapter 21), whether calculating a 100-year flood event or the long-term risks associated with a particular economic development. As Likens (1989) states, "Routine observations, often referred to as monitoring, are scorned by some. But routine observations, done carefully and with a purpose in mind, provide, in my opinion, the base of information and the necessary experience to develop meaningful or falsifiable hypotheses or questions."

Summary

Environmental change occurs across a variety of different spatial and temporal scales. For ecosystems, environmental change can either occur rapidly, such as with a forest fire, or slowly, as in climate change. In most cases, repeated measurements, often extending over several years, are necessary to generate the basic information needed as a context for short-term data. Although the collection of such data is often considered as being outside "research" (Likens 1983, 1989; Franklin 1989), it is worth noting that many environmental problems were recognized only as a result of such measurements. Examples include the identification of climate change, desertification, acidic precipitation, stratospheric ozone depletion, and losses of biological diversity.

When monitoring environmental change, there are appropriate spatial and temporal scales for measurements. A variety of techniques are available that can be used to minimize scale problems, but the establishment of a sound theoretical basis for the measurements is a critical first step (Bradshaw, chapter 11). This requires a priori knowledge of the different processes of interest in relation to the specific aims of the study. Such a theoretical framework determines the boundary conditions for the study. At the same time, it may become apparent which processes or parameters need to be measured, and which will involve scaling.

The extrapolation of process measurements from one scale to another is extremely difficult and can lead to substantial errors. In many cases, the different processes may be measured at a physiological level, and they will then have to be scaled to the forest canopy (Amthor 1994) or ecosystem level (figure 19.9) (Hinckley et al., chapter 15). Further extrapolations present even greater problems, and there have been relatively few attempts to extrapolate from the physiological scale to the regional or global scale, with Nikolov and Fox (1993) and Schulze et al. (1994) being interesting exceptions. These extrapolations are generally based on the development of suitable models: sufficient information is usually available for one of the scales (e.g., leaf respiration), but extrapolating to the global level requires the use of modeling. Naturally, the risk of error increases substantially when doing so, and it is important to identify constraints that can be used to assess the reliability of the model.

Many of the problems with extrapolation stem from the presence of nonlinear relationships between processes and scale (King 1991; Rastetter et al. 1992; White and Running 1994). Some processes may operate only at certain scales, becoming redundant at other scales. The recognition and verification of the domains within which each process operates represent an important research area attached to environmental monitoring. In particular, the stability of these domains needs to be tested in view of the possibilities that anthropogenic disturbances may be inducing changes in them (May 1989).

Interactive individual-based models appear to offer some potential for reconciling different scales of processes (Perestrello de Vasconcelos et al. 1993). Various proposals have been made to use one set of models to provide a link between two others (e.g., succession models to link physiological models to landscape models) (Huston 1991; Luxmoore et al. 1991; Taylor et al. 1994). These have met with varying degrees of success, but further developments are likely, and the resolution and accuracy of such approaches is steadily improving.

Information gathered from long-term observations, either as environmental variables or from the responses of ecosystem components to environmental change, forms a fundamental basis for decision making at all scales. At the scale of individual ecosystems, habitat management requires a knowledge of the likely impact of any intervention. At national and international levels, policy decisions are most likely to

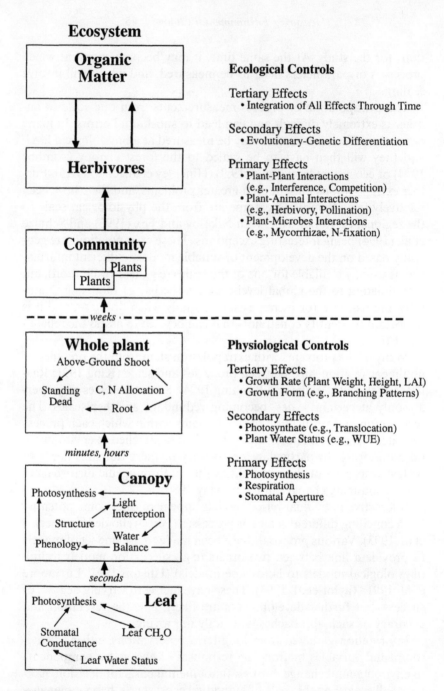

FIGURE 19.9

Physiological and ecological controls on translating effects of elevated CO_2 from leaf-level processes up to ecosystems.

From *Scaling Ecophysiological Processes: Leaf to Globe*, by J. R. Ehleringer and C. B. Field. Copyright © 1993 by Academic Press. Reprinted with permission of the publisher.

result in environmental improvements when they are based on credible science. The interaction of scientists and policy makers is essential. In relation to environmental change, the Intergovernmental Panel on Climatic Change provides a particularly good example of what can happen when scientists, supported by detailed observational data, and policy makers interact (Watson et al. 1996).

In interpreting environmental change from the viewpoint of management policy, an important concept is that of "limits of acceptable change" (Stankey et al. 1985). Similar ideas are implicit in concepts such as critical loads for air pollution, in which a policy decision is stated on the level (threshold) of acceptable damage. Quantifying acceptability implies knowledge of the processes in question, and long-term environmental measurements often form an essential part of the knowledge acquisition. Such measurements may also be critical in determining the effectiveness of any policy to reduce (adverse) changes. In such cases, the problems associated with scale are generally related to top-down issues: how will actions taken at the national or international level impact individual ecosystems?

The difficulties associated with the interpretation of some indicators, whether biological or physical, highlight the need to coordinate environmental measurements with experimental studies and modeling. Many authors have emphasized this (Taylor 1989; Heal 1991), yet it is surprising how seldom the research (including environmental measurements that are an integral part of research) is coordinated. Tilman (1989) has argued that observations on their own have very limited explanatory powers, but that observations become extremely valuable when linked to other types of research:

> Ecological research requires a synthetic approach in which observational, experimental, and theoretical approaches are pursued in a simultaneous, coordinated, interactive manner. (Tilman 1989)

Acknowledgments

I am grateful to Dan Fagre and Thomas Stohlgren for many thought-provoking comments on the manuscript. Michèle Kaennel provided valuable assistance with the literature survey for this chapter.

20

MANAGING ECOLOGICAL SYSTEMS
AND PROCESSES

Richard J. Hobbs

All ecosystems on earth are subject to some degree of human impact, and in many cases they have been shaped by that impact (McDonnell and Pickett 1993). Impacts range from subtle, inadvertent changes, through varying degrees of modification due to harvesting and exploitation, to complete destruction by mining, acute pollution, or warfare. The management of ecosystems and processes is in essence an attempt by humans to control these human impacts and thus to produce favorable outcomes, while minimizing negative effects. The type of outcome deemed favorable depends on the type of management being undertaken. For instance, conservation management may seek to maintain the biodiversity of a region by ensuring the continuation of essential ecological processes and interdependencies and by minimizing the impacts of degrading agents such as invading species and pollutants. Management for agricultural, forest, fisheries, or other types of production, on the other hand, seeks to maximize the sustainable yield of the harvested commodity while minimizing the impacts of agents likely to have a negative impact on that yield. Increasingly, however, management in a particular area is carried out not for mutually exclusive single goals, but to achieve multiple goals.

In this chapter, the importance of scale in the management of ecosystems and ecosystem processes is explored. First, there is an examination of the interface between management and ecological research and theory. Then, the importance of scale is discussed from three perspectives: (1) ecological levels of organization, (2) spatial scales, and (3) temporal scales. Examples are used to examine the importance of

correctly matching the scale of management activities with the appropriate ecological scales and the current barriers to achieving this
match. Finally, the implications of all these factors for ecological research are discussed.

The Interface Between Management and Research

Management and *research* are both broad terms, and a multitude of
activities and objectives exists within each. Examining the interface
between the two presents some problems. Broad management objectives can include the maintenance of a region's biodiversity, the maintenance of a sustainable supply of some product (timber, grain, fish, etc.),
and the maintenance of environmental quality. Within each of these,
however, lie a large number of more narrowly focused objectives. For
instance, the maintenance of biodiversity theoretically calls for the
maintenance of all its aspects, from genetic diversity to ecosystem and
landscape diversity. This demands management objectives that focus
on the maintenance of particular species, habitat management, restoration activities, and mitigating the impacts of human activities on elements of the biodiversity. Again, within each of these objectives, a multitude of more focused objectives must be developed. Habitat management may thus involve management of fire, invasive species, diseases,
people, and many other factors that influence the structure and dynamics of the system. Management is thus a complex process that often has multiple objectives. This is made more complex where objectives are mutually exclusive and tradeoffs become necessary.

Research is an equally complex process, with a similar range of
broad objectives with many narrower objectives. Objectives range from
acquiring fundamental knowledge to applying this knowledge to particular problems. Narrower objectives include developing an understanding of a given subject. This may require the use of survey, statistical and mathematical analysis and modeling, experimentation, and the
development of theory. At any given time, however, the level of knowledge and understanding of a given subject is liable to be incomplete,
but it improves as more information becomes available or new theoretical advances are made. Thus, applying knowledge always takes place
within the current constraints and level of understanding.

It is within this context that the interface between research and
management is found. Much research may have little immediate application to management problems, especially where the level of fundamental knowledge is still low. Application thus depends on the prior

development of sound theory and principles, and adequate levels of knowledge and understanding. It is also true that research can be directed more towards applied problems by ensuring that theory and principles have a sound empirical foundation and by addressing questions likely to be important in a management framework.

Until recently, a major gulf has existed in ecology between so-called pure research and applied research. This has evolved to the extent that a kind of academic apartheid exists, such that the different branches rarely communicate, theory is developed with little reference to real systems, and much ecological research has little apparent relevance to environmental problems. Research scientists are often reluctant to cross the line between scientific endeavor and the application of the knowledge gained by it. This gulf spills over into natural resource management and planning, where decisions are often made on an ad hoc, site-by-site basis, without the benefit of sound ecological advice. Indeed, environmental, industry, and other lobby groups probably have a greater impact on many natural resource management decisions than do scientists.

There are a number of reasons for the gulf between research and management. First, the subject matter of ecology is complex and ecological systems are inherently unpredictable. This makes the prospect of translating research results into management prescriptions daunting. If an attempt to do this is made, the scientist feels obliged to hedge the recommendations with a large number of "ifs and buts" to encompass the range of uncertainties involved. On the other hand, the manager and policy maker feel the need to operate with a higher degree of definitiveness; they perform a certain management operation or they do not. Thus, researchers who recognize the inherent complexity and uncertainty of their subject matter may choose not to explore the resource management implications of their results. On the other hand, a researcher may attempt to simplify things and end up trivializing the research. When faced with complex systems, researchers and managers alike may respond by ignoring the complexity and seeking a simple solution, or by hiding behind the complexity and stating that a solution is not possible. Fortunately, researchers and managers are now more aware of the complexity of the systems they study and manage, so more appropriate research and more realistic management can be implemented.

The complexity of ecology has led to a vast literature on the subject. The difficulty that individual researchers have in keeping up with all but a fraction of the literature provides a partial reason for increasing specialization and fragmentation of ecology (Hobbs and Saunders 1995). In turn, managers rarely have time to keep up with current scientific literature or to assess the relative merits of different viewpoints. The research scientist often makes little effort to condense or synthesize research results so that they are more accessible to managers. The manager observes the outcome of the scientific process as an uncertain morass of (usually unconnected) details, which are often either unlinked to any particular theory or linked to theoretical constructs that have a dubious genealogy (figure 20.1). Any current ecological theory or principle will find enthusiastic support from some and enthusiastic damnation from others. Even the nature of the basic building blocks of ecology is still vigorously debated in the ecological literature. Discussions have appeared in the past few years on how to define populations, communities, ecosystems, and landscapes, and there is still active debate on whether or not communities and ecosystems even exist as units. Although such debate is healthy from an academic viewpoint, it does little to enhance the applicability of ecology to real-life management, policy, and planning problems. As Porritt (1994) observes, "scientists are constantly hypothesizing, testing, rejecting, amending, positively thriving in between the 'grey areas': politicians like it black and white or not at all." Managers are faced with the problem of sorting out which current theories and approaches are likely to be important, which are simply fads, and which are just plain wrong.

Part of the problem lies with the dramatic shift underway in some ecological paradigms, thereby creating a range of perspectives on how ecologists view the world and hence in the way they would advise on management issues (Botkin 1990). Difficulties with the way ecology has been conducted in the past are being recognized, as is the need for change to a more integrated perspective relevant to management issues (Botkin 1990; Underwood 1995). Franklin (1995) has pointed out that "many (most?) important ecological questions are actually emerging from our efforts to apply existing knowledge in a real world setting, rather than in academic isolation." The split between pure and applied research is becoming fuzzier, and moves are afoot to break down subdisciplinary barriers. Recent initiatives, including the development of journals devoted to the application of ecology (e.g., *Ecological Applications*) and to the assessments of the broad directions in which ecological research should develop (Lubchenco et al. 1991), indicate an increasing awareness of the importance of closer linkages between research and management.

The reassessment of ecological paradigms cuts across many differ-

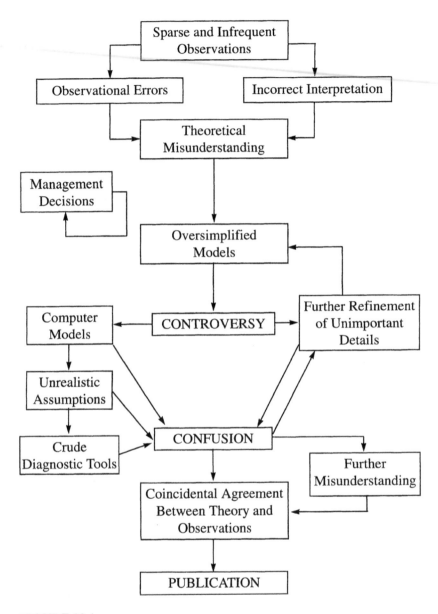

FIGURE 20.1

The cynic's view of the interface between ecological research and management. Note that the primary products from the research are controversy, confusion, and publications. Often this is not too far from the truth. Academic ecologists and theoreticians delight in vigorous debate on anything and everything, particularly in the absence of relevant data. They also enjoy developing models that are internally elegant but frequently bear no relation to the real world. The aim of the research is to produce papers in scientific journals, rather than find answers to particular problems or aid in the understanding of complex natural systems.

From an original idea by Sue Briggs.

ent issues, including those of scale. Questions concerning the spatial and temporal scales and levels of organization at which processes operate are central to many important management questions. In the remainder of this chapter, these issues are explored in more detail.

Levels of Organization

Numerous levels of organization can be recognized in ecological systems, ranging from individual organisms, through populations, communities, ecosystems, and landscapes, to the global level. Although it can be argued whether these levels are hierarchical or simply different perspectives on the same system (O'Neill et al. 1986; Allen and Hoekstra 1990; King 1993), their differentiation can be important in a management perspective. The recognition that natural systems exist at multiple levels of organization is important in directing management efforts. Figure 20.2 shows the different levels of organization hierarchically because this is often how they are dealt with operationally.

Arguments over the existence of particular levels of organization are a hindrance to the pragmatic application of ecological principles to natural resource management. One can argue about what constitutes a population, or whether communities or ecosystems actually exist, but the world will not stop while we resolve these arguments. A pragmatic definition of populations as interacting groups of individuals of the same species, or of communities as repeatable combinations of species, is essential if we are to sensibly manage the biotic diversity of the world. This does not exclude important questions, such as whether communities are composed of strongly interacting species or are the simple co-occurrence of species with largely individualistic responses. It also does not exclude using levels of organization as conceptual constructs not necessarily related to particular spatial scales; for example, a leaf surface might be construed as an ecosystem in some contexts, whereas the relevant unit may be the forest stand or the catchment in other contexts. But it does force us to be more practical. Management should be conducted on spatially explicit entities, not poorly defined concepts.

At the same time, managers need to appreciate the difficulties involved in resolving many issues. Some uncertainty is inherent in all decisions. Decisions involving processes and concepts about which there is still active debate must be made with uncertainty. Increased appreciation of this uncertainty, and the adoption of methods that deal with it, will greatly enhance the decision-making process (Rykiel,

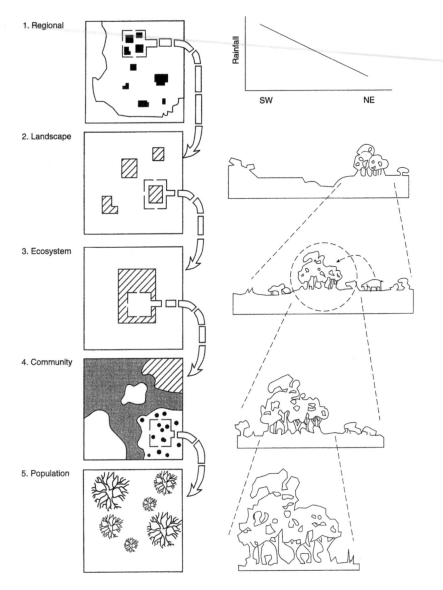

FIGURE 20.2

A hierarchical view of organizational levels relevant to management for bio-diversity conservation. Management needs to be directed at all levels, from individual populations, through communities and ecosystems [ecosystems are shown here in the way they are being regarded in the current terminology of "ecosystem management (i.e., encompassing individual reserves *and* their surroundings), and taking broad-scale processes into account]. In addition, landscape and regional perspectives also must be incorporated into the management framework.

Reprinted from *Biological Conservation*, vol. 64, by D. A. Saunders et al., pp. 185–192, Copyright © 1993, with kind permission from Elsevier Science Ltd., The Boulevard, Landford Land, Kidlington OX5 1GB UK.

chapter 21). For instance, decision analysis techniques (Raiffa 1968; Wilman et al. 1989) can be employed, and the precautionary principle can be adopted more widely (Costanza and Cornwell 1992; O'Riordan and Cameron 1994; Dovers and Handmer 1995).

Management has often concentrated on one particular level of organization while ignoring other levels. A significant example is management and policy resulting from the United States Endangered Species Act. Considerable argument has arisen as to the value of a species-based approach to conservation, versus an approach that embraces ecosystem- and landscape-level considerations (Kohm 1991; Franklin 1993b). The resolution of this argument derives from the recognition that an adequate approach must take multiple levels of organization into account. The prevailing trend increasingly recognizes that a species or community approach to conservation that dwells entirely on single-site issues is not sufficient, and that a larger-scale perspective that incorporates ecosystem, landscape, and regional perspectives must be adopted (Saunders et al. 1991). Noss (1990) indicates that a hierarchy of scales is important in the conservation of biodiversity. He argues that biodiversity needs to be assessed at a number of levels of organization, ranging from genes to landscapes. At each level, structural, compositional, and functional aspects need to be considered. In many cases, simultaneous consideration of different levels will lead to a more integrated management approach, with the needs of one level dependent on the other levels. However, this need not always be the case, and Baker (1989) has suggested that management practices that focus on the landscape, community, and species levels may indeed conflict.

Resource managers generally equate the terms *ecosystem management* and *landscape management* as implying management at a larger spatial scale than is implied by *single-species management*. The concept of *greater ecosystems* involves the amalgamation of parcels of land previously managed in different ways, usually by different organizations. This arises because of the need for spatially explicit management units (see later), but it runs contrary to the desire expressed by some (Allen and Hoekstra 1992) to distinguish between organizational and spatial hierarchies. In part, this is simply a language problem: terminology used for organizational levels is used to describe different spatial scales. On the other hand, the assumption of different spatial scales for different levels of organization does not always hold true, as discussed previously in relation to leaves and forest stands. Of more direct importance to management, some species have broad spatial requirements that may incorporate a range of "ecosystems." Blending different organizational levels across different spatial scales need not be a problem from either a research or a management perspective, provided it is explicitly recognized and does not confuse things unnecessarily.

Spatial Scales of Management

Salwasser (1991) noted that a number of spatial scales must be considered in ecosystem management (figure 20.3). These spatial scales are translated into administrative levels in management agencies. For example, five levels of decision making can be recognized within the United States Department of Agriculture (USDA) Forest Service (table 20.1) (Cawrse 1994). Decisions at one level have the potential to influence other levels, but this interlinking of levels is often not recognized. Allen (1994) has discussed in detail the interrelationships between management at the local site–scale patch and at the level of the landscape mosaic (made up of numerous local patches). Most management activities are directed at the level of the individual patch, but these activities must be planned in the context of the landscape level. The spatial and temporal distribution of patches with different characteristics is manipulated with reference to a landscape-level design or plan (which might incorporate size of areas to be logged, prescribed fire rotation, corridor retention, etc.) (Lertzman and Fall 1997, chapter 16). Successful management thus requires the blending of site-specific information into a landscape or regional context (Allen 1994).

The spatial scales at which individual species operate can also be important from a management perspective, because species use of the landscape may or may not correspond with particular landscape units (figure 20.4). Management directed at particular target taxa (e.g., endangered species) must consider the spatial requirements of these species. Concentrating on the traditional approach of site-by-site management is inappropriate for species with requirements spanning several different individual management units. This multisite approach has led to a stronger focus on landscape- and broad-scale management (Hudson 1991; Rodiek and Bolen 1991; Salwasser 1991; Grumbine 1992).

The incorporation of spatial requirements for particular species into management strategies is still problematic, because these requirements are often not known in detail. Recent studies have indicated that different species are liable to perceive the same patch in a number of different ways (Young et al. 1995). Currently, habitat and landscape descriptions developed for use in wildlife management operate entirely

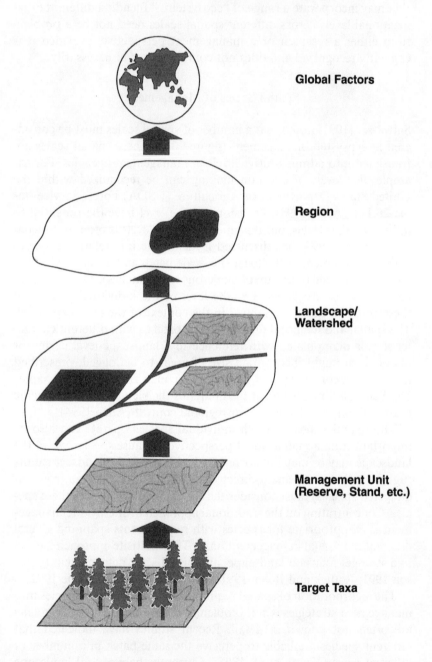

Global Factors

Region

**Landscape/
Watershed**

**Management Unit
(Reserve, Stand, etc.)**

Target Taxa

on our perception of how the landscape is patterned and structured, and this perception may be entirely inappropriate for species that operate at different spatial scales from humans. Many currently used measures of habitat structure and heterogeneity may be relatively useless in assessing the quality of different areas because they generalize relationships that may be quite different from species to species (Cale and Hobbs 1994). The same may be true for different ecosystem and landscape processes. This is a case in which our knowledge of scale-dependent factors is insufficient to provide good metrics for use in guiding management decisions. However, this insufficiency has not

TABLE 20.1

Hierarchy for levels of classification and their relation to planning and management within the United States Forest Service

Organizational level	Land division	Related laws and regulations
National or Congressional	Domain, division, or province	Environmental laws for national parks, forests, and wildernesses
Regional	Section	Regional guidelines; recovery plans
Forest	Landtype association	Forestwide guidelines and standards; objectives for the area
District	Landtype (watershed)	When and where harvesting occurs; delineation of old growth; road closure
Project	Landtype phase or site	Type of harvesting; marking guidelines; whether a stand is old growth

From Cawrse (1994), with permission of Island Press.

FIGURE 20.3

Spatial scales relevant to management. The basic management unit has been the individual land parcel (e.g., reserve, forest stand, etc.). These individual units exist within a broader landscape framework, which is often divided on the basis of catchments or watersheds. In turn, these aggregate into a broader region. Natural regions may exist within, or straddle, political boundaries, but management decisions are usually made only within individual political units (e.g., states or countries). Finally, the influence of global processes on management needs to be recognized; these include global environmental factors such as pollution and human population growth, as well as global economic and trade influences. Target taxa are included because they are frequently part of management goals. They can exist, as shown, within a particular management unit, but often species requirements must be considered at a range of spatial scales. See also figure 20.4.

been adequately recognized, and such measures have been proposed for use in management and planning without adequate assessment of their performance (Li et al. 1993).

A failure to recognize the importance of scale in management prescriptions is evident in discussions on fragmented systems (Lord and Norton 1991). The literature on the "single large or several small"

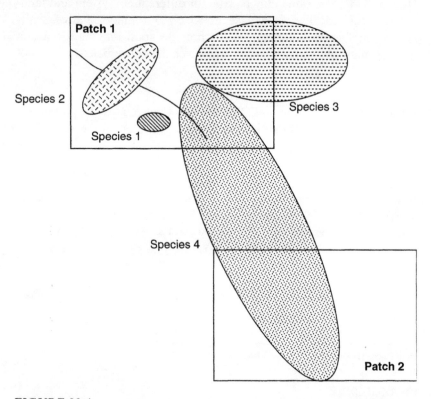

FIGURE 20.4

Scales at which management of species with different spatial requirements must be considered. Represented diagrammatically are two patches, which can represent resource-rich patches within a natural matrix, or habitat fragments within an altered matrix. Patch 1 has two areas of different resource type (e.g., different plant communities or stand ages). Species 1 has a very limited home range and depends entirely on one resource type within patch 1. Species 2 has a broader range, and a requirement to move between two resource types. Species 3 is a generalist that utilizes both the patch and the surrounding matrix. Species 4 is a wide-ranging species that needs to move between patches 1 and 2 to fulfill its resource requirements. The management requirements for each of these species differ markedly.

(SLOSS) debate on reserve design focuses on island biogeographic processes, but it frequently fails to account for factors such as the effects of size on habitat heterogeneity and edge effects (Soulé and Simberloff 1986; Saunders et al. 1991; Lomolino 1994). Factors such as changes in microenvironmental conditions, weed invasion, and nutrient inputs from the surrounding altered matrix are all likely to have a greater impact in small reserves than in large ones, simply as a result of the larger edge-to-area ratio of smaller reserves (Hobbs 1993a; Scougall et al. 1993; Sisk and Margules 1993). Nevertheless, size need not determine the quality of the reserve, because other factors such as degree of degradation or level of habitat heterogeneity may be more important. Small, undegraded fragments may be an important part of a reserve network, particularly if conservation objectives include the retention of species that operate on small spatial scales. Such species may persist in small areas either independently or as part of a metapopulation (Gotelli 1991; Opdam 1991).

A further example of the failure to recognize the scale issue in fragmented systems is the current debate over the utility of corridors in maintaining connectivity and reducing extinctions in fragmented systems (McEuen 1993; Mann and Plummer 1995). Unfortunately, this debate is hampered by lack of data with which to assess corridor function (Hobbs 1992). The relative benefits of corridors need to be assessed against other conservation options (Simberloff and Cox 1987; Simberloff et al. 1992).

The scale of the corridor in relation both to the rest of the landscape and to the requirements of the biota are clearly important in determining their potential utility. However, some of the corridors discussed by Simberloff, particularly in the Florida context, are often kilometers wide and provide for movement of a largely intact fauna, including large mammals. In contrast, corridors in the Western Australian wheatbelt are composed of roadside vegetation—often only tens of meters wide—running between small fragments that retain only a subset of the original biota. These corridors, although small in scale, frequently constitute a relatively large proportion of the land remaining in native vegetation in the area, and they appear to play an important function by providing habitat and movement conduits for components of the remaining fauna (Hobbs et al. 1993; Lynch et al. 1995).

Temporal Scales

The biota currently present in any given area is there as a result of processes operating at a variety of time scales, from geological to seasonal. This gives rise to the idea of slow and fast variables operating in any given system. In addition, individual components of the biota have life spans varying from days to centuries, and they respond to these slow and fast variables in different ways. Individual species may respond to different variables at different stages of their life cycle. For instance, *Sequoiadendron giganteum* (giant sequoia) seed germinates after the passage of a fire (a fast variable), and it subsequently persists as an adult through decadal- and century-scale climatic fluctuations. Holling (1986, 1995) has explored the dynamics of systems in relation to the alternation of rapid and slow changes (figure 20.5). As Holling suggests, "During this cycle, time flows unevenly."

Consideration of ecosystems and ecosystem processes in the context of figure 20.5 has clear implications for management of these systems and processes. Management interventions and goals are likely to be quite different at different phases in the cycle, and they may be designed to drive components of the cycle [e.g., by designed disturbance (Luken 1990)]. Management needs to recognize that system dynamics proceed at different rates at different phases of the cycle. Much conservation management is still based on the concept that natural systems are either static or change only gradually, whereas it is becoming clear that rapid change is a normal component of many systems (Botkin 1990; Hobbs 1994a). This presents something of a conundrum for managers, because they need definable entities to manage. How then does one manage loosely defined entities that are subject to constant change? This is an important question both for the management of natural ecosystems and for restoration ecology. In ecological restoration, the goal often is to restore aspects of the natural system in degraded areas. Achieving successful restoration depends partially on the recognition of the dynamic nature of natural systems and the inappropriateness of setting rigid goals that relate to particular reference states (Hobbs and Norton 1996).

It is also important to note that management is regulated by another set of time scales more related to social and economic aspects than to natural systems: the constraints imposed by human history and the development of human organizational and political frameworks (figure 20.6). Indeed, political and management time horizons are frequently mismatched with the time frames of the ecosystems that are

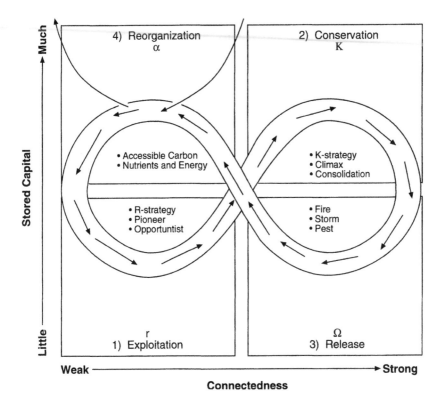

FIGURE 20.5

The dynamics of ecosystems viewed as the four functions of exploitation, conservation, release, and reorganization. The *arrows* show the speed of events in the ecosystem cycle; arrows close to each other indicate a rapidly changing situation, and arrows far from each other indicate a slowly changing situation. The cycle reflects changes in (1) the *y*-axis, which represents the amount of accumulated capital (nutrients, carbon) stored in variables that are currently important in the system, and (2) the *x*-axis, which is the degree of connectedness among the variables. Exploitation involves rapid colonization of recently disturbed areas. Conservation involves the slow accumulation and storage of energy and material. Release occurs when the tightly bound accumulation of biomass and nutrients is suddenly released by agents such as fires or pests. Reorganization involves the minimization and reorganization of nutrients to become available for the next phase of exploitation. The exit from the cycle indicated at the left of the figure suggests the stage at which a transition is most likely, into a less or more productive and organized system.

being managed. Political time horizons are very short compared with the dynamics of most ecosystems, and most management agencies have short histories or brief life spans as a result of frequent reorganizations.

Current management practices in the majority of the world's ecosystems have histories measuring only decades, with some exceptions in places where traditional management practices spanning centuries are maintained. In addition, the global context within which these management practices are carried out is changing rapidly. Most current management practices have short histories with which to assess their impacts. This is true for all of modern agriculture, forestry, and fishing, and for most current ideas on the conservation of biodiversity.

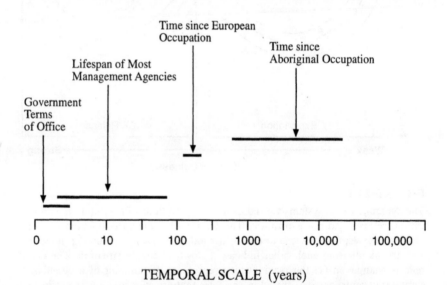

TEMPORAL SCALE (years)

FIGURE 20.6

Human timescales relevant to resource management issues. Humans have been present in many parts of the world for long periods of time and have modified the landscape through management activities such as hunting, burning, and cultivation. In many parts of the world, a dramatic change in land management occurred when they were colonized by Europeans. Most of the agencies currently responsible for management of nature conservation and other lands have been in existence for under a century, often considerably less. Finally, the major driving force behind many land management policies is the political time frame, which revolves around the recurrence of elections every 3 to 5 years. Compare these time frames with those given elsewhere in this volume for natural processes (see also Delcourt and Delcourt 1991).

In addition to the short duration of the management practices themselves, the histories of research into their impacts is even shorter. Even for basic climatic data, the longest data sets are less than 200 years, and much less for many parts of the world. Baselines against which to measure change are thus incredibly short. Similarly, few ecological data span more than a few years. Most ecological projects last the 2 to 4 years of a Ph.D., postdoctoral, or agency-funded program (Tilman 1989). Although rapid change is a characteristic of many natural systems, these changes may occur only episodically over relatively long time periods. Short-term research therefore runs the risk of missing these episodic changes and, hence, of misinterpreting the dynamics of the system. For instance, Hobbs and Mooney (1996) have illustrated the impact of episodic high rainfall events on a variety of ecosystem types in California. Similarly, Hobbs and Mooney (1995) have discussed how research on annual grassland in California conducted at one time may yield results that contradict those obtained at another time. These contradictions can be explained by relating the results to the long-term changes in precipitation that have occurred. However, without longer-term studies, this link would not be apparent.

It is apparent that management and research time horizons are frequently much shorter than those relevant to the systems being managed and studied. This has important ramifications when concepts such as sustainability are considered. Implicit in the term *sustainability* is a temporal component, with the parameters to be sustained (yield, productivity, etc.) persisting into the future. Although sustainability is now a policy buzz word, the long time frame involved is not accepted universally. What does sustainability actually mean when applied to systems that are likely to be temporally and spatially dynamic? How do we measure sustainability in a system that is subject to large, and often unpredictable, changes? The shift in paradigm from the "balance of nature" to the "flux of nature" (Pickett et al. 1992) necessitates a careful review of management objectives and of how we measure achievement of those objectives.

Species redundancy helps to indicate how a consideration of time scale is important in setting management responses to theoretical approaches. The concept of species redundancy has been the center of much debate in conservation biology. Originally proposed as a mechanism for assessing the functional importance of species within an ecosystem and potentially for setting conservation priorities (Solbrig 1991b; Walker 1992; Lawton and Brown 1993), the concept has been attacked as a potential means of consigning "useless" species to the extinction scrap heap (Blake 1993; Martinez and Dunne 1997, chapter 10). Nevertheless, it is useful for assessing the extent to which species are complementary in terms of ecosystem function and hence contribute to system resilience (Frost et al. 1995; Walker 1995). However, the assessment of redundancy must include a time component, because species that appear unimportant at one time may become important at another time or under different environmental conditions. For instance, Hobbs and Mooney (1995), studying an annual grassland in California, found that a species of annual forb, which was present in very low abundances at the start of the study, increased rapidly in abundance and was a dominant component of the grassland 10 years later. The importance of that species for ecosystem function would be perceived quite differently depending on when it was assessed. Our view of how natural systems work thus must incorporate the potential for long-term changes.

The problem of long-term change has wider ramifications for our ability to deal with current environmental problems. Humans are good at recognizing and responding to sudden and rapid change, but less able to recognize and deal with gradual change (Ornstein and Ehrlich 1989). Most of our current pressing environmental problems are caused by slow changes (in atmospheric composition, land use, etc.) that occur gradually and over time horizons longer than the usual human attention span. However, these gradual changes can also result in rapid changes in response variables because of likely nonlinearities and threshold effects (May 1977). Gradual changes, which humans tend to ignore, are thus likely to result in more rapid changes that arrive as somewhat a surprise and may have drastic consequences (Holling 1986, 1995). Developing an appreciation of the linkages between variables changing over relatively long time scales and those likely to undergo rapid change is an important challenge for ecologists.

Matching Management to Appropriate Scales

It is critical that scale considerations are adequately incorporated in conservation and resource management planning and decision making. There are several major hurdles to overcome before this can be achieved. Important factors include the need to gather information at different scales, to develop greater congruence between natural and management boundaries (with a concomitant change in the way management agencies are organized), and to address scale mismatches (spatial and temporal) between systems and processes and their management. Collecting information at the relevant scale is essential for informed management decision making. This can involve obtaining information at very fine spatial and temporal scales in some cases, and broad regional and global scales in other cases. Ecosystem, landscape, and regional management increasingly uses remote sensing and geographic information system (GIS) technologies (Sample 1994). Walker and Walker (1991) have illustrated the way in which a hierarchical GIS can be utilized to gather information relevant to management issues (figure 20.7). Their study also illustrates the value of assessing the spatial and temporal scale of natural and human disturbances, and it provides a good example of patterns and impacts across numerous spatial and

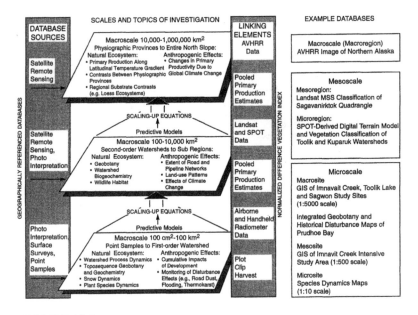

FIGURE 20.7

Organization of the information needed for management of Arctic tundra systems, within a hierarchical geographic information system. This illustrates the multiplicity of scales that have to be considered, and the types of data that can be useful at each scale.

From *Journal of Ecology*, vol. 28, pp. 244–276, by D. A. Walker and M. D. Walker. Copyright © 1991 by Blackwell Science Ltd. Reprinted with permission of the publisher.

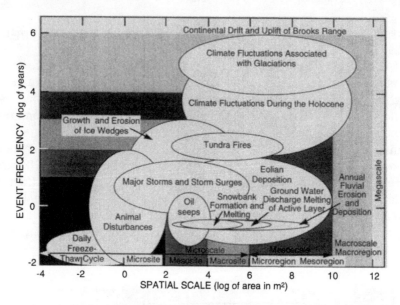

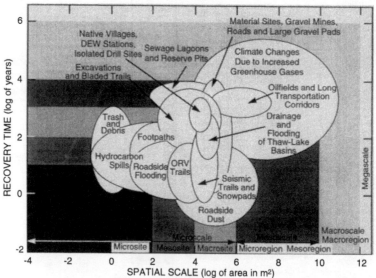

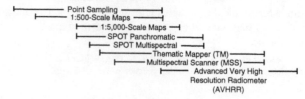

Spatial Data Bases

Point Sampling	
1:500-Scale Maps	
1:5,000-Scale Maps	
SPOT Panchromatic	
SPOT Multispectral	
Thematic Mapper (TM)	
Multispectral Scanner (MSS)	
Advanced Very High Resolution Radiometer (AVHRR)	

temporal scales (figure 20.8). There will often not be a single "correct" scale that needs to be considered in management applications.

Lee (1993) has proposed that most environmental problems are driven by mismatches of scale between human responsibility and natural interactions. These mismatches can be spatial, such as when pollution is released from a point source and spreads throughout the environment; functional, as in the allocation of water in response to specialized human activities rather than in an overall systems-perspective; and temporal, as in the unsustainable harvesting of natural populations (e.g., when short-term fish catches are in excess of long-term reproductive rates).

Scale mismatches show up in many other detailed aspects of natural resource management. The distribution of management units across the landscape has often developed in an arbitrary way. For example, geometric patterns of roads and field boundaries are often superimposed on the landscape without necessarily following any natural feature. The distribution of vegetation fragments in such landscapes thus depends more on the superimposed geometry than any underlying natural pattern (Saunders et al. 1993). The boundary between conservation and production land (e.g., national park versus national forest) may also be equally arbitrary. Landscape processes such as water and material fluxes operate within the context of the natural landscape units, such as catchments. Similarly, movement of biota is not necessarily constrained by management boundaries (Goodwin and Fahrig, chapter 9; Stern, chapter 14). The frequent mismatch between legal and biotic boundaries and the need for management units to represent some sensible biotic or landform unit should receive greater emphasis (Pickett and Thompson 1978; Newmark 1985).

The question of the appropriate scale of management relative to the natural scales of the managed system and the biota has been examined extensively, as in the effects of fragmentation (Harris 1984; Angelstam

FIGURE 20.8

An assessment of the spatial and temporal scales of natural disturbances (*top diagram*) versus human-induced disturbances (*bottom diagram*) in an Arctic tundra ecosystem, with the relevant databases indicated at the bottom. Analyses such as this are essential if the correct scales for research and management are to be identified. Concentration on one particular scale is not sufficient.

From *Journal of Ecology*, vol. 28, pp. 244–276, by D. A. Walker and M. D. Walker. Copyright © 1991 by Blackwell Science Ltd. Reprinted with permission of the publisher.

1992; Spies et al. 1994; Andrén 1995). The temporal and spatial patterns of management activity (e.g., logging) alter the spatial configuration of the landscape, such that the grain, connectivity, edge-to-area ratio and many other parameters are affected. These in turn have the capacity to impact the biota of the region, in terms of both altering the availability of habitat and altering movement patterns. As discussed above, the precise effects may be difficult to understand and measure.

Fragmentation and superimposition of a human-induced geometry to the landscape can also alter landscape processes. The mismatch between natural systems and management units can be particularly important in terms of hydrology. For example, the value of particular reserves or aquatic systems is sometimes compromised by larger-scale hydrologic changes (Barendregt et al. 1995; George et al. 1995; Light et al. 1995). In each case, it has been recognized that local management of individual parcels of land could not ensure the persistence of the systems under study, and that a regional-scale approach was necessary. The context within which individual units are positioned is critical.

Light et al. (1995) also recognized a number of scale-related barriers to the development of adequate management approaches, including the lack of any agency that had a system-wide perspective or jurisdiction. Indeed, management at the appropriate scale is often likely to involve numerous agencies or landholders, many of whom have conflicting goals and interests. Regional-scale approaches involve increased degrees of cooperation between agencies and landholders, or, alternatively, suitable means of mediating conflicting uses. Regional land-use allocation models are one possible tool that could be useful in extending rational land management beyond traditional single management units (Foran and Wardle 1995). However, because of the dynamic nature of natural systems, allocation models must not produce static results but must be able to produce dynamic solutions that respond to changing conditions.

Implications for Ecological Research

Ecological research has traditionally focused on particular levels of organization, and it has generally been conducted over small spatial and temporal scales. There are obvious exceptions to this, and excellent long-term and large-scale studies exist [e.g., the Hubbard Brook and Rothampstead experiments, and numerous long-term population studies (Likens et al. 1977; Bormann and Likens 1994; Johnston 1991)]. The importance of large-scale ecological investigation is being increasingly recognized (Edwards et al. 1995). The establishment and use of long-term ecological research sites (Callahan 1991; Magnuson et al. 1991) also promotes the increase in time frame of ecological studies. Initiatives such as the International Geosphere Biosphere Program (IGBP 1990) promote larger-scale, cross-disciplinary studies.

The subdivision of ecology into scale- and organizational-level-related subdisciplines has led to a fragmented subject with little communication between subdisciplines (Pickett et al. 1994; Jones and Lawton 1995). The complexity of ecological systems has led to a reductionist approach that aims to take manageable chunks of natural systems and to deal with them in a scientifically rigorous manner. Unfortunately, this has resulted in an inward spiral of looking at details while failing to tackle broader issues. Some single-level detailed research has been useful in a management context, but there is a need for increased emphasis on research that straddles a number of scales and approaches (Pickett et al. 1994; Gunderson et al. 1995).

Ecological experiments have been conducted mostly at small spatial scale and over short time frames (Tilman 1989). The relevance of this type of experimentation to management depends on the type of management being considered. For instance, fertilizer trials carried out in small plots can be extrapolated easily to larger areas because the processes operating within the small plot (plant growth and competition) are likely to be replicated at the larger scale. However, problems arise when small-scale experiments are conducted in an attempt to understand processes that normally occur at larger spatial scales.

The problems involved in relating small-scale research to possible management implications are illustrated by the criticism of fragmentation studies conducted by Robinson and Quinn (1988). These studies, which aimed to elucidate the effects of fragmentation on plant communities, experimentally fragmented small areas of annual grassland and concluded that regional diversity increased with increasing degrees of habitat subdivision, a result contrary to most conventional wisdom. Murphy (1989) pointed out that the experiments were carried out at a scale of square meters, which was entirely inappropriate to the question and yielded results at odds with other studies carried out at larger spatial scales. This highlights the problem of extrapolating from isolated small-scale studies to larger systems, a problem that remains unaddressed in Robinson's subsequent work (Robinson et al. 1992, 1995). As discussed above, this problem also arises in terms of temporal scale.

How relevant is the average graduate or postdoctoral project to the management of systems' multicentury dynamics (Likens 1989; Lertzman and Fall, chapter 16; Parker and Pickett, chapter 8)?

There are concomitant problems of studying large-scale systems at the appropriate scale, especially in terms of achieving the desired degree of replication and statistical confidence in the experimental outcome. Although there have been repeated calls in the literature for increased rigor in ecological experimentation (Hurlbert 1984; Underwood 1990), the practicalities of experimentation at large spatial scales are rarely considered. Indeed, examining the question of whether corridors could aid in the recolonization of isolated fragments, Nicholls and Margules (1991) suggest that it could well prove impossible to carry out experimentation in real landscapes. They concluded that covarying and confounding factors would be likely to swamp any attempt to obtain statistically meaningful results (see also Inglis and Underwood 1992). Hargrove and Pickering (1992) suggest that classical experimentation is virtually impossible at larger spatial scales, and that the quest for statistical robustness has led to increasing reductionism and a move away from the larger scale that is supposed to be the perspective being studied. Dealing with systems holistically is virtually impossible using standard reductionist analytical techniques (Hammer, chapter 6), and it is increasingly recognized that information on individual parts does not equal an understanding of the whole.

How do we conduct experiments and acquire information relevant to management at larger scales? We need a more interactive approach to ecological research, in which management activities are viewed as experiments, and managers become part of the research process (Walters and Holling 1990; Underwood 1995). In this way, experimentation occurs by default at the scale most relevant to management, and information can also be gleaned in ways other than by classical experimentation (Hargrove and Pickering 1992). Experimentation needs to take its place in a research strategy that embraces observational, theoretical, modeling, and experimental approaches. Information also needs to be collected and presented at the appropriate scales, as discussed earlier.

A shifting of emphasis from "pristine" systems to managed and degraded systems will also encourage increased linkages between research and management. This shift is already occurring, and much can be learned about ecological processes in damaged and managed systems. Indeed, many basic ecological principles have been derived from agricultural systems, in which the simplicity of the system facilitates interpretation. Agroecosystems provide important opportunities to test many ideas, especially in relation to the function of various components of biodiversity. Questions of scale are as important in agroecosystems as they are in natural systems. For example, Lowrance et al. (1986) and Lefroy et al. (1993) explore major influences on the farming enterprise, ranging from the individual-field scale to the farm scale to catchment and regional scales.

The emerging field of restoration ecology is also a testing ground for many ecological theories, because ecological restoration seeks to put systems back together again. This is impossible without a clear understanding of how systems work in the first place. There is a pressing need for restoration ecology to consider larger scales (Hobbs and Norton 1996). Until now, most restoration has been on individual, localized sites. However, many systems have been degraded at a much broader scale, and landscape- or regional-scale approaches to restoration are required. This broader-scale approach is being applied in the Western Australian wheatbelt (Hobbs 1993b), and in Florida (Cummins and Dahm 1995; Dahm et al. 1995).

Although good science will undoubtedly continue to be driven primarily by the curiosity and vigor of the individual scientist, we should not choose our research topics purely from the standpoint of personal curiosity. There is a myriad of pressing research needs in management-related issues, and this should be enough to whet the curiosity of even the most selective researcher. If ecologists do research because of their fascination with their subject matter, they also have a responsibility to ensure that research facilitates the long-term persistence of desired topics of study. Spatial and temporal scales of research that match scales important from a management perspective are a vital component of ensuring the utility of ecology. Changing ecological paradigms, which recognize the complex, dynamic, and multiscale nature of ecosystems, have made the task of setting management and policy goals more difficult. The same changes have also made conducting effective ecological research more complex and difficult. It is essential that research and management become more interrelated so that each can benefit from the other.

Acknowledgments

I thank two anonymous reviewers for insightful reviews of the draft manuscript, and my colleagues, particularly Robert Lambeck and Denis Saunders, for many discussions that shaped the ideas in this paper.

21

RELATIONSHIPS OF SCALE TO POLICY AND DECISION MAKING

Edward J. Rykiel Jr.

Facts are facts, but perception is reality.
— *Conventional political wisdom*

The study of scale in ecology has progressed to the point that entire books, including this volume, are devoted to elaborating its intricacies (Rosswall et al. 1988; Allen and Hoekstra 1992; Schneider 1994a). This great interest has developed because ecological systems embody processes that operate at many spatial and temporal scales. It is a grand scientific challenge to understand how the interactions of these processes produce the dynamic patterns and behaviors we observe at every level of biological organization. The central question has always been how multilevel organization emerges from the interactions of processes operating over such a wide range of spatial and temporal scales. In response, ecologists have given considerable attention to scientific problems of spatial and temporal scale (without making much of a dent in the central question). Some ecologists have suggested that "the problem" of scale is the central problem for ecology (Allen and Starr 1982; Levin 1992). The problem is how to relate processes that occur on different scales of space and time to ecological patterns and dynamic behaviors observed on other scales of space and time, in other words, relating phenomena across spatiotemporal scales (King 1991; Ehleringer and Field 1993).

Relatively little attention has been given to a rigorous evaluation of the influence of ecological scale on decision making and the formulation of policy. A notable exception, a course titled "Problems of Temporal and Spatial Scaling in Environmental Policy Formation," is being taught at the Georgia Institute of Technology, School of Public Policy (Bryan Norton, personal communication). Issues of environmental policy are even appearing on the World Wide Web (Levy 1996). This chapter is a call for ecologists and policy researchers to work together to develop a deeper understanding of the proper role of ecological scale considerations in policy- and decision-making processes. This chapter is also an introductory overview, not an exhaustive review. As the human population grows, collisions between ecological scale and public policy will become more frequent.

The relationship of scale to policy is made more complex by the interaction of "value-neutral" science with "value-laden" policy. The ground between the value-neutral and the value-laden is an area where scientists often fear to tread, lest their reputations be diminished by a perceived loss of objectivity associated with making value judgments about ecological outcomes (Wagner 1996); yet, value judgments are the essence of policy- and decision-making processes. New disciplines, such as conservation biology and ecological economics, that base their philosophical and ethical foundations on explicit assertions of the value of nature have emerged over the last decade. Conservation biology has become one of the principal drivers in the development of value-based scientific study, yet the index of an important textbook lacks a reference to scale (Primack 1993). Ecological economics (Costanza 1991) does better in explicitly recognizing scale issues (Norton 1995a,b). In a recent paper, Norton and Ulanowicz (1992) link scale and hierarchy theory in a discussion of biodiversity policy.

Any relationship between scale and policy must bridge the gap between scientific judgment and value judgment. In a simplistic sense, science deals with true and false, whereas society deals with good and bad. Science can delineate the possibilities and describe the system that is likely to result from a policy, but it cannot decide if the resulting system is good or bad. Consequently, the relationship between scale and policy, if any consistent relationship exists, must involve a value judgment about whether a selected scale leads to an appropriate decision. Scale enters at least implicitly into decision making, and the goal must be to make it an explicit consideration. Humans are inherently constrained in their view of scale by their own spatial and temporal scales of operation (tables 21.1 and 21.2). The human viewpoint is just one of a plethora of interacting ecological scales. The first task is to understand something about what these scales are and how they might interact to formulate policy.

It may initially appear that there is no consistent relationship between ecological scale and public policy, because policy making can ignore scientific facts and theories. A problem that is ecologically insignificant in scale can receive a great deal of attention if people perceive it to be important, whereas a large-scale ecological problem can be blithely ignored if its solution has a price people are unwilling to pay. The cost of ignoring the scientific facts may become apparent only after a policy has failed or led to irreparable damage. The effort to understand the complex relationships of scale and policy is in large part an effort to establish a linkage between human perceptions and ecological realities. Scale must be recognized as a critical concern in the policy-making process and must play a significant role in formulating and executing environmental policies, if these policies are to provide real solutions to environmental problems.

Scale

Scale by itself has no special ecological meaning. Ecologists have responded by pouring ecological content into the term and exploring the richness of scale concepts in ecology (Allen and Hoekstra 1992; Schneider 1994a; Allen, chapter 3; Gardner, chapter 2; Schneider, chapter 12). Allen and Hoekstra (1992) indicate that scale pertains to

TABLE 21.1

Examples of some human temporal scales

Characteristic	Temporal scale
Attention span	Seconds to minutes
Memory	Seconds to decades
Individual activity	Minutes to days
Birthdays	Years
Marriage	Months to decades
Child rearing	Seconds to two decades
Lifetime	Years to a century
Generation	25 years
Intergenerational time	Decades to centuries
Calendar time	Days to years
Work week	Four to seven days
Career	Months up to 50 years
Season	Days to months
Political time	Months to decades

TABLE 21.2
Examples of some human spatial scales

Characteristic	Spatial scale
Home	100 m^2 to 1000 m^2
Neighborhood	0.05 km^2 to 1 km^2
City/ town/ village	1 km^2 to 1000 km^2
Hectareage (acreage)/section	0.005 km^2 to 2.6 km^2
Management unit	0.1 km^2 to 1000 km^2
Irrigation district	1000 km^2
Travel: foot, vehicle	1 km to 1000 km
Distance to nearest mountain	
State	
Region	
Country	
International sector	
Continent	
Earth	

size, and size is determined by measurement. Schneider (1994) suggests a definition of scale that denotes the resolution within the range of a measured quantity. Range corresponds to the extent over which measurement is made, and resolution corresponds to the grain of the measurement as defined by Turner and Gardner (1991b).

Ecologists have generally defined scale in terms of the spatial and temporal dimensions of an object or process (Delcourt and Delcourt 1988) characterized by both grain and extent, where grain is the unit of spatial and temporal resolution, and extent is the spatial size and temporal duration (Turner and Gardner 1991b). Details of spatial and temporal scales are dealt with elsewhere in this volume. For a preliminary assessment of the role of scale in policy and decision making, a relatively simple view of scale is adequate. In a general context, any system of measurement based on proportion or characteristic-length intervals is a scale, or, in other words, a system of grouping or classifying in a series of steps or degrees according to a standard of relative size, amount, rank, etc. The underlying sense of scale used here is that it is a "yardstick" for measurement. Space and time are not scales until they are divided into segments that can be used for measurement. Although ecological scales have generally been conceived as space–time scales in the majority of recent literature, many other scales. both quantitative and qualitative, are relevant to the policy- and decision-making context. In particular, measurement scales that reveal public perceptions are crucial to developing the relationships between ecological scale and policy.

Psychometric Scales

Some scales are qualitative and cannot be quantified except arbitrarily. For example, reviewers for the United States National Science Foundation are asked to rate proposals on a scale that ranges from "excellent" to "poor." As one might suspect, this rating system is not objective: it is a psychometric or judgment rating scale (McDaniels et al. 1995). From a scientific perspective, an intrinsic problem with such a scale is that the grain and extent are qualitative (some might say irrational) and subjective. How is "excellent" distinguished from "very good" on this scale? Although numbers can be assigned to permit calculations, the numbers still represent psychometric variables and are not quantitative in any sense related to physical science.

The pseudoquantitative nature of numbers in a psychometric context introduces another problem concerning the useful grain size for achieving the objective of the rating. Is a proposal rated 3.5 out of 5.0 better than one rated 3.4? Psychometric scales have a profound influence on the policy-making process and are cast in terms of value judgments (Clark 1994).

Another illustration of judgment scales is provided by McDaniels et al. (1995). They used a questionnaire to study perception of ecological risk. They devised 31 judgment scales to evaluate the perceived impact of 65 items suggested to pose some level of ecological threat. One of these scales was species loss, ranging from "no species" to "many species," and ranked numerically from 1 to 7 (the extent is 7 and the grain size 1). The questionnaire asked the respondents to judge ecological risk using the species-loss scale: "Please rate the impacts of each event in terms of any potential for loss of animal or plant species." The events included, for example, nuclear war, climate change, society's desire for continued economic growth, and scuba diving. Ecological scale is implicit in comparing items such as climate change to scuba diving, but it was not explicitly considered. The point here is that there is a clear need to understand and evaluate the linkage between psychometric scales used to measure environmental perceptions and ecological scales relevant to policy and decision making. This is necessary because surveys of human perceptions of environmental conditions, such as visual air quality (Pryor 1996), are increasingly being used to set standards and influence policy.

Policy

According to Clark (1994), the policy sciences focus on problems. Dovers (1995) defines a policy problem as "a sub-issue, issue or suite of issues perceived to require resolution in some way, thus posing the challenge of choosing the optimum policy response(s)." To identify or understand a problem requires a broad qualitative interpretation of the relevant context, which may include biological, socioeconomic, and political concerns. This context could also include issues of ecological scale, but that is not explicit at present.

Stakeholders are central to policy and decision making. A stakeholder is anyone with an identifiable interest in an issue or problem. That interest can run the gamut from an individual whose personal well-being is directly and immediately affected, to someone who just wants to know that autumn *Oncorhynchus tshawytscha* (chinook salmon) still spawn in the Hanford Reach of the Columbia River (Washington State). Stakeholders are all those who feel they have a stake in the matter, and who come forward to express and defend their interest.

Policy is a principle, plan, or course of action adopted to accomplish a specified purpose. Policies are strategic statements that constitute solutions to problems, and programs are the tactical measures that implement the strategy (Clark 1994). Statements that are defined to be policy by the issuing authority run the gamut from ethereal (the pursuit of happiness) to mundane (no shoes, no service). Although we all have our own personal policies for dealing with life's exigencies and opportunities, policy is often considered a governmental or organizational function, and the job of high-level officials is viewed as "setting policy." For example, consider the U.S. Department of Energy ecosystem management policy, with emphasis on land and facility use:

It is the Department of Energy policy to manage all of its land and facilities as valuable national resources. Our stewardship will be based on the principles of ecosystem management and sustainable development. We will integrate mission, economic, ecologic, social and cultural factors in a comprehensive plan for each site that will guide land and facility decisions. Each comprehensive plan will consider the site's larger regional context and be developed with stakeholder participation. This policy will result in land and

facility uses which support the Department's critical missions, stimulate the
economy, and protect the environment.
>
> (Memorandum from H. R. O'Leary, Secretary of Energy,
> to Secretarial Officers and Operations Office Managers,
> 21 December, 1994)

Policy also provides guidance for situations that have not yet arisen.
When a situation arises for which there is no policy, one is invented.
When a situation arises in which the application of current policy appears inappropriate to some stakeholder, the policy may be reevalu-
ated, and modified or replaced with a new policy. The ideal situation
occurs when a compromise among competing positions establishes a
policy that satisfies all stakeholders. However, the policy-making process is a political undertaking that often involves a choice between
alternative, and sometimes mutually exclusive, courses of action.

Policy is developed based on what is believed to be good, rather
than on rational and objective evaluation of scientific phenomena.
This distinction has sometimes been stated as the difference between
"what ought to be" and "what is" (Loomis 1993). Deciding what ought
to be and how to achieve it inevitably involves politics, and this is a
point at which many scientists get up and leave the table. Policy is
generally embodied in (1) laws and acts at any political level, (2) administrative directives (e.g., executive orders and explicit agency policy), (3) agreed standards, and (4) decision criteria. Scientific quantities
enter into the policy-making process because a numerical value leads
to a normative value. Thus, in the policy-making arena, numbers are
psychometric variables.

Policy-relevant scales reflect human concerns. Dovers (1995) developed a framework for including scale in the policy-making process
(table 21.3). In the policy realm, spatial and temporal scales often have
values different from those employed for ecological research, and they
are often a mixture of spatiotemporal and psychometric scales.

Examples of the Influence of Scale

Ecosystem Management

The preceding quotation by H. R. O'Leary illustrates a high-level government policy statement adopting ecosystem management of United
States federal lands. Fitzsimmons (1994) argues against the general
adoption of ecosystem management as government policy, on the

TABLE 21.3

Problem and response attributes for resource management policy

Attributes	Scale or associated characteristic
Problem attributes	
Spatial scale	Local, regional, national, international, global
Temporal scale	1. Timing of occurrence: near (months to years), medium (years to decades), long (decades to centuries).
	2. Duration: short, medium, long
Magnitude of impacts	Minor, moderate, severe, catastrophic
Reversibility	1. Easy, inexpensive, quick
	2. Difficult, expensive, long
	3. Irreversible
Understanding	Well-known, probable, uncertain, unknown
Complexity	Discrete, linear, nonlinear
Response attributes	
Nature of causes	1. Discrete, simple
	2. Complex, multifactor
Relevance	Direct, indirect, beyond jurisdiction, irrelevant
Tractability	1. Availability of means: fully sufficient, research required, beyond current technology
	2. Acceptability of means: acceptable, barriers, insurmountable
Public concern	1. Low, moderate, high
	2. Widely shared, moderate disagreement, no consensus
Goals	Clearly stated, generally stated, absent

Adapted from Dovers (1995).

grounds that ecosystems have no fixed spatial scale: "The ecosystem concept is unavoidably characterized by geographic fog when actually applied on the ground and should not be used to garb policies in the cloak of science." He concludes that ecosystem management could therefore be a disguise for a federal land grab: "Providing the federal government with the authority to centrally manage ecosystems would entail a massive transfer of power from the individual to the state. . . . The key to minimizing 'train wrecks' is the abandonment of existing policies that elevate environmental protection above the pursuit of human welfare." The contrast of human welfare with environmental protection misses the point that ecosystems have real physical limits that are not determined by human values but by ecological and environmental constraints. Thus, it is important to understand what kind of system can result from human values and decision making within the context of biological reality. There must be an interplay between human aspirations and desires and what is ecologically possible (Ecological Society of America 1995). Policy discussions often ignore this point, and stakeholders cannot understand why their unrealistic desires cannot be met. Although Fitzsimmons repeatedly refers to private property *rights,* he does not suggest that private property *responsibilities* also exist. The rights seem generally to apply to relatively small land areas, whereas the neglected responsibilities seem to arise in consideration of the larger-scale impacts of exercising ownership rights.

Fitzsimmons unwittingly identifies an important cross-scale problem reminiscent of the tragedy of the commons: under what circumstances do actions that are beneficial at the individual level become destructive at the ecosystem level? How does the ensemble of individual behaviors become the integrated behavior of the whole system? This issue also crosses temporal scales because a clearly negative impact may not appear for many years.

Norton (1995a) identifies two scales of spatiotemporal dynamics, which he describes as "individual choices affecting small subcomponents of a system that are reversible in one human lifetime, and decisions that threaten to create change that affects whole ecosystems. If these large-scale changes cannot be reversed within a single human lifetime, then they must be treated as decisions affecting intergenerational equity." He indicates a fundamental difference between small-scale economic decision making and large-scale impacts:

> Scale issues are therefore at the heart of environmental problems in two related senses. First, . . . environmental problems involve an asymmetrical problem in decision scale—actions that are rational from an individual viewpoint lead inexorably to the destruction of a public, community-level value that emerges on a larger and more long-term scale . . . [which is] a social trap inherent in the scalar discontinuity between the scale of individual concern and the scale at which landscape-scale environmental problems emerge. Second, since environmental problems result from cross-scale impacts of trends in individual decisions, resulting in changes in the larger, physical variables that form the environment—these changes are multigenerational impacts.

Global Change: The MINK Study

The aim of the so-called MINK Study was to develop and test a process for analyzing the impact of global warming at the spatial scale of a region (Rosenberg et al. 1993). The study region includes the states of Missouri, Iowa, Nebraska, and Kansas (MINK). General circulation models (GCM) generally predict higher temperatures and decreased precipitation for this region. One of the principal motivations for the study was to investigate the effects of natural spatial and temporal variability in climate. In previous studies, climate change was based on the spatial scale of GCM grid cells and was applied uniformly across this large spatial domain. To include the effects of spatial and temporal variability, the MINK researchers used weather records for the period from 1931 to 1940 as a surrogate for global warming because this period was the hottest and driest on record for the region. Weather records for the period from 1951 to 1980 were used as a control climate for comparisons.

The results of a warmer-climate simulation indicate that a general decrease in agricultural production would be expected for the region as a whole. However, there would be variation in growth across the region, with some areas having increased crop yields. Thus, predictions of climate-change effects at the spatial scale of a farm could differ dramatically from predictions made at the spatial scale of a GCM grid cell. Furthermore, adequate evaluation of climatic effects on water resources required a larger spatial scale than the region, because water balances outside the region influence the streamflow into it. As proposed in hierarchy theory (O'Neill 1988), a minimal assessment of scale effects requires examining a level below and a level above the level of interest. In this case, the level of interest is the region, farms represent a level below the region, and the larger hydrologic system represents the level above. Different conclusions about the effects of climate change arise at these different levels because they are structured to represent different spatial and temporal scales within the hierarchy.

Human Perceptions and Impacts

At least some segments of the human population do not perceive direct connections between themselves and ecological systems, even when they value nature highly. Studies of perceptions of ecological risk show that people believe that events that have great impacts on species also have little impact on humans. Data analyses by McDaniels et al. (1995) show orthogonal axes for impacts on species and ecosystems versus impacts on humans. In other words, the population segment they sampled, although well educated, perceives humans as disconnected from nature. Thus, large impacts on nature do not translate to impacts on humans. The highest values are always accorded to impacts that directly and immediately affect individual humans. A significant policy implication of studies such as this is that impacts that are not immediate and deadly are always perceived as less important, even if in the long term the whole civilization collapses (Hillel 1991). It is critical to understand the scale at which a problem or issue is being perceived and the ecological scale at which a solution or policy is likely to be effective.

The spatiotemporal dimensions of human scales are generally small compared to the wide range of ecological processes and structures. The time scale of humans is minutes to decades (see table 21.2) and the spatial scale is meters to kilometers (see table 21.1). Ecological effects that fall outside these scales are likely to rank low on judgment scales. In addition, ecological effects at any scale that do not have direct effects on individual humans are likely to rank low. Clearly, evolutionary time scales ranging from hundreds to millions of years have no meaning to the average human. The impacts of ecological structures and processes not directly observable by unaided humans are essentially invisible and therefore without value. The structure of an old-growth forest is visible and can be valued, but the process that creates an old-growth forest takes so long that it is incomprehensible. Consequently, when the last of the old growth is destroyed, the old-growth forest has no value. It would take a commitment never before observed in humans to set aside land to reconstitute old-growth forests.

Perhaps without realizing it, we have entered a new phase in the history of earth, a phase in which a single species determines the fate of millions of other species; by design or accident, this species is creating the ecosystems with which future generations must live (Kates et al. 1990). This transition from a wilderness planet to a tame planet started about 10,000 years ago with the rise of agriculture, and the magnitude of change mirrors the growth of the human population. Human habitations are no longer islands in a sea of wilderness; instead, wilderness islands float in a sea of humanity. A critical role of ecological scale is to identify policy options that can be sustainable for long periods of time in a tame world managed within the framework of our conflicting value systems (Mangel et al. 1996; Wagner 1996).

"There is no more powerful statement of government policy than budget decisions" (Press 1995). These decisions tell us where our values are, at least where they are politically and perhaps only temporarily. Nevertheless, studies of ecological scale will have little effect on government policy until scale effects are made an explicit part of the policy- and decision-making process. That requires government to (1) fund research on the relationship between scale and policy, and (2) ensure that scale is an explicitly considered factor in environmental assessments, impact studies, and policy and decision making. The ecological future of earth is now being determined by the human authors of our environmental and economic policies.

Summary

To be policy-relevant, ecological scales must be interpreted in terms of human perceptions of space and time. For example, an average human-lifetime (HLT) unit may be a better expression of ecological time in a policy context than simply stating an effect in years. Thus, if old-growth forests were eliminated from the Pacific Northwest region of North America, it would take almost seven HLTs to regenerate a forest with similar old-growth structure and function (assuming an HLT of 75 years, a regrowth period of 500 years, and the requisite ecological resources).

Human perceptions and value judgments must be informed by ecological scale. It is necessary to educate citizens about the relevance of scales—especially of ecological structures and processes—that are outside their normal range of experience. For example, fire is a major factor in maintaining the aquatic character of the Okefenokee Swamp (Georgia). However, the recurrence interval of significant fire events is estimated at about 50 years (Rykiel 1984). Thus, the importance of fire is easily missed or dismissed.

In addition, policy and decision making must be evaluated at multiple spatial and temporal scales before implementation occurs (Holling and Meffe 1996). Solving a problem at one scale can create a problem at another scale. The basic principle of hierarchy theory that requires examination of levels above and below the level of interest should be applied to policy- and decision-making procedures. For example, one of the major conclusions of the MINK study was that future work should take an approach to include impacts associated with the larger hydrologic system. Scientifically acceptable correspondences between ecological scale and human psychometric scales need to be developed to effectively communicate the ecological consequences of policies and decisions. Scientific knowledge in and of itself is insufficient (Pastor 1995).

The value of ecological *processes* must be emphasized in addition to the value of structure. The political firestorm that accompanied the Yellowstone (Wyoming) fires of 1988 are an instructive example. People were educated to believe that the *structure* of the Yellowstone ecosystem had been preserved in a national park, so they were unprepared to deal with the processes that generate the structure and spatiotemporal scales over which those processes operate. Human activities are also ecological processes that influence ecological patterns, stability, and sustainability.

Acknowledgment

This study was supported by the U.S. Department of Energy under contract DE-AC06-76RLO 1830.

22

DIMENSIONS OF SCALE IN ECOLOGY, RESOURCE MANAGEMENT, AND SOCIETY

David L. Peterson and V. Thomas Parker

Nature is sufficiently surprising that I, at least, am not smart enough to have deduced the patterns described herein and their causes without a framework that allowed nature to point the direction for me.

(Holling 1992)

As we near the end of the twentieth century, we are entering an age of increasing awareness of the complexity of the biological world. We are also aware that the integrity and quality of ecological systems have been dramatically altered by human impacts. The "environmental era" of the 1970s provided a bridge from a period of descriptive ecology to our current grappling with biological diversity and attempts to manage ecosystems that contain thousands of components and connections. More ecologists are now studying species and systems with an eye toward spatial and temporal dynamics (Peterson et al. 1997), although this effort is not without some frustration. Consider the problems inherent in understanding the long-term growth patterns of a thousand-year-old *Pinus aristata* (bristlecone pine), the annual 19,000-km migration route of *Eschrictius robustus* (California gray whale) (Stern 1997, chapter 14), and networks of mycorrhizae that include 100 species and 10,000 m of hyphae per square meter.

We should continue to describe the fascinating complexity of nature, but we need to merge concepts of space, time, and other aspects of scale toward a deeper understanding of how ecological systems work (Allen, chapter 3; O'Neill and King, chapter 1). As suggested by Holling (1992), we need a framework that allows phenomena of the natural world to reveal themselves. Until that framework and a basic understanding of biological interactions are in place, it seems presumptuous to think that we can "manage" ecosystems. We believe that scale concepts are a key component of a framework for addressing ecological problems and improving resource management.

Concepts, Constraints, and Uses of Scale in Ecology

One of the principal goals of this book is to emphasize issues about dimensionality and scale. A number of ecologists have raised these concepts, but their incorporation into general usage has been limited. However, the linkage between *pattern, process,* and *scale* has become more common in recent years, and these words are often found in titles of articles or books. Nonetheless, terms such as *scale* are used with multiple meanings by ecologists. Ecologists generally refer to either temporal or spatial measurement, but scale also has been used to refer to other types of concepts, such as the type of measurement scale or metric being used, or to the level of organization of a system (Rahel 1990). The former measures referring to grain, extent, or size all seem appropriate uses, whereas the latter (level of organization) correlates only with size [but with many exceptions (Allen and Hoekstra 1992; O'Neill and King, chapter 1)].

Multidimensional Aspects of Ecological Entities and Processes

Ecologists now are comfortable viewing systems as embedded in larger systems and having smaller components that can be studied independently. This hierarchical perspective is easy for ecologists to incorporate into their research. More difficult is the realization that all systems and processes have more dimensions than the one articulated by the research question. Of course, the dimensionality of systems often must be reduced for ecologists to ask clear questions. Unfortunately, there is a considerable difference between reducing dimensions for the convenience of study and acting as if systems really are so simple. Few articles or ecological texts mention this, much less emphasize it.

Ecological measurements are distinct from many other types of physical measurements. During the past 30 years, ecologists have changed their metrics for a number of physical measurements to conform with international standards. In such cases, changes in metrics are a matter of convenience or comparison, and such changes do not affect ecological interpretations of processes. But ecological processes and entities consistently have multiple dimensions, and each dimension represents another metric that can probe for patterns in ecological systems (figure 22.1). Multidimensionality results from ecological systems being ongoing historical entities with dynamic spatial and temporal dimensions (Parker and Pickett, chapter 8). This is true of the biological components, regardless of the level of organization investigated, and consequently it is true of processes that influence them. A clear example is the use of the term *regime* to describe multiple dimensions of fire or other disturbances. Fire is a process that contains dimensions of frequency, seasonality, intensity, area, and type. Any of these dimensions represents different metrics for investigating effects of fire on ecosystems. This multidimensionality is a characteristic of

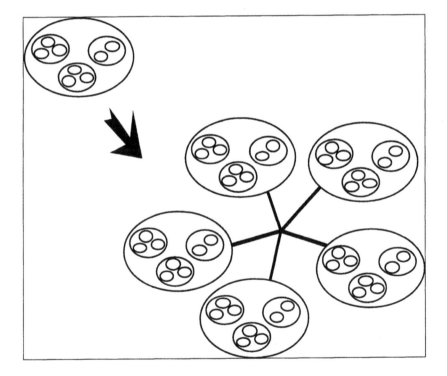

FIGURE 22.1

The left portion of the figure represents a classical hierarchical structure containing nested subsets. The larger structure on the right represents the same hierarchy but represents the different dimensions contained by ecological entities (e.g., frequency, magnitude, size, etc.). Each dimension represents an alternate scale or metric of analysis.

all ecological processes and entities, particularly as scale is increased in measurements.

Consequently, the metric chosen in many ecological studies is critical because of the number of different metrics possible. The choice of metric can determine significant differences in interpretation. Arguments about theoretical concepts have sometimes developed from different scales (in the sense of "size" of investigation) [reviewed by Allen and Starr (1982)]. Some disagreements are based on the use of different metrics as well.

Ecologists investigate a diversity of systems, some focusing on physiological properties (Hinckley et al., chapter 15), others on populations or communities of organisms (Goodwin and Fahrig, chapter 9; Kolasa and Waltho, chapter 4; Martinez and Dunne, chapter 10), and others on mineral or energy dynamics. The entities of interest in these examples are not equivalent and are subject to different types of interpretive problems. One of the significant interpretive issues is the dimensionality of the entities investigated; this is particularly true if conclusions are scaled up or down. In some cases, the entity investigated (e.g., nitrogen) may remain the same as one changes scales, and in other cases (e.g., organism dynamics), the "currency" changes from individuals to populations to guilds or communities as one increases scale, and the dimensionality of the entity changes as a consequence.

Organisms vary in life history characteristics and patterns. Bacteria are relatively simple, whereas multicellular organisms express a broad array of life history stages that may vary greatly in size, duration, impact on the environment, and sensitivity to environmental processes. Because of these differences in scale among stages, without considering any other organismic characteristic, organisms respond differentially to each other and to abiotic environmental processes. Ecologists have a history of building interactive models that tend to simplify life histories into a single response. Over the last few decades, ecologists have developed matrix models, which either have explicit life history stages (Horn 1975) or emphasize the role of stages that have been ignored (Grubb 1977; Parker et al. 1989).

Multidimensionality of organismal systems reflects not only differences among component organisms but the processes that impact them. For example, seasonality and other temporal aspects emphasize the multifaceted aspects of systems with organisms in them. They change through time, the order of additions and losses can be significant (Drake 1991), and their history emphasizes the contingency of their current conditions (Parker and Pickett, chapter 8).

The hierarchical nature of systems has been emphasized by a number of researchers, as well as the different interpretations required as additional processes become important. In addition, ecological systems and influential processes have multiple dimensions at all scales (Pahl-Wostl, chapter 9). Ecologists must deliberately and consciously integrate not only multiple scales but multiple dimensions.

How Scale Is Used in Ecology

Some authors have noted that ecologists are limited in their usage of scale concepts, perception of scale-based issues, and spatial scale of sampling regimes (Kareiva and Anderson 1989; May 1994). Although there is some truth to this statement, ecologists sometimes have little choice but to ask realistic questions of limited scope in order to address large-scale problems. An examination of the literature suggests that for all ecologists, a focus on spatial or temporal issues has a long history. Before continental-plate movement was recognized, there were arguments about dispersal and the biogeographic relationships of continents and island biota. Huffaker's (1958) experiments with predator–prey systems illustrated spatial concepts in maintaining species relationships well before explicit metapopulation models were developed (Levins 1969, 1970; Hanski 1991; Hanski and Gilpin 1991; Taylor 1991). Ecologists working in marine pelagic systems are an excellent example of those who have been grappling with questions concerning the scale of processes and the response of biological systems, mainly because they have no other choice (Steele 1978b; Giller et al. 1994; Pahl-Wostl, chapter 7; Stern, chapter 14).

The focus on scale concepts in ecology is extensive and has explored a number of directions. The compilation in table 22.1 is incomplete but indicates the diversity of scale concepts already being incorporated into ecological science. The table is arbitrarily divided into several categories that designate areas in which concepts about scale have developed. Ecological pattern ties basic descriptions of ecological systems into considerations of the scale of patterns involved (table 22.1). At what scale of measurement do we discover patterns? Sometimes, simply maintaining consistent measurements of systems over a long enough time period begins to reveal new patterns. Familiar examples of this include the recognition of increases in atmospheric CO_2,

TABLE 22.1
Examples of ecological investigations exploring different aspects of ecological scale

I. Scale

  A. Ecological pattern

    1. Environmental heterogeneity

      a. Patchiness, patch dynamics

        (1) Effects of spatial and temporal scales of habitat heterogeneity on regional abundance (Fahrig 1992)

        (2) Hierarchical model of habitat or patch structure (Kolasa 1989; Kolitar and Wiens 1990)

        (3) Mosaic model of patch structure (Wiens et al. 1993)

        (4) Effect of patch spatial scale on forager movement and foraging success (Crist and Wiens 1994)

      b. Habitat fragmentation: effect of spatial scale on animal dispersal (Doak et al. 1992)

      c. Habitat persistence (Fahrig 1992)

    2. Species-area relationship (Gaston and Lawton 1990)

      a. Effect of internal spatial dynamics of "islands" and of inhabiting species (Holt 1992)

      b. Clusters of species at distinct hierarchical habitat levels (Kolasa 1989)

    3. Fractal geometry of landscape

      a. Relation to allometric herbivory ("scale-dependent foraging activity") (Milne et al. 1992)

      b. General [Milne 1991b (general introduction on fractal applications to landscape patterns); Wiens and Milne 1989]

    4. Hierarchical organization of food webs (Pahl-Wostl 1993a)

  B. Ecological process

    1. Mechanisms

      a. Disturbance

        (1) Effect of localized disturbance on competition/succession dynamics at larger spatial scales (Moloney et al. 1992)

        (2) "Disturbance" impacts change at different hierarchical organizational levels (Allen and Wyleto 1983)

        (3) Scale of disturbance determines perceived "stability/equilibrium" of lands/habitat (Turner et al. 1993)

        (4) Disturbance, succession and hierarchy (Pickett et al. 1989)

      b. Competition

        (1) Interspecific competition outcome modified by fractal geometry of landscape (Milne et al. 1992)

        (2) Time scale of resource/consumer dynamics affects consumer coexistence (Loreau 1992)

      c. Dispersal: are population-level dynamics chaotic or stable? (Ruxton 1995)

      d. Predator-prey interactions: Lotka-Volterra diffusion model with trophic hierarchy (McLaughlin and Roughgarden 1992)

      e. Herbivory impacts at different hierarchical organizational levels (Brown and Allen 1989)

      f. Foraging (Campbell et al. 1991; Crist and Wiens 1994)

TABLE 22.1

Continued

2. Multiple mechanisms at multiple scales (Levin 1992)
   a. Size-dependent spatial structure affects island species-area relationship (Holt 1992)
   b. Age structure and spatial structures effects on density-dependent population dynamics (Hastings 1992)
   c. Review of models that incorporate all levels (Menge and Olson 1990)
   d. Time scales of resources and organisms affect trophic overlap and competition (Loreau 1992)
3. Processes in time and space vary as "operational ecosystem" changes [O'Neill et al. 1986; see Moloney et al. (1992) introduction]

C. Evolution
   1. Evolutionary forces act at different scales (Levin 1992)
   2. Decay of genetic correlations with distance [see Moloney et al. (1992) introduction]
   3. Evolutionary convergence of ecological traits of ecosystems does not happen because it requires longer than longest scale of environmental change (Ricklefs 1990)

D. Organism perception
   1. Organisms perceive environment differently (at different scales) at different life history stages (Levin 1992)
   2. Environmental heterogeneity is result of perceptual scale of organism (Fahrig 1992; Levin 1992; Milne et al. 1992; Wiens et al. 1993)

E. Human perception/observation/study
   1. Scale observed/measured is limited by human perceptual abilities (human choice always involved and inherently biased) (Levin 1992); ecologists implicitly scale processes and patterns to human perception (Hoekstra et al. 1991)
   2. Not necessarily (or absolutely not) one unique, natural, characteristic or correct scale of any given pattern or process; scale emerges from human perception and there are many appropriate/useful scales (Levin 1992)

F. Measurement
   1. Extent and grain of system (Kotliar and Wiens 1990)
   2. Determining scale of processes generating scale-dependent pattern (Horne and Schneider 1994)
   3. Determining appropriate scale to study vegetation (Carlile et al. 1989; Simmons et al. 1992)
   4. Measuring multiple, overlapping scales of a pattern
      a. Vegetation patch structure (Ver Hoef and Glenn-Lewin 1989)
      b. Nonparametric three-matrix permutation tests reveal scale of spatial heterogeneity (Gurevitch and Fortin 1991)
      c. Community mosaic analysis (Yoshioka and Yoshioka 1989)
   5. Scale dependence/independence of food-web analysis (Martinez 1991, 1992, 1993, 1994)
   6. Models (Levin 1992)
      a. Formal models allow explicit use of relative scale in order to generalize explicit ecological measurements (Fulton et al. 1993)
      b. Review of some models and conceptual frameworks for regulation of community structure (Menge and Olson 1990)

TABLE 22.1

Continued

II. Scaling: moving across scales, extrapolating data from one scale to another
   A. Scaling up to generate pattern at larger scale from smaller-scale information
      (Levin 1992; McKenzie et al. 1996)
      1. Structured population dynamics relating physiology of individuals to higher-order community dynamics (Cushing 1992)
      2. Individual-level mechanisms in heterogeneous environment (Wiens et al. 1993)
      3. Aggregating fine-scale data to model coarser-scale attributes (King et al. 1991; Rastetter et al. 1992; Crist and Wiens 1994)
      4. Individual-based models (DeAngelis and Post 1992)
   B. Scaling across (between different systems; e.g., land-sea) (Steele 1991)
   C. Model types
      1. Aggregation (Levin 1992)
      2. Random walk (Levin 1992)
      3. Individual-based models (DeAngelis and Post 1992)
      4. Experimental model systems (Wiens et al. 1993)
      5. Spatially explicit models (O'Neill et al. 1989; Moloney et al. 1992; O'Neill et al. 1992)
      6. Hierarchical simulation models: species abundance related to disturbance patterns (Moloney et al. 1992)
      7. Hierarchically organized systems of positive feedback loops (Pahl-Wostl 1993b)
III. Concepts and applications
   A. Vegetation dynamics theory and models
      1. Nonnested hierarchy (Acker 1990)
      2. General laws of vegetation dynamics (Brand and Parker 1995)
   B. Marine/aquatic environments: general overviews of importance/application of scale (Steele 1989; Ricklefs 1990)
   C. Landscape ecology
      1. Application to conservation biology (Hobbs 1994b)
      2. Different perspectives in chapters from Turner and Gardner (1991)
   D. Hierarchy theory
      1. General use in aquatic macrophyte ecology (Farmer and Adams 1991)
      2. Models
         a. Complexity of plant communities (Allen and Wyleto 1983)
         b. Comprehensive hierarchical framework for scaled ecological systems (O'Neill et al. 1989)
         c. Neutral model (hierarchically structured maps) for landscape analysis (O'Neill et al. 1992; Lavorel et al. 1993)
         d. Hierarchical maps used to model causes of plant species coexistence (Lavorel et al. 1994)
      3. Hierarchical structure of environment (Kolasa 1989; Waltho and Kolasa 1994)
   E. Network theory (Pahl-Wostl 1993a)
   F. Positive feedback loops (Perry et al. 1989; Pahl-Wostl 1993a)
   G. Complex ecological systems
      1. Axiomatic system to describe units of biological organization (Kolasa and Pickett 1989)
      2. Ecosystem as hierarchical self-organizing system of positive feedback cycles (Pahl-Wostl 1993b)

TABLE 22.1
Continued

H. Macroecology (Brown 1995)
I. Global applications
   1. Climatic change (Steele 1991)
   2. Conservation
      a. At larger scale than species (Hobbs 1994b)
      b. Reserve design and habitat management (Doak et al. 1992; Fahrig 1992; Wiens et al. 1993)
J. Fractals (Pahl-Wostl 1993a)
K. Prediction and/or guide for future experimental work (Acker 1990; Menge and Olson 1990; Doak et al. 1992; Levin 1992; Pahl-Wostl 1993b; Turner et al. 1993; Wiens et al. 1993)

---

increased acidity in rainfall, and the cycle of the El Niño Southern Oscillation. Other ecologists have considered how the change in scale of structural pattern impacts organisms or other system responses. The scale of processes involved in generating patterns differentially impacts systems, and this has led to a better articulation of how pattern reflects the scale of the processes producing it (Levins 1992; Wu and Loucks 1995). Other topics for which scale has been considered include evolution, the perception of organisms, the perception of observers, and how things are measured. Ecologists also have considerable interest in how interpretations at one scale can be used to make predictions at larger or smaller scales.

Extrapolation of Ecological Entities Across Scales

Scale-based constraints of interpretations or models have led ecologists to recognize that it is critical to consider such limitations. Generalized models of how systems work may be effective at one scale but not at others (Brand and Parker 1995). A number of models and approaches have attempted to directly incorporate concepts of scales (Bradshaw 1997, chapter 11; Dutilleul 1997, chapters 17 and 18; Gardner 1997, chapter 2; Schneider 1997, chapter 12). Landscape ecology is one outgrowth of the awareness of impacts of large-scale processes on ecosystems. Hierarchy theory takes the recognition that systems are embedded in larger systems and brings an explicit multiple-scale model into ecology.

The issue of ecological scale is a natural response to changes in how ecologists examine systems. A review of journal articles several decades ago would reveal a focus on organism interactions, productivity of systems, mineral nutrient pools, and other topics investigated at small scales. Some questions would be posed at large scales, but answers would still be sought at small scales. Nonetheless, increases in the scale of the question led to the awareness that the answers had to be similarly investigated at larger scales. Research in the last two decades has increasingly focused on larger questions related to atmospheric phenomena such as global climate change, the dynamics of fire over landscapes, and the impact of water-temperature shifts and hemispheric weather patterns on oceanic systems.

What has been discovered? Regardless of the level of organization or the scale of the study, ecological systems are not governed exclusively by local-scale processes. Instead, systems are linked by processes into larger systems, and by other processes into even larger systems. For example, because a variety of ecosystems actually may be interconnected, attempts to manage natural resources at a local scale can be confounded by larger constraints that must be taken into account. Recognition of these circumstances has occurred in a number of managed systems, such as the Greater Yellowstone Ecosystem (mostly in Wyoming) (Patten 1991).

Multidimensional Scale Issues: Impacts, Interpretations, and Societal Responses

The first 21 chapters of this book illustrate the integral nature of scale concepts in ecological theory and applications. Every ecological discipline and nearly every aspect of research and resource management need to consider how scale decisions affect data collection, analysis, and inference. Perhaps the greatest benefit of integrating the scale paradigm in the day-to-day work of ecologists is the potential for establishing a framework for addressing complex ecological and natural resource issues (Gardner, chapter 2; Schneider, chapter 12). Far from being a reductionist tool, scale concepts provide an organizational and conceptual approach for studying and managing ecosystem components.

To illustrate how scale can be applied to specific ecological issues, it is helpful to look at some environmental problems that are of concern to scientists and society. The examples discussed below address aquatic, terrestrial, and atmospheric components of ecosystems and their relationship to broader social issues. These examples include minimal detail because we assume most readers have at least some familiarity with them. The objective here is to illustrate the use of scale for assessing and solving problems.

Salmon in the North Pacific

The life cycle of *Oncorhynchus* (five salmon and two trout species) in the North Pacific region of North America is well known for having distinctive features related to spatial and temporal scales. These anadromous fish are born in freshwater streams or lakes, they swim downstream to spend the majority of their life in the Pacific Ocean and adjacent saltwater channels, and they return upstream to spawn and subsequently die. The seven species in this genus—*O. gorbuscha* (pink salmon), *O. keta* (chum salmon), *O. nerka* (sockeye salmon), *O. kisutch* (coho salmon), *O. tshawytscha* (chinook salmon), *O. mykiss* (steelhead trout), and *O. clarki* (sea-run cutthroat trout)—have different geographic ranges (with considerable overlap), migration routes, and spawning cycles. Some species reproduce at relatively constant time intervals (e.g., 2 years for *O. gorbuscha,* 3 years for *O. kisutch*). Others have quite variable reproduction intervals, but they occasionally spawn in streams other than the one in which they were born, thereby increasing the probability of at least some reproduction if their native stream is unavailable for spawning as a result of large-scale sedimentation or some other disturbance; they "hedge their bets" in space. Other species tend to be faithful to their native streams but have greater variation in the length of their life cycle (e.g., *O. nerka* and *O. tshawytscha,* 3 to 7 years), thereby increasing the probability of reproduction if their native stream is unavailable in a particular year; they "hedge their bets" in time. Each species has life history characteristics that provide adaptability to changing environmental conditions in space and time (Reimers 1973; Pauley 1991).

North Pacific *Oncorhynchus* (except *O. clarki*) range thousands of kilometers from inland spawning streams to the Gulf of Alaska and beyond (figure 22.2). Some species range (or have historically ranged) as far inland as south-central Idaho (Kaczynski and Palmisano 1993). Different mechanisms affect *Oncorhynchus* population dynamics at different spatial scales (Hanski and Gilpin 1991), from local breeding populations (e.g., in a creek), to metapopulations (e.g., in a river drainage), to geographical areas that contain many metapopulations (e.g., the west coast of the United States), to the total range of a species [e.g., *O. tshawytscha* ranges from the Kamchatka Peninsula (Russia) across the Bering Sea and North Pacific Ocean to northern California]

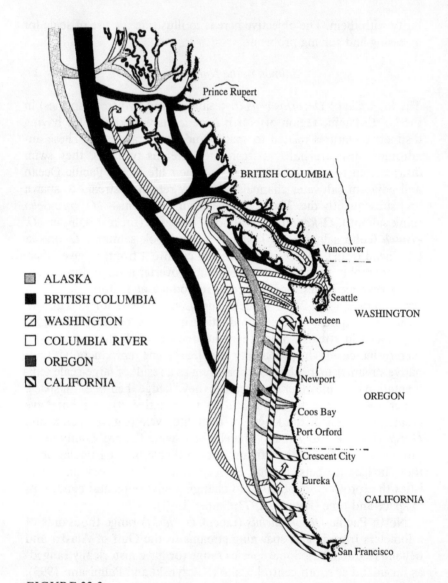

ALASKA

BRITISH COLUMBIA

WASHINGTON

COLUMBIA RIVER

OREGON

CALIFORNIA

Prince Rupert

BRITISH COLUMBIA

Vancouver

Seattle

Aberdeen WASHINGTON

Newport

Coos Bay

Port Orford

Crescent City

Eureka

OREGON

CALIFORNIA

San Francisco

FIGURE 22.2

Migration patterns of *Oncorhynchus kisutch* (coho salmon) by state and province stocks, illustrating variations in spatial scales. Width of lines has no quantitative meaning.

From U.S. National Marine Fisheries Service (1976).

(Healey 1991). Variation in distribution and abundance of *Oncorhynchus* populations is strongly affected by interannual variation in climate, ocean currents, and prey fish abundance (Beamish 1995); this natural variation must be considered when evaluating human impacts on fish population dynamics.

Pacific *Oncorhynchus* have disappeared from about 40% of their historical breeding ranges in Washington, Oregon, Idaho, and California, and many remaining populations are severely reduced (National Research Council 1995); stocks are also threatened in some areas of coastal British Columbia and Alaska. In less than a century, a genus that was once abundant throughout the North Pacific has dwindled to the point that many genetic stocks are extinct. The existence of fish that were—and continue to be—such an important part of Native [First Nation (Canada)] and non-Native American livelihood, culture, and art is now the focus of intensive study and political debate. What are the factors contributing to this natural resource crisis?

Reductions in native *Oncorhynchus* populations have been caused by (1) alteration of hydrological systems by dams and destruction of wetlands and estuaries, (2) mortality through fishing [commercial, recreational, and treaty (human subsistence)], (3) reduction in suitable spawning habitat by sedimentation of streams and increased stream temperature because of forest harvest practices, (4) loss of stream habitat as a result of the removal of vegetation cover and sedimentation caused by livestock grazing, (5) changes in stream velocity associated with urban runoff, water use, and channel alteration, and (6) propagation of various *Oncorhynchus* genetic stocks in fish hatcheries (Meehan 1991; National Research Council 1995). Each of these human activities has had impacts at different scales (table 22.2).

Effective restoration and management of *Oncorhynchus* must target specific scales of their population dynamics. For example, maintaining water volumes at appropriate levels within stream channels is critical for fish migration (table 22.2), but water releases must occur at the correct time for individual species. Successful manipulation of hydrological regimes at large spatial scales (Pahl-Wostl, chapter 7) could be ineffective unless habitat conditions are suitable at much smaller scales. Access to upstream spawning habitat is a smaller-scale concern fundamental to promotion of local populations.

The ongoing debate on how to manage the *Oncorhynchus* resource in the North Pacific has far-reaching economic and social implications (Castle 1993). Society and politicians are now looking at many complex issues, such as impacts of restoration strategies on the generation

TABLE 22.2
Scales at which human activities have affected *Oncorhynchus* populations in the North Pacific region

Human activity	Impacts on *Oncorhynchus* populations	Spatial scale (general estimates)	Temporal scale	Potential restoration activity
Dams and hydrological alterations	Metapopulations affected by large dams; local populations affected by small dams	10–10^3 km along streams	Dams, in place since late 1800s, affect upstream and downstream migration	Modify water release to promote fish migration; remove dams
Fishing	Metapopulations and groups of metapopulations affected by reduced numbers of spawning adults	10^2–10^3 km along streams; up to 10^4 km^2 in the ocean	Commercial fishing since mid-1800s; affects number of fish that can return upstream to reproduce	Reduce harvests spatially and temporally to allow stocks to recover; regulation varies greatly by species
Forest harvest	Local populations affected by damaged spawning habitat from sedimentation of streams 1–10 km, mostly along smaller streams	Since mid-1800s; affects reproductive phase of life cycle	Reduce forest harvest near streams; reduce sedimentation through different forest practices	
Grazing	Local populations and metapopulations affected by sedimentation of streams, loss of shading, higher water temperature	1–10^3 km along streams	Since mid-1800s; affects reproductive phase of life cycle	Reduce grazing; restrict grazing in riparian areas; restore riparian vegetation
Urbanization	Local populations and metapopulations affected by altered water velocity, sedimentation and altered channel structure	1–10 km along streams	Since about 1900; affects all phases of life cycle	Improve flow retardation mechanisms adjacent to and downstream of urban areas
Fish hatcheries	Local populations and metapopulations affected by mixing native and nonnative genetic stocks; nonnatives may have lower fitness	1–10^3 km along streams, with additional impacts on ocean populations at larger scales	Since about 1900; affects all phases of life cycle	Favor restoration of native stocks over hatchery propagation; restrict harvest of native stocks

of hydroelectric power and on the commercial fishing industry. The removal of dams from rivers has even been proposed, which represents a dramatic shift in attitudes regarding the protection of natural resources (U.S. National Park Service 1996). Conservation strategies must ensure that specific restoration approaches are compatible with appropriate scales of fish population dynamics. Coordination and integration of proposed restoration activities at all scales and among all stakeholders is critical for successful restoration efforts.

Air Pollution in Southern and Central California

The impacts of air pollution on ecosystems and human health in southern and central California are a striking example of how a local source can become a regional issue at various time scales. In the Los Angeles Basin (southern California), all the ingredients are present to create poor air quality: high human population (13 million), high emissions as a result of fossil fuel combustion (automobiles, industrial processes, and power generation), hot and dry summer weather, and a large valley encircled by mountains where air pollutants are trapped during atmospheric inversions. These conditions are ideal for the production of (1) photochemical haze, or smog, which includes a variety of gases, particulates and aerosols, and (2) photochemical oxidants such as ozone, which is colorless but potentially toxic to humans and vegetation at relatively low concentrations (Seinfeld 1989). The same phenomena, with less severe impacts, are also present in the Great Central Valley of California and western Sierra Nevada because of air pollutants produced in the San Francisco Bay region (Peterson et al. 1991).

While emissions are highest in the western Los Angeles Basin, air pollutant impacts are generally greater in the central and eastern part of the basin. Prevailing westerly winds transport air pollutants eastward to the San Bernardino Mountains, a distance of approximately 100 km from Los Angeles. Photochemical haze greatly obscures visibility in the eastern half of the basin, often reducing visual range to a few kilometers. During the summer, these pollutants are transported over mountain passes eastward to the Mojave Desert, and they contribute to reduced visibility as far away as the Grand Canyon (northern Arizona) (Shaver and Malm 1995), a distance of over 500 km from the primary source of pollution.

Perhaps more insidious than the haze are the high levels of ambient ozone that prevail for up to 6 months each year. Concentrations commonly exceed 100 parts per billion by volume (ppbv) and can reach as high as 250 ppbv. Prolonged exposure to elevated ozone concentrations reduces respiratory function in most humans, and sensitive individuals can suffer severe respiratory distress (Lippman 1989). Photosynthesis and productivity of some forest species and agricultural crops are reduced by prolonged exposures to greater than 80 ppbv (Reich and Amundson 1985).

Different scales of ozone production and dispersal have different impacts on vegetation. On a typical summer day in the eastern Los Angeles Basin, ozone is low in the morning, increases sharply in the mid afternoon, then decreases at night. Nighttime concentrations are near zero in the city (where abundant nitric oxide acts to break down ozone molecules), but they remain moderately high in outlying wildland areas. Ozone concentrations are much higher east of Los Angeles because ozone is continually synthesized from nitrogen oxides and volatile organic compounds as the air mass moves eastward in the presence of sunlight (specifically, ultraviolet light). As a result, ozone concentrations are particularly high in forest and shrubland ecosystems in the San Bernardino Mountains at the eastern end of the basin (Miller 1992).

Long-distance transport of ozone has affected the forest dynamics and competitive interactions of species in the mixed-conifer forests of the San Bernardino Mountains. Ozone reduces photosynthesis in sensitive species such as *Pinus ponderosa* (ponderosa pine) and *P. jeffreyi* (Jeffrey pine) (Bytnerowicz and Grulke 1992), causing chlorotic foliage, lower foliar biomass, reduced growth, and greater susceptibility to other stresses such as low soil moisture and bark beetles. Reduced vigor and susceptibility to other stresses are the result of chronic exposure to ozone over decades (since at least the 1950s in the Los Angeles Basin) (Miller et al. 1963). Mortality and lack of vigor in these sensitive species have allowed ozone-tolerant and shade-tolerant species such as *Libocedrus decurrens* (incense cedar) and *Abies concolor* (white fir) to attain greater dominance in some mixed-conifer forests of this region (Miller 1992). Furthermore, ozone stress has eliminated many sensitive lichen species from the San Bernardinos, resulting in a loss of diversity and function of the epiphytic component of the forests (Sigal and Nash 1983).

In the Sierra Nevada, the region of greatest impact is 300 km from the primary source of pollution, and symptomatic ozone injury to *P. ponderosa* and *P. jeffreyi* is found in the southernmost 500 km of the range. The most severe injury is found in the far southern part of the

Sierra Nevada, particularly in Sequoia National Park, where recent growth reductions (figure 22.3) have been measured (Peterson et al. 1991).

The severity of air pollution in southern and central California has prompted a substantial societal investment in monitoring and regulating air pollutants. The state of California has the largest pollutant monitoring network in the world and a large regulatory and scientific advisory agency (California Air Resources Board). The State restricts emissions from the combustion of fossil fuel (e.g., from passenger vehicles) and from volatilization of organic compounds (e.g., from gasoline pumps) at more restrictive levels than required by federal regulations. Emission controls on automobiles are especially strict because automobile emissions contribute such a large fraction of nitrogen oxides and volatile organic compounds (and hence formation of ozone).

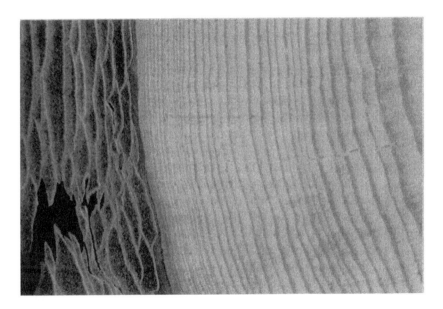

FIGURE 22.3

Cross section from the base of *Pinus jeffreyi*, showing a recent growth reduction. This pattern of growth reduction is common in mature *P. jeffreyi* and *P. ponderosa* with symptomatic ozone injury (chlorotic foliage and reduced foliar biomass) in the San Bernardino Mountains and southern Sierra Nevada. Statistical verification of growth reduction requires that time-series analysis be used to measure a small-scale spatial response (tree-stem basal area)—caused by a large-scale spatial phenomenon (atmospheric transport of air pollution)—over a temporal scale of several decades.

In addition, there are programs that encourage use of public transportation and multiple passengers per vehicle.

Efforts to reduce emissions at the source have had a significant, measurable impact in reducing ozone concentrations throughout the Los Angeles Basin, including the severely impacted eastern half of the basin. Ozone concentrations, including both mean and maximum values, are now considerably lower than they were 20 years ago. There is even evidence that the vigor (as measured by crown condition) of *P. ponderosa* has improved during that period of time (Miller 1992). Although air quality is still poor in southern and central California, it is clear in this case that regulatory actions instituted locally on relatively short time scales can have significant regional benefits at longer time scales.

The Dynamics of Fire Disturbance

Most plant communities are influenced to some extent by fire, and some are described by ecologists as "characterized by fire." The diversity of communities arises at a number of scales, from local heterogeneity to larger scale patterns. In the case of fire, the changing of scale in any particular region shifts our interpretation of influential processes or aspects of the fire regime, yet parallel shifts may not occur when comparing one region to another. Fires result from the interaction of climate and vegetation. Fuel accumulation patterns and vegetation structure lead to differential fire regimes in the context of appropriate climates. Both fuel accumulation and sources of ignition must be available at the time environmental conditions can maintain fire. Thus, fire exerts considerable influence on dry regions of western North America in most vegetation types. Because of sufficient ignition sources and fuels, fire also impacts conifer-dominated forests in the humid eastern United States along the coastal plains, from Massachusetts to Texas.

The community mosaic that develops in regions characterized by fire reflects climatic and landscape heterogeneities that influence or constrain fire regime patterns. Local vegetation dynamics are thus hierarchically nested within climate and landscape. Interactions of species adaptations with fire regime characteristics, and the predictability of fire patterns in the landscape, are an important feature of ecosystem biogeography in regions with high-intensity wildfire systems (Romme and Knight 1981; Myers 1985; Minnich 1988).

Vegetation dynamics in these fire-influenced systems result because of differential performance of species. Plant populations vary greatly in how they can survive a fire, but as a whole, species exhibit different combinations of characteristics. Fires vary considerably depending on whether they are ground, surface, or canopy fires, the intensity they exhibit, the area of the fire, the season the fire occurs, and the historical frequency of fires. All of these dimensions interact with plant characteristics and influence heterogeneous patterns at local to biogeographic scales.

At small spatial scales, patterns of vegetation result from local differences in fire characteristics such as intensity, ash accumulation, and chance events of which species happened to be available. Fire heat shifts predictably from gentle to steep slopes and with other environmental gradients (Minnich 1988), thus sorting species by their ability to survive such a temperature gradient, either as adults that might vegetatively recover, or as seed in soil (few species survive temperatures greater than 120°C). Fires that switch from surface to crown fire along landscape features create mosaics of vegetation types, such as adjacent chaparral and conifer forests in southern California mountains (Minnich 1988) or various sandhill and *Pinus clausa* (sand pine) scrub communities in Florida (Laessle 1942; Veno 1976; Myers 1985). Variablility in fire regimes also is influenced by site history, which increases the underlying landscape variability. Previous surface fires, patchy within the landscape, can break the continuity of subsequent crown fires within otherwise homogeneous forest (Wagle and Eakle 1979).

One significant characteristic that influences success among plants found in fire-prone regions is their population response after fires. Plants can be divided into a number of groups, but they easily sort into those that increase in population size after a fire and those that take a number of years to recover. Populations increasing after fire can be considered as fire-recruitment specialists, such as those with fire-stimulated dormant soil seed banks or conifers with closed cones, or as facultative postfire recruiters, such as pines that survive surface fires and disperse seed into recently opened habitat. If fire is sufficiently frequent, species that increase in population size after fire will dominate systems such as chaparral (Parker and Kelly 1989) and conifer communities (Romme and Knight 1981; Romme 1982).

As these postfire recruiters increase in dominance, so too does homogeneity of conditions, which leads to positive feedbacks to maintain the system. The positive feedbacks of these dominant species as they affect both environmental conditions and fire regime characteristics act to sort associated plants, animals, fungi, and characteristics of ecosystem processes. Such dominance reinforces fire regimes in these ecosystems because of the subsequent larger spatial scale and homogeneity of otherwise local processes. Large areas end up with single age classes, which leads to relatively uniform responses to environmental stresses, such as drought or insect attack, and results in similar amounts of fuel buildup. The relationship between fire, fungi, and insect outbreaks is well known in conifer systems. Within *Pinus contorta* (lodgepole pine) forest, a cycle of fire followed by fungal attack in older stands, subsequent beetle attack, tree mortality, and finally the return of fire is an example of how fire-recruitment dominance reinforces the fire regime of the system (Geiszler et al. 1980). Thus, fire becomes incorporated as a predictable aspect of fire-type vegetation, strongly linked by climate and vegetation characteristics. For example, fire suppression efforts in southern California chaparral have not altered point-frequencies of fires. A study comparing southern California wildfire history with that of adjacent Baja California determined that fire suppression may have changed fire events only from multiple, smaller fires to fewer, larger-scale fires, while not shifting the average frequency much for any site (Minnich 1983, 1989).

Humans interact with wildland fire in a variety of ways. Adjacent to urban areas, management of fire-prone vegetation includes fire suppression and fuel reduction by cutting or prescribed burning. Problems have arisen with prescribed burning in a number of locations because of conflicting objectives in managing vegetation. Managers have multiple objectives, and these include not only safety and fuel reduction but maintenance of watershed function and water quality, preservation of rare and threatened species, and conservation of recreational values. In meeting safety objectives, conditions appropriate to control prescribed fires often conflict with the natural fire regime (Parker 1989, 1990). Prescribed burning that is too frequent can cause long-term changes in species composition, and fires during seasons in which fire would not normally occur can differentially eliminate species. Other dimensions of the natural fire pattern (area, intensity, etc.) are modified by prescribed burning constraints and can significantly impact vegetation (Parker 1989, 1990).

In contrast to deliberately set prescribed burns, "prescribed natural fires" result from natural ignitions and are in locations and climatic conditions that managers feel do not threaten humans or structures. Under these conditions, managers allow the fire to continue burning while monitoring the fire's progress and any changes in weather. Humans often do not hold realistic attitudes about the dynamics of vegetation, and this can substantially influence how the general public assesses ecosystem management techniques (Parker and Pickett 1997). For example, the extensive fires of summer 1988 that burned through parts of Yellowstone National Park were not examples of prescribed natural fire. Nonetheless, the fires created a national debate on the use of prescribed natural fires in managing vegetation (Christensen et al. 1989; Knight 1991; Keiter and Boyce 1991).

Humans thus interact with fire in several ways, but overall their impact is to modify historical, natural fire regimes. The long-term result of such changes will be substantial shifts in vegetation composition and structure. The difficulty of managing wildfire-prone communities varies depending on the combination of species and the type of conditions and fire to which they are historically adapted. It is clear that conflicts between human development, management objectives, and the preservation of dynamic natural systems are increasing in number and scale.

Global Climatic Change: The Ultimate Scale Issue

The prospect of global-scale increases in temperature during the next century as a result of human-caused changes in the atmospheric environment poses a huge challenge for addressing physical and biological issues at large spatial and temporal scales. Modeling the earth's atmosphere has been addressed by a number of general circulation models (GCMs) (e.g., Manabe and Stouffer 1986) that calculate future climatic scenarios at a spatial scale of greater than 100,000 km^2. These models simulate temperature with some confidence but simulate precipitation unreliably. The models generally predict large spatial variations in future temperature, with much larger increases at higher latitudes (Ferguson 1995). Although future increases in atmospheric concentrations of greenhouse gases are certain, there is a high degree of uncertainty regarding predictions of future climate.

Even less certain are the potential impacts of climate change on terrestrial and aquatic systems, although this has not prevented a huge literature that predicts dramatic changes in ecosystem structure and function as a result of temperature increases (Watson et al. 1996). Unfortunately, many "prediction" studies contain a mismatch of scales. For example, claims of recent human-caused temperature increases have been made (cited in Schneider 1989; Kerr 1995) despite a short time series of reliable data (slightly over 100 years) compared to long-term climatic cycles (hundreds to thousands of years) (Schoonmaker 1997, chapter 5). In addition, climatic data at a spatial scale of 100 to

1000 km² are typically compared to vegetation data at a scale of less than 1 km² (Peterson et al. 1997), and GCMs (with high uncertainty) are now being linked to vegetation models (with additional uncertainty) (Gyalistras et al. 1994). How much confidence can we have in the final model output if there is no articulation of scales of different model components or of how data and processes are modeled across scales (Hinckley et al., chapter 15; Lertzman and Fall, chapter 16; McKenzie et al. 1996)? While large-scale modeling efforts have some heuristic value, the output has large confidence intervals for scientific and other applications.

Incompatible scales and a general lack of data make it difficult to accurately predict the potential impacts of climate change on natural resources (Innes, chapter 19). It is not our intent to harshly judge scientists who are addressing this critical issue. However, we strongly encourage researchers and modelers to more clearly state the spatial and temporal scales of their data and inferences, as well as how disparate scales are linked conceptually and quantitatively. If the scale paradigm is clearly integrated and the uncertainty of climate change impacts is stated, then the predictions will be more realistic for subsequent interpretation by resource managers and decision makers. Scientific predictions and model output of the past decade have been convincing enough that many nations in the world have taken steps to reduce emissions of carbon dioxide and other greenhouse gases and to take other actions to delay the onset of climate change or to mitigate potential impacts. This linkage with effective sociopolitical actions should be based on credible science (Rykiel, chapter 21), a linkage that can be facilitated by awareness of scale concepts in climate change research.

Using Scale to Solve Ecological Problems

Ecologists investigate such a diversity of issues that it is a challenge to develop frameworks or models that can effectively include all issues. One particular topic that has already been discussed is the concept of entities. Depending on how questions are phrased, some entities are readily scaled up or down to make other predictions. This seems relatively true when ecologists focus on entities that do not change from one scale to another, such as water in hydrological studies, forms of nitrogen from a small scale to a watershed study, or proportions of flora that are introduced species. In other cases, increases in scale of interpretation changes the analytical form of entities; consider the variation in assessing changes in numbers of individuals in a small plot study compared to species populations at larger scales, or compared to community types or vegetation structure at even larger scales. The dynamics and dimensionality of the entities at each scale change considerably, even though all are ultimately represented by success or failure of individual organisms.

Ecological entities are also susceptible to interpretive problems that change with the scale of analysis (Rahel 1990). Populations and communities are often poorly defined and have ambiguous boundaries. The species of which they are composed can also have poor taxonomic resolution, especially among taxa from different kingdoms or phyla. Approaching an ecological problem from multiple scales of investigation and from several scales of analysis is currently the only approach for helping ecologists avoid differences in opinions based on differently scaled studies (Allen and Starr 1982). Useful models have been developed that are neither constrained to a single level of investigation nor restricted in scale. Some are well known, such as hierarchical approaches (Allen and Starr 1982; O'Neill et al. 1989; Kolasa 1989; Pickett et al. 1989; Parker and Pickett, chapter 8). Such scale-independent models of systems allow comparative analysis among studies that may differ considerably with respect to organisms and spatial and temporal scale.

There is a strong tendency in ecological research to use reductionist approaches (Peters 1991; Hammer, chapter 6) by describing complex natural phenomena with simple concepts and mathematics (Martinez and Dunne, chapter 10). Reductionist ecology, short-term funding horizons for research, and emphasis on short-term productivity (i.e., publications) tend to produce scientific investigations of small spatial and temporal scales. Because most critical natural resource problems tend to occur at large spatial (a watershed or greater) and temporal (decades to centuries) scales, it is logical that more research and monitoring be conducted at these more appropriate scales (Magnuson et al. 1991; Innes, chapter 19; Vande Castle, chapter 13). Long-term studies [e.g., Hubbard Brook (New Hampshire)] have proven extremely valuable in quantifying ecosystem processes and can influence decision making and environmental regulations (Rykiel, chapter 21).

Scale is a nonreductionist unifying concept in ecology. Almost every type of ecological investigation exhibits scale limitations in its interpretation and in its application to other circumstances. Scientists and resource managers need to explicitly state the spatial, temporal, and other scales of their problems, data, analysis, and interpretation. We support the populist call for greater cooperation between scientists and resource mangers (Peterson 1996; Hobbs, chapter 20) and encourage the continued development of tools to facilitate this cooperation. Recognition of the scale paradigm in all ecological disciplines will make it more likely that the scales of application in resource management match relevant scales in research and monitoring.

Acknowledgments

We are grateful to many students and colleagues for providing inspiration and ideas for this chapter. We thank Katie Bode for compiling and summarizing most of the information in table 22.1. The section on specific impacts on natural resources benefited greatly from past discussions with and reviews by Paul Miller (air quality), Gil Pauley (salmon), and Ed Schreiner (salmon and biogeographic concepts).

REFERENCES

Aber, J. D., A. Magill, R. Boone, J. M. Melillo, P. Steudler, and R. Bowen. 1993. Plant and soil responses to chronic nitrogen additions at the Harvard Forest, Massachusetts. *Ecological Applications* 3:156–166.

Aber, J. D., J. Pastor, and J. M. Melillo. 1982. Changes in forest canopy structure along a site quality gradient in southern Wisconsin. *American Midland Naturalist* 108:256–265.

Ableiter, J. K. 1949. Soil classification in the United States. *Soil Science* 67: 183–191.

Abmeyer, W. and H. V. Campbell. 1970. Soil survey of Shawnee County, Kansas. In E. L. Stone, ed., *Forest Soils and Treatment Impacts, Proceedings,* 6th North American Forest Soils Conference, pp. 381–404. Madison, Wis.: Soil Science Society of America.

Aboul-Hosn, W., P. Dutilleul, and D. Boisclair. 1997. Use of spectral analysis to estimate short-term periodicities in growth rates of the brook trout *Salvelinus fontinalis. Canadian Journal of Fisheries and Aquatic Science* 54:1532–1541.

Acker, S. A. 1990. Vegetation as a component of a non-nested hierarchy: a conceptual model. *Journal of Vegetation Science* 1:683–690.

Adams, J. B., M. O. Smith, and A. R. Gillespie. 1993. Imaging spectrometry: interpretation based on spectral mixing analysis. In C. M. Pieters and P. Englert, eds., *Remote Geochemical Analysis: Elemental and Mineralogical Composition.* 7:145–166. New York: Cambridge University Press.

Addicott, J. F. 1978. The population dynamics of aphids on fireweed: a comparison of local populations and metapopulations. *Canadian Journal of Zoology* 56:2554–2564.

Addicott, J. F., J. M. Aho, M. R. Antolin, D. K. Padilla, J. S. Richardson, and D. A. Soluk. 1987. Ecological neighborhoods: scaling environmental patterns. *Oikos* 49:340–346.

Aebischer, N.J., P. A. Robertson, and R. E. Kenward. 1993. Compositional analysis of habitat use from animal radio-tracking data. *Ecology* 74:1313–1325.

Agee, J. K. and L. Smith. 1984. Subalpine tree establishment after fire in the Olympic Mountains, Washington. *Ecology* 65:810–819.

Ahl, V. A. and T. F. H. Allen. 1996. *Hierarchy Theory: A Vision, Vocabulary and Epistemology.* New York: Columbia University Press.

Ahrens, M. and R. H. Peters. 1991. Patterns and limitations in limnoplankton size spectra. *Canadian Journal of Fisheries and Aquatic Sciences* 48:1967–1978.

Aikman, D. P. 1992. Modeling of growth and competition in plant monocultures. In D. L. DeAngelis and L. J. Gross, eds., *Individual-based Models and Approaches in Ecology: Populations, Communities and Ecosystems,* pp. 472–491. New York: Chapman and Hall.

Allan, J. D. 1995. *Stream Ecology: Structure and Function of Running Waters.* London: Chapman and Hall.

Allen, C. D. 1994. Ecological perspective: linking ecology, GIS, and remote sensing to ecosystem management. In V. A. Sample, ed., *Remote Sensing and GIS in Ecosystem Management,* pp. 111–139. Washington, D.C.: Island Press.

Allen, T. F. H. and T. W. Hoekstra. 1990. The confusion between scale-defined levels and conventional levels of organization in ecology. *Journal of Vegetation Science* 1:5–12.

Allen, T. F. H. and T. W. Hoekstra. 1991. The role of heterogeneity in scaling of ecological systems under analysis. In J. Kolasa and S. T. A. Pickett, eds., *Ecological Heterogeneity. Ecological Studies 86,* pp. 47–68. New York: Springer-Verlag.

Allen, T. F. H. and T. W. Hoekstra. 1992. *Toward a Unified Ecology.* New York: Columbia University Press.

Allen, T. F H. and J. F. Koonce. 1973. Multivariate approaches to algal stratagems and tactics in systems analysis of phytoplankton. *Ecology* 54:1234–1246.

Allen, T. F. H. and T. B. Starr. 1982. *Hierarchy: Perspectives for Ecological Complexity.* Chicago: University of Chicago Press.

Allen, T. F. H. and E. P. Wyleto. 1983. A hierarchical model for the complexity of plant communities. *Journal of Theoretical Biology* 101:529–540.

Allen, T. F. H., S. M. Bartell, and J. F. Koonce. 1977. Multiple stable configurations in ordination of phytoplankton community change rates. *Ecology* 58:1076–1084.

Allen, T. F. H., A. W. King, B. Milne, A. Johnson, and S. Turner. 1993. The problem of scaling in ecology. *Evolutionary Trends in Plants* 7:3–8.

Allen, T. F. H., R. V. O'Neill, and T. W. Hoekstra. 1987. Interlevel relations in ecological research management: some working principles from hierarchy theory. *Journal of Applied Systems Analysis* 14:63–79.

Amen, A. E., D. L. Anderson, T. J. Hughes, and T. J. Weber. 1977. *Soil Survey of Logan County, Colorado.* USDA Soil Conservation Service in cooperation with Colorado Agricultural Experiment Station, Washington, D.C.: U.S. Government Printing Office.

Amthor, J. S. 1994. Scaling CO_2-photosynthesis relationships from the leaf to the canopy. *Photosynthesis Research* 39:321–350.

Andersen, S. T. 1984. Forests at Lovenholm, Djursland, Denmark, at present and in the past. *Biologiske Skrifter Danske Videnskabernes Selskab* 24:1–211.

Anderson, M. G. and T. P. Burt, ed. 1990. *Process Studies in Hillslope Hydrology.* New York: John Wiley and Sons.

Anderson, P. W. 1972. More is different. *Science* 177:393–396.

Anderson, R. C. 1982. An evolutionary model summarizing the roles of fire, climate, and grazing animals in the origin and maintenance of grasslands. In J. R. Estes, R. J. Tyrl, and J. N. Brunken, eds., *Grasses and Grasslands: Systematics and Ecology,* pp. 297–308. Norman: University of Oklahoma Press.

Andrén, H. 1995. Effects of habitat fragmentation on birds and mammals in landscapes with different proportions of suitable habitat: a review. *Oikos* 71: 355–66.

Andrewartha, H. G. and L. C. Birch. 1984. *The Ecological Web.* Chicago: University of Chicago Press.

Angelstam, P. 1992. Conservation of communities: the importance of edges, surroundings and landscapes mosaic structure. In L. Hansson, ed., *Ecological Principles of Nature Conservation: Application in Temperate and Boreal Environments,* pp. 9–69. London: Elsevier Applied Science.

Arnold, G. W., D. E. Steven, J. R. Weeldenburg, and E. A. Smith. 1993. Influences of remnant size, spacing pattern and connectivity on population boundaries and demography in Euros *Macropus robustus* living in a fragmented landscape. *Biological Conservation* 64:219–230.

Aubreville, A. 1932 (1933). La forêt de la Côte d'Ivoire: essai de géobotanique forestière. *Bulletin du Comité d'Études historiques et scientifiques de l'Afrique occidentale française* 15:20–61.

Axelrod, D. I. 1985. Rise of the grassland biome, central North America. *Biological Reviews* 51:163–201.

Bach, C. E. 1980. Effects of plant density and diversity on the population dynamics of a specialist herbivore, the striped cucumber beetle, *Acalymma vittata* (Fab.). *Ecology* 61:1515–1530.

Backman, A. E. and W. A. Patterson III. 1984. Vegetation and fire history of Miles Standish State Forest interpreted from sedimentary pollen and charcoal analyses. *Cooperative Forestry Research Unit, Report 2.* Amherst: University of Massachusetts.

Bailey, J. A. 1984. *Principles of Wildlife Management.* New York: John Wiley and Sons.

Baker, W. L. 1989. Landscape ecology and nature reserve design in the Boundary Waters Canoe Area, Minnesota. *Ecology* 70:23–35.

Baldocchi, D. 1994. A comparative study of mass and energy exchange over a closed C_3 (wheat) and an open C_4 (corn) canopy: I. The partitioning of available energy into latent and sensible heat exchange. *Agricultural and Forest Meteorology* 67:191–220.

Baldwin, M. and C. E. Kellogg. 1938. Soil classification. In *Soils and Men. Yearbook of Agriculture,* pp. 979–1001. U.S. Department of Agriculture. Washington, D.C.: U.S. Government Printing Office.

Barendregt, A., M. J. Wassen and P. P. Schot. 1995. Hydrological systems beyond a nature reserve, the major problem in wetland conservation of Naardermeer (The Netherlands). *Biological Conservation* 72:393–405.

Barnola, J. M., D. Raynaud, V. S. Korotkevitch, and C. Lorius. 1987. Vostok ice core provides 160,000 year record of atmospheric CO_2. *Nature* 329:408–414.

Bartell, S. M., A. L. Brenkert, R. V. O'Neill, and R. H. Gardner. 1988a. Temporal variation in regulation of production in a pelagic food web model. In S. R. Carpenter, ed., *Complex Interactions in Lake Communities*, pp. 101–118. New York: Springer-Verlag.

Bartell, S. M., W. G. Cale, R. V. O'Neill, and R. H. Gardner. 1988b. Aggregation error, research objectives and relevant model structure. *Ecological Modelling* 41:157–168.

Bartlett, M. S. 1978. Nearest neighbor models in the analysis of field experiments with large blocks. *Journal of the Royal Statistical Society Series B* 40:147–174.

Barton, C. V. M., H. S. J. Lee, and P. G. Jarvis. 1993. A branch bag and CO_2 control system for long-term CO_2 enrichment of mature Sitka spruce. *Plant, Cell and Environment* 16:1139–1148.

Batschelet, E. 1981. *Circular Statistics in Biology.* New York: Academic Press.

Battarbee, R. W., J. Mason, I. Tenberg, and J. F. Talling, eds. 1990. *Palaeolimnology and Lake Acidification.* London: The Royal Society.

Baumgartner, T., A. Soutar, and V. Ferreira. 1992. Reconstruction of the history of Pacific sardine and northern anchovy populations over the past two millennia from sediments of the Santa Barbara Basin, California. *California Cooperative Oceanic Fisheries Investigative Report* 33:24–40.

Bazzaz, F. A. 1993. Scaling in biological systems: population and community perspectives. In J. R. Ehleringer and C. B. Field, eds., *Scaling Physiological Processes: Leaf to Globe,* pp. 233–254. San Diego: Academic Press.

Beamish, R. J. 1995. Response of anadromous fish to climate change in the North Pacific. In D. L. Peterson and D. R. Johnson, eds., *Human Ecology and Climate Change: People and Resources in the Far North,* pp. 123–136. Washington, D.C.: Taylor and Francis.

Bell, S. S., E. D. McCoy, and H. R. Mushinsky, eds. 1991. *Habitat Structure. The Physical Arrangement of Objects in Space.* London: Chapman and Hall.

Belsky, A. J. 1995. Spatial and temporal landscape patterns in arid and semi-arid African savannas. In L. Hansson, L. Fahrig, and G. Merriam, eds., *Mosaic Landscapes and Ecological Processes,* pp. 31–56. London: Chapman and Hall.

Bennett, K. D. 1990. Milankovitch cycles and their effects on species in ecological and evolutionary time. *Paleobiology* 16:11–21.

Bergeron, Y. 1991. The influence of island and mainland lakeshore landscapes on boreal forest fire regimes. *Ecology* 72:1980–1992.

Berglund, B. E., ed., 1986. *Handbook of Holocene Paleoecology and Paleohydrology.* New York: John Wiley and Sons.

Berry, J. K. 1993. *Beyond Mapping: Concepts, Algorithms, and Issues in GIS.* Fort Collins, Colo.: GIS World Inc.

Bertram, B. 1979. Serengeti predators and their social systems. In A. R. E. Sinclair and M. Norton-Griffiths, eds., *Serengeti: Dynamics of an Ecosystem*, pp. 221–248. Chicago: University of Chicago Press.

Bessie, W. C. and E. A. Johnson. 1995. The relative importance of fuels and weather on fire behavior in subalpine forests. *Ecology* 76:747–762.

Betancourt, J. L., T. R. Van Devander, and P. S. Martin, eds. 1990. *Packrat Middens: The Last 40,000 Years of Biotic Change*. Tucson: University of Arizona Press.

Beyschlag, W., F. Kresse, R. J. Ryel, and H. Pfanz. 1994. Stomatal patchiness in conifers: experiments with *Picea abies* (L.) Karst. and *Abies alba* Mill. *Tree* 8:132–138.

Binford, M. W. and B. Leyden. 1987. Ecosystems, paleoecology and human disturbance in subtropical and tropical America. *Quaternary Science Review* 6: 115–128.

Birks, H. J. B. and H. H. Birks. 1980. *Quaternary Paleoecology*. London: Edward Arnold.

Blake, G. 1993. A response to Brian Walker. *Conservation Biology* 7:5–6.

Bobisud, L. E. and W. L. Voxman. 1979. Predator response to variation in prey density in a patchy environment: a model. *American Naturalist* 114:63–75.

Bock, C. E. and R. E. Ricklefs. 1983. Range size and local abundance of some North American songbirds: a positive correlation. *American Naturalist* 122:463–466.

Bonner, J. T. 1965. *Size and Cycle*. Princeton, N.J.: Princeton University Press.

Boose, E. R., D. R. Foster, and F. Marcheterre Fluet. 1994. Hurricane impacts to tropical and temperate forest landscapes. *Ecological Monographs* 64:369–400.

Borcard, D., P. Legendre, and P. Drapeau. 1992. Partialling out the spatial component of ecological variation. *Ecology* 73:1045–1055.

Borchers, J. G. 1995. A hierarchical context for sustaining ecosystem health. In R. Jaindl and T. Quigley, eds., *Search for a Solution*, pp. 63–80. Washington, D.C.: American Forestry Association.

Bormann, B. T., M. H. Brookes, F. E. Ford, R. A. Kiester, C. D. Oliver, and J. F. Weigand. 1994. Volume V: A framework for sustainable-ecosystem management. *USDA Forest Service General Technical Report PNW-GTR-331*. Portland, Ore.: Pacific Northwest Research Station.

Bormann, F. H. and G. E. Likens. 1994. *Pattern and Process in a Forested Ecosystem. Disturbance, Development and the Steady State Based on the Hubbard Brook Ecosystem Study*. New York: Springer-Verlag.

Botkin, D. B. 1990. *Discordant Harmonies. A New Ecology for the Twenty-First Century*. Oxford: Oxford University Press.

Botkin, D. B., J. M. Mellilo, and L. S. Wu. 1981. How ecosystem processes are linked to large mammal population dynamics. In C. W. Fowler and T. D. Smith, eds., *Dynamics of Large Mammal Populations*, pp. 373–388. New York: John Wiley and Sons.

Bouma, J. 1989. Land qualities in space and time. In J. Bouma and K. Bregt, eds., *Land Qualities in Space and Time*, pp. 3–13. Wageningen, Netherlands: Pudoc.

Bowman, W. D., T. A. Theodose, J. C. Schardt, and R. T. Conant. 1993. Constraints of nutrient availability on primary production in two alpine tundra communities. *Ecology* 74:2085–2097.

Box, G. E. P. and C. R. Cox. 1964. An analysis of transformations. *Journal of the Royal Statistical Society Series B* 26:211–243.

Box, G. E. P. and G. M. Jenkins. 1976. *Time Series Analysis: Forecasting and Control,* revised ed. San Francisco: Holden-Day.

Box, G. E. P., G. M. Jenkins, and G. C. Reinsel. 1994. *Time Series Analysis: Forecasting and Control,* 3rd edition. Englewood Cliffs, N.J.: Prentice-Hall.

Boyd, R. G., ed. 1997. *Indians, Fire and the Land in the Pacific Northwest.* Corvallis: Oregon State University Press.

Bradley, R. S. 1985. *Quaternary Paleoclimatology.* Boston: Allen and Unwin.

Bradshaw, A. D. 1987. Restoration: an acid test for ecology. In W. R. Jordan, M. E. Gilpin, and J. D. Aber, eds., *Restoration Ecology: A Synthetic Approach to Ecological Research,* pp. 23–30. Cambridge: Cambridge University Press.

Bradshaw, G. A. 1991. Analysis of hierarchical pattern and process in Douglas-fir forests using wavelet analysis. *Ph.D. Dissertation.* Corvallis: Oregon State University.

Bradshaw, G. A. and S. L. Garman. 1994. Detecting fine-scale disturbance in forested ecosystems as measured by large-scale landscape patterns. In W. K. Michener, J. W. Brunt, and S. G. Stafford, eds., *Environmental Information Management and Analysis: Ecosystem to Global Scales,* pp. 535–550. London: Taylor and Francis.

Bradshaw, G. A. and B. McIntosh. 1993. Detecting climate-induced change using wavelet analysis. *Environmental Pollution* 83:135–142.

Bradshaw, R. W. H. 1981. Quantitative reconstruction of local woodland vegetation using pollen analysis from a small basin in Norfolk, England. *Journal of Ecology* 69:941–955.

Bradshaw, R. H. W. 1988. Spatially precise studies of forest dynamics. In B. Huntley and T. Webb III, eds., *Vegetation History,* pp. 725–751. The Hague: Kluwer.

Brand, T. and V. T. Parker. 1995. Scale and general laws of vegetation dynamics. *Oikos* 73:375–380.

Bratton, S. P. 1975. The effect of the European wild boar (*Sus scrofa*) on gray beech forest in the Great Smoky Mountains. *Ecology* 56:1356–1366.

Breininger, D. R., V. L. Larson, B. W. Duncan, R. B. Smith, D. M. Oddy, and M. F. Goodchild. 1995. Landscape patterns of Florida scrub jay habitat use and demographic success. *Conservation Biology* 9:1442–1453.

Briand, F. 1983. Environmental control of food web structure. *Ecology* 64: 253–263.

Briand F. and J. E. Cohen. 1984. Community food webs have scale-invariant structure. *Nature* 307:264–266.

Briand F. and J. E. Cohen. 1987. Environmental correlates of food chain length. *Science* 238:956–960.

Brillinger, D. R. 1973. The analysis of time series collected in an experimental design. In P. R. Krishnaiah, ed., *Multivariate Analysis III,* pp. 241–256. London: Academic Press.

Brillinger, D. R. 1980. Analysis of variance and problems under time series models. In P. R. Krishnaiah, ed., *Handbook of Statistics: Analysis of Variance*, pp. 237–278. New York: North-Holland.

Brillinger, D. R. 1981. *Time Series: Data Analysis and Theory*, expanded edition. New York: McGraw-Hill.

Brillinger, D. R. 1983. The finite Fourier transform of a stationary process. In D. R. Brillinger and P. R. Krishnaiah, eds., *Handbook of Statistics: Time Series in the Frequency Domain*, pp. 21–37. New York: North-Holland.

British Columbia Government. 1995. *Forest Practices Code of British Columbia.* Victoria, B.C.

Brockwell, P. J. and R. A. Davis. 1991. *Time Series: Theory and Methods.* New York: Springer-Verlag.

Brooks, J. R. 1987. Foliage respiration of *Abies amabilis.* Master's thesis. Seattle: University of Washington.

Brooks, J. R., T. M. Hinckley, E. D. Ford, and D. G. Sprugel. 1990. Foliage respiration in *Abies amabilis:* scaling measurements from twigs to a whole stand. *Bulletin of the Ecological Society of America* 71:104.

Brooks, J. R., T. M. Hinckley, E. D. Ford, and D. G. Sprugel. 1991. Dark respiration in *Abies amabilis:* variation within the canopy. *Tree Physiology* 9:325–338.

Brooks, J. R., T. M. Hinckley, and D. G. Sprugel. 1994. Acclimation responses of mature *Abies amabilis* sun foliage to shading. *Oecologia* 100:316–324.

Brooks, J. R., T. M. Hinckley, and D. G. Sprugel. 1996. The effects of light acclimation during and after foliage expansion on photosynthetic function of *Abies amabilis* foliage within the canopy. *Oecologia* 107:21–32.

Broschart, M. R., C. A. Johnston, and R. J. Naiman. 1989. Predicting beaver colony density in boreal landscapes. *Journal of Wildlife Management* 53:929–934.

Brown, B. J. and T. F. H. Allen. 1989. The importance of scale in evaluating herbivory impacts. *Oikos* 54:189–194.

Brown, J. 1994. Grand challenges in scaling up environmental research. In W. K. Michener, J. W. Brunt, and S. G. Stafford, eds., *Environmental Information Management and Analysis: Ecosystem to Global Scales*, pp. 21–26. London: Taylor and Francis.

Brown, J. H. 1984. On the relationship between abundance and distribution of species. *American Naturalist* 124:255–279.

Brown, J. H. 1995. *Macroecology.* Chicago: Chicago University Press.

Brown, J. H. and A. Kodric-Brown. 1977. Turnover rates in insular biogeography: effect of immigration on extinction. *Ecology* 58:445–449.

Brown, J. H. and B. A. Maurer. 1986. Body size, ecological dominance and Cope's rule. *Nature* 324:248–250.

Brown, J. H. and B. A. Maurer. 1989. Macroecology: the division of food and space among species on continents. *Science* 243:1145–1150.

Brown, J. H. and P. F. Nicoletto. 1991. Spatial scaling of species composition: body masses of North American land mammals. *American Naturalist* 138:1478–1512.

Brubaker, L. B. 1975. Postglacial forest patterns associated with till and outwash in northcentral upper Michigan. *Quaternary Research* 5:499–527.

Brugam, R. B. 1978. Pollen indicators of land-use change in southern Connecticut. *Quaternary Research* 9:349–362.

Brunsden, D. 1980. Applicable models of long term landform evolution. *Zeitschrift für Geomorphologie Supplement* 36:16–26.

Brunsden, D. and J. B. Thornes. 1979. Landscape sensitivity and change. *Transactions of the Institute of British Geography* NS4:463–484.

Buckland, S. T., D. R. Anderson, K. P. Burnham, and J. L. Laake. 1993. *Distance Sampling.* London: Chapman and Hall.

Bunnell, F. J. 1995. Forest-dwelling vertebrate faunas and natural fire regimes in British Columbia: patterns and implications for conservation. *Conservation Biology* 9:636–644.

Buol, S. W. and F. D. Hole. 1959. Some characteristics of clay skins on peds in the B horizon of a gray-brown podzolic soil. *Soil Science Society of America Proceedings* 23:239–241.

Burel, F. 1989. Landscape structure effects on carabid spatial patterns in western France. *Landscape Ecology* 2:215–226.

Burke, I. C., E. T. Elliott, and C. V. Cole. 1995a. Influence of macroclimate, landscape position, and management on soil organic matter in agroecosystems. *Ecological Applications* 5:124–131.

Burke, I. C., W. K. Lauenroth, and D. P. Coffin. 1995b. Soil organic matter recovery in semiarid grasslands: implications for the Conservation Reserve Program. *Ecological Applications* 5:793–801.

Burrough, P. A. 1981. Fractal dimensions of landscapes and other environmental data. *Nature* 294:241–242.

Burrough, P. A. 1983. Multiscale sources of spatial variation in soil. *Journal of Soil Science* 34:577–597.

Burrough, P. A. 1986. *Principles of Geographical Information Systems for Land Resources Assessment.* Oxford: Oxford University Press.

Busing, R. T. and P. S. White. 1993. Effects of area on old-growth forest attributes: implications for the equilibrium landscape concept. *Landscape Ecology* 8:119–126.

Butler, B. E. 1959. Periodic phenomena in a landscape as a basis for soil studies. *Soil Publication 14.* Melbourne: C.S.I.R.O.

Bycroft, C. M., N. Nicolaou, B. Smith, and J. B. Wilson. 1993. Community structure (niche limitation and guild proportionality) in relation to the effect of spatial scale, in a Nothofagus forest sampled with a circular transect. *New Zealand Journal of Ecology* 17:95–101.

Byers, H. G., C. E. Kellogg, M. S. Anderson, and J. Thorp. 1938. Formation of soil. In *Soils and Men. Yearbook of Agriculture,* pp. 948–978. U.S. Department of Agriculture. Washington, D.C.: U.S. Government Printing Office.

Bytnerowicz, A. and N. E. Grulke. 1992. Physiological effects of air pollutants on Western trees. In R. K. Olson, D. Binkley, and M. Böhm, eds., *Response of Western Forests to Air Pollution,* pp. 183–233. New York: Springer-Verlag.

Calder, W. A. 1984. *Size, Function and Life History.* Cambridge: Harvard University Press.

Calder, W. A., III. 1983. Ecological scaling: mammals and birds. *Annual Review of Ecology and Systematics* 14:213–230.

Caldwell, M. M., P. A. Matson, C. Wessman, and J. Gamon. 1993. Prospects for scaling. In J. R. Ehleringer and C. B. Field, eds., *Scaling Physiological Processes: Leaf to Globe,* pp. 223–230. San Diego: Academic Press.

Cale, P. and R. J. Hobbs. 1994. Landscape heterogeneity indices: problems of scale and applicability, with particular reference to animal habitat description. *Pacific Conservation Biology* 1:183–193.

Cale, W. G. 1995. Model aggregation: ecological perspectives. In B. C. Patten and S. E. Jorgenson, eds., *Complex Ecology,* pp. 230–241. Englewood Cliffs, N.J.: Prentice-Hall.

Cale, W. G. and P. L. Odell. 1979. Concerning aggregation in ecosystem modeling. In E. Halfon, ed., *Theoretical Systems Ecology,* pp. 55–77. New York: Academic Press.

Cale, W. G. and P. L. Odell. 1980. Behavior of aggregate state variables in ecosystem models. *Mathematical Bioscience* 49:121–137.

Cale, W. G., R. V. O'Neill, and R. H. Gardner. 1983. Aggregation error in nonlinear ecological models. *Journal of Theoretical Biology* 100:539–550.

Callahan, J. T. 1991. Long-term ecological research in the United States: a federal perspective. In P. G. Risser, ed., *Long-Term Ecological Research: An International Perspective,* pp. 9–22. Chichester: John Wiley and Sons.

Callaway, R. M., E. H. DeLucia, and W. H. Schlesinger. 1994. Biomass allocation of montane and desert ponderosa pine: an analog for response to climate change. *Ecology* 75:1474–1481.

Campbell, B. D., J. P. Grime, and J. M. L. Mackey. 1991. A trade-off between scale and precision in resource foraging. *Oecologia* 87:532–538.

Campbell, D. J. and D. J. Clarke. 1971. Nearest neighbour tests of significance for non-randomness in the spatial distribution of singing crickets (Walker). *Animal Behaviour* 19:750–756.

Campbell, I. D. and J. H. Andrews. 1990. Forest disequilibrium caused by rapid Little Ice Age cooling. *Nature* 366:336–338.

Campbell, K. E., Jr and D. Frailey. 1984. Holocene flooding and species diversity in southwestern Amazonia. *Quaternary Research* 21:369–375.

Cappuccino, N. 1987. Comparative population dynamics of two goldenrod aphids: spatial patterns and temporal constancy. *Ecology* 68:1634–1646.

Cappuccino, N. and P. Kareiva. 1985. Coping with a capricious environment: a population study of a rare pierid butterfly. *Ecology* 66:152–161.

Carey, A. B. 1985. A summary of the scientific basis for spotted owl management. In R. J. Gutiérrez and A. B. Carey, eds., *Ecology and Management of the Spotted Owl in the Pacific Northwest,* pp. 100–114. USDA Forest Service General Technical Report PNW-GTR-185. Portland, Ore.: Pacific Northwest Research Station.

Carey, A. B. and M. L. Johnson. 1995. Small mammals in managed, naturally young, and old-growth forests. *Ecological Applications* 5:336–352.

Carlile, D. W., J. R. Skalski, J. E. Baker, J. M. Thomas, and V. I. Cullinan. 1989. Determination of ecological scale. *Landscape Ecology* 2:203–213.

Carpenter, S. and J. Kitchell. 1992. Trophic cascade and biomanipulation: interface of research and management; a reply to the comment of DeMelo et al. *Limnology and Oceanography* 37:208–213.

Carpenter, S. R. 1990. Large-scale permutations: opportunities for innovation. *Ecology* 71:2038–2043.

Carpenter, S. R. and J. E. Chaney. 1983. Scales of spatial patterns: four methods compared. *Vegetatio* 53:153–160.

Carpenter, S. R. and J. F. Kitchell. 1987. Plankton community structure and limnetic primary production. *American Naturalist* 124:159–172.

Carpenter, S. R. and J. F. Kitchell. 1988. Consumer control of lake productivity. *BioScience* 38:764–769.

Castle, E. N. 1993. A pluralistic, pragmatic and evolutionary approach to natural resource management. *Forest Ecology and Management* 56:279–295.

Caswell, H. 1976. Community structure: a neutral model hypothesis. *Ecological Monographs* 46:327–354.

Cawrse, D. 1994. Resource management perspective: managing to minimize fragmentation of native hardwood forests. In V. A. Sample, ed., *Remote Sensing and GIS in Ecosystem Management,* pp. 192–200. Washington, D.C.: Island Press.

Cermák, J. 1989. Solar equivalent leaf area: an efficient biometrical parameter of individual leaves, trees and stands. *Tree Physiology* 5:269–289.

Cermák, J., J. Ulehla, J. Kucera, and M. Penka. 1982. Sap flow rate and transpiration dynamics in the full-grown oak (*Quercus robur* L.) in floodplain forest exposed to seasonal floods as related to potential evapotranspiration and tree dimensions. *Biologia Plantarum* 24:446–460.

Chappellaz, J., T. Blunier, D. Raynaud, J. M. Barnola, J. Schwander, and R. Stauffer. 1993. Synchronous changes in atmospheric CH_4 and Greenland climate between 40 and 8 kyr B. P. *Nature* 366:443–445.

Charles, D. F. 1990. Effects of acidic deposition on North American lakes: palaeolimnological evidence from diatoms and chrysophytes. *Philosophical Transactions of the Royal Society of London B* 327:403–412.

Charnov, E. L. 1993. *Life History Invariants: Some Explorations of Symmetry in Evolutionary Ecology.* New York: Oxford University Press.

Chatters, J. C. 1996. Taking the long view: geologic, paleontological, and archeological evidence of salmon population dynamics during the Holocene. In C. Steward and D. MacDonald, eds., *Towards Sustainable Fisheries: Balancing Conservation and Use of Salmon and Steelhead in the Pacific Northwest, Conference Proceedings and Abstracts,* p. 120. Bothell, Wash.: Sustainable Fisheries Foundation.

Chesson, P. 1991. Stochastic population models. In J. Kolasa and S. T. A. Pickett, eds., *Ecological Heterogeneity,* pp. 123–143. New York: Springer-Verlag.

Chorley, R. J. and B. A. Kennedy. 1971. *Physical Geography: A Systems Approach.* London: Prentice-Hall.

Christensen, N. L., J. K. Agee, P. F. Brussard, J. Hughes, D. H. Knight, G. W. Marshall, J. M. Peek, S. J. Pyne, F. J. Swanson, J. W. Thomas, S. Wells, S. E.

Williams, and H. A. Wright. 1989. Interpreting the Yellowstone fires of 1988. *BioScience* 39:678–685.

Cienciala, E. and A. Lindroth. 1995. Gas-exchange and sap flow measurements of *Salix viminalis* trees in short-rotation forest. I. Transpiration and sap flow. *Tree* 9:289–294.

Clark, J. S. 1988. Effect of climate change on fire regimes in northwestern Minnesota. *Nature* 334:233–235.

Clark, J. S. 1990. Fire and climate during the last 750 years in northwestern Minnesota. *Ecological Monographs* 60:135–159.

Clark, T. W. 1994. Creating and using knowledge for species and ecosystem conservation: science, organizations, and policy. In R. E. Grumbine, ed., *Environmental Policy and Biodiversity*, pp. 335–364. Washington, D.C.: Island Press.

Clements, F. E. 1916. *Plant Succession.* Carnegie Institute Publication 242. Washington, D.C.: Carnegie Institute.

Cliff, A. D. and J. K. Ord. 1973. *Spatial Autocorrelation.* London: Pion.

Cliff, A. D. and J. K. Ord. 1981. *Spatial Processes: Models and Applications.* London: Pion.

Cline, M. G. 1949. Basic principles of soil classification. *Soil Science* 67:81–91.

Cline, M. G. 1961. The changing model of soil. *Soil Science Society of America Proceedings* 25:442–446.

Closs, G. P. and P. S. Lake. 1994. Spatial and temporal variation in the structure of an intermittent-stream food web. *Ecological Monographs* 64:1–21.

Coffey, G. N. 1912. A study of the soils of the United States. *U.S. Agricultural Bureau Soils Bulletin,* 5. Washington, D.C.

Cohen, J. E. and F. Briand. 1984. Trophic links of community food webs. *Proceedings of the National Academy of Sciences* 81:4105–4109.

Cohen, J. E. and C. M. Newman. 1985. When will a large complex system be stable? *Journal of Theoretical Biology* 113:153–156.

Cohen, J. E., F. Briand, and C. M. Newman. 1990. *Community Food Webs: Data and Theory.* Berlin: Springer-Verlag.

Cohen, W. B., T. A. Spies, and G. A. Bradshaw. 1990. Semivariograms of digital imagery for analysis of conifer canopy structure. *Remote Sensing of the Environment* 34:167–192.

Cohen, W. B., T. A. Spies, F. J. Swanson, and D. O. Wallin. 1995. Land cover on the western slopes of the central Oregon Cascade Range. *International Journal of Remote Sensing* 16:595–596.

COHMAP Members. 1988. Climate changes of the last 18,000 years: observations and model simulations. *Science* 241:1043–1052.

Coile, T. S. 1952. Soil and the growth of forests. *Advances in Agronomy* 4:239–398.

Colinvaux, P. 1986. *Ecology.* New York: John Wiley and Sons.

Collins, S. L. and S. C. Barber. 1985. Effects of disturbance on diversity in mixed-grass prairie. *Vegetatio* 64:87–94.

Collins, S. L. and S. M. Glenn. 1991. Importance of spatial and temporal dynamics in species regional abundance and distribution. *Ecology* 72:654–664.

Comerford, N. B., G. Kidder, and A. V. Mollitor. 1984. Importance of subsoil fertility to forest and non-forest plant nutrition. In E. L. Stone, ed., *Forest Soils* and Treatment Impacts, Proceedings, 6th North American Forest Soils Conference, pp. 381–404. Madison, Wis.: Soil Science Society of America.

Connell, J. H. and W. P. Sousa. 1983. On the evidence needed to judge ecological stability, or persistence. *American Naturalist* 121:789–824.

Cooper, W. S. 1923. The recent ecological history of Glacier Bay, Alaska: III. Permanent quadrats at Glacier Bay: an initial report upon a long-period study. *Ecology* 4:355–365.

Costanza, R., ed. 1991. *Ecological Economics.* New York: Columbia University Press.

Costanza, R. and L. Cornwell. 1992. The 4P approach to dealing with scientific uncertainty. *Environment* 34:12–20.

Costanza, R. and T. Maxwell. 1994. Resolution and predictability: an approach to the scaling problem. *Landscape Ecology* 9:47–57.

Coughenour, M. B. and J. E. Ellis. 1993. Landscape and climatic control of woody vegetation in a dry tropical ecosystem: Turkana District, Kenya. *Journal of Biogeography* 22:107–122.

Coulson, R. N., C. N. Lovelady, R. O. Flamm, S. L. Spradling, and M. C. Saunders. 1991. Intelligent geographic information system for natural resource management. In M. G. Turner and R. H. Gardner, eds., *Quantitative Methods in Landscape Ecology,* pp. 153–172. New York: Springer-Verlag.

Covington, W. W. and L. F. DeBano, tech. coords., 1994. *Sustainable Ecological Systems: Implementing an Ecological Approach to Land Management.* USDA Forest Service General Technical Report GTR-RM-247. Fort Collins, Colo.: Rocky Mountain Forest Research Station.

Cressie, N. A. C. 1993. *Statistics for Spatial Data.* New York: Wiley.

Crist, T. O. and J. A. Wiens. 1994. Scale effects of vegetation on forager movement and seed harvesting by ants. *Oikos* 69:37–46.

Crompton, E. 1962. Soil formation. *Outlook on Agriculture* 3:209–218.

Cronan, C. S. and D. F. Grigal. 1995. Use of calcium/aluminum rations as indicators of stress in forest ecosystems. *Journal of Environmental Quality* 24:209–226.

Cropper, W. P. and H. L. Gholz. 1990. Modeling the labile carbon dynamics of a Florida slash pine plantation. *Silva Carelica* 15:121–130.

Crowder, M. J. and D. J. Hand. 1990. *Analysis of Repeated Measures.* London: Chapman and Hall.

Cummins, K. W. and C. N. Dahm. 1995. Restoring the Kissimmee. *Restoration Ecology* 3:147–148.

Curtis, J. T. 1959. *The Vegetation of Wisconsin.* Madison: University of Wisconsin Press.

Cushing, J. M. 1992. A discrete model for competing stage-structured species. *Theoretical Population Biology* 41:372–387.

Cwynar, L. C. 1987. Fire and the forest history of the North Cascade Range. *Ecology* 68:791–802.

Dahm, C. N., K. W. Cummins, H. M. Valett, and R. L. Coleman. 1995. An ecosystem view of the restoration of the Kissimmee River. *Restoration Ecology* 3:225–238.

Dale, M. R. T. and D. A. MacIsaac. 1989. New methods for the analysis of spatial pattern in vegetation. *Journal of Ecology* 77:78–91.

Dale, V. H., R. H. Gardner, and M. G. Turner. 1989. Predicting across scales: comments of the guest editors of *Landscape Ecology*. *Landscape Ecology* 3:147–151.

Damman, A. W. H. 1979. The role of vegetation analysis in land classification. *Forestry Chronicle* 55:175–182.

Daniels, L. D., P. L. Marshal, R. E. Carter, and K. Klinka. 1995. Age structure of *Thuja plicata* in the tree layer of old-growth stands near Vancouver, British Columbia. *Northwest Science* 69:175–183.

Daniels, R. B. and R. D. Hammer. 1992. *Soil Geomorphology*. New York: John Wiley and Sons.

Daniels, R. B., J. F. Brasfield, and F. F. Riecken. 1962. Distribution of sodium hydrofulfite extractable manganese in some Iowa profiles. *Soil Science Society of America Proceedings* 26:75–78.

Davidson, E. A., P. A. Matson, P. M. Vitousek, R. Riley, K. Dunkin, G. Garcia-Mendez, and J. M. Maas. 1993. Processes regulating soil emissions of NO and N_2O in a seasonally dry tropical rainforest. *Ecology* 74:130–139.

Davis, M. B. 1976. Pleistocene biogeography of temperate deciduous forests. *Geoscience and Man* 13:13–26.

Davis, M. B. 1981. Quaternary history and the stability of forest communities. In D.C. West, H. H. Shugart, and D. B. Botkin, eds., *Forest Succession,* pp. 132–153. New York: Springer-Verlag.

Davis, M. B. 1983. Quaternary history of deciduous forests of eastern North America and Europe. *Annals of the Missouri Botanical Garden* 70:550–563.

Davis, M. B. 1986. Climatic instability, time lags, and community disequilibrium. In J. Diamond and T. J. Case, eds., *Community Ecology,* pp. 269–284. New York: Harper and Row.

Davis, M. B. 1987. Paleolimnological diatom studies of acidification of lakes by acid rain: an application of Quaternary science. *Quaternary Science Review* 6:147–164.

Davis, M. B. 1994. Ecology and paleoecology begin to merge. *Trends in Ecology and Evolution* 9:357–358.

Davis, M. B., S. Sugita, R. R. Calcote, J. B. Ferrari, and L. E. Frelich. 1994. Historical development of alternate communities in a hemlock-hardwood forest in northern Michigan, USA. In R. M. May, N. Webb, and D. P. Edwards, eds., *Large Scale Ecology and Conservation Biology,* pp. 19–39. Oxford: Blackwell.

Dawson, T. E. and L. C. Bliss. 1993. Plants as mosaics: leaf-, ramet-, gender-specific influences on the physiology of dwarf willow. *Functional Ecology* 7:293–304.

Dawson, T. E. and J. R. Ehleringer. 1993. Isotopic enrichment of water in the "woody" tissues of plants: implications for plant water source, water uptake, and other studies which use the stable isotopic composition of cellulose. *Geochimica Cosmochimica Acta* 57:3487–3492.

Dayton, P. D. and M. J. Tegner. 1984. The importance of scale in community ecology: a kelp forest example with terrestrial analogs. In P. W. Price, C. M. Slobodchikoff, and W. S. Gaud, eds., *A New Ecology: Novel Approaches to Interactive Systems,* pp. 457–481. New York: John Wiley and Sons.

Dayton, P. K., M. J. Tegner, P. E. Parnell, and P. B. Edwards. 1992. Temporal and spatial patterns of disturbance and recovery in a kelp forest community. *Ecological Monographs* 62:421–445.

De Boer, D. H. 1992. Hierarchies and spatial scale in process geomorphology: a review. *Geomorphology* 4:303–318.

De Cola, L. 1994. Simulating and mapping spatial complexity using multi-scale techniques. *International Journal of Geographical Information Systems* 8:411–427.

de Gruijter, J. J. and C. J. F. ter Braak. 1990. Model-free estimation from spatial samples: a reappraisal of classical sampling theory. *Mathematical Geology* 22:407–415.

de Vries, W., J. Kros, R. M. Hootsmans, J. G. van Uffelen, and J. C. H. Voogd. 1992. Critical loads for Dutch forest soils. In T. Schneider, ed., *Acidification Research: Evaluation and Policy Applications,* pp. 307–318. Amsterdam: Elsevier Science.

DeAngelis, D. L. and L. J. Gross. 1992, eds. *Individual-based Models and Approaches in Ecology: Populations, Communities and Ecosystems.* New York: Chapman and Hall.

Deb, D. 1995. Scale-dependence of food web structures: tropical ponds as paradigm. *Oikos* 72:242–262.

Décamps, H. and M. Fortuné. 1991. Long-term ecological research and fluvial landscapes. In Risser, P. G., ed., *Long-Term Ecological Research: An International Perspective,* pp. 135–151. New York: Wiley.

Deegan, L. A. 1993. Nutrient and energy transport between estuaries and coastal marine ecosystems by fish migration. *Canadian Journal of Fisheries and Aquatic Sciences* 50:74–79.

Delcourt, H. R. and P. A. Delcourt. 1988. Quaternary landscape ecology: relevant scales in space and time. *Landscape Ecology* 2:45–61.

Delcourt, H. R. and P. A. Delcourt. 1991. *Quaternary Ecology: A Paleoecological Perspective.* New York: Chapman and Hall.

Delcourt, H. R., P. A. Delcourt, and T. Webb. 1983. Dynamic plant ecology: the spectrum of vegetational change in space and time. *Quaternary Science Review* 1:153–175.

Delcourt, P. A. and H. R. Delcourt. 1987. *Long-Term Forest Dynamics of the Temperate Zone.* New York: Springer-Verlag.

DeLong, S. C. and D. Tanner. 1996. Managing the pattern of forest harvest: lessons from wildfire. *Biodiversity and Conservation* 5:1191–1205.

DeMelo, R., R. France, and D. McQueen. 1992. Biomanipulation: hit or myth? *Limnology and Oceanography* 37:192–207.

Deming, W. E. 1982. *Out of the Crisis.* Cambridge, Mass.: Massachusetts Institute of Technology, Center for Advanced Engineering Study.

den Boer, P. J. 1981. On the survival of populations in a heterogeneous and variable environment. *Oecologia* 50:39–53.

Denman, K. 1994. Scale-determining biological-physical interactions in oceanic food webs. In P. Giller, A. Hildrew, and D. Rafaelli, eds., *Aquatic Ecology: Scale, Pattern, and Process*, pp. 377–439. Oxford: Blackwell Scientific Publications.

Denslow, J. S. 1980. Gap partitioning among tropical rainforest trees. *Biotropica* 12(supplement):47–55.

DeRuiter, P. C., A.-M. Neutel, and J. C. Moore. 1995. Energetics, patterns of interaction strength, and stability in real ecosystems. *Science* 269:1257–1260.

Deswysen, A. G., P. Dutilleul, and W. C. Ellis. 1989. Quantitative analysis of nycterohemeral eating and ruminating patterns in heifers with different voluntary intakes and effects of monensin. *Journal of Animal Science* 67:2751–2761.

Deswysen, A. G., P. Dutilleul, J.-P. Godfrin, and W. C. Ellis. 1993. Nycterohemeral eating and ruminating patterns in heifers fed grass and corn silage: Analysis by finite Fourier transform. *Journal of Animal Science* 71:2739–2747.

Deutchmann, D. H., G. A. Bradshaw, W. Childress, K. Daly, D. Grunbaum, M. Pascual, N. Schumaker, and J. Wu. 1993. Some thoughts on how patches form. In S. A. Levin, T. M. Powell, and J. H. Steele, eds., *Patch Dynamics*, pp. 184–209. Berlin: Springer-Verlag.

Deutsch, C. V. and A. G. Journel. 1992. *GSLIB: Geostatistical Software Library and User's Guide*. New York: Oxford University Press.

Diawara, A., D. Loustau, and P. Berbigier. 1991. Comparison of two methods for estimating the evaporation of a *Pinus pinaster* (Ait.) stand: sap flow and energy balance with sensible heat flux measurements by an eddy covariance method. *Agricultural and Forest Meteorology* 54:49–66.

Diggle, P. 1990. *Time Series: A Biostatistical Introduction*. Oxford: Clarendon Press.

Doak, D. F., P. C. Marino, and P. M. Kareiva. 1992. Spatial scale mediates the influence of habitat fragmentation on dispersal success: implications for conservation. *Theoretical Population Biology* 41:315–336.

Dovers, S. R. 1995. A framework for scaling and framing policy problems in sustainability. *Ecological Economics* 12:93–106.

Dovers, S. R. and J. W. Handmer. 1995. Ignorance, the precautionary principle and sustainability. *Ambio* 24:92–97.

Drake, D. R. and D. Mueller-Dombois. 1993. Population development of rain forest trees on a chronosequence of Hawaiian lava flows. *Ecology* 74:1012–1019.

Drake, J. A. 1991. Community-assembly mechanics and the structure of an experimental species ensemble. *American Naturalist* 137:1–26.

Drosdoff, M., R. B. Daniels, and J. J. Nicholaides. 1978. *Diversity of Soils in the Tropics*. Special Publication 34. Madison, Wis.: American Society of Agronomy.

Drury, W. H. and I. C. T. Nisbet. 1973. Succession. *Arnold Arboretum Journal* 54:331–368.

Dufrene, E., J.-Y. Pontailler, and B. Saugier. 1993. A branch bag technique for simultaneous CO_2 enrichment and assimilation. *Plant, Cell and Environment* 16:1131–1138.

Dunning, J. B., Jr., D. L. Stewart, B. Danielson, B. Noon, T. Root, R. Lamberson, and E. Stevens. 1995. Spatially explicit population models: current forms and future uses. *Ecological Applications* 5:3–11.

Dutilleul, P. 1988. On the information available in a modified periodogram at the frequencies corresponding to a non integer number of cycles over the observation period. Contributed paper at *The XIVth International Biometric Conference,* Namur, Belgium, July 18–22.

Dutilleul, P. 1989. Estimation par moindres carrés dans une classe particulière de modèles non linéaires. Application à la détection de périodicités dans une série chronologique. Paper presented at *Ajustements non linéaires,* Meeting organized by The Adolphe Quetelet Society, Louvain-la-Neuve, Belgium, November 17. Abstract published in *Biométrie-Praximétrie* 29:188–189.

Dutilleul, P. 1990. Apport en Analyse Spectrale d'un Périodogramme Modifié et Modélisation des Séries Chronologiques avec Répétitions en Vue de Leur Comparaison en Fréquence. *D. Sc. Dissertation.* Louvain-la-Neuve, Belgium: Université catholique de Louvain.

Dutilleul, P. 1993a. Modifying the *t* test for assessing the correlation between two spatial processes. *Biometrics* 49:305–314.

Dutilleul, P. 1993b. Spatial heterogeneity and the design of ecological field experiments. *Ecology* 74:1646–1658.

Dutilleul, P. 1994. Maximum likelihood estimation for the matrix normal distribution. *Report 94-09, Department of Mathematics and Statistics.* Montreal: McGill University.

Dutilleul, P. 1995. Rhythms and autocorrelation analysis. *Biological Rhythm Research* 26:173–193.

Dutilleul, P. and P. Legendre. 1993. Spatial heterogeneity against heteroscedasticity: an ecological paradigm versus a statistical concept. *Oikos* 66:152–171.

Dutilleul, P. and B. Pinel-Alloul. 1996. A doubly multivariate model to deal with the spatiotemporal autocovariance structure of environmental data in ANOVA and correlation analysis. *Environmetrics* 7:551–566.

Dutilleul, P. and C. Potvin. 1995. Among-environment heteroscedasticity and genetic autocorrelation: implications for the study of phenotypic plasticity. *Genetics* 139:1815–1829.

Dutilleul, P. and C. Till. 1992. Evidence of periodicities related to climate and planetary behaviors in ring-width chronologies of Atlas cedar (*Cedrus atlantica*) in Morocco. *Canadian Journal of Forest Research* 22:1469–1482.

Dye, P. J. and B. W. Olbrich. 1993. Estimating transpiration from 6-year-old *Eucalyptus grandis* trees: development of a canopy conductance model and comparison with independent sap flux measurements. *Plant, Cell and Environment* 16:45–53.

Eberhard, K. E. and P. M. Woodard. 1987. Distribution of residual vegetation associated with large fires in Alberta. *Canadian Journal of Forest Research* 17: 1207–1212.

Ecological Society of America. 1995. *The Scientific Basis for Ecosystem Management.* Washington, D.C.: Ecological Society of America.

Edwards, P. J., R. M. May, and N. R. Webb, eds. 1994. *Large-Scale Ecology and Conservation Biology.* Oxford: Blackwell.

Ehleringer, J. R. and C. B. Field, eds. 1993. *Scaling Physiological Processes: Leaf to Globe.* New York: Academic Press.

Ekeberg, D., A. M. Jablonska, and G. Ogner. 1995. Phytol as a possible indicator of ozone stress by *Picea abies. Environmental Pollution* 89:55–58.

Ellenberg, H. 1971. *Integrated Experimental Ecology: Methods and Results of Ecosystem Research in the German Solling Project.* Berlin: Springer-Verlag.

Ellenberg, H., Mayer, R., and J. Schauermann, eds. 1986. *Ökosystemforschung-Ergebnisse des Sollings-Projekts.* Stuttgart: Ulmer Verlag.

Elwood, J. W., J. D. Newbold, R. V. O'Neill, and W. Van Winkle. 1983. Resource spiraling: an operational paradigm for analyzing lotic ecosystems. In T. D. Fontaine, III and S. M. Bartell, eds., *Dynamics of Lotic Ecosystems,* pp. 3–27. Ann Arbor, Mich.: Ann Arbor Science.

Engen, S. 1978. *Stochastic Abundance Models, with Emphasis on Biological Communities and Species Diversity.* London: Chapman and Hall.

Engstrom, D. R. and B. C. S. Hansen. 1985. Postglacial vegetational change and soil development in southeastern Labrador as inferred from pollen and chemical stratigraphy. *Canadian Journal of Botany* 63:543–561.

Engstrom, D. R. and H. E. Wright. 1984. Chemical stratigraphy of lake sediments as a record of environmental history. In E. Y. Haward and W. G. Lund, eds., *Lake Sediments and Environmental History,* pp. 11–67. Minneapolis: University of Minnesota Press.

Ewel, J. J. 1987. Restoration is the ultimate test of ecological theory. In W. R. Jordan, M. E. Gilpin, and J. D. Aber, eds., *Restoration Ecology: A Synthetic Approach to Ecological Research,* pp 31–33. Cambridge: Cambridge University Press.

Facelli, J. M. and S. T. A. Pickett. 1990. Markovian chains and the role of history in succession. *Trends in Ecology and Evolution* 5:27–29.

Fahrig, L. 1991. Simulation methods for developing general landscape-level hypotheses of single-species dynamics. In M. G. Turner and R. H. Gardner, eds., *Quantitative Methods in Landscape Ecology,* pp. 417–442. New York: Springer-Verlag.

Fahrig, L. 1992. Relative importance of spatial and temporal scales in a patchy environment. *Theoretical Population Biology* 41:300–314.

Fahrig, L. 1993. Effect of fish movement and fleet spatial behaviour on management of fish substocks. *Natural Resource Modeling* 7:37–56.

Fahrig, L. and K. Freemark. 1995. Landscape-scale effects of toxic events for ecological risk assessment. In J. Cairns, Jr. and B. R. Niederlehner, eds., *Ecological Toxicity Testing: Scale, Complexity, and Relevance,* pp. 193–208. Boca Raton: Lewis Publishers.

Fahrig, L. and G. Merriam. 1985. Habitat patch connectivity and population survival. *Ecology* 66:1762–1768.

Fahrig, L. and G. Merriam. 1994. Conservation of fragmented populations. *Conservation Biology* 8:50–59.

Fahrig, L. and J. Paloheimo. 1988. Determinants of local population size in patchy habitats. *Theoretical Population Biology* 34:194–213.

Fall, J. and A. Fall. 1996. SELES: A spatially explicit landscape event simulator. In *Proceedings of the NCGIA Third International Conference on GIS and Environmental Modeling.* Fort Collins, Colo.: GIS World Inc. (also located at World Wide Web site: //www.ncgia.ucsb.edu/conf/santa_fe.html).

Farmer, A. M. and M. S. Adams. 1991. The nature of scale and the use of hierarchy theory in understanding the ecology of aquatic macrophytes. *Aquatic Botany* 41:252–261.

Fastie, C. L. 1995. Causes and ecosystem consequences of multiple pathways of primary succession at Glacier Bay, Alaska. *Ecology* 76:1899–1916.

Feder, J. 1988. *Fractals.* New York: Plenum Press.

Federer, C. A., L. D. Flynn, C. W. Martin, J. W. Hornbeck, and R. S. Pierce. 1990. Thirty years of hydrometeorologic data at the Hubbard Brook Experimental Forest, New Hampshire. *USDA Forest Service General Technical Report GTR-NE-141.* Radnor, Penn.: Northeastern Forest Research Station.

Feibleman, J. K. 1954. Theory of integrative levels. *British Journal of the Philosophical Society* 5:59–66.

Ferguson, S. A. 1995. Potential climate change in northern North America. In D. L. Peterson and D. R. Johnson, eds., *Human Ecology and Climate Change: People and Resources in the Far North,* pp. 15–30. Washington, D.C.: Taylor and Francis.

Finney, M. A. 1996. *FARSITE Fire Area Simulator Version 2.0 User's Guide and Technical Documentation.* Missoula, Mont.: Systems for Environmental Management.

Fisher, R. F. 1995. Soil organic matter: clue or conundrum? In W. W. McFee and J. M. Kelley, ed., *Carbon Forms and Functions in Forest Soils, Proceedings,* 8th North American Forest Soils Conference, pp. 1–11. Madison, Wis.: Soil Science Society of America.

Fitzsimmons, A. K. 1994. Federal ecosystem management: a "train wreck" in the making. *Policy Analysis no. 217.* Washington, D.C.: CATO Institute.

Folse, L. J., H. E. Mueller, and A. D. Whittaker. 1990. Object-oriented simulation and geographic information systems. *AI Applications* 4:41–47.

Folse, L. J., J. M. Packard, and W. E. Grant. 1989. AI modeling of animals movements in a heterogeneous habitat. *Ecological Modelling* 46:57–72.

Foran, B. and K. Wardle. 1995. Transitions in land use and the problems of planning: a case study from the mountainlands of New Zealand. *Journal of Environmental Management* 43:97–127.

Ford, E. D. and E. Renshaw. 1984. The interpretation of process from pattern using a two-dimensional spectral analysis. *Vegetatio* 56:113–123.

Ford, E. D., A. Avery, and R. Ford. 1990. Simulation of branch growth in the Pinaceae: interactions of morphology, phenology, foliage productivity, and the requirement for structural support on the export of carbon. *Journal of Theoretical Biology* 146:13–36.

Forest Ecosystem Management Assessment Team (FEMAT). 1993. *Forest Ecosys-*

tem Management: An Ecological, Economic and Social Assessment. Washington, D.C.: U.S. Departments of Agriculture, Commerce, and Interior, and the Environmental Protection Agency.

Forman, R. T. T. 1995. *Land Mosaics: The Ecology of Landscapes and Regions.* Cambridge: Cambridge University Press.

Forsman, E. D., E. C. Meslow, and M. J. Strub. 1977. Spotted owl abundance in young versus old-growth forests, Oregon. *Wildlife Society Bulletin* 5:43–47.

Forsman, E. D., E. C. Meslow, and H. M. Wight. 1984. Distribution and biology of the spotted owl in Oregon. *Wildlife Society Monograph 87.*

Fortin, M.-J., P. Drapeau, and P. Legendre. 1989. Spatial autocorrelation and sampling design in plant ecology. *Vegetatio* 83:209–222.

Foster, D. R., P. K. Schoonmaker, and S. T. A. Pickett. 1990. Insights from paleoecology to community ecology. *Trends in Ecology and Evolution* 5:119–122.

Foster, M. S. 1990. Organization of macroalgal assemblages in the Northeast Pacific: the assumption of homogeneity and the illusion of generality. *Hydrobiologia* 192:21–33.

Foster, M. S. 1991. Rammed by the Exxon Valdez: a reply to Paine. *Oikos* 62: 93–96.

Francis, M. M., R. J. Naiman, and J. M. Melillo. 1985. Nitrogen fixation in subarctic streams influenced by beaver (*Castor canadensis*). *Hydrobiologia* 121: 193–202.

Francour, P., C. F. Boudouresque, J. G. Harmelin, M. L. Harmelin-Vivien, and J. P. Quignard. 1994. Are the Mediterranean waters becoming warmer? Information from biological indicators. *Marine Pollution Bulletin* 28:523–526.

Franklin, J. F. 1989. Importance and justification of long-term studies in ecology. In G. E. Likens, ed., *Long-Term Studies in Ecology: Approaches and Alternatives,* pp. 3–19. New York: Springer-Verlag.

Franklin, J. F. 1990. Biological legacies: a critical management concept from Mount St. Helens. *North American Wildlife and Natural Resources Conference* 55:216–219.

Franklin, J. F. 1993a. Lessons from old-growth. *Journal of Forestry* 91:10–13.

Franklin, J. F. 1993b. Preserving biodiversity: species, ecosystems or landscapes? *Ecological Applications* 3:202–205.

Franklin, J. F. 1993c. The fundamentals of ecosystem management with applications in the Pacific Northwest. In G. Aplet, N. Johnson, J. Olson, and V. Sample, eds., *Defining Sustainable Forestry,* pp. 127–144. Washington, D.C.: The Wilderness Society and Island Press.

Franklin, J. F. 1995. Why link species conservation, environmental protection, and resource management? In C. G. Jones and J. H. Lawton, eds., *Linking Species and Ecosystems,* pp. 326–35. New York: Chapman and Hall.

Franklin, J. F. and T. Spies. 1991. Composition, function and structure of old-growth Douglas-fir forests. In L. F. Ruggiero, K. B. Aubry, A. B. Carey, and M. H. Huff, eds., *Wildlife and Vegetation of Unmanaged Douglas-fir Forests,* pp. 71–80. General Technical Report PNW-GTR-285. Portland, Ore.: USDA Forest Service, Pacific Northwest Research Station.

Franklin, J. F., J. A. MacMahon, F. J. Swanson, and J. R. Sedell. 1985. Ecosystem responses to the eruption of Mount St. Helens. *National Geographic Research* Spring 1985:198–216.

Franklin, J. F., T. Spies, D. Perry, M. Harmon and A. McKee. 1986. Modifying Douglas-fir management regimes for non-timber objectives. In C. D. Oliver, D. P. Hanley, and J. A. Johnson, eds., *Douglas-fir: Stand Management for the Future: Symposium Proceedings,* pp.373–379. Seattle: College of Forest Resources, University of Washington.

Frelich, L. E. and P. B. Reich. 1995. Spatial patterns and succession in a Minnesota southern-boreal forest. *Ecological Monographs* 65:325–356.

Fretwell, S. D. 1987. Food chain dynamics: the central theory of ecology? *Oikos* 50:291–301.

Frissell, C. A., W. J. Liss, C. Warren, and M. D. Hurley. 1986. A hierarchical framework for stream habitat classification: viewing streams in a watershed context. *Environmental Management* 10:199–214.

Fritts, H. C. 1976. *Tree Rings and Climate.* London: Academic Press.

Frontier, S. 1985. Diversity and structure in aquatic ecosystems. In M. Barnes, ed., *Oceanography and Marine Biology—An Annual Review,* pp. 253–312. Aberdeen: Aberdeen University Press.

Frontier, S. 1987. Applications of fractal theory to ecology. In P. Legendre and L. Legendre, eds., *Developments in Numerical Ecology,* pp. 335–378. Berlin: Springer-Verlag.

Frost, T. M., S. R. Carpenter, F. Ives, and T. K. Kratz. 1995. Species compensation and complementarity in ecosystem function. In Jones, C. G. and J. H. Lawton, eds., *Linking Species and Ecosystems,* pp. 224–239. New York: Chapman and Hall,.

Frost, T. M., D. L. DeAngelis, S. M. Bartell, D. J. Hall, and S. H. Hurlbert. 1988. Scale in the design and interpretation of aquatic community research. In S. R. Carpenter, ed., *Complex Interactions in Lake Communities,* pp. 229–258. New York: Springer-Verlag.

Fulton, M. R., E. J. Rykiel Jr, T. F. H. Allen, and A. W. King. 1993. The distinctive character of ecological theory: scale and criteria in formal models. *Bulletin of the Ecological Society of America* 74:243.

Furness, R. W. 1993. Birds as monitors of pollutants. In R. W. Furness and J. J. D. Greenwood, eds., *Birds as Monitors of Environmental Change,* pp. 86–143. London: Chapman and Hall.

Gaedke, U. 1992. The size distribution of plankton biomass in a large lake and its seasonal variability. *Limnology and Oceanography* 37:1202–1220.

Gaedke, U. 1995. *Struktur und Funktion des pelagischen Nahrungsnetzes im Bodensee.* Habilitationsschrift, University of Constance.

Gale, M. R., D. F. Grigal, and R. B. Harding. 1991. Soil productivity index: predictions of site quality for white spruce pantations. *Soil Science Society of America Journal* 55:1701–1708.

Gallant, A. R. 1987. *Nonlinear Statistical Models.* New York: Wiley.

Garcia-Molinar, G., C. Greene, B. Li, A. Lobo, D. Mason, G. Bradshaw. 1993.

Description and analysis of spatial patterns. In S. A. Levin, T. M. Powell, and J. H. Steele, eds., *Patch Dynamics*, pp. 70–89. Berlin: Springer-Verlag.

Gardner, R. H., W. G. Cale, and R. V. O'Neill. 1982. Robust analysis of aggregation error. *Ecology* 63:1771–1779.

Gardner, R. H., B. T. Milne, M. G. Turner, and R. V. O'Neill. 1987. Neutral models for the analysis of broad-scale landscape pattern. *Landscape Ecology* 1: 19–28.

Gardner, R. H. and R. V. O'Neill. 1991. Pattern, process and predictability: the use of neutral models for landscape analysis. In M. G. Turner and R. H. Gardner, eds., *Quantitative Methods in Landscape Ecology*, pp. 289–307. New York: Springer-Verlag.

Garrod, D. J. 1977. The North Atlantic cod. In J. A. Gulland, ed., *Fish Population Dynamics*, pp. 216–239. Chichester: Wiley.

Gaston, K. J. 1994. *Rarity.* London: Chapman and Hall.

Gaston, K. J., ed. 1996a. *Biodiversity: A Biology of Numbers and Difference.* Oxford: Blackwell.

Gaston, K. J. 1996b. Species richness and measurement. In K. J. Gaston, ed., *Biodiversity: a Biology of Numbers and Difference*, pp. 77–113. Oxford: Blackwell.

Gaston, K. J. and J. H. Lawton. 1990. Effects of scale and habitat on the relationship between regional distribution and local abundance. *Oikos* 58:329–335.

Gaston, K. J. and B. H. McArdle. 1993. Measurement of variation in the size of populations in space and time: some points of clarification. *Oikos* 68:357–360.

Gaudreau, D.C. 1986. Late-quaternary history of the Northeast: paleoecological implications of topographic patterns in pollen distributions. *Ph.D. Dissertation.* New Haven, Conn.: Yale University.

Gavin, D., K. P. Lertzman, L. Brubaker, and E. Nelson. 1996. Long-term fire histories in a coastal temperate rainforest. *Bulletin of the Ecological Society of America* 77(3):157.

Geiszler, D. R., R. I. Gara, C. H. Driver, V. F. Gallucci, and R. E. Martin. 1980. Fire, fungi, and beetle influences on a lodgepole pine ecosystem of south-central Oregon. *Oecologia* 46:239–243.

George, R. J., D. J. McFarlane, and R. J. Speed. 1995. The consequences of a changing hydrologic environment for native vegetation in south Western Australia. In D. A. Saunders, J. Craig, and L. Mattiske, eds., *Nature Conservation 4: The Role of Networks*, pp. 9–22. Chipping Norton, NSW, Australia: Surrey Beatty and Sons.

Gerrard, A. J. 1990. Soil variations on hillslopes in humid temperate systems. *Geomorphology* 3:225–244.

Getis, A. and J. Franklin. 1987. Second-order neighborhood analysis of mapping point patterns. *Ecology* 68:473–477.

Getty, T. and H. R. Pulliam. 1991. Random prey detection with pause-travel search. *American Naturalist* 138:1459–1477.

Gholz, H. L. and R. F. Fisher. 1984. The limits to productivity: fertilization and nutrient cycling in coastal plain slash pine forests. In E. L. Stone, ed., *Forest*

Soils and Treatment Impacts, Proceedings, 6th North American Forest Soils Conference, pp. 105–120. Madison, Wis.: Soil Science Society of America.

Gill, D. E. 1978. The metapopulation ecology of the red-spotted newt, *Notophthalmus viridescens* (Rafinesque). *Ecological Monographs* 48:145–166.

Giller, P. S. 1984. *Community Structure and the Niche.* New York: Chapman and Hall.

Giller, P. S., A. G. Hildrew, and D. G. Raffaelii. 1994. *Aquatic Ecology: Scale, Pattern and Process.* Oxford: Blackwell Scientific Publications.

Girling, A. J. 1995. Periodograms and spectral estimates for rhythm analysis. *Biological Rhythm Research* 26:149–172.

Gleason, H. A. 1926. The individualistic concept of the plant association. *Bulletin of the Torrey Botanical Club* 53:7–26.

Glitzenstein, J. S., P. A. Harcombe, and D. R. Streng. 1986. Disturbance, succession, and maintenance of species diversity in an east Texas forest. *Ecological Monographs* 56:243–258.

Glover, R. K., L. D. Zavesky, W. R. Swafford, and Q. L. Markley. 1975. *Soil Survey of Ellis County, Kansas.* USDA Soil Conservation Service in cooperation with Kansas Agricultural Experiment Station, Washington, D.C.: U.S. Government Printing Office.

Goldwasser, L. and J. Roughgarden. 1993. Construction of a large Caribbean food web. *Ecology* 74:1216–1233.

Gordon, A. D. and H. J. B. Birks. 1985. *Numerical Methods in Paleoecology.* New York: John Wiley and Sons.

Gordon, N. D., T. A. McMahon and B. L. Finlayson. 1992. *Stream Hydrology.* Chichester: Wiley.

Gosz, J. R. 1992. Gradient analysis of ecological change in time and space: implications for forest management. *Ecological Applications* 2:248–261.

Gotelli, N. J. 1991. Metapopulation models: the rescue effect, the propagule rain, and the core-satellite hypothesis. *American Naturalist* 138:768–776.

Gotelli, N. J. and G. R. Graves. 1996. *Null Models in Ecology.* Washington, D.C.: Smithsonian Institution.

Gould, S. J. 1989. *Wonderful Life: The Burgess Shale and the Nature of History.* New York: Norton.

Grace, J., J. Lloyd, J. McIntyre, A. Miranda, P. Meir, H. Miranda, J. Moncrieff, J. Massheder, I. Wright, and J. Gash. 1995. Fluxes of carbon dioxide and water vapour over an undisturbed tropical forest in south-west Amazonia. *Global Change Biology* 1:1–12.

Granier, A. and D. Loustau. 1994. Measuring and modeling the transpiration of a maritime pine canopy from sap-flow data. *Agricultural and Forest Meteorology* 71:61–81.

Granier, A., R. Huc, and S. T. Barigah. 1996. Transpiration of natural rain forest and its dependence on climatic factors. *Agricultural and Forest Meteorology* 78:19–29.

Grant, B. R. and P. R. Grant. 1993. Evolution of Darwin's finches caused by a rare climatic event. *Proceedings of the Royal Society of London B* 251:111–117.

Grant, P. R. 1986. *Ecology and Evolution of Darwin's Finches.* Princeton, N.J.: Princeton University Press.

Graves, S., J. Maldanado, and J. O. Wolff. 1988. Use of ground and arboreal microhabitats by *Peromyscus leucopus* and *Peromyscus maniculatus. Canadian Journal of Zoology* 66:277–278.

Graybill, F. A. 1983. *Matrices with Applications in Statistics,* 2nd ed. Belmont: Wadsworth.

Green, D. G. 1982. Fire and stability in the post-glacial forests of southwest Nova Scotia. *Journal of Biogeography* 9:29–40.

Green, D. G. and G. S. Dolman. 1988. Fine resolution pollen analysis. *Journal of Biogeography* 15:685–701.

Green, R. H. 1967. *Sampling Design and Statistical Methods for Environmental Biologists.* New York: Wiley.

Greenhouse, S. W. and S. Geisser. 1959. On the methods in the analysis of profile data. *Psychometrika* 24:95–112.

Greig-Smith, P. 1952. The use of random and contiguous quadrats in the study of plant communities. *Annals of Botany, New Series* 16:293–316.

Greig-Smith, P. 1979. Pattern in vegetation. *Journal of Ecology* 67:755–759.

Greig Smith, P. 1983. *Quantitative Plant Ecology,* 3rd edition. Berkeley: University of California Press.

Griffith, D. A. 1987. *Spatial Autocorrelation: A Primer.* Washington, D.C.: Association of American Geographers.

Grimm, E. C. 1983. Chronology and dynamics of vegetation change in the prairie-woodland region of southern Minnesota, U.S.A. *New Phytologist* 93:311–335.

Groom, M. J. and N. Schumaker. 1993. Evaluating landscape change: patterns of worldwide deforestation and local fragmentation. In P. M. Kareiva, J. G. Kingsolver, and R. B. Huey, eds., *Biotic Interactions and Global Change,* pp. 24–44. Sunderland, MA: Sinaur Associates.

Groot, C. and L. Margolis. 1991. *Pacific Salmon Life Histories.* Vancouver: University of British Columbia Press.

Grootes, P. M., M. Stuiver, J. W. C. White, S. Johnsen, and J. Jouzel. 1993. Comparison of oxygen isotope records from the GISP2 and GRIP Greenland ice cores. *Nature* 366:552–554.

Grubb, P. J. 1977. The maintenance of species-richness in plant communities: the importance of the regeneration niche. *Biological Reviews* 52:107–145.

Grumbine, R. E. 1992. *Ghost Bears: Exploring the Biodiversity Crisis.* Washington, D.C.: Island Press.

Gunderson, L. H., C. S. Holling and S. S. Light. 1995. *Barriers and Bridges to the Renewal of Ecosystems and Institutions.* New York: Columbia University Press.

Gurevitch, J. and M. J. Fortin. 1991. Field experiments in community ecology: new perspectives on old problems of scale and structure. *Bulletin of the Ecological Society of America* 72:131.

Gurney, W. S. C. and R. M. Nisbet. 1978. Single-species population fluctuations in patchy environments. *American Naturalist* 112:1075–1090.

Gutiérrez, R. J. 1994. Conservation planning: lessons from the spotted owl. In

W. W. Covington and L. F. DeBano, tech. coords., *Sustainable Ecological Systems: Implementing an Ecological Approach to Land Management,* pp. 51–58. USDA Forest Service General Technical Report RM-GTR-247. Fort Collins, Colo.: Rocky Mountain Research Station.

Gyalistras, D., H. von Storch, A. Fischlin, and M. Beniston. 1994. Linking GCM-simulated climatic changes to ecosystem models: case studies of statistical downscaling in the Alps. *Climate Research* 4:167–189.

Hack, J. T. and J. G. Goodlett. 1960. Geomorphology and forest ecology of a mountain region in the central Appalachians. *U.S. Geological Survey Professional Paper 347.*

Haefner, J. W. 1980. Two metaphors of the niche. *Synthese* 43:123–153.

Haidinger, T. L. and J. E. Keeley. 1993. Role of high fire frequency in destruction of mixed chaparral. *Madroño* 40:41–147.

Haigh, M. J. 1987. The holon: hierarchy theory and landscape research. *Catena Supplement* 10:181–192.

Haines-Young, R. 1994. Remote sensing and environmental change. In N. Roberts, ed., *The Changing Global Environment,* pp. 22–43. Oxford: Basil Blackwell.

Haines-Young, R., D. R. Green, and S. H. Cousins, eds. 1993. *Landscape Ecology and GIS.* London: Taylor and Francis.

Haining, R. 1990. *Spatial Data Analysis for the Social and Environmental Sciences.* Cambridge: Cambridge University Press.

Hall, F. G., D. B. Botkin, D. E. Strebel, K. D. Woods, and S. J. Goetz. 1991a. Large-scale patterns of forest succession as determined by remote sensing. *Ecology* 72:628–640.

Hall, F. G., D. E. Stebel, J. E. Nickeson, and S. J. Goetz. 1991b. Radiometric rectification: toward a common radiometric response among multidate, multisensor images. *Remote Sensing of the Environment* 35:11–27.

Hamburg, S. P. and R. L. Sanford, Jr. 1986. Disturbance, *Homo sapiens,* and ecology. *Bulletin of the Ecological Society of America* 67:169–171.

Hammer, R. D., G. S. Henderson, R. Udawatta, and D. K. Brandt. 1995. Soil organic carbon in the Missouri forest-prairie ecotone. In W. W. McFee and J. M. Kelley, ed., *Carbon Forms and Functions in Forest Soils, Proceedings,* 8th North American Forest Soils Conference, pp. 201–231. Madison, Wis.: Soil Science Society of America.

Hammer, R. D., R. G. O'Brien, and R. J. Lewis. 1987. Temporal and spatial soil variability on three forested land types on the mid-Cumberland Plateau. *Soil Science Society of America Journal* 51:1320–1326.

Hansen, A. and D. Urban. 1992. Avian response to landscape pattern: the role of species life histories. *Landscape Ecology* 7:163–180.

Hansen, A., D. Urban, B. Marks. 1992. Avian community dynamics: the interplay of landscape trajectories and species' life histories. In A. H. Hansen and F. Di Castri, eds., *Landscape Boundaries: Consequences for Biodiversity and Ecological Flows,* pp. 170–195. New York: Springer-Verlag.

Hansen, A. J., W. McComb, R. Vega, M. Raphael, and M. Hunter. 1995. Bird habitat relationships in natural and managed forests in the west Cascades of Oregon. *Ecological Applications* 5:555–569.

Hansen, A. J., T. A. Spies, F. J. Swanson, and J. L. Ohmann. 1991. Conserving biodiversity in managed forests: lessons from natural forests. *BioScience* 41: 382–392.

Hansen, B., P. Bjørnsen and P. J. Hansen. 1994. The size ratio between planktonic predators and their prey. *Limnology and Oceanography* 39:395–403.

Hanski, I. 1982. Dynamics of regional distribution: the core and satellite species hypothesis. *Oikos* 38:210–221.

Hanski, I. 1991. Single-species metapopulation dynamics: concepts, models and observations. *Biological Journal of the Linnaean Society* 42:17–38.

Hanski, I. 1993. Dynamics of small mammals on islands. *Ecography* 16:372–375.

Hanski, I. 1994a. A practical model of metapopulation dynamics. *Journal of Animal Ecology* 63:151–162.

Hanski, I. 1994b. Spatial scale, patchiness and population dynamics on land. *Philosophical Transactions of the Royal Society of London B* 343:19–25.

Hanski, I. and M. Gilpin. 1991. Metapopulation dynamics: brief history and conceptual domain. *Biological Journal of the Linnean Society* 42:3–16.

Hanski, I., J. Kouki, and A. Halkka. 1993. Three explanations of the positive relationship between distribution and abundance of species. In R. E. Ricklefs and D. Schluter, eds., *Species Diversity in Ecological Communities*, pp. 108–116. Chicago: University of Chicago Press.

Hanson, G. C., P. M. Groffman, and A. J. Gold. 1994. Symptoms of nitrogen saturation in a riparian wetland. *Ecological Applications* 4:750–756.

Hansson, L., L. Fahrig, and G. Merriam, eds. 1995. *Mosaic Landscapes and Ecological Processes*. London: Chapman and Hall.

Harestad, A. S. and F. L. Bunnell, F. L. 1979. Home range and body weight—a reevaluation. *Ecology* 60:389–402.

Hargrove, W. W. 1994. *Simulating Fire Patterns in Heterogeneous Landscapes*. Located on the World Wide Web at http://juniper.esd.ornl.gov/ern/embyr/embyr.html:3.

Hargrove, W. W. and J. Pickering. 1992. Pseudoreplication: a sine qua non for regional ecology. *Landscape Ecology* 6:251–258.

Harper, J. 1977. *The Population Biology of Plants*. New York: Academic Press.

Harris, G. P. 1980. Temporal and spatial scales in phytoplankton ecology: mechanisms, methods, models, and management. *Canadian Journal of Fisheries and Aquatic Sciences* 37:877–900.

Harris, L. D. 1984. *The Fragmented Forest. Island: Biogeographic Theory and the Preservation of Biotic Diversity*. Chicago: University of Chicago Press.

Harris, M. P. and D. Osborn. 1981. Effect of a polychlorinated biphenyl on the survival and breeding of puffins. *Journal of Applied Ecology* 18:471–479.

Harrison, S., D. D. Murphy, and P. R. Ehrlich. 1988. Distribution of the bay checkerspot butterfly, *Euphydryas editha baynesis:* evidence for a metapopulation model. *American Naturalist* 132:360–382.

Hartgerink, A. P. and F. A. Bazzaz. 1984. Seedling-scale environmental heterogeneity influences individual fitness and population structure. *Ecology* 65:198–206.

Hastings, A. 1988. Food web theory and stability. *Ecology* 69:1655–1668.

Hastings, A. 1992. Age dependent dispersal is not a simple process: density dependence, stability, and chaos. *Theoretical Population Biology* 41:388–400.

Hastings, A. and G. Sugihara. 1993. *Fractals, a User's Guide for the Natural Sciences.* Oxford: Oxford University Press.

Hatton, J. C. and N. O. E. Smart. 1984. The effect of long-term exclusion of large herbivores on soil nutrient status in Murchison Falls National Park, Uganda. *African Journal of Ecology* 22:23–30.

Haury, L. R., J. S. McGowan, and P. Wiebe. 1978. Patterns and processes in the time-space scales of plankton distributions. In J. Steele, ed., *Spatial Pattern in Plankton Communities,* pp. 277–327. New York: Plenum Press.

Havens, K. 1992. Scale and structure in natural food webs. *Science* 257:1107–1109.

Haydon, D. 1994. Pivotal assumptions determining the relationship between stability and complexity: an analytical synthesis of the stability-complexity debate. *American Naturalist* 144:14–29.

He, F., P. Legendre, C. Bellehumeur, and J. V. LaFrankie. 1994. Diversity and spatial scale: a study of a tropical rain forest of Malaysia. *Environmental and Ecological Statistics* 1:265–284.

Heal, O. W. 1991. The role of study sites in long-term ecological research: A UK experience. In P. G. Risser, ed., *Long-Term Ecological Research: An International Perspective,* pp. 23–44. Chichester: John Wiley and Sons.

Healey, M. C. 1991. Life history of chinook salmon (*Oncorhynchus tshawytscha*). In C. Groot and L. Margolis, eds., *Pacific Salmon Life Histories,* pp. 311–394. Vancouver: University of British Columbia Press.

Hebda, R. J. and C. Whitlock. 1996. Environmental history of the coastal temperate rain forest. In P. K. Schoonmaker, B. von Hagen, and E. C. Wolf, eds., *The Rain Forests of Home: Profile of a North American Bioregion,* pp. 300–339. Washington, D.C.: Island Press.

Heide, K. 1984. Holocene pollen stratigraphy from a lake and small hollow in north-central Wisconsin, USA. *Palynology* 8:3–20.

Heil, R. D. and G. J. Buntley. 1965. A comparison of the characteristics of the ped faces and ped interiors of the B horizon in a Chestnut soil. *Soil Science Society of America Proceedings* 29:583–587.

Heinselman, M. L. 1973. Fire in the virgin forests of the Boundary Waters Canoe Area, Minnesota. *Quaternary Research* 3:329–382.

Hemstrom, M. A. and J. F. Franklin. 1982. Fire and other disturbances of the forests in Mount Rainier National Park. *Quaternary Research* 18:32–51.

Henderson, M. T., G. Merriam, and J. Wegner. 1985. Patchy environments and species survival: chipmunks in an agricultural mosaic. *Biological Conservation* 31:95–105.

Henderson-Sellers, A., M. F. Wilson, and G. Thomas. 1985. The effect of spatial resolution on archives of land cover type. *Climate Change* 7:391–402.

Henebry, G. M. and H. J. H. Kux. 1995. Lacunarity as a texture measure for SAR imagery. *International Journal of Remote Sensing* 16:565–571.

Henry, J. D. and J. M. A. Swan. 1974. Reconstructing forest history from live and dead plant material: an approach to the study of forest succession in southwest New Hampshire. *Ecology* 55:772–783.

Herbold, B. and P. B. Moyle. 1986. Introduced species and vacant niches. *American Naturalist* 128:751–760.

Hillel, D. J. 1991. *Out of the Earth: Civilization and the Life of the Soil.* New York: The Free Press, Macmillan.

Hinckley, T. M., J. R. Brooks, J. Cermák, R. Ceulemans, J. Kucera, F. C. Meinzer, and D. A. Roberts. 1994. Water flux in a hybrid poplar stand. *Tree Physiology* 14:1005–1018.

Hinckley, T. M., H. Imoto, K. M. Lee, S. Lacker, Y. Morikawa, K. A. Vogt, C. C. Grier, M. R. Keyes, R. O. Teskey, and V. A. Seymour. 1984. Impact of tephra deposition on growth in conifers: the year of the eruption. *Canadian Journal of Forest Research* 14:731–739.

Hobbs, R. J. 1992. Corridors for conservation: solution or bandwagon? *Trends in Ecology and Evolution* 7:389–392.

Hobbs, R. J. 1993a. Can revegetation assist in the conservation of biodiversity in agricultural areas? *Pacific Conservation Biology* 1:29–38.

Hobbs, R. J. 1993b. Effects of landscape fragmentation on ecosystem processes in the Western Australian wheatbelt. *Biological Conservation* 64:193–201.

Hobbs, R. J. 1994a. Dynamics of vegetation mosaics: can we predict responses to global change? *Ecoscience* 1:346–356.

Hobbs, R. J. 1994b. Landscape ecology and conservation: moving from description to application. *Pacific Conservation Biology* 1:170–76.

Hobbs, R. J. and H. A. Mooney. 1995. Spatial and temporal variability in California annual grassland: results from a long-term study. *Journal of Vegetation Science* 6:43–57.

Hobbs, R. J. and H. A. Mooney. 1996. Effects of episodic rainfall events on Mediterranean-climate ecosystems. In J. Roy, J. Aronson, and F. di Castri, eds., *Timescales in Biological Responses to Water Constraints,* pp 71–85. Amsterdam: SPB Academic Publishing.

Hobbs, R. J. and D. A. Norton. 1996. Towards a conceptual framework for restoration ecology. *Restoration Ecology* 4:93–110.

Hobbs, R. J. and D. A. Saunders. 1995. Conversing with aliens: do scientists communicate with each other well enough to solve complex environmental problems? In D. A. Saunders, J. Craig, and L. Mattiske, eds., *Nature Conservation 4: The Role of Networks,* pp. 195–198. Chipping Norton, NSW, Australia: Surrey Beatty and Sons.

Hobbs, R. J., D. A. Saunders, and G. W. Arnold. 1993. Integrated landscape ecology: a Western Australian perspective. *Biological Conservation* 64:231–238.

Hoekstra, T. W., T. F. H. Allen, and C. Flather. 1991. Implicit scaling in the choice of organism and ecological concept: on when to make studies of mice and men. *BioScience* 41:148–154.

Holling, C. S. 1986. Resilience of ecosystems: local surprise and global change. In W. C. Clark, and R. E. Munn, eds., *Sustainable Development of the Biosphere,* pp. 292–317. Cambridge: Cambridge University Press.

Holling, C. S. 1992. Cross-scale morphology, geometry, and dynamics of ecosystems. *Ecological Monographs* 62:447–502.

Holling, C. S. 1995. What barriers? What bridges? In L. H. Gunderson, C. S. Holling, and S. S. Light, eds., *Barriers and Bridges to the Renewal of Ecosystems and Institutions*, pp. 3–34. New York: Columbia University Press.

Holling, C. S. and G. K. Meffe. 1996. Command and control and the pathology of natural resource management. *Conservation Biology* 10:328–337.

Hollinger, D. Y., F. M. Kelliher, J. N. Byers, J. E. Hunt, T. M. McSeveny, and P. L. Weir. 1994. Carbon dioxide exchange between an undisturbed old-growth temperate forest and the atmosphere. *Ecology* 75:134–150.

Holt, R. D. 1992. A neglected facet of island biogeography: the role of internal spatial dynamics in area effects. *Theoretical Population Biology* 41:354–371.

Holt, R. D., S. W. Pacala, T. W. Smith, and J. Liu. 1995. Linking contemporary vegetation models with spatially explicit animal population models. *Ecological Applications* 5:20–27.

Hopkins, M. S., J. Ash, A. W. Graham, J. Head, and R. K. Hewett. 1993. Charcoal evidence of the spatial extent of the Eucalyptus woodland expansions and rainforest contractions in North Queensland during the late Pleistocene. *Journal of Biogeography* 20:357–372.

Horn, H. S. 1975. Markovian properties and forest succession. In M. L. Cody and J. M. Diamond, eds., *Ecology and Evolution of Communities*, pp. 196–211. Cambridge, Mass.: Harvard University Press.

Horne, J. K. and D.C. Schneider. 1994. Analysis of scale-dependent processes with dimensionless ratios. *Oikos* 70:201–211.

Horsley, S. B. 1977. Allelopathic inhibition of black cherry by fern, grass, goldenrod, and aster. *Canadian Journal of Forest Research* 7:205–216.

Houpis J. L. J., M. P. Costella, and S. Cowles. 1991. A branch exposure chamber for fumigating ponderosa pine to atmospheric pollution. *Journal of Environmental Quality* 20:467–474.

Huberman, B. A. and T. Hogg. 1988. The behaviour of computational ecologies. In B. A. Huberman, ed., *The Ecology of Computation*, pp. 71–115. Amsterdam: North-Holland.

Hudson, W. E. 1991. *Landscape Linkages and Biodiversity*. Washington, D.C.: Island Press.

Huff, M. H. 1995. Forest age structure and development following wildfires in the western Olympic Mountains, Washington. *Ecological Applications* 5:471–483.

Huffaker, C. B. 1958. Experimental studies on predation: dispersion factors and predator-prey oscillations. *Hilgardia* 27:343–383.

Huggett, R. J. 1991. *Climate, Earth Processes, and Earth History*. Berlin: Springer-Verlag.

Hughes, R. G. 1986. Theories and models of species abundance. *American Naturalist* 128:879–899.

Hunter, M. D. and P. W. Price. 1992. Playing chutes and ladders: heterogeneity and the relative roles of bottom-up and top-down forces in natural communities. *Ecology* 73:724–732.

Hunter, M. L. Jr., G. L. Jacobson, Jr., and T. Webb III. 1988. Paleoecology and the coarse filter approach to maintaining biological diversity. *Conservation Biology* 2:375–385.

Huntley, B. and H. J. B. Birks. 1983. *An Atlas of Past and Present Pollen Maps for Europe: 0–13,000 Years Ago.* Cambridge: Cambridge University Press.

Huntley, B. and T. Webb III. 1988. *Vegetation History.* The Hague: Kluwer Academic Publishers.

Hurlbert, S. H. 1984. Pseudoreplication and the design of ecological field experiments. *Ecological Monographs* 54:187–211.

Hurlbert, S. H. 1990. Spatial distribution of the montane unicorn. *Oikos* 58: 257–271.

Huston, M., D. DeAngelis, and W. Post. 1988. New computer models unify ecological theory. *BioScience* 38:682–691.

Huston, M. A. 1991. Use of individual-based forest succession models to link physiological whole-tree models to landscape-scale ecosystem models. *Tree Physiology* 9:293–306.

Hutchinson, G. E. 1957. Concluding remarks. Cold Spring Harbor *Symposium on Quantitative Biology* 22:415–427.

Hutchinson, G. E. 1959. Homage to Santa Rosalia. *American Naturalist* 93: 145–159.

Hutto, R. L. 1995. Composition of bird communities following stand-replacement fires in Northern Rocky Mountain (U.S.A.) conifer forests. *Conservation Biology* 9:1041–1059.

Huynh, H. and S. Feldt. 1970. Conditions under which mean square ratios in repeated measurements designs have exact F-distributions. *Journal of the American Statistical Association* 65:1582–1589.

Huynh, H. and S. Feldt. 1976. Estimation of the Box correction for degrees of freedom for sample data in randomized block and split-plot designs. *Journal of Educational Statistics* 1:69–82.

Hyman, J. B., J. B. McAninch, and D. L. DeAngelis. 1991. An individual-based simulation model of herbivory in a heterogeneous landscape. In M. G. Turner and R. H. Gardner, eds., *Quantitative Methods in Landscape Ecology*, pp. 443–475. New York: Springer-Verlag.

IGBP. 1990. *The International Geosphere Biosphere Program: A Study of Global Change. The Initial Core Projects.* Report No. 12. Stockholm: IGBP.

Imboden, D. 1990. Mixing and transport in lakes: mechanisms and ecological relevance. In M. Tilzer and C. Serruya, eds., *Large Lakes: Ecological Structures and Functions*, pp. 47–80 Berlin: Springer-Verlag.

Imboden, D. and A. Wüest. 1995. Mixing Mechanisms in Lakes. In A. Lerman, D. Imboden, and J. Gat, eds., *Physics and Chemistry of Lakes*, pp. 83–135. Berlin: Springer-Verlag.

Imbrie, J. and K. P. Imbrie. 1986. *Ice Ages.* Cambridge: Harvard University Press.

Indorante, S. J., L. R. Follmer, R. D. Hammer, and P. G. Koenig. 1990. Particle size analysis by a modified pipette procedure. *Soil Science Society of America Journal* 54:560–563.

Inglis, G. and A. J. Underwood. 1992. Comments on some designs proposed for experiments on the biological importance of corridors. *Conservation Biology* 6:581–586.

Innes, J. L. 1993. *Forest Health: Its Assessment and Status.* Wallingford, U.K.: CAB International.

Innes, J. L. 1994. Design of an intensive monitoring system for Swiss forests. In M. Beniston, ed., *Mountain Environments in Changing Climates,* pp. 281–300. London: Routledge.

Innes, J. L. 1995. Theoretical and practical criteria for the selection of ecosystem monitoring plots in Swiss forests. *Environmental Monitoring and Assessment* 36:271–294.

Isaaks, E. H. and R. M. Srivastava. 1989. *Applied Geostatistics.* New York: Oxford University Press.

Iversen, J. 1973. The development of Denmark's nature since the last glacial. *Danmarks Geologiske Undersogelse,* volume 7-C.

Iwao, S. 1972. Application of the m-m method to the analysis of spatial patterns by changing the quadrat size. *Researches on Population Ecology* 14:97–128.

Iwasa, Y., V. Andreasen, and S. A. Levin. 1987. Aggregation in model ecosystems: I. Perfect aggregation. *Ecological Modelling* 37:287–302.

Iwasa, Y., S. A. Levin, and V. Andreasen. 1989. Aggregation in model ecosystems: II. Approximate aggregation. *IMA Journal of Mathematics Applied in Medicine and Biology* 6:1–23.

Jackson, S. T. 1989. *Postglacial Vegetational Changes along an Elevational Gradient in the Adirondack Mountains (New York): A Study of Plant Macrofossils.* Bulletin 465. Albany: New York State Museum and Science Service.

Jackson, S. T. 1990. Pollen source area and representation in small lakes of the northeastern United States. *Review of Paleobotany and Palynology* 63:53–76.

Jackson, S. T. 1994. Pollen and spores in Quaternary lake sediments as sensors of vegetation composition: theoretical models and empirical evidence. In A. Traverse, ed., *Sedimentation of Organic Particles,* pp. 253–266. Cambridge: Cambridge University Press.

Jackson, S. T. and A Wong. 1994. Using forest patchiness to determine pollen source areas of closed-canopy pollen assemblages. *Journal of Ecology* 82:89–99.

Jacobson, G. L. and R. H. W. Bradshaw. 1981. The selection of sites for paleovegetational studies. *Quaternary Research* 16:80–96.

Jacobson, G. L. and E. C. Grimm. 1986. A numerical analysis of Holocene forest and prairie vegetation in central Minnesota. *Ecology* 67:958–966.

Jacobson, G. L., T. Webb, and E. C. Grimm. 1987. Patterns and rates of vegetation change during the deglaciation of eastern North America. In W. F. Ruddiman and H. E. Wright, eds., *The Geology of North America,* volume K-3, pp. 277–288. Boulder, Colo.: Geological Society of America.

James F. C., R. F. Johnson, N. O. Warner, G. J. Niemi, and W. J. Boecklen. 1984. The Grinnellian niche of the wood thrush. *American Naturalist* 124:17–30.

Jardine, P. M., G. V. Wilson, and R. J. Luxmoore. 1990. Unsaturated solute transport through a forest soil during rain storm events. *Geoderma* 46:103–118.

Jarvis, P. G. 1995. Scaling processes and problems. *Plant, Cell and Environment* 18:1079–1089.

Jarvis, P. G., G. B. James, and J. J. Landsberg. 1976. Coniferous forest. In J. L. Monteith, ed., *Vegetation and the Atmosphere,* Volume II, Case Studies, pp. 171–240. London: Academic Press.

Jenny, H. 1941. *Factors of Soil Formation: A System of Quantitative Pedology.* New York: McGraw Hill.

Jensen, J. R. 1986. *Introductory Digital Image Processing: A Remote Sensing Perspective.* Englewood Cliffs, N.J.: Prentice-Hall.

Johnson, A. R., B. T. Milne, and J. A. Wiens. 1992. Diffusion in fractal landscapes: simulations and experimental studies of tenebrionid beetle movements. *Ecology* 73:1968–1983.

Johnson, D. W. and R. I. Van Hook, eds. 1989. *Analysis of Biogeochemical Cycling Processes in Walker Branch Watershed.* New York: Springer-Verlag.

Johnson, D. W., W. T. Swank, and J. M. Vose. 1995a. Effects of liming on soils and streamwaters in a deciduous forest: comparison of field results and simulations. *Journal of Environmental Quality* 24:1104–1117.

Johnson, D. W., R. F. Walker, and J. T. Ball. 1995b. Lessons from lysimeters: soil N release from disturbance compromises controlled environment study. *Ecological Applications* 5:395–400.

Johnson, E. and C. Larsen. 1991. Climatically induced change in fire frequency in the Southern Canadian Rockies. *Ecology* 72:194–201.

Johnson, E. and D. R. Wowchuk. 1993. Wildfires in the southern Canadian Rocky Mountains and their relationship to mid-tropospheric anomalies. *Canadian Journal of Forest Research* 23:1213–1222.

Johnston, A. E. 1991. Benefits of long-term ecosystem research: some examples from Rothamsted. In P. G. Risser, ed., *Long-Term Ecological Research: An International Perspective,* pp. 89–114. Chichester: John Wiley and Sons.

Johnston, C. A. and R. J. Naiman. 1987. Boundary dynamics at the aquatic-terrestrial interface: the influence of beaver and geomorphology. *Landscape Ecology* 1:47–57.

Johnston, C. A. and R. J. Naiman. 1990. The use of a geographic information system to analyze long-term landscape alteration by beaver. *Landscape Ecology* 4:5–19.

Jones, B. and M. G. Kenward. 1988. *Design and Analysis of Cross-Over Trials.* London: Chapman and Hall.

Jones, C. G. and J. H. Lawton. 1995. *Linking Species and Ecosystems.* New York: Chapman and Hall.

Jones, K. B. 1995. Phylogeography of the desert horned lizard (*Phrynosoma platyrhinos*) and the short-horned lizard (*Phrynosoma douglassi*): patterns of divergence and diversity. *Ph.D. Dissertation.* Las Vegas: University of Nevada.

Jongman, R. H. G., C. J. F. ter Braak, and O. F. R. van Tongeren, eds. 1987. *Data Analysis in Community and Landscape Ecology.* Wageningen: Pudoc.

Judson, O. P. 1994. The rise of individually-based models in ecology. *Trends in Ecology and Evolution* 9:9–14.

Kachanoski, R. G. 1988. Processes and soils—from pedon to landscape. In T. Rosswall, R. G. Woodmansee, and P. G. Risser, eds., *Scales and Global Change,* pp. 153–177. New York: John Wiley and Sons.

Kaczynski, V. W. and J. F. Palmisano. 1993. *Oregon's Wild Salmon and Steelhead Trout: A Review of the Impact of Management and Environmental Factors.* Salem: Oregon Forest Industries Council.

Kareiva, P. 1990. Population dynamics in spatially complex environments: theory and data. *Philosophical Transactions of the Royal Society of London B* 330: 175–190.

Kareiva, P. and M. Andersen. 1986. Spatial aspects of species interactions: the wedding of models and experiments. In A. Hastings, ed., *Community Ecology: Lecture Notes in Biomathematics 77*, pp. 35–50. Berlin: Springer-Verlag.

Kareiva, P. and M. Anderson. 1989. Spatial aspects of species interactions: the wedding of models and experiments. In A. Hastings, ed., *Community Ecology*, pp. 35–50. New York: Springer-Verlag.

Kareiva, P. M. and N. Shigesada. 1983. Analyzing insect movement as a correlated random walk. *Oecologia* 56:234–238.

Kareiva, P. and U. Wennergren. 1995. Connecting landscale pattern to ecosystem and population processes. *Nature* 373:299–302.

Kasischke, E., N. L. Christensen, Jr., and B. Stocks. 1995. Fire, global warming, and the carbon balance of boreal forests. *Ecological Applications* 5:437–451.

Kates, R. W., B. L. Turner II, and W. C. Clark. 1990. The great transformation. In B. L. Turner, W. C. Clark, R. W. Kates, J. F. Richards, J. T. Matthews, and W. B. Meyer, eds., *The Earth as Transformed by Human Action*, pp.1–17. New York: Cambridge University Press.

Kauffman, S. A. 1993. *The Origins of Order*. New York: Oxford University Press.

Keeley, J. E. and P. Zedler. 1978. Reproduction of chaparral shrubs after fire: a comparison of sprouting and seeding strategies. *American Midland Naturalist* 99:142–161.

Keiter, R. B. and M. S. Boyce. 1991. *The Greater Yellowstone Ecosystem*. New Haven, Conn.: Yale University Press.

Kelliher, F. M., B. M. M. Kostner, D. Y. Hollinger, J. N. Byers, J. E. Hunt, T. M. McSeveny, R. Meserth, P. L. Weir, and E.-D. Shulze. 1992. Evaporation, xylem sap flow, and tree transpiration in a New Zealand broad-leaved forest. *Agricultural and Forest Meteorology* 62:53–73.

Kellogg, C. E. 1961. A challenge to American soil scientists: on the occasion of the 25th anniversary of the Soil Science Society of America. *Soil Science Society of America Proceedings* 25:419–423.

Kelt, D. A., P. L. Meserve, and B. K. Lang. 1994. Quantitative habitat associations of small mammals in a temperate rainforest in southern Chile: empirical patterns and the importance of ecological scale. *Journal of Mammalogy* 75: 890–904.

Kenny, D. and C. Loehle. 1991. Are food webs randomly connected? *Ecology* 72:1794–1799.

Kerr, R. A. 1995. It's official: first glimmer of greenhouse warming seen; news item. *Science* 270:1565–1567.

Kilgore, B. M. and D. Taylor. 1979. Fire history of a sequoia-mixed conifer forest. *Ecology* 60:129–142.

King, A. W. 1991. Translating models across scales in the landscape. In M. G. Turner and R. H. Gardner, eds., *Quantitative Methods in Landscape Ecology: The Analysis and Interpretation of Landscape Heterogeneity*, pp. 479–517. New York: Springer-Verlag.

King, A. W. 1993. Considerations of scale and hierarchy. In S. Woodley, J. Kay, and G. Francis, eds., *Ecological Integrity and the Management of Ecosystems,* pp. 19–46. Delray Beach, Fla.: Saint Lucie Press.

King, A. W., D. L. DeAngelis, and W. M. Post. 1987. The seasonal exchange of carbon dioxide between the atmosphere and the terrestrial biosphere: extrapolation from site-specific models to regional models. *ORNL/TM-10570.* Oak Ridge, Tenn.: Oak Ridge National Laboratory.

King, A. W., W. R. Emanuel, and R. V. O'Neill. 1990. Linking mechanistic models of tree physiology with models of forest dynamics: problems of temporal scale. In R. K. Dixon, R. S. Meldahl, G. A. Ruark, and W. G. Warren, eds., *Process Modeling of Forest Growth Responses to Environmental Stress,* pp. 241–248. Portland, Ore.: Timber Press.

King, A. W., A. R. Johnson, and R. V. O'Neill. 1991. Transmutation and functional representation of heterogeneous landscapes. *Landscape Ecology* 5:239–253.

King, A. W., R. V. O'Neill, and D. L. DeAngelis. 1989. Using ecosystem models to predict regional CO_2 exchange between the atmosphere and the terrestrial biosphere. *Global Biogeochemical Cycles* 3:337–361.

Kiniry, L. N., C. L. Scrivner, and M. E. Keener. 1983. A soil productivity index based upon predicted water depletion and root growth. *University of Missouri Agricultural Experiment Station Research Bulletin* 1051.

Kirkby, M. J., ed. 1978. *Hillslope Hydrology.* New York: John Wiley and Sons.

Kitchell, J. F., R. V. O'Neill, D. Webb, G. W. Gallepp, S. M. Bartell, J. F. Koonce, and B. S. Ausmus. 1979. Consumer regulation of nutrient cycling. *BioScience* 29:28–34.

Kleiber, M. 1961. *The Fire of Life.* New York: John Wiley and Sons.

Knick, S. T. and J. T. Rotenberry. 1995. Landscape characteristics of fragmented shrubsteppe habitats and breeding passerine birds. *Conservation Biology* 9: 1059–1071.

Knight, D. H. 1991. The Yellowstone fire controversy. In R. B. Keiter and M. S. Boyce, eds., *The Greater Yellowstone Ecosystem,* pp. 87–104. New Haven, Conn.: Yale University Press.

Knopf, F. L. and F. B. Samson. 1994. Scale perspectives on avian diversity in western riparian ecosystems. *Conservation Biology* 8:669–676.

Kohm, K. A. 1991. *Balancing on the Brink of Extinction: The Endangered Species Act and Lessons for the Future.* Washington, D.C.: Island Press.

Kolasa, J. 1989. Ecological systems in hierarchical perspective: breaks in community structure and other consequences. *Ecology* 70:36–47.

Kolasa J. and J. A. Drake. 1997. Abundance and range relationship in a fragmented landscape: connections and contrasts between competing models. *Coenoses* (in review).

Kolasa, J. and S. T. A. Pickett. 1989. Ecological systems and the concept of biological organization. *Proceedings of the National Academy of Science* 86:8837–8841.

Kolasa, J. and C. D. Rollo. 1991. The heterogeneity of heterogeneity: a glossary. In J. Kolasa and S. T. A. Pickett, eds., *Ecological Heterogeneity,* pp. 1–23. New York: Springer-Verlag.

Kolasa, J. and D. Strayer. 1988. Patterns of the abundance of species: a comparison of two hierarchical models. *Oikos* 53:235–241.

Kolitar, N. B. and J. A. Wiens. 1990. Multiple scales of patchiness and patch structure: a hierarchical framework for the study of heterogeneity. *Oikos* 59:253–260.

Koopman, B. 1980. *Search and Screening*. New York: Pergamon Press.

Krajina, V. J. 1965. Biogeoclimatic zones and classification of British Columbia. *Ecology of Western North America* 1:1–17.

Krishnaiah, P. R. and C. R. Rao, eds. 1988. *Handbook of Statistics: Sampling*. New York: North-Holland.

Krummel J. R., R. H. Gardner, G. Sugihara, R. V. O'Neill, and P. R. Coleman. 1987. Landscape patterns in a disturbed environment. *Oikos* 48:321–324.

Kurtz, L. T. and S. W. Melsted. 1973. Movement of chemicals in soils by water. *Soil Science* 115:231–239.

Kurz, W. A. and M. J. Apps. 1996. Retrospective assessment of carbon flows in Canadian boreal forests. In M. J. Apps and D. T. Price, eds., *Forest Ecosystems, Forest Management and the Global Carbon Cycle*, pp. 173–182. *NATO ASI Series 1: Global Environmental Change*, volume 40. Heidelberg: Springer-Verlag.

Laan, R. and B. Verboom. 1990. Effects of pool size and isolation on amphibian communities. *Biological Conservation* 54:251–262.

Laessle, A. M. 1942. The origin and successional relationship of sandhill vegetation and sand-pine scrub. *Ecological Monographs* 28:361–387.

Lahti, T. 1995. Understorey vegetation as an indicator of forest site potential in southern Finland. *Acta Forestalia Fennica* 246:1–68.

Lair, H. 1987. Estimating the location of focal center in red squirrel home ranges. *Ecology* 68:1092–1101.

Lande, R. 1988. Genetics and demography in biological conservation. *Science* 241:1455–1459.

Landres, P. B., J. Verner, and J. W. Thomas. 1988. Ecological uses of vertebrate indicator species: a critique. *Conservation Biology* 2:316–328.

Larson, D. W. 1996. Curing the incurable. *American Scientist* 84:7–10.

Lauenroth, W. K., O. E. Sala, D. P. Coffin, and T. B. Kirchner. 1994. The importance of soil water in the recruitment of *Bouteloua gracilis* in the shortgrass steppe. *Ecological Applications* 4:741–749.

Lavorel, S., R. H. Gardner, and R. V. O'Neill. 1993. Analysis of patterns in hierarchically structured landscapes. *Oikos* 67:521–528.

Lavorel, S., R. V. O'Neill, and R. H. Gardner. 1994. Spatio-temporal dispersal strategies and annual plant species coexistence in a structured landscape. *Oikos* 71:75–88.

Law, R. and J. C. Blackford. 1992. Self-assembling food webs: a global viewpoint of coexistence of species in Lotka-Volterra communities. *Ecology* 73:567–579.

Laws, R. M. 1981. Experiences in the study of large mammals. In C. W. Fowler and T. D. Smith, eds. *Dynamics of Large Mammal Populations*, pp. 19–46. New York: John Wiley and Sons.

Lawton, J. H. 1989. Food webs. In J. M. Cherrett, ed., *Ecological Concepts*, pp. 43–78. Oxford: Blackwell Scientific.

Lawton, J. H. 1995. Webbing and WIWACS. *Oikos* 72:305–306.

Lawton, J. H. and V. K. Brown. 1993. Redundancy in ecosystems. In E.-D. Schulze and H. A. Mooney, eds., *Biodiversity and Ecosystem Function,* pp. 255–270. Heidelberg: Springer-Verlag.

Lawton, J. H., S. Nee, A. J. Letcher, and P. H. Harvey. 1994. Animal distribution: patterns and processes. In P. J. Edwards, R. M. May, and N. R. Webb, eds., *Large-Scale Ecology and Conservation,* pp. 41–58. New York: Blackwell Science.

Leatherland, J. F., K. J. Fabridge, and T. Bouchard. 1992. Lunar and semi-lunar rhythms in fish. In M. A. Ali, ed., *Rhythms in Fishes, NATO ASI Series,* Vol. A-236, pp. 83–107. New York: Plenum Press.

Leck, M. A., V. T. Parker, and R. L. Simpson, eds. 1989. *Ecology of Soil Seed Banks.* San Diego: Academic Press.

Leduc, A., P. Drapeau, Y. Bergeron, and P. Legendre. 1992. Study of spatial components of forest cover using partial Mantel tests and path analysis. *Journal of Vegetation Science* 3:69–78.

Lee, K. N. 1993. Greed, scale mismatch, and learning. *Ecological Applications* 3:560–564.

Lefkovitch, L. P. and L. Fahrig. 1985. Spatial characteristics of habitat patches and population survival. *Ecological Modelling* 30:297–308.

Lefroy, E. C., J. Salerian, and R. J. Hobbs. 1993. Integrating economic and ecological considerations: a theoretical framework. In R. J. Hobbs and D. A. Saunders, eds., *Reintegrating Fragmented Landscapes. Towards Sustainable Production and Nature Conservation,* pp. 209–244. New York: Springer Verlag.

Legendre, L. and S. Demers. 1984. Towards dynamic biological oceanography and limnology. *Canadian Journal of Fisheries and Aquatic Science* 41:2–19.

Legendre, L. and P. Legendre. 1983. *Numerical Ecology.* Amsterdam: Elsevier Scientific Publishing Company.

Legendre, P. 1993. Spatial autocorrelation: Trouble or new paradigm? *Ecology* 74:1659–1673.

Legendre, P. and D. Borcard. 1994. Rejoinder. *Environmental and Ecological Statistics* 1:57–61.

Legendre, P. and P. Dutilleul. 1992. Introduction to the analysis of periodic phenomena. In M. A. Alli, ed., *Rhythms in Fishes, NATO ASI Series,* Vol. A-236, pp. 11–25. New York: Plenum Press.

Legendre, P. and M.-J. Fortin. 1989. Spatial pattern and ecological analysis. *Vegetatio* 80:107–138.

Legendre, P. and L. Legendre. 1997. *Numerical Ecology,* 2nd English ed. Amsterdam: Elsevier.

Legendre, P. and A. Vaudor. 1991. *The "R package": Multidimensional Analysis, Spatial Analysis.* Montréal: Département de sciences biologiques, Université de Montréal.

Legendre, P., N. Oden, R. R. Sokal, A. Vaudor, and J. Kim. 1990. Approximate analysis of variance of spatially autocorrelated regional data. *Journal of Classification* 7:53–75.

Leibold, M. A. 1995. The niche concept revisited: mechanistic models and community context. *Ecology* 76:1371–1382.

Lertzman, K., T. Spies, and F. Swanson. 1997. From ecosystem dynamics to eco-system management. In P. Schoonmaker, B. von Hagen, and E. Wolf, eds., *The Rainforests of Home: Profile of a North American Bioregion,* pp. 361–382. Washington, D.C.: Island Press.

Lertzman, K. P. 1992. Patterns of gap-phase replacement in a subalpine, old-growth forest. *Ecology* 73:657–669.

Lertzman, K. P. 1995. Forest dynamics, differential mortality and variable transition probabilities. *Journal of Vegetation Science* 6:191–204.

Lertzman, K. P. and C. L. Gass. 1983. Alternative models of pollen transfer. In C. Jones and C. Little, eds., *The Handbook of Experimental Pollination Biology,* pp. 474–489. New York: Van Nostrand Rheinhold.

Lertzman, K. P., G. Sutherland, A. Inselberg, and S. Saunders. 1996. Canopy gaps and the landscape mosaic in a temperate rainforest. *Ecology* 77:1254–1270.

Leverenz, J., J. D. Deans, E. D. Ford, P. G. Jarvis, R. Milne, and D. Whitehead. 1982. Systematic spatial variation of stomatal conductance in a Sitka spruce plantation. *Journal of Applied Ecology* 19:835–851.

Levin, S. A. 1978. Pattern formation in ecological communities. In J. Steele, ed., *Spatial Pattern in Plankton Communities,* pp. 433–465. New York: Plenum Press.

Levin, S. A. 1986. Pattern, scale, and variability: an ecological perspective. In A. Hastings, ed., *Community Ecology: Lecture Notes in Biomathematics 77,* pp. 1–12. Berlin: Springer-Verlag.

Levin, S. A. 1992. The problem of pattern and scale in ecology. *Ecology* 73:1943–1967.

Levin, S. A. 1993. Concepts of scale at the local level. In J. R. Ehleringer and C. B. Field, eds., *Scaling Physiological Processes: Leaf to Globe,* pp. 7–19. San Diego: Academic Press.

Levin, S. A. and L. Buttel. 1986. *Measures of Patchiness in Ecological Systems.* Publication ERC-130. Ithaca: Ecosystem Research Center, Cornell University.

Levins, R. 1969. Some demographic and genetic consequences of environmental heterogeneity for biological control. *Bulletin of the Entomological Society of America* 15:237–240.

Levins, R. 1970. Extinction. In M. Gerstenhaber, ed., *Some Mathematical Problems in Biology,* pp. 77–107. Providence, R.I.: American Mathematical Society.

Levy, M. 1996. *Global Environmental Policy Research Tools.* Located on the World Wide Web at http://www.princeton.edu/mlevy/.

Lhomme, J.-P. 1992. The energy balance of heterogeneous terrain: averaging the controlling parameters. *Agricultural and Forest Meteorology* 61:11–21.

Li, B-L. and C. Loehle. 1995. Wavelet analysis of multiscale permeabilities in the subsurface. *Geophysical Research Letters* 22:3123–3126.

Li, H., J. F. Franklin, F. J. Swanson, and T. A. Spies. 1993. Developing alternative forest cutting patterns: a simulation approach. *Landscape Ecology* 8:63–75.

Light, S. S., L. H. Gunderson, and C. S. Holling. 1995. The Everglades: evolution of management in a turbulent ecosystem. In L. H. Gunderson, C. S. Holling, and S. S. Light, eds., *Barriers and Bridges to the Renewal of Ecosystems and Institutions,* pp. 103–168. New York: Columbia University Press.

Likens, G. E. 1983. A priority for ecological research. *Bulletin of the Ecological Society of America* 64:234–243.

Likens, G. E., ed. 1989. *Long-Term Studies in Ecology: Approaches and Alternatives.* New York: Springer-Verlag.

Likens, G. E. 1989. Preface. In G. E. Likens, ed., *Long-Term Studies in Ecology: Approaches and Alternatives,* pp. ix-xi. New York: Springer-Verlag.

Likens, G. E. and F. H. Bormann. 1995. *Biogeochemistry of a Forested Ecosystem.* New York: Springer-Verlag.

Lillesand, T. M. and R. W. Kiefer. 1994. *Remote Sensing and Image Interpretation.* New York: John Wiley and Sons.

Lindstedt, S. L., B. J. Miller, and S. W. Buskirk. 1986. Home range, time and body size in mammals. *Ecology* 67:413–418.

Lippman, M. 1989. Health effects of ozone: a critical review. *Journal of the Air and Waste Management Association* 39:672–695.

Liu, S. and R. O. Teskey. 1995. Responses of foliar gas exchange to long-term elevated CO_2 concentrations in mature loblolly pine trees. *Tree Physiology* 15:351–359.

Lodge, D., J. Barko, D. Strayer, J. Melack, G. Mittelbach, R. Howarth, B. Menge, and J. Titus. 1988. Spatial heterogeneity and habitat interactions in lake communities. In S. Carpenter, ed., *Complex Interactions in Lake Communities,* pp. 229–260. New York: Springer-Verlag.

Loehle, C. and G. Wein. 1994. Landscape habitat diversity: a multiscale information theory approach. *Ecological Modelling* 73:311–329.

Lomolino, M. V. 1994. An evaluation of alternative strategies for building networks of nature reserves. *Biological Conservation* 69:243–249.

Loomis, J. B. 1993. *Integrated Public Lands Management.* New York: Columbia University Press.

Lord, J. M. and D. A. Norton. 1990. Scale and the spatial concept of fragmentation. *Conservation Biology* 4:197–201.

Loreau, M. 1992. Time scale of resource dynamics and coexistence through time partitioning. *Theoretical Population Biology* 41:401–412.

Lorimer, C. G. 1989. Relative effects of small and large disturbances on temperate hardwood forest structure. *Ecology* 70:565–567.

Lovejoy, S. and D. Schertzer. 1986. Scale invariance, symmetries, fractals, and stochastic simulations of atmospheric phenomena. *Bulletin of the American Meteorological Society* 67:21–32.

Lowrance, R., P. F. Hendrix, and E. P. Odum. 1986. A hierarchical approach to sustainable agriculture. *American Journal of Alternative Agriculture* 1:169–173.

Lubchenco, J., A. M. Olson, L. B. Brubaker, S. R. Carpenter, M. M. Holland, S. P. Hubbell, S. A. Levin, J. A. MacMahon, P. A. Matson, J. M. Melillo, H. A. Mooney, C. H. Peterson, R. Pulliam, L. A. Real, P. J. Regal, and P. G. Risser. 1991. The Sustainable Biosphere Initiative: an ecological research agenda. A report from the Ecological Society of America. *Ecology* 2:371–412.

Luckyanov, N. K. 1995. Model aggregation: mathematical perspectives. In B. C. Patten and S. E. Jorgenson, eds., *Complex Ecology,* pp. 242–261. Englewood Cliffs, N.J.: Prentice-Hall.

Luckyanov, N. K., Y. M. Svirezhev, and O. V. Voronkova. 1983. Aggregation of variables in simulation models of water ecosystems. *Ecological Modelling* 18:235–240.

Ludwig, J. A. and J. F. Reynolds. 1988. *Statistical Ecology.* New York: John Wiley and Sons.

Lugo, A. E. and S. Brown. 1986. Steady state terrestrial ecosystems and the global carbon cycle. *Vegetatio* 68:83–90.

Luken, J. O. 1990. *Directing Ecological Succession.* New York: Chapman and Hall.

Luxmoore, R. J., P. M. Jardine, G. V. Wilson, J. R. Jones, and L. W. Zelazny. 1990. Physical and chemical controls of preferred path flow through a forested hill-slope. *Geoderma* 46:139–154.

Luxmoore, R. J., A. W. King, and M. L. Tharp. 1991. Approaches to scaling up physiologically based soil-plant models in space and time. *Tree Physiology* 9:281–292.

Lynch, J. F., W. F. Carmen, D. A. Saunders, and P. Cale. 1995. Short-term use of vegetated road verges and habitat patches by four bird species in the central wheatbelt of Western Australia. In D. A. Saunders, J. Craig, and L. Mattiske, eds., *Nature Conservation 4: The Role of Networks,* pp. 34–42. Chipping Norton, NSW, Australia: Surrey Beatty and Sons.

MacArthur, R. H. 1955. Fluctuation of animal populations and a measure of community stability. *Ecology* 36:533–536.

MacArthur, R. H. 1957. On the relative abundance of bird species. *Proceedings of the National Academy of Science* 43:293–295.

MacDonald, G. A., A. T. Abbott, and F. L. Peterson. 1984. *Volcanoes in the Sea: The Geology of Hawaii,* 2nd edition. Honolulu: University of Hawaii Press.

MacDonald, N. 1979. Simple aspects of food web complexity. *Journal of Theoretical Biology* 80:577–588.

Magnuson, J. J., T. K. Kratz, T. M. Frost, C. J. Bowser, B. J. Benson, and R. Nero. 1991. Expanding the temporal and spatial scales of ecological research and comparison of divergent ecosystems: roles for LTER in the United States. In P. G. Risser, ed., *Long-Term Ecological Research: An International Perspective,* pp. 45–70. West Sussex, U.K.: John Wiley and Sons.

Magurran, A. E. 1988. *Ecological Diversity and Its Measurement.* Princeton, N.J.: Princeton University Press.

Mallat, S. G. 1988. Review of multi-frequency channel decomposition of images and wavelet models. *Courant Institute of Mathematical Sciences Robotic Report 178.* New York: New York University.

Manabe, S. and R. J. Stouffer. 1986. Sensitivity of a global climate to an increase in CO_2 concentration in the atmosphere. *Journal of Geophysical Research* 85: 5529–5554.

Mandelbrot, B. B. 1977. *Fractals—Form, Chance, and Dimension.* San Francisco: Freeman.

Mandelbrot, B. B. 1983. *The Fractal Geometry of Nature.* New York: Freeman.

Mangel, M. and C. W. Clark. 1988. *Dynamic Modeling in Behavioral Ecology.* Princeton, N.J.: Princeton University Press.

Mangel, M., L. M. Talbot, G. K. Meffe et al. 1996. Principles for the conservation of wild living resources: another perspective. *Ecological Applications* 6:338–362.

Mann, C. C. and M. L. Plummer. 1995. Are wildlife corridors the right path? *Science* 270:1428–1430.

Mantel, N. 1967. The detection of disease clustering and a generalized regression approach. *Cancer Research* 27:209–220.

Marbut, C. F. 1922. Soil classification. *American Soil Survey Association Bulletin* 3:24–32.

Marcot, B. G., M. J. Wisdom, H. W. Li, and G. C. Castillo. 1994. Managing for featured, threatened, endangered, and sensitive species and unique habitats for ecosystem sustainability. *USDA Forest Service General Technical Report PNW-GTR-329.* Portland, Ore.: Pacific Northwest Research Station.

Margalef, R. 1995. Information theory and complex ecology. In B. C. Patten and S. E. Jorgensen, eds., *Complex Ecology,* pp. 40–50. Englewood Cliffs, N.J.: Prentice Hall.

Martin, T. A., K. J. Brown, J. Cermák, R. Ceulemans, J. Kucera, F. C. Meinzer, J. S. Rombold, D. G. Sprugel, and T. M. Hinckley. 1997. Crown conductance and tree and stand transpiration in a second-growth *Abies amabilis* forest. *Canadian Journal of Forest Research* 27:797–808.

Martinez, N. D. 1991. Artifacts or attributes? Effects of resolution on the Little Rock Lake food web. *Ecological Monographs* 61:367–392.

Martinez, N. D. 1992. Constant connectance in community food webs. *American Naturalist* 139:1208–1218.

Martinez, N. D. 1993a. Effect of scale on food web structure. *Science* 260:242–243.

Martinez, N. D. 1993b. Effects of resolution on food web structure. *Oikos* 66:403–412.

Martinez, N. D. 1994. Scale-dependent constraints on food-web structure. *American Naturalist* 144:935–953.

Martinez, N. D. 1995. Unifying ecological subdisciplines with ecosystem food webs. In C. J. Jones and J. H. Lawton, eds., *Linking Species and Ecosystems,* pp. 166–175. London: Chapman and Hall.

Martinez, N. D. 1996. Defining and measuring functional aspects of biodiversity. In K. J. Gaston, ed., *Biodiversity: A Science of Numbers and Difference,* pp. 114–148. Oxford: Blackwell.

Martinez, N. D. and J. H. Lawton. 1995. Scale and food-web structure—from local to global. *Oikos* 73:148–154.

Maser, C. and J. R. Sedell. 1994. *From the Forest to the Sea: The Ecology of Wood in Streams, Rivers, Estuaries, and Oceans.* Delray Beach, Fla.: St. Lucie Press.

Matheron, G. 1971. *The Theory of Regionalized Variables and Its Applications.* Cahiers du Centre de Morphologie Mathématique de Fontainebleau 5.

Matson, P. A. and P. M. Vitousek. 1990. Ecosystem approach to a global nitrous oxide budget. *BioScience* 40:667–672.

Matthews, J. A. 1978. Plant colonization patterns on a gletschervorfeld, southern Norway: a meso-scale geographical approach to vegetation change and phytometric dating. *Boreas* 7:155–178.

Matthews, J. A. 1992. *The Ecology of Recently-Deglaciated Terrain. A Geoecologi-*

cal Approach to Glacier Forelands and Primary Succession. Cambridge: Cambridge University Press.

Matyasovszky, I. and I. Bogardi. 1994. Comparison of two general circulation models to downscale temperature and precipitation under climate change. *Water Resources Research* 30:3437–3448.

Maurer, B. A. 1990. The relationship between distribution and abundance in a patchy environment. *Oikos* 58:181–189.

May, R. M. 1974. *Stability and Complexity in Model Ecosystems.* Princeton, N.J.: Princeton University Press.

May, R. M. 1975. Patterns of species abundance and diversity. In M. L. Cody and J. M. Diamond, eds., *Ecology and Evolution of Communities,* pp. 81–120. Cambridge: Belknap Press.

May, R. M. 1977. Thresholds and breakpoints in ecosystems with a multiplicity of stable states. *Nature* 269:471–4777.

May, R. M. 1983. The structure of food webs. *Nature* 301:566–568.

May, R. M. 1988. How many species are there on earth? *Science* 241:1441–1449.

May, R. M. 1989. Levels of organization in ecology. In J. M. Cherrett, ed., *Ecological Concepts: The Contribution of Ecology to an Understanding of the Natural World,* pp. 339–363. Oxford: Blackwell Scientific Publications.

May, R. M. 1994. The effects of spatial scale on ecological questions and answers. In R. J. Edwards, R. M. May, and N. R. Webb, eds., *Large-Scale Ecology and Conservation Biology,* pp. 1–17. Boston: Blackwell Scientific Publications.

Mayr, E. 1976. *Evolution and the Diversity of Life: Selected Essays.* Cambridge, Mass.: Belknap Press.

Mayr, E. 1984. *The Growth of Biological Thought.* Cambridge: Harvard University Press.

McArdle, B. H., K. J Gaston, and J. H. Lawton. 1990. Variation in the size of animal populations: patterns, problems and artifacts. *Journal of Animal Ecology* 59:439–454.

McComb, W. C., T. A. Spies, and W. H. Emmingham. 1993. Douglas-fir forests: managing for timber and mature-forest habitat. *Journal of Forestry* 91:31–42.

McConnaughay, K. D. M. and F. A. Bazzaz. 1987. The relationship between gap size and performance of several colonizing annuals. *Ecology* 68:411–416.

McCoy, E. D. and S. S. Bell. 1991. "Habitat structure" in ecology. In S. S. Bell, E. D. McCoy, and H. R. Mushinsky, eds., *Habitat Structure,* pp. 3–27. New York: Chapman and Hall.

McCullagh, P. and J. A. Nelder. 1989. *Generalised Linear Models.* London: Chapman and Hall.

McCune, B. and T. F. H. Allen. 1985. Will similar forests develop on similar sites? *Canadian Journal of Botany* 63:367–376.

McDaniels, T., L. J. Axelrod, and P. Slovic. 1995. Characterizing perception of ecological risk. *Risk Analysis* 15:575–588.

McDonald, J. A. and J. H. Brown. 1992. Using montane mammals to model extinction due to global change. *Conservation Biology* 6:409–415.

McDonnell, M. J. and S. T. A. Pickett. 1993. *Humans as Components of Ecosys-*

tems: The Ecology of Subtle Human Effects and Populated Areas. New York: Springer-Verlag.

McEuen, A. 1993. The wildlife corridor controversy: a review. *Endangered Species Update* 10:1–7.

McGarigal, K. and B. Marks. 1994. Fragstats: a spatial pattern analysis program for quantifying landscape structure. Unpublished software. Corvallis, Ore.: Department of Forest Sciences, Oregon State University.

McKenzie, D., D. L. Peterson, and E. Alvarado. 1996. Extrapolation problems in modeling fire effects at large spatial scales: a review. *International Journal of Wildland Fire* 6:165–176.

McLaughlin, J. F. and J. Roughgarden. 1992. Predation across spatial scales in heterogeneous environments. *Theoretical Population Biology* 41:277–299.

McMahon, T. 1973. Size and shape in biology. *Science* 179:1201–1204.

McMillan, C. 1959. Nature of the plant community. V. Variation within the true prairie community-type. *American Journal of Botany* 46:418–424.

McNab, B. K. 1963. Bioenergetics and home range size. *American Naturalist* 97:133–140.

McNaughton, K. G. 1994. Effective stomatal and boundary-layer resistances of heterogeneous surfaces. *Plant, Cell and Environment* 17:1061–1068.

McNaughton, S. J. 1979. Grassland-herbivore dynamics. In A. R. E. Sinclair and M. Norton-Griffiths, eds., *Serengeti: Dynamics of an Ecosystem,* pp. 46–81. Chicago: University of Chicago Press.

McSweeney, K., P. E. Gessler, B. Slater, R. D. Hammer, J. Bell, and G. W. Petersen. 1994. Towards a new framework for modeling the soil-landscape continuum. In R. Amundson, J. Harden, and M. Singer, eds., *Factors of Soil Formation: A Fiftieth Anniversary Retrospective,* pp. 127–145. Special Publication 33. Madison, Wis.: Soil Science Society of America.

Mead, R. 1988. *The Design of Experiments. Statistical Principles for Practical Application.* Cambridge: Cambridge University Press.

Meehan, W. R. 1991. Influences of Forest and Rangeland Management on Salmonid Fishes and their Habitats. *American Fisheries Society Special Publication 19.* Bethesda, Md.: American Fisheries Society.

Meentemeyer, V. and E. O. Box. 1987. Scale effects in landscape studies. In M. G. Turner, ed., *Landscape Heterogeneity and Disturbance,* pp. 15–34. New York: Springer-Verlag.

Meese, D. A., A. J. Gow, P. Grootes, P. A. Mayewski, M. Ram, M. Stuiver, K. C. Taylor, E. D. Waddington, and G. A. Zielinski. 1994. The accumulation record from the GISP2 core as an indicator of climate change throughout the Holocene. *Science* 266:1680–1682.

Meinzer, F. C., G. Goldstein, N. M. Holbrook, P. Jackson, and J. Cavelier. 1993. Stomatal and environmental control of transpiration in a lowland tropical forest tree. *Plant, Cell and Environment* 16:429–436.

Menge, B. A. and A. M. Olson. 1990. Role of scale and environmental factors in regulation of community structure. *Tree* 5:52–57.

Menge, B. A., E. L. Berlow, C. A. Blanchette, S. A. Navarrete, and S. B. Yamada.

1994. The keystone species concept: variation in interaction strength in a rocky intertidal habitat. *Ecological Monographs* 64:249–286.

Mercer, W. B. and A. D. Hall. 1911. The experimental error of field trials. *Journal of Agricultural Science* 4:107–132.

Merriam, G. 1984. Connectivity: a fundamental ecological characteristic of landscape pattern. In *Proceedings of the First International Seminar on Methodology in Landscape Ecological Research and Planning,* Theme 1:5–15. Roskilde, Denmark: International Association for Landscape Ecology.

Merriam, G. 1988. Landscape dynamics in farmland. *Trends in Ecology and Evolution* 3:16–20.

Michener, W. K., J. W. Brunt, and S. G. Stafford, eds. 1994. *Environmental Information Management and Analysis: Ecosystem to Global Scales.* London: Taylor and Francis.

Millar, C. E., L. M. Turk, and H. D. Foth. 1965. *Fundamentals of Soil Science,* 4th edition. New York: John Wiley and Sons.

Miller, P. R. 1992. Mixed conifer forests of the San Bernardino Mountains, California. In R. K. Olson, D. Binkley, and M. Böhm, eds., *Response of Western Forests to Air Pollution,* pp. 461–497. New York: Springer-Verlag.

Miller, P. R., J. R. Parmeter, O. C. Taylor, and E. A. Cardiff. 1963. Ozone injury to the foliage of *Pinus ponderosa. Phytopathology* 53:1072–1076.

Mills, L. S., M. E. Soulé, and D. F. Doak. 1993. The keystone-species concept in ecology and conservation. *BioScience* 43:219–224.

Milne, B. T. 1988. Measuring the fractal geometry of landscapes. *Applied Mathematical Computing* 27:67–79.

Milne, B. T. 1991a. Heterogeneity as a multiscale characteristic of landscapes. In J. Kolasa and S. T. A. Pickett, eds., *Ecological Heterogeneity,* pp. 69–84. New York: Springer-Verlag.

Milne, B. T. 1991b. Lessons from applying fractal models to landscape patterns. In M. G. Turner and R. H. Gardner, eds., *Quantitative Methods in Landscape Ecology,* pp. 199–235. New York: Springer-Verlag.

Milne, B. T. 1992. Spatial aggregation and neutral models in fractal landscape. *American Naturalist* 139:32–57.

Milne, B. T. and A. R. Johnson. 1993. Renormalization relations for scale transformation in ecology. In R. H. Gardner, ed., *Some Mathematical Questions in Biology: Theoretical Approaches for Predicting Spatial Effects in Ecological Systems,* pp. 109–128. Providence, R.I.: American Mathematical Society.

Milne, B. T., A. R. Johnson, T. H. Keitt, C. A. Hatfield, J. D. David, and P. T. Hraber. 1996. Detection of critical densities associated with piñon-juniper woodland ecotones. *Ecology* 77:805–821.

Milne, B. T., M. G. Turner, J. A. Wiens, and A. R. Johnson. 1992. Interactions between the fractal geometry of landscapes and allometric herbivory. *Theoretical Population Biology* 41:337–353.

Milne, G. 1935. Some suggested units of classification and mapping particularly for East African soils. *Soil Research* 4.

Minnich, R. A. 1983. Fire mosaics in southern California and northern Baja California. *Science* 219:1287–1294.

Minnich, R. A. 1988. The biogeography of fire in the San Bernardino Mountains of California. *University of California Publications in Geography* 28:1–122.

Minnich, R. A. 1989. California fire history in San Diego County and adjacent northern Baja California: an evaluation of natural fire regimes and the effects of suppression management. In S. C. Keeley, ed., *The California Chaparral: Paradigms Reexamined*, pp. 37–47. Los Angeles: Natural History Museum of Los Angeles.

Minta, S. C. 1992. Tests of spatial and temporal interaction among animals. *Ecological Applications* 2:178–188.

Mladenoff, D. J., M. A. White, T. R. Crow, and J. Pastor. 1994. Applying principles of landscape design and management to integrate old-growth forest enhancement and commodity use. *Conservation Biology* 8:752–762.

Mladenoff, D. J., M. A. White, J. Pastor, and T. R. Crow. 1993. Comparing spatial pattern in unaltered old-growth and disturbed forest landscapes for biodiversity design and management. *Ecological Applications* 3:293–305.

Moloney K. A., S. A. Levin, N. R. Chiariello, and L. Buttel. 1992. Pattern and scale in a serpentine grassland. *Theoretical Population Biology* 41:257–276.

Moloney, K. A., A. Morin, and S. A. Levin. 1991. Interpreting ecological pattern generated through simple stochastic processes. *Landscape Ecology* 5:163–174.

Monteith, J. L. 1965. Evaporation and environment. *Symposium of the Society for Experimental Biology* 19:205–234.

Mooney, H. A., E. R. Fuentes, and B. I. Kronberg, eds. 1993. *Earth System Response to Global Change: Contrasts between North and South America*. New York: Academic Press.

Moore, J. C., P. C. DeRuiter, and H. W. Hunt. 1993. Influence of productivity on the stability of real and model ecosystems. *Science* 261:906–908.

Morgan, E. and D. S. Minors. 1995. The analysis of biological time-series data: some preliminary considerations. *Biological Rhythm Research* 26:124–148.

Morin, P. J. 1984. Odonate guild composition: experiments with colonization history and fish predation. *Ecology* 65:1866–1873.

Morisita, M. 1964. Application of the Iδ index to sampling techniques. *Researches on Population Ecology* 6:46–53.

Morisita, M. 1971. Composition of the Iδ index. *Researches on Population Ecology* 13:1–27.

Morris, D. W. 1987. Ecological scale and habitat use. *Ecology* 68:362–369.

Morrison, D. F. 1990. *Multivariate Statistical Methods*, 3rd ed. New York: McGraw-Hill.

Morrison, M. L., B. G. Marcot, and R. W. Mannan. 1992. *Wildlife-Habitat Relationships: Concepts and Applications*. Madison: University of Wisconsin Press.

Morrison, P. H. and F. J. Swanson. 1990. Fire history and pattern in a Cascade Range landscape. *USDA Forest Service General Technical Report PNW-GTR-254*. Portland, Ore.: USDA Forest Service, Pacific Northwest Research Station.

Morse, D. R., J. H. Lawton, M. M. Dodson, and M. H. Williamson. 1985. Fractal distribution of vegetation and the distribution of arthropod body lengths. *Nature* 314:731–733.

Motomura, I. 1932. On the statistical treatment of communities. *Japanese Journal of Zoology* 44:379–383 (cited in May 1975).

Muller, R. N. and F. H. Bormann. 1976. Role of *Erythronium americanum* Ker. in energy flow and nutrient dynamics of a northern hardwood forest ecosystem. *Science* 193:1126–1128.

Munn, R. E. 1988. The design of integrated monitoring systems to provide early indications of environmental changes. *Environmental Monitoring and Assessment* 11:203–217.

Murphy, D. D. 1989. Conservation and confusion: wrong species, wrong scale, wrong conclusions. *Conservation Biology* 3:82–4.

Murphy, E. J., D. J. Morris, J. L. Watkins, and J. Priddle. 1988. Scales of interaction between Antarctic krill and the environment. In D. Sahrhage, ed., *Antarctic Ocean and Resource Variability*, pp. 120–130. Berlin: Springer-Verlag.

Murphy, G. I. 1966. Population biology of the Pacific sardine (*Sardinops caerulea*). *Proceedings of the California Academy of Sciences* 34:1–34.

Murray, B. G. 1986. The structure of theory, and the role of competition in community dynamics. *Oikos* 46:145–158.

Murray, D. R. 1986. *Seed Dispersal.* Orlando, Fla.: Academic Press.

Murtaugh, P. A. 1994. Statistical analysis of food webs. *Biometrics* 50:1199–1202.

Myers, R. L. 1985. Fire and the dynamic relationship between Florida sandhill and sand pine scrub. *Bulletin of the Torrey Botanical Club* 112:241–252.

Naiman, R. J. 1988. Animal influences on ecosystem dynamics. *BioScience* 38:750–752.

Naiman, R. J., C. A. Johnston, and J. C. Kelly. 1988. Alteration of North American streams by beaver. *BioScience* 38:753–762.

Naiman, R. J. and J. M. Melillo. 1984. Nitrogen budget of a subarctic stream altered by beaver (*Castor canadensis*). *Oecologia* 62:150–155.

Naiman, R. J., J. M. Melillo, and J. E. Hobbie. 1986. Ecosystem alteration of boreal forest streams by beaver (*Castor canadensis*). *Ecology* 67:1254–1269.

National Aeronautics and Space Administration Advisory Council, Earth System Sciences Committee, eds. 1988. *Earth System Science: A Closer View.* Washington, D.C.: National Aeronautics and Space Administration.

National Research Council. 1995. *Upstream: Salmon and Society in the Pacific Northwest.* Washington, D.C.: National Academy Press.

Nee, S., P. H. Harvey, and R. M. May. 1991. Lifting the veil on abundance patterns. *Proceedings of the Royal Society of London B* 243:161–163.

Neilson, R. P. and L. H. Wullstein. 1983. Biogeography of two southwest American oaks in relation to atmospheric dynamics. *Journal of Biogeography* 10:275–297.

Nelder, J. A. and R. W. M. Wedderburn. 1972. Generalised linear models. *Journal of the Royal Statistical Society* A135:370–384.

Nellis, D. M and J. M. Briggs. 1988. SPOT satellite data for pattern recognition on the North American tall-grass prairie Long-term Ecological Research Site. *Geocarto International* 3:37–40.

Newbold, J. D., J. W. Elwood, R. V. O'Neill, and A. L. Sheldon. 1983. Phosphorus dynamics in a woodland stream ecosystem: a study of nutrient spiraling. *Ecology* 64:1249–1265.

Newbold, J. D., J. W. Elwood, R. V. O'Neill, and W. Van Winkle. 1981. Measuring nutrient spiraling in streams. *Canadian Journal of Fisheries and Aquatic Sciences* 38:860–863.

Newbold, J. D., P. J. Mulholland, J. W. Elwood, and R. V. O'Neill. 1982a. Organic carbon spiraling in stream ecosystems. *Oikos* 38:266–272.

Newbold, J. D., R. V. O'Neill, J. W. Elwood, and W. Van Winkle. 1982b. Nutrient spiraling in streams: implications for nutrient limitation and invertebrate activity. *American Naturalist* 120:628–652.

Newmark, W. D. 1985. Legal and biotic boundaries of western North American national parks, a problem of congruence. *Biological Conservation* 33:197–208.

Nials, F. L., E. E. Deeds, M. E. Moseley, S. G. Pozorski, T. G. Pozorski, and R. A. Feldman. 1979. El Niño: the catastrophic flooding of coastal Peru. *Bulletin of the Field Museum of Natural History* 50(7): 4–14; 50(8): 4–10.

Nicholls, A. O. and C. R. Margules. 1991. The design of studies to demonstrate the biological importance of corridors. In D. A. Saunders and R. J. Hobbs, eds., *Nature Conservation 2: The Role of Corridors,* pp. 49–61. Chipping Norton, NSW, Australia: Surrey Beatty and Sons.

Niemelè, J. K. and J. R. Spence. 1994. Distribution of forest dwelling carabids (Coleoptera): spatial scale and the concept of communities. *Ecography* 17: 166–175.

Nikiforoff, C. C. 1959. Reappraisal of the soil. *Science* 129:186–196.

Nikolov, N. T. and D. G. Fox. 1993. A coupled carbon-water-energy-vegetation model to assess responses of temperate forest ecosystems to changes in climate and atmospheric CO_2. Part 1. Model concept. *Environmental Pollution* 83: 251–262.

Norman, J. M. 1993. Scaling processes between leaf and canopy levels. In J. R. Ehleringer and C. B. Field, eds., *Scaling Physiological Processes: Leaf to Globe,* pp. 41–76. San Diego: Academic Press.

Northcote, T. G. 1988. Fish in the structure and function of freshwater systems: a "top-down" view. *Canadian Journal of Fisheries and Aquatic Sciences* 45: 361–379.

Norton, B. G. and R. E. Ulanowicz. 1992. Scale and biodiversity policy: a hierarchical approach. *Ambio* 21:244–249.

Norton, B. G. 1995a. Ecological integrity and social values: at what scale? *Ecosystem Health* 1:228–241.

Norton, B. G. 1995b. Evaluating ecosystem states: two competing paradigms. *Ecological Economics* 14:113–127.

Noss, R. F. 1990. Indicators for monitoring biodiversity: a hierarchical approach. *Conservation Biology* 4:355–364.

Odum, E. P. 1971. *Fundamentals of Ecology,* 3rd edition. Philadelphia: W. B. Saunders.

Oksanen, L., S. D. Fretwell, J. Arruda, and P. Niemela. 1981. Exploitation ecosystems in gradients of primary productivity. *American Naturalist* 118:240–261.

Okubo, A. 1980. *Diffusion and Ecological Problems: Mathematical Models.* Berlin: Springer-Verlag.

Oliver, C. D. and E. P. Stephens. 1978. Reconstruction of a mixed-species forest in central New England. *Ecology* 58:562–572.

Oliver, M. A. and R. Webster. 1986. Combining nested and linear sampling for determining the scale and form of spatial variation of regionalized variables. *Geographic Analysis* 18:227–242.

O'Neill, R. V. 1979. Transmutation across hierarchical levels. In G. S. Innis and R. V. O'Neill, eds., *Systems Analysis of Ecosystems,* pp. 59–78. Fairlands, Md.: International Cooperative Publishing House.

O'Neill, R. V. 1988. Hierarchy theory and global change. In T. Rosswall, R. G. Woodmansee, and P. G. Risser, eds., *Scales and Global Change,* pp. 29–45. New York: John Wiley and Sons.

O'Neill, R. V. 1989. Perspectives in hierarchy and scale. In J. Roughgarden, R. M. May, and S. A. Levin, eds., *Perspectives in Ecological Theory,* pp. 140–156. Princeton, N.J.: Princeton University Press.

O'Neill, R. V. 1996. Recent developments in ecological theory: hierarchy and scale. In J M. Scott, T. H. Tear, and F. W. Davis, eds., *GAP Analysis: A Landscape Approach to Biodiversity Planning,* pp. 7–14. Bethesda, Md.: American Society of Photogrammetry and Remote Sensing.

O'Neill, R. V., D. L. DeAngelis, J. B. Waide, and T. F. H. Allen. 1986. *A Hierarchical Concept of Ecosystems.* Princeton, N.J.: Princeton University Press.

O'Neill, R. V., J. W. Elwood, and S. G. Hildebrand. 1979. Theoretical implications of spatial heterogeneity in stream ecosystems. In G. S. Innis and R. V. O'Neill, eds., *Systems Analysis of Ecosystems,* pp. 79–101. Fairlands, Md.: International Cooperative Publishing House.

O'Neill, R. V., R. H. Gardner, B. T. Milne, M. G. Turner, and B. Jackson. 1991a. Heterogeneity and spatial hierarchies. In J. Kolasa and S. T. A. Pickett, eds., *Ecological Heterogeneity,* pp. 85–96. New York: Springer-Verlag.

O'Neill, R. V., R. H. Gardner, and M. G. Turner. 1992. A hierarchical neutral model for landscape analysis. *Landscape Ecology* 7:55–61.

O'Neill, R. V., A. R. Johnson, and A. W. King. 1989. A hierarchical framework for the analysis of scale. *Landscape Ecology* 3:193–205.

O'Neill, R. V., J. R. Krummel, R. H. Gardner, G. Sugihara, B. Jackson, D. L. DeAngelis, B. T. Milne, M. G. Turner, B. Zygmunt, S. W. Christensen, V. H. Dale, and R. L. Graham. 1988. Indices of landscape pattern. *Landscape Ecology* 1:153–162.

O'Neill, R. V., S. J. Turner, V. I. Cullinan, D. P. Coffin, T. Cook, W. Conley, J. Brunt, J. M. Thomas, M. R. Conley, and J. Gosz. 1991b. Multiple landscape scales: an intersite comparison. *Landscape Ecology* 5:137–144.

Opdam, P. 1991. Metapopulation theory and habitat fragmentation: a review of holarctic breeding bird studies. *Landscape Ecology* 5:93–106.

Opdam, P., D. van Dorp, and C. J. F. ter Braak. 1984. The effect of isolation on the number of woodland birds in small woods in the Netherlands. *Journal of Biogeography* 11:473–478.

Ord, J. K. 1975. Estimation methods for models of spatial interaction. *Journal of the American Statistical Association* 70:120–126.

O'Riordan, T. and J. Cameron, ed. 1994. *Interpreting the Precautionary Principle.* London: Cameron and May.

Ornstein, R. and P. R. Ehrlich. 1989. *New World, New Mind.* New York: Doubleday.

Pagel, M. D., P. H. Harvey and H. C. J. Godfray. 1991. Species-abundance, biomass, and resource-use distributions. *American Naturalist* 138:836–850.

Pahl-Wostl, C. 1993a. Food webs and ecological networks across spatial and temporal scales. *Oikos* 66:415–432.

Pahl-Wostl, C. 1993b. The hierarchical organization of the aquatic ecosystem: an outline of how reductionism and holism may be reconciled. *Ecological Modelling* 66:81–100.

Pahl-Wostl, C. 1995. *The Dynamic Nature of Ecosystems: Chaos and Order Entwined.* Chichester, U.K.: Wiley.

Pahl-Wostl, C. 1997. Dynamic structure of a food web model: comparison with a food chain model. *Ecological Modelling* (in press).

Pahl-Wostl, C. and D. Imboden. 1990. DYPHORA: dynamic model for the rate of photosynthesis of algae. *Journal of Plankton Research* 12:1207–1221.

Pahl-Wostl, C., M. Van Asselt, C. Jaeger, S. Rayner, C. Schaer, D. Imboden, and A. Vickovski. 1997. Integrated assessment of climate change and the problem of indeterminacy. In P. Cebon, U. Dahinden, H. Davies, D. Imboden, and C. Jaeger, eds., *A View from the Alps: Regional Perspectives on Climate Change.* Cambridge, Mass.: MIT-Press (in press).

Paine, R. T. 1966. Food web complexity and species diversity. *American Naturalist* 100:65–75.

Paine, R. T. 1974. Intertidal community structure: experimental studies on the relationship between a dominant competitor and its principal predator. *Oecologia* 15:93–120.

Paine, R. T. 1980. Food webs: linkage, interaction strength and community infrastructure. *Journal of Animal Ecology* 49:667–685.

Paine, R. T. 1984. Ecological determinism in the competition for space. *Ecology* 65:1339–1348.

Paine, R. T. 1988. Food webs: road maps of interactions or grist for theoretical development? *Ecology* 69:1648–1654.

Paine, R. T. 1992. Food-web analysis through field measurement of per capita interaction strength. *Nature* 355:73–75.

Palmer, M. W. 1992. The coexistence of species in fractal landscapes. *American Naturalist* 139:375–397.

Palmer, M. W. and P. S. White. 1994. On the existence of ecological communities. *Journal of Vegetation Science* 5:279–282.

Pantastico-Caldas, M. and D. L. Venable. 1993. Competition in two species of desert annuals along a topographic gradient. *Ecology* 74:2192–2203.

Parker, V. T. 1984. Correlation of physiological divergence with reproductive mode in chaparral shrubs. *Madroño* 31:231–242.

Parker, V. T. 1989. Maximizing vegetation response on management burns by identifying fire regimes. In *Fire and Watershed Management,* pp. 87–91. Sacramento, Calif.: Watershed Management Council.

Parker, V. T. 1990. Problems encountered while mimicking nature in vegetation management: an example from a fire-prone vegetation. In R. S. Mitchell, C. J. Sheviak and D. J. Leopold, eds., *Ecosystem Management: Rare Species and Significant Habitats, Proceedings* of the 15th Annual Natural Areas Conference. New York State Museum Bulletin 471:231–234.

Parker, V. T. and V. R. Kelly. 1989. Seed banks in California chaparral and other Mediterranean climate shrublands. In M. A. Leck, V. T. Parker, and R. L. Simpson, eds., *Ecology of Soil Seed Banks,* pp. 231–255. San Diego: Academic Press.

Parker, V. T. and S. T. A. Pickett. 1997. Restoration as an ecosystem process: implications of the modern ecological paradigm. In K. M. Urbanska, N. R. Webb, and P. J. Edwards, eds., *Restoration Ecology and Sustainable Development.* Cambridge: Cambridge University Press (in press).

Parker, V. T., M. A. Leck, and R. L. Simpson. 1989. Pattern and process in seed banks dynamics. In M. A. Leck, V. T. Parker, and R. L. Simpson, eds., *Ecology of Soil Seed Banks,* pp. 367–384. New York: Academic Press.

Parrish, J. A. D. and F. A. Bazzaz. 1976. Niche differences in use of pollinators between plant species of an early successional community. *Bulletin of the Ecological Society of America* 57:35.

Pastor, J. 1995. Ecosystem management, ecological risk, and public policy. *BioScience* 45:286–288.

Pastor, J. and M. Broschart. 1990. The spatial pattern of a northern conifer-hardwood landscape. *Landscape Ecology* 4:55–68.

Pastor, J., J. D. Aber, and C. A. McClaugherty. 1984. Aboveground production and N and P cycling along a nitrogen mineralization gradient on Blackhawk Island, Wisconsin. *Ecology* 65:256–268.

Pastor, J., B. Dewey, R. J. Naiman, P. F. McInnes, and Y. Cohen. 1993. Moose browsing and soil fertility in the boreal forests of Isle Royale National Park. *Ecology* 74:467–480.

Pastor, J., R. J. Naiman, B. Dewey, and P. McInnes. 1988. Moose, microbes and the boreal forest. *BioScience* 38:770–777.

Patrick, R. and D. M. Palavage. 1994. The value of species as indicators of water quality. *Proceedings of the Academy of Natural Sciences of Philadelphia* 145:55–92.

Pattee, H. H. 1973. The physical basic and origin of hierarchical control. In H. H. Pattee, ed., *Hierarchy Theory: The Challenge of Complexity,* pp. 71–108. New York: Braziller.

Patten, D. T. 1991. Defining the Greater Yellowstone Ecosystem. In R. B. Keiter and M. S. Boyce, eds., *The Greater Yellowstone Ecosystem,* pp. 19–26. New Haven, Conn.: Yale University Press.

Pauley, G. B. 1991. Anadromous trout. In J. Stolz and J. Schnell, eds., *Trout,* pp. 96–104. Harrisburg, Penn.: Stackpole Books.

Paulson, L. J. and J. R. Baker. 1981. *Nutrient Interactions among Reservoirs on the*

Colorado River, Symposium on Surface Water Impoundments. New York: American Society of Civil Engineers.

Pearson, D. L. 1994. Selecting indicator taxa for the quantitative assessment of biodiversity. *Philosophical Transactions of the Royal Society of London B* 345: 75–79.

Peitgen, H.-O., H. Jurgens, and D. Saupe. 1992. *Chaos and Fractals: New Frontiers of Science*. Berlin: Springer-Verlag.

Perestrello de Vasconcelos, M. J. 1993. Modeling multi-scale spatial ecological processes under the discrete event systems paradigm. *Landscape Ecology* 8:273–286.

Perlin, J. 1989. *A Forest Journey: The Role of Wood in the Development of Civilization*. New York: W. W. Norton and Company.

Perry, D. A. 1995. Self-organizing systems across scales. *Trends in Ecology and Evolution* 10:241–244.

Perry, D. A., M. P. Amaranthus, J. G. Borchers, S. L. Borchers, and R. E. Brainerd. 1989. Bootstrapping in ecosystems. *BioScience* 39:230–237.

Perry, D. A., R. Molina, and M. P. Amaranthus. 1987. Mycorrhizae, mycorrhizospheres, and reforestation: current knowledge and research needs. *Canadian Journal of Forest Research* 17:929–940.

Peters, R. H. 1983. *The Ecological Implications of Body Size*. Cambridge: Cambridge University Press.

Peters, R. H. 1991. *A Critique for Ecology*. Cambridge: Cambridge University Press.

Peterson, D. L. 1996. Research in parks and protected areas: forging the link between science and management. In R. G. Wright, ed., *National Parks and Protected Areas*, pp. 417–434. Cambridge, Mass.: Blackwell Science.

Peterson, D. L. and G. L. Rolfe. 1982. Nutrient dynamics of herbaceous vegetation in upland and floodplain forest communities. *American Midland Naturalist* 107:325–339.

Peterson, D. L., M. J. Arbaugh, and L. J. Robinson. 1991. Regional growth changes in ozone-stressed ponderosa pine (*Pinus ponderosa*) in the Sierra Nevada, California, USA. *The Holocene* 1:50–61.

Peterson, D. L., D. L. Schmoldt, J. M. Eilers, R. W. Fisher, and R. D. Doty. 1992. Guidelines for evaluating air pollution impacts on class I wilderness areas in California. *USDA Forest Service General Technical Report*, PSW-GTR-136. Albany, Calif.: Pacific Southwest Research Station.

Peterson, D. L., E. G. Schreiner, and N. M. Buckingham. 1997. Gradients, vegetation, and climate: spatial and temporal dynamics in the Olympic Mountains, USA. *Global Ecology and Biogeography Letters* 6:7–17.

Phillips, J. D. 1987. Choosing the level of detail for depicting two-variable spatial relationships. *Mathematical Geology* 19:539–547.

Phillips, J. D. 1988. Nonpoint source pollution and spatial aspects of risk assessment. *Annals of the Association of American Geographers* 78:611–623.

Phillips, J. D. 1988. The role of spatial scale in geomorphic systems. *Geographic Analysis* 20:359–368.

Phillips, J. D. 1995. Biogeomorphology and landscape evolution: the problem of scale. *Geomorphology* 13:337–347.

Phipps, M. J. 1992. From local to global: the lesson of cellular automata. In D. L. DeAngelis and L. J. Gross, eds., *Individual-based Models and Approaches in Ecology: Populations, Communities and Ecosystems*, pp. 165–187. New York: Chapman and Hall.

Pianka E. R. 1976. Competition and niche theory. In R. M. May, ed., *Theoretical Ecology: Principles and Applications*, pp. 292–314. Oxford: Blackwell.

Pickett, S. T. A. and M. J. McDonnell. 1989. Changing perspectives in community dynamics: a theory of successional forces. *Trends in Ecology and Evolution* 4:241–245.

Pickett, S. T. A. and J. N. Thompson. 1978. Patch dynamics and the design of nature reserves. *Biological Conservation* 13:27–37.

Pickett S. T. A. and P. S. White. 1985. *The Ecology of Natural Disturbance and Patch Dynamics*. San Diego: Academic Press.

Pickett, S. T. A., S. L. Collins, and J. J. Armesto. 1987a. A hierarchical consideration of causes and mechanisms of succession. *Vegetatio* 69:109–114.

Pickett, S. T. A., S. L. Collins, and J. J. Armesto. 1987b. Models, mechanisms and pathways of succession. *Botanical Review* 53:335–371.

Pickett, S. T. A., J. Kolasa, J. J. Armesto, and S. L. Collins. 1989. The ecological concept of disturbance and its expression at various hierarchical levels. *Oikos* 54:129–136.

Pickett, S. T. A., J. Kolasa, and C. G. Jones. 1984. *Ecological Understanding: The Nature of Theory and the Theory of Nature*. San Diego: Academic Press.

Pickett, S. T. A., V. T. Parker, and P. Fielder. 1992. The new paradigm in ecology: implications for conservation biology above the species level. In P. Fielder, and S. Jain, eds., *Conservation Biology: The Theory and Practice of Nature Conservation*, pp. 65–88. New York: Chapman and Hall.

Pielou, E. C. 1969. *An Introduction to Mathematical Ecology*. New York: Wiley-Interscience.

Pienitz, R. and J. P. Smol. 1993. Diatom assemblages and their relationship to environmental variables in lakes from the boreal forest-tundra ecotone near Yellowknife, Northwest Territories, Canada. *Hydrobiologia* 269/270:391–404.

Pienitz, R., J. P. Smol, and H. J. B. Birks. 1995. Assessment of freshwater diatoms as quantitative indicators of past climatic change in the Yukon and Northwest Territories, Canada. *Journal of Paleolimnology* 13:21–49.

Pierce, L. L. and S. W. Running. 1995. The effects of aggregating sub-grid land surface variation on large-scale estimates of net primary production. *Landscape Ecology* 10:239–253.

Pimm, S. L. 1982. *Food Webs*. London: Chapman and Hall.

Pimm, S. L. and J. H. Lawton. 1977. Number of trophic levels in ecological communities. *Nature* 268:329–331.

Pimm, S. L., J. H. Lawton, and J. E. Cohen. 1991. Food web patterns and their consequences. *Nature* 350:669–674.

Pisek, A. and E. Winkler. 1958. Assimilationsvermögen und Respiration der Fichte (*Picea excelsa* Link.) in verschiedener Höhenlage und der Zirbe (*Pinus cembra* L.) an der alpinen Waldgrenze. *Planta* 51:518–543.

Platt, T. 1981. Thinking in terms of scale. Introduction to dimensional analysis.

In T. Platt, K. H. Mann, and R. E. Ulanowicz, eds., *Mathematical Models in Biological Oceanography,* pp. 112–121. Paris: UNESCO Press.

Platt, T. R. and K. L. Denman. 1975. Spectral analysis in ecology. *Annual Review of Ecology and Systematics* 6:189–210.

Platt, W. J. 1975. The colonization and formation of equilibrium plant species associations on badger disturbances in a tallgrass prairie. *Ecological Monographs* 45:285–305.

Platt, W. J. and I. Weis. 1977. Resource partitioning and competition within a guild of fugitive prairie plants. *American Naturalist* 111:479–513.

Plotnick, R. E., R. H. Gardner, and R. V. O'Neill. 1993. Lacunarity indices as a measure of landscape texture. *Landscape Ecology* 8:201–211.

Plotnick, R. E. and K. L. Prestegaard. 1993. Fractal analysis of geologic time series. In N. Lam and L. DeCola, eds., *Fractals in Geography,* pp. 207–224. Englewood Cliffs, N.J.: Prentice Hall.

Poff, N. L. and J. V. Ward. 1989. Implications of streamflow variability and predictability for lotic community structure: a regional analysis of streamflow patterns. *Canadian Journal of Fisheries and Aquatic Science* 46:1805–1818.

Polis, G. A. 1991. Complex desert food webs: an empirical critique of food web theory. *American Naturalist* 138:123–155.

Polis, G. A. and K. O. Winemiller, eds. 1996. *Food Webs: Integration of Structure and Dynamics.* New York: Chapman and Hall.

Porritt, J. 1994. Translating ecological science into practical policy. In P. J. Edwards, R. M. May, and N. R. Webb, eds., *Large-Scale Ecology and Conservation Biology,* pp. 345–353. Oxford: Blackwell.

Porter, J. H. and J. L. Dooley. 1993. Animal dispersal patterns: a reassessment of simple mathematical models. *Ecology* 74:2436–2443.

Post, J. and L. Rudstam. 1992. Fisheries management and the interactive dynamics of walleye and perch populations. In J. Kitchell, *Food Web Management: A Case Study of Lake Mendota,* pp. 381–406. New York: Springer-Verlag.

Potvin, C. 1993. ANOVA: Experiments in controlled environments. In S. M. Scheiner and J. Gurevitch, eds., *Design and Analysis of Ecological Experiments,* pp. 46–68. New York: Chapman and Hall.

Potvin, C., M. J. Lechowicz, and S. Tardif. 1990. The statistical analysis of ecophysiological response curves obtained from experiments involving repeated measures. *Ecology* 71:1389–1400.

Powell, T. M. 1989. Physical and biological scales of variability in lakes, estuaries, and the coastal ocean. In J. Roughgarden, R. M. May, and S. A. Levin, eds., *Perspectives in Ecological Theory,* pp. 157–180. Princeton, N.J.: Princeton University Press.

Power, M. E. 1992. Top-down and bottom-up forces in food webs: do plants have primacy? *Ecology* 73:733–746.

Power, M. E. and L. S. Mills. 1995. The Keystone cops meet in Hilo. *Trends in Ecology and Evolution* 10:182–184.

Prentice, I. C. 1985. Pollen representation, source area, and basin size: toward a unified theory of pollen analysis. *Quaternary Research* 23:76–86.

Prentice, I. C. 1992. Climate change and long-term vegetation dynamics. In D.C.

Glenn-Lewin, R. K. Peet, and T. V. Veblen, eds., *Plant Succession: Theory and Prediction*, pp. 293–339. London: Chapman and Hall.

Press, F. 1995. Needed: coherent budgeting for science and technology. *Science* 270:1448–1450.

Preston, F. W. 1948. The commonness and rarity of species. *Ecology* 29:254–283.

Price, J. C. 1987. Calibration of satellite radiometers and the comparison of vegetation indices. *Remote Sensing of the Environment* 21:15–27.

Priestley, M. B. 1981. *Spectral Analysis and Time Series.* London: Academic Press.

Primack, R. J. 1993. *Essentials of Conservation Biology.* Sunderland, Mass.: Sinauer Associates.

Pryor, S. C. 1996. Assessing public perception of visibility for standard setting exercises. *Atmospheric Environment* 30:2705–2716.

Putman, R. J. 1994. *Community Ecology.* London: Chapman and Hall.

Pyne, S. J. 1982. *Fire in America.* Princeton, N.J.: Princeton University Press.

Rahel, F. J. 1990. The hierarchical nature of community persistence: a problem of scale. *American Naturalist* 136:328–344.

Raiffa, H. 1968. *Decision Analysis: Introductory Lectures on Choices under Uncertainty.* New York: Random House.

Rastetter, E. B., A. W. King, B. J. Cosby, G. M. Hornberger, R. V. O'Neill, and J. E. Hobbie. 1992. Aggregating fine-scale ecological knowledge to model coarse-scale attributes of ecosystems. *Ecological Applications* 2:55–70.

Raupach, M. R. 1991. Vegetation-atmosphere interaction in homogeneous and heterogeneous terrain. *Vegetatio* 91:105–120.

Raupach, M. R. 1995. Vegetation-atmosphere interaction and surface conductance at leaf, canopy and regional scales. *Agricultural and Forest Meteorology* 73:151–179.

Reich, P. B. and R. G. Amundson. 1985. Ambient levels of ozone reduce net photosynthesis in tree and crop species. *Science* 230:566–570.

Reijnders, P. J. H. 1986. Reproductive failure in common seals (*Phoca vitulina*) feeding on fish from polluted coastal waters. *Nature* 324:456–457.

Reimers, P. E. 1973. The length of residence of juvenile fall chinook salmon in Sixes River, Oregon. *Oregon Fish Commission Research Report,* 4(2). Portland, Ore.: Oregon Department of Fish and Wildlife.

Reiners, W. A., A. F. Bouwman, W. F. J. Parsons, and M. Keller. 1994. Tropical rain forest conversion to pasture: changes in vegetation and soil properties. *Ecological Applications* 4:363–377.

Reiners, W. A., I. A. Worley, and D. B. Lawrence. 1971. Plant diversity in a chronosequence at Glacier Bay, Alaska. *Ecology* 52:55–69.

Rejmanek, M. and J. Jenik. 1975. Niche, habitat, and related ecological concepts. *Acta Biotheoretica* 24:100–107.

Rejmanek, M. and P. Stary. 1979. Connectance in real biotic communities and critical values for stability of model ecosystems. *Nature* 280:311–313.

Remillard, M. M., G. K. Gruendling, and D. J. Bogucki. 1987. Disturbance by beaver (*Castor canadensis* Kuhl) and increased landscape heterogeneity. In M. G. Turner, ed., *Landscape Heterogeneity and Disturbance,* pp 103–123. New York: Springer-Verlag.

Reynolds, C. 1994. The ecological base for the successful biomanipulation of aquatic communities. *Archiv für Hydrobiologie* 130:1–33.

Reynolds, C. S. 1984. Phytoplankton periodicity: the interaction of form, function and environmental variability. *Freshwater Biology* 14:111–142.

Reynolds, C. S. 1984. *The Ecology of Freshwater Phytoplankton*. Cambridge: Cambridge University Press.

Reynolds, J. F., D. W. Hilbert, and P. R. Kemp. 1993. Scaling ecophysiology from the plant to the ecosystem: a conceptual framework. In J. R. Ehleringer and B. B. Field, eds., *Scaling Physiological Processes: Leaf to Globe*, pp. 127–140. San Diego: Academic Press.

Richie, J. C. 1994. Current trends in studies of long-term plant community dynamics. *Journal of Ecology* (supplement) 83:469–494.

Ricklefs, R. E. 1987. Community diversity: relative roles of local and regional processes. *Science* 235:167–171.

Ricklefs, R. E. 1990. Scaling pattern and process in marine ecosystems. In K. Sherman, L. M. Alexander, and B. D. Gold, eds., *Large Marine Ecosystems: Patterns, Processes and Yields*, pp. 169–178. Washington, D.C.: American Association for the Advancement of Science.

Rigler, F. H. and R. H. Peters. 1995. *Science and Limnology*. Excellence in Ecology 6. Oldendorf/Luhe, Germany: Ecology Institute.

Riitters, K. H., R. V. O'Neill, C. T. Hunsaker, J. D. Wickham, D. H. Yankee, S. P. Timmins, K. B. Jones, and B. L. Jackson. 1995. A factor analysis of landscape pattern and structure metrics. *Landscape Ecology* 10:23–39.

Rind, O., R. Goldberg and R. Reudy. 1989. Change in climatic variability in the 21st century. *Climatic Change* 13:5–39.

Ripley, B. D. 1981. *Spatial Statistics*. New York: Wiley.

Ripley, B. D. 1978. Spectral analysis and the analysis of pattern in plant communities. *Journal of Ecology* 66:965–981.

Rippey, B. 1990. Sediment chemistry and atmospheric contamination. *Philosophical Transactions of the Royal Society of London B* 327:311–317.

Roberts, D. A., J. B. Adams, and M. O. Smith. 1990. Predicted distribution of visible and near-infrared radiant flux above and below a transmittant leaf. *Remote Sensing of the Environment* 34:1–17.

Roberts, D. A., R. O. Green, and J. B. Adams. 1997. Temporal and spatial patterns in vegetation and atmospheric properties from AVIRIS. *Remote Sensing of the Environment* 62:233–240.

Roberts, D. A., M. O. Smith, and J. B. Adams. 1993a. Green vegetation, non-photosynthetic vegetation, and soils in AVIRIS data. *Remote Sensing of the Environment* 44:255–269.

Roberts, J., O. M. R. Cabral, G. Fisch, L. C. B. Molion, C. J. Moore, and W. J. Shuttleworth. 1993b. Transpiration from an Amazonian rainforest calculated from stomatal conductance measurements. *Agricultural and Forest Meteorology* 65:175–196.

Roberts, N. 1994. The global environmental future. In N. Roberts, ed., *The Changing Global Environment*, pp. 3–21. Oxford: Basil Blackwell.

Robinson, G. R. and J. F. Quinn. 1988. Extinction, turnover, and species diversity in an experimentally fragmented California annual grassland. *Oecologia* 76:71–82.

Robinson, G. R. and J. F. Quinn. 1992. Habitat fragmentation, species diversity, extinction, and design of nature reserves. In S. K. Jain and L. W. Botsford, eds., *Applied Population Biology*, pp. 223–248. Boston: Kluwer Academic Publishers.

Robinson, G. R., R. D. Holt, M. S. Gaines, S. P. Hamburg, M. L. Johnson, H. S. Fitch, and E. A. Martinko. 1992. Diverse and contrasting effects of habitat fragmentation. *Science* 257:524–526.

Robinson, S. K., F. R. Thompson, III, Donovan, T. M., D. R. Whitehead, and J. Faaborg. 1995. Regional forest fragmentation and the nesting success of migratory birds. *Science* 267:1987–1990.

RockWare Inc. 1991. *MacGRIDZO™: The Contour Mapping Program for the Macintosh*. Wheat Ridge: RockWare, Inc.

Rodiek, J. E. and E. G. Bolen. 1991. *Wildlife and Habitats in Managed Landscapes*. Washington D.C.: Island Press.

Rodríguez, J. F. and M. M. Mullin. 1986. Relation between biomass and body weight of plankton in a steady state oceanic ecosystem. *Limnology and Oceanography* 31:361–370.

Roese, J. H., K. L. Risenhoover, and L. J. Folse. 1991. Habitat heterogeneity and foraging efficiency: an individual based model. *Ecological Modelling* 57:133–143.

Rogers, C. A. 1993. Describing landscapes: indices of structure. Unpublished report for the Masters in Resource Management program. Burnaby, British Columbia: Simon Fraser University.

Roller, N. E. G. and J. E. Colwell. 1989. Coarse-resolution satellite data for ecological surveys. *BioScience* 36:468–475.

Romme, W. H. 1982. Fire and landscape diversity in sub-alpine forests of Yellowstone National Park. *Ecological Monographs* 52:199–211.

Romme, W. H. and D. Despain. 1989. The Yellowstone fires. *Scientific American* 261:37–46.

Romme, W. H. and D. H. Knight. 1981. Fire frequency and subalpine forest succession along a topographic gradient in Wyoming. *Ecology* 62:319–326.

Root, T. L. and S. H. Schneider. 1995. Ecology and climate: research strategies and implications. *Science* 269:334–341.

Rose, G. A. and W. C. Legget. 1990. The importance of scale to predator-prey spatial correlations: an example of Atlantic fishes. *Ecology* 71:33–43.

Rosen, R. 1989. Similitude, similarity, and scaling. *Landscape Ecology* 3:207–216.

Rosen, R. 1991. *Life Itself.* New York: Columbia University Press.

Rosenberg, N.J., P. R. Crosson, K. D. Frederick, W. E. Easterling, III, M. S. McKenney, M. D. Bowes, R. A. Sedjo, J. Darmstadter, L. A. Katz, and K. M. Lemon. 1993. The MINK methodology: background and baseline. *Climate Change* 24:7–22.

Rosenzweig, M. L. 1995. *Species Diversity in Space and Time.* Cambridge: Cambridge University Press.

Rossi, R. E., D. J. Mulla, A. G. Journel, and E. H. Franz. 1992. Geostatistical tools for modeling and interpreting ecological spatial dependence. *Ecological Monographs* 62:277–314.

Rosswall, T., R. G. Woodmansee, and P. G. Risser, eds. 1988. *Scales and Global Change.* New York: John Wiley and Sons.

Rouanet, H. and D. Lépine. 1970. Comparison between treatments in a repeated-measures design: ANOVA and multivariate methods. *British Journal of Mathematical and Statistical Psychology* 23:147–163.

Rouhani, S. and H. Wackernagel. 1990. Multivariate geostatistical approach to space-time data analysis. *Water Resources Research* 10:585–591.

Rowe, J. S. 1961. The level-of-integration concept and ecology. *Ecology* 42: 420–427.

Rowe, J. S. 1984a. Forestland classification: limitations of the use of vegetation. In J. G. Bockheim, ed., *Proceedings, Symposium on Forest Land Classification: Experience, Problems, Perspectives,* pp. 132–147. Madison: Department of Soil Science, University of Wisconsin.

Rowe, J. S. 1984b. Understanding forest landscapes: what you conceive is what you get. Leslie L. Schaffer Lectureship in Forest Science, Oct. 24, 1984. Vancouver: University of British Columbia.

Royama, T. 1992. *Analytical Population Dynamics.* London: Chapman and Hall.

Ruesink, J. L. 1996. Macrophyte, epiphyte, and grazer relationships: an experimental study in an intertidal assemblage. Ph.D. Dissertation. Seattle: University of Washington.

Ruggiero, L. F., K. B. Aubry, A. B. Carey, and M. H. Huff, eds. 1991. Wildlife and Vegetation of Unmanaged Douglas-fir Forests. *General Technical Report PNW-GTR-285.* Portland, Ore.: USDA Forest Service, Pacific Northwest Research Station.

Ruggiero, L. F., G. D. Hayward, and J. R. Squires. 1994. Viability analysis in biological evaluations: concepts of population viability analysis, biological population, and the ecological scale. *Conservation Biology* 8:364–372.

Ruhe, R. V. 1956. Geomorphic surfaces and the nature of soils. *Soil Science* 82:441–445.

Running, S. W., C. O. Justice, V. Salomonson, D. Hall, J. Barker, Y. J. Kaufman, A. H. Strahler, A. R. Huete, J.-P. Muller, V. Vanderbilt, Z. M. Wan, P. Teillet, and D. Carneggie. 1994. Terrestrial remote sensing science and algorithm planned for EOS/MODIS. *International Journal of Remote Sensing* 15:3587–3620.

Ruxton, G. D. 1995. Temporal scales and the occurrence of chaos in coupled populations. *Tree* 10:141–142.

Ryan, M. G. 1991. A simple method for estimating gross carbon budgets for vegetation in forest ecosystems. *Tree Physiology* 9:255–266.

Rykiel, E. J., Jr. 1984. Okefenokee Swamp watershed: water balance and nutrient budgets. In K. C. Ewel and H. T. Odum, eds., *Cypress Swamps,* pp. 374–385. Gainesville, Fla.: University of Florida Press.

Sagova, M. and M. S. Adams. 1993. Aggregation of numbers, size and taxa of benthic animals at four levels of spatial scale. *Archiv für Hydrobiologie* 128: 329–352.

Sala, O. E. 1992. Long-term soil water dynamics in the shortgrass steppe. *Ecology* 73:1175–1181.

Saldarriaga, J. G. and D.C. West. 1986. Holocene fires in the northern Amazon basin. *Quaternary Research* 26:358–366.

Salo, J., R. Kalliola, I. Hakkinen, Y. Makinen, P. Niemela, M. Pahukka, and P. Coley. 1986. River dynamics and the diversity of Amazon lowland forest. *Nature* 322:254–258.

Salwasser, H. 1991. In search of an ecosystem approach to endangered species conservation. In K. A. Kohm, ed., *Balancing on the Brink of Extinction: The Endangered Species Act and Lessons for the Future,* pp. 247–265. Washington, D.C.: Island Press.

Sample, V. A. 1994. *Remote Sensing and GIS in Ecosystem Management.* Washington, D.C.: Island Press.

Sarnelle, O. 1997. *Daphnia* effects on microzooplankton: comparisons of enclosure and whole-lake responses. *Ecology* 78:913–928.

SAS Institute Inc. 1988. *SAS/ETS User's Guide: Version 6,* 1st ed. Cary, N.C.: SAS Institute Inc.

SAS Institute Inc. 1989. *SAS/STAT User's Guide: Version 6,* 4th ed. Cary, N.C.: SAS Institute Inc.

SAS Institute Inc. 1994. *SAS/STAT Software, Changes and Enhancements, Release 6.10.* Cary, N.C.: SAS Institute Inc.

Saunders, D. A. and J. A. Ingram. 1987. Factors affecting survival of breeding populations of Carnaby's cockatoo *Calyptorhyncus funereus latirostris* in remnants of native vegetation. In D. A. Saunders, G. W. Arnold, A. A. Burbridge, and A. J. M. Hopkins, eds., *Nature Conservation: The Role of Remnants of Native Vegetation,* pp. 249–258. Chipping Norton, N.S.W., Australia: Surrey Beatty and Sons.

Saunders, D. A., G. W. Arnold, A. A. Burbridge, and A. J. M. Hopkins, eds. 1987. *Nature Conservation: The Role of Remnants of Native Vegetation.* Chipping Norton, N.S.W., Australia: Surrey Beatty and Sons.

Saunders, D. A., R. Hobbs, and G. W. Arnold. 1993. The Kellerberrin project on fragmented landscapes: a review of current information. *Biological Conservation* 64:185–92.

Saunders, D. A., R. J. Hobbs, and C. R. Margules. 1991. Biological consequences of ecosystem fragmentation: a review. *Conservation Biology* 5:18–32.

Schaap, W. 1992. Use of branch and whole tree exposure systems to evaluate ozone impacts on forest trees. Ph.D. Dissertation. Seattle: University of Washington.

Schaffer, F. M. 1981. Ecological abstraction: the consequences of reduced dimensionality in ecological models. *Ecological Monographs* 51:383–401.

Schelstraete, I., E. Knaepen, P. Dutilleul, and M. H. Weyers. 1992. Maternal behaviour in the Wistar rat under atypical Zeitgeber. *Physiological Behavior* 52:189–193.

Schelstraete, I., P. Dutilleul, and M. H. Weyers. 1995. Effect of atypical Zeitgeber on the maturation and the expression in adulthood of motor and drinking activities in rats. Contributed paper at *The World Conference on Chronobiology and Chronotherapeutics,* Ferrara, Italy, September 5–10. Abstract published in *Biological Rhythm Research* 26:441.

Schelstraete, I., E. Knaepen, P. Dutilleul, and M. H. Weyers. 1993. Simulation chez le rat des caractéristiques chronobiologiques et ontogénétiques des troubles de l'humeur. In *Rythmes Biologiques: De la Molécule à l'Homme, Actes du Congrès du GERB 92,* pp. 169–178. Paris: Polytechnica.

Schimel, D. S., T. G. Kittel, A. K. Knapp, T. R. Seastedt, W. J. Parton, and V. B. Brown. 1991. Physiological interactions along resource gradients in a tallgrass prairie. *Ecology* 72:672–684.

Schlesinger, W. H., E. H. DeLucia, and W. D. Billings. 1989. Nutrient-use efficiency of woody plants on contrasting soils in the western Great Basin, Nevada. *Ecology* 70:105–113.

Schmidt-Nielsen, K. 1972. *How Animals Work.* Cambridge: Cambridge University Press.

Schmidt-Nielsen, K. 1984. *Scaling—Why Is Animal Size So Important?* Cambridge: Cambridge University Press.

Schneider, D.C. 1985. Migratory shorebirds: resource depletion in the tropics? *Ornithological Monographs* 36:546–557.

Schneider, D.C. 1994a. *Quantitative Ecology. Spatial and Temporal Scaling.* San Diego: Academic Press.

Schneider, D.C. 1994b. Scale-dependent patterns and species interactions in marine nekton. In P. S. Giller, A. G. Hildrew, and D. G. Raffaelli, eds., *Aquatic Ecology: Scale, Pattern and Process,* pp. 441–467. London: Blackwell Scientific Publications.

Schneider, D.C. and J. F. Piatt. 1986. Scale dependent correlation of seabirds with schooling fish in a coastal ecosystem. *Marine Ecology Progress Series* 32: 237–246.

Schneider, S. H. 1989. *Global Warming: Are We Entering the Greenhouse Century?* San Francisco: Sierra Club Books.

Schoener, T. W. 1989. Food webs from the small to the large. *Ecology* 70:1559–1589.

Schoenly, K. and J. E. Cohen. 1991. Spatial variation in food web structure—16 empirical cases. *Ecological Monographs* 61:267–298.

Schoenly, K., R. A. Beaver, and T. A. Heumier. 1991. On the trophic relations of insects—a food-web approach. *American Naturalist* 137:597–638.

Scholes, R. J. 1991. The influence of soil fertility on the ecology of southern African dry savannas. In P. A. Werner, ed., *Savanna Ecology and Management,* pp. 71–75. Oxford: Basil Blackwell.

Schoonmaker, P. K. 1992. Long-term vegetation dynamics in southwestern New Hampshire. Ph.D. Dissertation. Cambridge: Harvard University.

Schrader-Frechette, K. and E. D. McCoy. 1993. *Method in Ecology: Strategies for Conservation.* Cambridge: Cambridge University Press.

Schulte, P. J., R. O. Teskey, T. M. Hinckley, R. G. Stevens, and D. A. Leslie. 1985. The effect of tephra deposition and planting treatment on soil oxygen levels and water relations of newly planted seedlings. *Forest Science* 31:109–116.

Schultz, E.-D. and H. A. Mooney, eds. 1993. *Biodiversity and Ecosystem Function.* Berlin: Springer-Verlag.

Schulze, E.-D., F. M. Kelliher, C. Körner, J. Lloyd, and R. Leuning. 1994. Relationships among maximum stomatal conductance, ecosystem surface conductance, carbon assimilation rate, and plant nitrogen nutrition: a global ecology scaling exercise. *Annual Review of Ecology and Systematics* 25:629–660.

Schumm, S. 1988. Variability of the fluvial system in space and time. In T. Rosswall, R. Woodmansee, and P. Risser, eds., *Scales and Global Change,* pp. 225–250. New York: Wiley.

Schumm, S. A. and R. W. Lichty. 1965. Time, space, and causality in geomorphology. *American Journal of Science* 263:110–119.

Schuster, A. 1898. On the investigation of hidden periodicities with application to a supposed 26-day period of meteorological phenomena. *Terrestrial Magnetism, Atmosphere and Electricity* 3:13–41.

Schwinghamer, P. 1981. Characteristic size distributions of intgral benthic communities. *Canadian Journal of Fisheries and Aquatic Science* 38:1255–1263.

Scientific Panel for Sustainable Forest Practices in Clayoquot Sound. 1995. *Report 5: Sustainable Ecosystem Management in Clayoquot Sound: Planning and Practices.* Victoria, British Columbia.

Scougall, S. A., J. D. Majer, and R. J. Hobbs. 1993. Edge effects in grazed and ungrazed Western Australian wheatbelt remnants in relation to ecosystem reconstruction. In D. A. Saunders, R. J. Hobbs, and P. R. Ehrlich, eds., *Nature Conservation 3: Reconstruction of Fragmented Ecosystems, Global and Regional Perspectives,* pp. 163–178. Chipping Norton, NSW, Australia: Surrey Beatty and Sons.

Searle, S. R. 1971. *Linear Models.* New York: Wiley.

Sedell, J. R., F. J. Swanson, and S. V. Gregory. 1984. Evaluating fish response to woody debris. In T. J. Hassler, ed., *Proceedings of the Pacific Northwest Stream Habitat Management Workshop,* pp. 222–245. Arcata, Calif.: American Fisheries Society, Western Division, and Cooperative Fisheries Unit, Humboldt State University.

Segura, G., L. B. Brubaker, J. F. Franklin, T. M. Hinckley, D. A. Maguire, and G. Wright. 1994. Recent mortality and decline in mature *Abies amabilis:* the interaction between site factors and tephra deposition from Mount St. Helens. *Canadian Journal of Forest Research* 24:1112–1122.

Segura, G., T. M. Hinckley, and C. D. Oliver. 1995a. Stem growth responses of declining mature *Abies amabilis* trees after tephra deposition from Mount St. Helens. *Canadian Journal of Forest Research* 25:1493–1502.

Segura, G., T. M. Hinckley, and L. B. Brubaker. 1995b. Variations in radial growth of declining old-growth stands of *Abies amabilis* after tephra deposition from Mount St. Helens. *Canadian Journal of Forest Research* 25:1484–1492.

Seifried, A. 1993. Die benthische Besiedlung der obersten Gauchach unter besonderer Berücksichtigung des Versauerungsgradienten. *Diplome thesis,* University of Constance.

Seinfeld, J. H. 1989. Urban air pollution: state of the science. *Science* 243:645–752.

Sellers, P. J. 1985. Canopy reflectance, photosynthesis and transpiration. *International Journal of Remote Sensing* 6:1335–1372.

Sellers, P. J. 1987. Canopy reflectance, photosynthesis and transpiration. II. The role of biophysics in the linearity of their interdependence. *Remote Sensing of the Environment* 21:143–183.

Sellers, P. J., J. A. Berry, G. J. Collatz, C. B. Field, and F. G. Hall. 1992. Canopy reflectance, photosynthesis, and transpiration. III. A reanalysis using improved leaf models and a new canopy integration scheme. *Remote Sensing of the Environment* 42:187–216.

Senft, R. L., M. B. Coughenour, D. W. Bailey, L. R. Rittenhouse, O. E. Sala, and D. M. Swift. 1987. Large herbivore foraging and ecological hierarchies. *BioScience* 37:789–799.

Serem, V. K., C. A. Madramootoo, G. T. Dodds, P. Dutilleul, and G. Mehuys. 1997. Nitrate-N and water movement in soil columns as influenced by tillage and corn residues. *Transactions of the American Society of Agricultural Engineering* 40:1001–1012.

Seymour, V. A., T. M. Hinckley, Y. Morikawa, and J. F. Franklin. 1983. Foliage damage in coniferous trees following volcanic ashfall from Mt. St. Helens. *Oecologia* 59:339–343.

Shapiro, J., W. T. Edmondson, and D. E. Allison. 1971. Changes in the chemical composition of Lake Washington, 1958–70. *Limnology and Oceanography* 16:437–452.

Shaver, C. L. and W. C. Malm. 1995. Air quality in Grand Canyon. In W. L. Halvorson and G. E. David, eds., *Science and Ecosystem Management in the National Parks,* pp. 229–250. Tucson: University of Arizona Press.

Shugart, H. H., ed. 1978. *Time Series and Ecological Processes.* Philadelphia: Society for Industrial and Applied Mathematics.

Shumway, R. H. 1983. Replicated time series regression: an approach to signal estimation and detection. In D. R. Brillinger and P. R. Krishnaiah, eds., *Handbook of Statistics: Time Series in the Frequency Domain,* pp. 383–408. New York: North-Holland.

Sigal, L. L. and T. H. Nash. 1983. Lichen communities on conifers in southern California mountains: an ecological survey relative to oxidant air pollution. *Ecology* 64:1343–1354.

Silvert W. 1994. The revolution in ecology and population dynamics. Presentation at a Symposium in Honour of Peter J. Wangersky, Dalhousie University, July 1994. Available on the World Wide Web at http://www.maritimes.dfo.ca/science/mesd/he/staff//silvert/pjw.html.

Simberloff, D. and J. Cox. 1987. Consequences and costs of conservation corridors. *Conservation Biology* 1:63–71.

Simberloff, D., J. A. Farr, J. Cox, and D. W. Mehlman. 1992. Movement corridors: conservation bargains or poor investments? *Conservation Biology* 6:493–504.

Simmons, M. A., V. I. Cullinan, and J. M. Thomas. 1992. Satellite imagery as a tool to evaluate ecological scale. *Landscape Ecology* 7:77–85.

Simon Moffat, A. 1996. Biodiversity is a boon to ecosystems, not species. *Science* 271:1497.

Simon, H. A. 1962. The architecture of complexity. *Proceedings of the American Philosophical Society* 106:467–482.

Simonson, R. W. 1951. Lessons from the first half century of soil survey: I. Classification of soils. *Soil Science* 73:249–257.

Simonson, R. W. 1959. Outline of a generalized theory of soil genesis. *Soil Science
Society of America Proceedings* 23:152–156.

Sims, P. L., J. S. Singh, and W. K. Lauenroth. 1978. The structure and function of
ten western North American grasslands. I. Abiotic and vegetational characteristics. *Journal of Ecology* 66:983–1009.

Sisk, R. D. and C. R. Margules. 1993. Habitat edges and restoration: methods for
quantifying edge effects and predicting the results of restoration efforts. In
D. A. Saunders, R. J. Hobbs, and P. R. Ehrlich, eds., *Nature Conservation 3:
Reconstruction of Fragmented Ecosystems, Global and Regional Perspectives*, pp.
57–69. Chipping Norton, NSW, Australia: Surrey Beatty and Sons.

Sjögren, P. 1991. Extinction and isolation gradients in metapopulations: the case
of the pool frog (*Rana lessonae*). *Biological Journal of the Linnaean Society*
42:135–147.

Smith, P. E. 1978. Biological effects of ocean variability: time and space scales
of biological response. *Journal du Conseil International pour l'Exploration du
Mer* 173:117–127.

Smith, T. M. and D. L. Urban. 1988. Scale and resolution of forest structural pattern. *Vegetatio* 74:143–150.

Soil Survey Staff. 1975. *Soil Taxonomy: A Basic System of Soil Classification for
Making and Interpreting Soil Surveys.* U.S. Department of Agriculture, Soil
Conservation Service, Agriculture Handbook 436. Washington, D.C.: U.S.
Government Printing Office.

Sokal, R. R. and F. J. Rohlf. 1995. *Biometry: The Principles and Practice of Statistics in Biological Research,* 3rd edition. New York: Freeman.

Solbreck, C. and B. Sillén-Tullberg. 1990. Population dynamics of a seed feeding
bug, *Lygaeus equestris.* 1. Habitat patch structure and spatial dynamics. *Oikos*
58:199–209.

Solbrig, O., ed. 1991a. Savanna modelling for global change. *Biology International* 24:1–47.

Solbrig, O. T., ed. 1991b. *From Genes to Ecosystems: A Research Agenda for Biodiversity.* Paris: IUBS.

Solow, A. R. 1989. Bootstrapping sparsely sampled spatial point patterns. *Ecology* 70:379–382.

Sorrensen-Cothern, K. A., E. D. Ford, and D. G. Sprugel. 1993. A model of competition incorporating plasticity through modular foliage and crown development. *Ecological Monographs* 63:277–304.

Soulé, M. E. and D. S. Simberloff. 1986. What do genetics and ecology tell us
about the design of nature reserves? *Biological Conservation* 35:19–40.

Spears, C. F., A. E. Amen, L. A. Fletcher, and L. Y. Healey. 1968. *Soil Survey of
Morgan County, Colorado.* USDA Soil Conservation Service in cooperation
with Colorado Agricultural Experiment Station. Washington, D.C.: U.S. Government Printing Office.

Spellerberg, I. F. 1991. *Monitoring Ecological Change.* Cambridge: Cambridge
University Press.

Spies, T. A. and J. F. Franklin. 1989. Gap characteristics and vegetation response in coniferous forests of the Pacific Northwest. *Ecology* 70:543–545.

Spies, T. A., W. Ripple, and G. A. Bradshaw. 1994. Dynamics and pattern of managed coniferous forest landscapes in Oregon. *Ecological Applications* 4:555–568.

Sprugel, D. G. 1976. Dynamic structure of wave-regenerated *Abies balsamea* forests in the northeastern United States. *Journal of Ecology* 64:889–911.

Sprugel, D. G. 1991. Disturbance, equilibrium, and environmental variability: What is "natural" vegetation in a changing environment? *Biological Conservation* 58:1–18.

Sprugel, D. G., T. M. Hinckley, and W. Schaap. 1991. The theory and practice of branch autonomy. *Annual Review of Ecology and Systematics* 22:309–334.

Stankey, G. H., D. N. Cole, R. C. Lucas, M. E. Petersen, and S. S. Frissell. 1985. The limits of acceptable change (LAC) system for wilderness planning. *USDA Forest Service General Technical Report* INT-GTR-176. Ogden, Utah: Intermountain Research Station.

Stauffer, D. and A. Aharony. 1992. *Introduction to Percolation Theory,* 2nd edition. London: Taylor and Francis.

Steele, J. H. 1978a. Some comments on plankton patches. In J. H. Steele, ed., *Spatial Pattern in Plankton Communities,* pp. 11–20. New York: Plenum Press.

Steele, J. H. 1978b. *Spatial Pattern in Plankton Communities.* New York: Plenum Press.

Steele, J. H. 1989a. Scale and coupling in ecological systems. In J. Roughgarden, R. M. May, and S. A. Levin, eds., *Perspectives in Ecological Theory,* pp. 177–180. Princeton, N.J.: Princeton University Press.

Steele, J. H. 1989b. The ocean "landscape." *Landscape Ecology* 3:185–192.

Steele, J. H. 1991. Can ecological theory cross the land-sea boundary? *Journal of Theoretical Biology* 153:425–436.

Steinauer, E. M and S. L. Collins. 1995. Effects of urine deposition on small-scale patch structure in prairie vegetation. *Ecology* 76:1195–1205.

Stern, S. J. 1992. Surfacing rates and surfacing patterns of minke whales (*Balaenoptera acutorostrata*) off central California and the probability of a whale surfacing with visual range. *Report of the International Whaling Commission* 42: 379–385.

Stevens, G. C. 1989. The latitudinal gradient in geographical range: how so many species coexist in the tropics. *American Naturalist* 133:240–256.

Stevens, S. S. 1946. On the theory of scales of measurement. *Science* 103:677–680.

Stevenson, F. J. 1982. *Humus Chemistry: Genesis, Composition, Reactions.* New York: John Wiley and Sons.

Stevenson, F. J. 1986. *Cycles of Soil: Carbon, Nitrogen, Phosphorus, Sulfur, Micronutrients.* New York: John Wiley and Sons.

Stockner, J. G. and W. W. Benson. 1967. The succession of diatom assemblages in the recent sediments of Lake Washington. *Limnology and Oceanography* 12: 513–532.

Stockwell, J. D. 1996. Spatial distribution of zooplankton biomass in Lake Erie. *Doctoral Dissertation,* Department of Zoology. Missisauga: University of Toronto, Erindale College.

Stockwell, J. D. and W. G. Sprules. 1995. Spatial and temporal patterns of zoo-plankton biomass in Lake Erie. *ICES Journal of Marine Sciences* 52:557–564.

Stommel, H. 1963. Varieties of oceanographic experience. *Science* 139:572–576.

Stone, E. L. 1975. Effects of species on nutrient cycles and soil change. *Philosophical Transactions of the Royal Society of London B* 271:149–162.

Stone, E. L. and P. J. Kalisz. 1991. On the maximum extent of tree roots. *Forest Ecology and Management* 46:59–102.

Stott, P. 1994. Savanna landscapes and global environmental change. In N. Roberts, ed., *The Changing Global Environment*, pp. 287–303. Oxford: Basil Blackwell.

Strahler, A. N. 1952. Dynamic basis of geomorphology. *Bulletin of the Geological Society of America* 63:923–928.

Strang, D. 1994. Wavelets. *American Scientist* 82:250–255.

Strayer, D., J. S. Glitzenstein, C. G. Jones, J. Kolasa, G. E. Likens, M. J. McDonnell, G. G. Parker, and S. T. A. Pickett. 1986. *Long-Term Ecological Studies: An Illustrated Account of Their Design, Operation, and Importance to Ecology.* Occasional Publication of the Institute for Ecosystem Studies, Number 2. Millbrook, N.Y.: Institute for Ecosystem Studies.

Student (W. S. Gosset). 1914. The elimination of spurious correlation due to position in time or space. *Biometrika* 10:179–180.

Suchanek, T. H. 1979. The *Mytilus californianus:* studies on the composition, structure, organization, and dynamics of a mussel bed. Ph.D. Dissertation. Seattle: University of Washington.

Sugihara, G. 1980. Minimal community structure: an explanation of species abundance patterns. *American Naturalist* 116:770–787.

Sugihara, G. and R. M. May. 1990. Applications of fractals in ecology. *Trends in Ecology and Evolution* 5:79–86.

Sugihara, G., K. Schoenly, and A. Trombla. 1989. Scale invariance in food web properties. *Science* 245:48–52.

Sullivan, A. L. and M. L. Shaffer. 1975. Biogeography of the megazoo. *Science* 189:13–17.

Swain, A. M. 1978. Environmental changes during the past 2000 years in north-central Wisconsin: analysis of pollen, charcoal, and seeds from varied lake sediments. *Quaternary Research* 10:55–68.

Swanson, F. J. and J. F. Franklin. 1992. New forestry principles from ecosystem analysis of Pacific Northwest forests. *Ecological Applications* 2:262–274.

Swanson, F. J., J. A. Jones, D. O. Wallin, and J. H. Cissel. 1993. Natural variability: implications for ecosystem management. In M. E. Jensen and P. S. Bougeron, eds., *Eastside Forest Ecosystem Health Assessment, volume II, Ecosystem Management: Principles and Applications,* pp. 89–104. Portland, Ore.: USDA Forest Service.

Swanson, F. J., T. K. Kranz, N. Caine, and R. G. Woodmansee. 1988. Landform effects on ecosystem patterns and processes. *BioScience* 38:92–98.

Swetnam, T. W. 1993. Fire history and climate change in giant sequoia groves. *Science* 262:885–889.

Swihart, R. K. and N. A. Slade. 1985a. Influence of sampling interval on estimates of home range size. *Journal of Wildlife Management* 49:1019–1025.

Swihart, R. K. and N. A. Slade. 1985b. Testing for independence of observations in animal movements. *Ecology* 66:1176–1184.

Swihart, R. K. and N. A. Slade. 1986. The importance of statistical power for testing independence in animal movement. *Ecology* 67:255–258.

Swihart, R. K., N. A. Slade, and B. J. Bergstrom. 1988. Relating body size to the rate of home range use in mammals. *Ecology* 69:393–399.

Szacki, J. and A. Liro. 1991. Movements of small mammals in the heterogeneous landscape. *Landscape Ecology* 5:219–224.

Szacki, J., J. Babinska-Werka, and A. Liro. 1993. The influence of landscape spatial structure on small mammal movements. *Acta Theriologica* 38:113–123.

Taggert, C. T. and K. T. Frank. 1990. Perspectives on larval fish ecology and recruitment processes: probing the scales of relationships. In K. Sherman and L. M. Alexander, eds., *Patterns, Processes and Yields of Large Marine Ecosystems, AAAS Selected Symposium*. Washington, D.C.: American Association for the Advancement of Science.

Tardif, J., P. Dutilleul, and Y. Bergeron. 1998. Variations in the periodicities of the ring width of black ash (*Fraxinus nigra* Marsh.) in relation to flooding and ecological site factors at Lake Duparquet in northwestern Québec. *Biological Rhythm Research* 29:1–29.

Tavares-Cromar, A. F. and D. D. Williams. 1996. The importance of temporal resolution in food web analysis: evidence from a detritus-based stream. *Ecological Monographs* 66:91–113.

Taylor, A. D. 1988a. Large-scale spatial structure and population dynamics in arthropod predator-prey systems. *Annales Zoologici Fennici* 25:63–74.

Taylor, A. D. 1991. Studying metapopulation effects in predator-prey systems. *Biological Journal of the Linnean Society* 42:305–323.

Taylor, G. E., D. W. Johnson, and C. P. Andersen. 1994. Air pollution and forest ecosystems: a regional to global perspective. *Ecological Applications* 4:662–689.

Taylor, L. R. 1989. Objective and experiment in long-term research. In G. E. Likens, ed., *Long-Term Studies in Ecology: Approaches and Alternatives*, pp. 20–70. New York: Springer-Verlag.

Taylor, P. D. and G. Merriam. 1995. Wing morphology of a forest damselfly is related to landscape structure. *Oikos* 73:43–48.

Taylor, P. D., L. Fahrig, K. Henein, and G. Merriam. 1993. Connectivity is a vital element of landscape structure. *Oikos* 68:571–573.

Taylor, P. J. 1988b. Consistent scaling and parameter choice for linear and generalized Lotka-Volterra models used in community ecology. *Journal of Theoretical Biology* 135:543–568.

Taylor, R. A. J. 1992. Simulating populations obeying Taylor's power law. In D. L. DeAngelis and L. J. Gross, eds., *Individual-Based Models and Approaches in Ecology*, pp. 295–311. New York: Chapman-Hall.

Teskey, R. O. 1995. A field study of the effects of elevated CO_2 on carbon assimilation, stomatal conductance and leaf and branch growth of *Pinus taeda* trees. *Plant, Cell and Environment* 18:565–573.

Teskey R. O., P. M. Dougherty, and A. E. Wieslogel. 1991. Design and performance of branch chambers suitable for long term ozone fumigation of foliage in large trees. *Journal of Environmental Quality* 20:591–596.

Teskey, R. O., C. C. Grier, and T. M. Hinckley. 1984. Change in photosynthesis and water relations with age and season in *Abies amabilis. Canadian Journal of Forest Research* 14:77–84.

Thiebaux, M. L. and L. M. Dickie. 1993. Structure of the body-size spectrum of the biomass in aquatic ecosystems: a consequence of allometry in predator-prey interactions. *Canadian Journal of Fisheries and Aquatic Science* 50:1308–1317.

Thomas, C. D. 1994. Difficulties in deducing dynamics from static distributions. *Trends in Ecology and Evolution* 9:300.

Thompson, D. W. 1961. *On Growth and Form* (an abridged edition edited by J. T. Bonner). Cambridge: Cambridge University Press.

Thompson, L. G., E. Mosley-Thompson, M. E. Davis, P.-N. Lin, K. A. Henderson, J. Cole-Dai, J. F. Bolsan, and K. G. Liu. 1995. *Late Glacial and Holocene tropical ice core records from Huascaran,* Peru. *Science* 269:46–50.

Thorhallsdottir, T. E. 1991. The dynamics of a grassland community: a simultaneous investigation of spatial and temporal heterogeneity at various scales. *Journal of Ecology* 78:884–908.

Thorp, J, J. G. Cady, and E. E. Gamble. 1959. Genesis of Miami silt loam. *Soil Science Society of America Proceedings* 23:65–70.

Tilman, D. 1982. *Resource competition and community structure.* Princeton, N.J.: Princeton University Press.

Tilman, D. 1988. *Plant Strategies and the Dynamics and Structure of Plant Communities.* Princeton, N.J.: Princeton University Press.

Tilman, D. 1989. Ecological experimentation: strengths and conceptual problems. In G. E. Likens, ed., *Long-Term Studies in Ecology: Approaches and Alternatives,* pp. 136–157. New York: Springer-Verlag.

Tilman, D. and J. Downing. 1994. Biodiversity and stability in grasslands. *Nature* 367:363–365.

Tilman, D. and H. Olff. 1991. An experimental study of the effects of pH and N on grassland vegetation. *Acta Oecologica* 12:427–441.

Tilman, D., D. Wedin, and J. Knops. 1996. Productivity and sustainability influenced by biodiversity in grassland ecosystems. *Nature* 379:718–720.

Timm, N. H. 1980. Multivariate analysis of variance of repeated measurements. In P. R. Krishnaiah, ed., *Handbook of Statistics: Analysis of Variance,* pp. 41–87. New York: North-Holland.

Tinley, K. L. 1982. The influence of soil moisture balance in ecosystem pattern in southern Africa. In B. J. Huntley and B. H. Walker, eds., *Ecology of Tropical Savannas,* pp. 175–92. Berlin: Springer-Verlag.

Tokeshi, M. 1990. Niche apportionment or random assortment: species abundance patterns revisited. *Journal of Animal Ecology* 59:1129–1146.

Tokeshi, M. 1993. Species abundance patterns and community structure. *Advances in Ecological Research* 24:111–186.

Townsend, C. R. 1989. The patch dynamics concept of stream community ecology. *Journal of the North American Benthological Society* 8:36–50.

Transeau, E. N. 1935. The prairie peninsula. *Ecology* 16:23–437.

Troeh, F. R. and L. M. Thompson. 1993. *Soils and Soil Fertility,* 5th edition. New York: Oxford University Press.

Tucker, C. J. 1979. Red and photographic infrared linear combinations for monitoring vegetation. *Remote Sensing of the Environment* 8:127–150.

Turner, J. and S. P. Gessel. 1990. Forest productivity in the southern hemisphere with particular emphasis on managed forests. In S. P. Gessel, D. S. Lacae, G. F. Weetman, and R. F. Powers, eds., *Sustained Productivity of Forest Soils, Proceedings, Seventh North American Forest Soils Conference,* pp. 23–39. Vancouver: Faculty of Forestry, University of British Columbia.

Turner, M. G. 1989. Landscape ecology: the effect of pattern on process. *Annual Review of Ecology and Systematics* 20:171–198.

Turner, M. G. and R. H. Gardner, eds. 1991a. *Quantitative Methods in Landscape Ecology: The Analysis and Interpretation of Landscape Heterogeneity.* New York: Springer-Verlag.

Turner, M. G. and R. H. Gardner. 1991b. Quantitative methods in landscape ecology: an introduction. In M. G. Turner and R. H. Gardner, eds., *Quantitative Methods in Landscape Ecology,* pp. 3–14. New York: Springer-Verlag.

Turner, M. G., R. H. Gardner, V. H. Dale, and R. V. O'Neill. 1989. Predicting the spread of disturbance across heterogeneous landscapes. *Oikos* 55:121–129.

Turner, M. G., W. W. Hargrove, R. H. Gardner, and W. H. Romme. 1994. Effects of fire on landscape heterogeneity in Yellowstone National Park, Wyoming. *Journal of Vegetation Science* 5:731–742.

Turner, M. G. and R. V. O'Neill. 1994. Exploring aggregation in space and time. In C. G. Jones and J. H. Lawton, eds., *Linking Species and Ecosystems,* pp. 194–208. New York: Chapman and Hall.

Turner, M. G., R. V. O'Neill, R. H. Gardner, and B. T. Milne. 1989. Effects of changing spatial scale on the analysis of landscape pattern. *Landscape Ecology* 3:153–162.

Turner, M. G. and W. H. Romme. 1994. Landscape dynamics in crown fire ecosystems. *Landscape Ecology* 9:59–77.

Turner, M. G., W. H. Romme, and R. H. Gardner. 1994. Landscape disturbance models and the long-term dynamics of natural areas. *Natural Areas Journal* 14:3–11.

Turner, M. G., W. H. Romme, R. H. Gardner, R. V. O'Neill, and T. K. Kratz. 1993. A revised concept of landscape equilibrium: disturbance and stability on scaled landscapes. *Landscape Ecology* 8:213–227.

Turner, S. J. 1995. Scale, observation and measurement: critical choices for biodiversity research. In T. J. B. Boyle and B. Boontawee, eds., *Measuring and Monitoring Biodiversity in Tropical and Temperate Forests,* pp. 97–111. Bogor: Centre for International Forestry Research.

Turner, S. J., R. V. O'Neill, W. Conley, M. R. Conley, and H. C. Humphries. 1991. Pattern and scale: statistics for landscape ecology. In M. G. Turner and R. H. Gardner, eds., *Quantitative Methods in Landscape Ecology,* pp. 17–41. New York: Springer-Verlag.

Udawatta, R. P. and R. D. Hammer. 1995. Variation in soil organic carbon due to surface shape and topography in forested watersheds. *Agronomy Abstracts,* p. 307.

Ugland, K. I. and J. S. Gray. 1982. Lognormal distributions and the concept of community equilibrium. *Oikos* 39:171–178.

Uhl, C., K. Clark, N. Dezzeo, and P. Maquirino. 1988. Vegetation dynamics in Amazon treefall gaps. *Ecology* 69:751–763.

Ulanowicz, R. E. 1997. *Ecology, the Ascendant Perspective.* New York: Columbia University Press.

Ulanowicz, R. E. 1986. *Growth and Development: Ecosystems Phenomenology.* New York: Springer-Verlag.

Underwood, A. J. 1990. Experiments in ecology and management: their logics, functions and interpretations. *Australian Journal of Ecology* 15:365–389.

Underwood, A. J. 1995. Ecological research and (and research into) environmental management. *Ecological Applications* 5:232–247.

Urban, D. L., R. V. O'Neill, and H. H. Shugart, Jr. 1987. Landscape ecology. *Bio-Science* 37:119–127.

USDA Forest Service. 1993. *Forest Ecosystem Management: An Ecological, Economic, and Social Assessment.* Report of the Forest Ecosystem Management Assessment Team (FEMAT). Washington, D.C.: USDA Forest Service.

U.S. National Marine Fisheries Service. 1976. *Final Environmental Impact Statement/Preliminary Fishery Management Plan: Trawl Fishery of the Washington, Oregon and California Region.* Washington, D.C.: National Oceanic and Atmospheric Administration.

U.S. National Park Service. 1996. *Elwha River Ecosystem Restoration Implementation: Draft Environmental Impact Statement.* Denver, Colo.: National Park Service.

Usher, M. B. 1969. The relation between mean square and block size in the analysis of similar patterns. *Journal of Ecology* 57:505–514.

Ustin, S. L., M. O. Smith, and J. B. Adams. 1993. Remote sensing of ecological processes: a strategy for developing and testing ecological models using spectral mixture analysis. In J. R. Ehleringer and C. B. Field, eds., *Scaling Physiological Processes: Leaf to Globe,* pp. 339–357. New York: Academic Press.

Ustin, S. L., C. A. Wessman, B. Curtiss, E. Kasischeke, J. Way, and V. C. Vanderbilt. 1991. Opportunities for using the EOS imaging spectrometers and synthetic aperture radar in ecological models. *Ecology* 72:1934–1945.

Valentine, J. W. 1973. *The Evolutionary Paleoecology of the Marine Biosphere.* New York: Prentice-Hall.

Valentini, R., G. Scarascia Mugnozza, P. DeAngelis, and G. Matteucci. 1995. Coupling water sources and carbon metabolism of natural vegetation at integrated time and space scales. *Agricultural and Forest Meteorology* 73:297–306.

Vande Castle, J. R. 1995. Integration of spatial analysis in long-term ecological studies. In T. M. Powell and J. H. Steele, eds., *Ecological Time Series,* pp. 48–53. New York: Chapman and Hall.

Vande Castle, J. R. and E. Vermote. 1996. Operational remote sensing data for comparative ecological research: application of atmospheric correction using automated sun photometers. In *Proceedings, Eco-Inofrma '96 — Global*

Networks for Environmental Information. Lake Buena Vista, Fla.: ERIM (in press).

Veit, R. R., E. D. Silverman, and I. Iversen. 1993. Aggregation patterns of pelagic predators and their principal prey, Antarctic krill, near South Georgia. *Journal of Animal Ecology* 62:551–564.

Veno, P. A. 1976. Successional relationships of five Florida plant communities. *Ecology* 57:498–508.

Vepraskis, M. J. 1992. Redoximorphic features for identifying aquatic conditions. *North Carolina Agricultural Research Service Technical Bulletin* 301. Raleigh: North Carolina State University.

Ver Hoef, J. M. and N. A. C. Cressie. 1993. Spatial statistics: analysis of field experiments. In S. M. Scheiner and J. Gurevitch, eds., *Design and Analysis of Ecological Experiments,* pp. 319–341. New York: Chapman and Hall.

Ver Hoef, J. M. and D.C. Glenn-Lewin. 1989. Multiscale ordination: a method for detecting pattern at several scales. *Vegetatio* 82:59–67.

Verhagen, J. H. 1994. Modeling phytoplankton patchiness under the influence of wind-driven currents in lakes. *Limnology and Oceanography* 39:1551–1565.

Villard, M. A., K. E. Freemark, and G. Merriam. 1992. Metapopulation dynamics as a conceptual model for neotropical migrant birds: an empirical investigation. In J. M. Hagan and D. W. Johnston, eds., *Ecology and Conservation of Neotropical Migrant Landbirds,* pp. 474–482. Washington, D.C.: Smithsonian Institution Press.

Vitousek, P. M. 1992. Factors controlling ecosystem structure and function. In R. Amundson, J. Harden, and M. Singer, eds., *Factors of Soil Formation: A Fiftieth Anniversary Retrospective,* pp. 87–97. Special Publication 33. Madison, Wis.: Soil Science Society of America.

Vitousek, P. M. 1993. Global dynamics and ecosystem processes: scaling up or scaling down? In J. R. Ehleringer and C. B. Field, eds., *Scaling Physiological Processes: Leaf to Globe,* pp. 169–177. San Diego: Academic Press.

Vitousek, P. M. 1994. Beyond global warming: ecology and global change. *Ecology* 75:1861–1876.

Vitousek, P. M. and R. W. Howarth. 1991. Nitrogen limitation on land and in the sea: how can it occur? *Biogeochemistry* 13:87–115.

von Ende, C. N. 1993. Repeated-measures analysis: growth and other time-dependent measures. In S. M. Scheiner and J. Gurevitch, eds., *Design and Analysis of Ecological Experiments,* pp. 113–137. New York: Chapman and Hall.

von Post, L. 1916. (M. B. Davis and K Faegri, translators, 1967). Forest pollen in south Swedish bog deposits. *Pollen et Spores* 9:375–401.

Vreeken, W. J. 1984. Soil-landscape chronograms for pedochronological analysis. *Geoderma* 34:149–164.

Wackernagel, H., R. Webster, and M. A. Oliver. 1988. A geostatistical method for segmenting multivariate sequences of soil data. In H. H. Block, ed., *Classification and Related Methods of Data Analysis,* pp. 641–650. Amsterdam: Elsevier.

Wagle, R. F. and T. W. Eakle. 1979. A controlled burn reduces the impact of a subsequent wildfire in a ponderosa pine vegetation type. *Forest Science* 25: 123–129.

Wagner, F. H. 1996. Principles for the conservation of wild living resources: another perspective. *Ecological Applications* 6:365–367.

Walker, B. H. 1992. Biodiversity and ecological redundancy. *Conservation Biology* 6:18–23.

Walker, B. H. 1995. Conserving biological diversity through ecosystem resilience. *Conservation Biology* 9:47–752.

Walker, D. A. and M. D. Walker. 1991. History and pattern of disturbance in Alaskan Arctic terrestrial ecosystems: a hierarchical approach to analyzing landscape change. *Journal of Applied Ecology* 28:244–276.

Wallace, L. L., M. G. Turner, W. H. Romme, R. V. O'Neill, and Y. Wu. 1995. Scale of heterogeneity of forage production and winter foraging by elk and bison. *Landscape Ecology* 10:75–83.

Wallin, D. O., F. J. Swanson, and B. Marks. 1994. Landscape pattern response to changes in pattern generation rules: land-use legacies in forestry. *Ecological Applications* 4:569–580.

Walters, C. J. and C. S. Holling. 1990. Large-scale management experiments and learning by doing. *Ecology* 71:2060–2068.

Waltho, N. and J. Kolasa. 1994. Organization of instabilities in multispecies systems: a test of hierarchy theory. *Proceedings of the National Academy of Science* 91:1682–1685.

Waltho, N. and J. Kolasa. 1996. Stochastic determinants of assemblage patterns in coral reef fishes: a quantification by means of two models. *Environmental Biology of Fishes* 47:255–267.

Wangersky, P. J. 1972. Evolution and the niche concept. *Transactions of the Connecticut Academy of Arts and Sciences* 44:369–376.

Ward, J. V. 1989a. *Riverine-Wetland Interactions: Freshwater Wetlands and Wildlife.* Oak Ridge, Tenn., U.S. Department of Energy, Office of Scientific and Technical Information.

Ward, J. V. 1989b. The four-dimensional nature of lotic ecosystems. *Journal of the North American Benthological Society* 8:2–8.

Ward, J. V. and J. A. Stanford. 1982. Thermal responses in the evolutionary ecology of aquatic insects. *Annual Review of Entomology* 27:97–117.

Ward, J. V. and J. A. Stanford. 1995a. The serial discontinuity concept: extending the model to floodplain rivers. *Regulated Rivers: Research and Management* 10:159–168.

Ward, J. V. and J. A. Stanford. 1995b. Ecological connectivity in alluvial river ecosystems and its disruption by flow regulation. *Regulated Rivers: Research and Management* 11:105–119.

Warren, P. H. 1989. Spatial and temporal variation in the structure of a freshwater food web. *Oikos* 55:299–311.

Warren, P. H. 1990. Variation in food-web structure: the determinants of connectance. *American Naturalist* 136:689–700.

Warren, P. H. 1994. Making connections in food webs. *Trends in Ecology and Evolution* 9:136–141.

Warwick, R. 1984. Species size distributions in marine benthic communities. *Oecologia* 61:32–41.

Warwick, R. and I. Joint. 1987. The size distribution of organisms in the Celtic Sea: from bacteria to Metazoa. *Oecologia* 73:185–191.

Watson, R. T., M. C. Zinyowera, R. H. Moss, and D. J. Dokken, eds. 1996. *Climate Change 1995. Impacts, Adaptations and Mitigation of Climate Change: Scientific-Technical Analyses.* Cambridge: Cambridge University Press.

Watson, V. and O. L. Loucks. 1979. An analysis of turnover times in a lake ecosystem and some implications for system properties. In E. Halfon, ed., *Theoretical Systems Ecology,* pp. 355–383. New York: Academic Press.

Watt, A. S. 1947. Pattern and process in the plant community. *Journal of Ecology* 13:27–73.

Weaver, J. C. 1995. Indicator species and scale of observation. *Conservation Biology* 9:939–942.

Webb, T., III. 1986. Is vegetation in equilibrium with climate? How to interpret late-Quaternary pollen data. *Vegetatio* 67:75–96.

Webster, R. and M. A. Oliver. 1990. *Statistical Methods in Soil and Land Resource Survey.* Oxford: Oxford University Press.

Wegner, J. and G. Merriam. 1990. Use of spatial elements in a farmland mosaic by a woodland rodent. *Biological Conservation* 54:263–276.

Wells, P. V. 1969. The relation between mode of reproduction and extent of speciation in woody genera of the California chaparral. *Evolution* 23:264–267.

Wennergren, U., M. Ruckelshaus, and P. Kareiva. 1995. The promise and limitations of spatial models in conservation biology. *Oikos* 74:349–356.

Werner, P. A. 1979. Competition and coexistence of similar species. In O. T. Solbrig, S. Jain, G. B. Johnson, and P. H. Raven, eds., *Topics in Plant Population Biology,* pp. 287–310. New York: Columbia University Press.

Werner, P. A. and W. J. Platt. 1976. Ecological relationships of co-occurring goldenrods (*Solidago:* Compositae). *American Naturalist* 110:959–971.

Wessman, C. 1992. Spatial scales and global change: bridging the gap from plots to GCM grid cells. *Annual Review of Ecology and Systematics* 23:175–200.

White, J. D. and S. W. Running. 1994. Testing scale dependent assumptions in regional ecosystem simulations. *Journal of Vegetation Science* 5:687–702.

Whitlock, C. and S. H. Millspaugh. 1996. Testing the assumptions of fire history studies: an examination of modern charcoal accumulation in Yellowstone National Park, USA. *The Holocene* 6:7–15.

Whitmore, T. C. 1975. *Tropical Rainforests of the Far East.* Oxford: Clarendon Press.

Whitmore, T. C. 1988. The influence of tree population dynamics on forest species composition. In A. J. Davy, M. J. Hutchings, and A. R. Watkinson, eds., *Plant Population Ecology,* pp. 271–291. Oxford: Blackwell Scientific Publications.

Whittaker, R. H. 1965. Dominance and diversity in land plant communities. *Science* 147:250–260.

Whittaker, R. H. and S. A. Levin. 1977. The role of mosaic phenomena in natural communities. *Theoretical Population Biology* 12:117–139.

Whittaker, R. H., S. A. Levin, and R. B. Root. 1973. Niche, habitat, and ecotope. *American Naturalist* 107:321–338.

Whittaker, R. J. 1985. Plant community and plant population studies of a successional sequence: Storbreen glacier foreland, Jotunheimen, Norway. *Ph.D. Dissertation.* Cardiff: University of Wales.

Whittaker, R. J. 1991. The vegetation of Storbreen gletschervorfeld, Jotunheimen, Norway. IV. Short-term vegetation change. *Journal of Biogeography* 18:41–52.

Wiegert, R. G. 1962. The selection of an optimum quadrat size for sampling the standing crop of grasses and forbs. *Ecology* 43:125–129.

Wiens, J. A. 1976. Population responses to patchy environments. *Annual Review of Ecology and Systematics* 7:81–120.

Wiens, J. A. 1977. On competition and variable environments. *American Scientist* 65:590–597.

Wiens, J. A. 1984. Resource systems, populations and communities. In P. W. Price, C. N. Slobodchikoff, and W. S. Gaud, eds., *A New Ecology: Novel Approaches to Interactive Systems,* pp. 397–436. New York: Wiley and Sons.

Wiens, J. A. 1989. Spatial scaling in ecology. *Functional Ecology* 3:385–397.

Wiens, J. A. 1995. Landscape mosaics and ecological theory. In L. Hansson, L. Fahrig, and G. Merriam, eds. *Mosaic Landscapes and Ecological Processes,* pp. 1–26. London: Chapman and Hall.

Wiens, J. A. and B. T. Milne. 1989. Scaling of "landscapes" in landscape ecology, or, landscape ecology from a beetle's eye perspective. *Landscape Ecology* 3:87–96.

Wiens, J. A., J. F. Addicott, T. J. Case, and J. Diamond. 1986. Overview: the importance of spatial and temporal scales in ecological investigations. In J. Diamond and T. J. Case, eds., *Community Ecology,* pp. 145–153. New York: Harper and Row.

Wiens, J. A., N. C. Stenseth, B. Van Horne, and R. A. Ims. 1993. Ecological mechanisms and landscape ecology. *Oikos* 66:369–380.

Wilding, L. P. 1992. Factors of soil formation: contributions to pedology. In R. Amundson, J. Harden, and M. Singer, eds., *Factors of Soil Formation: A Fiftieth Anniversary Retrospective,* pp. 15–30. Special Publication 33. Madison, Wis.: Soil Science Society of America.

Wilkinson, G. N., S. R. Eckert, T. W. Hancock, and O. Mayo. 1983. Nearest neighbour (NN) analysis of field experiments. *Journal of the Royal Statistical Society Series B* 45:152–212.

Williams, D. L. 1991. A comparison of spectral reflectance properties at the needle, branch, and canopy level for selected conifer species. *Remote Sensing of the Environment* 35:79–93.

Wilman, E. A., P. N. V. Tu, and R. Cullan. 1989. Decision-making under uncertainty: an application to wildlife management. *Journal of Environmental Management* 28:243–253.

Wilson, G. V. and R. J. Luxmoore. 1988. Infiltration, macroporosity, and misoporisity distributions on two forested watersheds. *Soil Science Society of America Journal* 52:329–335.

Wilson, G. V., P. M. Jardine, R. J. Luxmoore, and J. R. Jones. 1990. Hydrology of a forested hillslope during storm events. *Geoderma* 46:119–138.

Winemiller, K. O. 1989. Must connectance decrease with species richness? *American Naturalist* 134:960–968.

Winemiller, K. O. 1990. Spatial and temporal variation in tropical fish trophic networks. *Ecological Monographs* 60:331–367.

Winer, B. J. 1971. *Statistical Principles in Experimental Design.* New York: McGraw-Hill.

Winer, B. J., D. R. Brown, and K. M. Michels. 1991. *Statistical Principles in Experimental Design,* 3rd ed. New York: McGraw-Hill.

With, K. A., R. H. Gardner, and M. G. Turner. 1997. Landscape connectivity and population distributions in heterogeneous environments. *Oikos* 78:151–169.

Woodcock, C. E. and A. H. Strahler. 1987. The factor of scale in remote sensing. *Remote Sensing of the Environment* 21:311–332.

Woodcock, C. E., A. H. Strahler, and D. L. B. Jupp. 1988a. The use of variograms in remote sensing. I. Scene models and simulated images. *Remote Sensing of the Environment* 25:323–348.

Woodcock, C. E., A. H. Strahler, and D. L. B. Jupp. 1988b. The use of variograms in remote sensing. II. Scene models and simulated images. *Remote Sensing of the Environment* 25:467–479.

Woodyard, S. O., R. E. Preator, R. E. Moreland, and M. B. McCulough. 1973. *Soil Survey of Baca County, Colorado.* USDA Soil Conservation Service in cooperation with Colorado Agricultural Experiment Station Washington, D.C.: U.S. Government Printing Office.

Wright, D. H., D. J. Currie, and B. A. Maurer. 1993a. Energy supply and patterns of species richness on local and regional scales. In R. E. Ricklefs and D. Schluter, eds., *Species Diversity in Ecological Communities,* pp. 66–74. Chicago: University of Chicago Press.

Wright, H. A. and A. W. Bailey. 1982. *Fire Ecology.* New York: John Wiley and Sons.

Wright, H. E., Jr. 1974. Landscape development, forest fires, and wilderness management. *Science* 186:487–495.

Wright, H. E., Jr., J. E. Kutzbzch, T. Webb III, F. A. Street-Perrot, and P. J. Bartlein, eds. 1993b. *Global Climates since the Last Glacial Maximum.* Minneapolis: University of Minnesota Press.

Wu, J. and O. L. Loucks. 1995. From balance of nature to hierarchical patch dynamics: a paradigm shift in ecology. *The Quarterly Review of Biology* 70:439–466.

Wu, J. and J. L. Vankat. 1995. Island biogeography, theory and applications. *Encyclopedia of Environmental Biology* 2:371–379.

Wulff, F., J. Field, and K. Mann, eds. 1989. *Network Analysis in Marine Ecology.* Berlin: Springer-Verlag.

Wyatt, R. 1992. *Ecology and Evolution of Plant Reproduction: New Approaches.* New York: Chapman and Hall.

Yaalon, D. H. 1983. Climate, time and soil development. In L. P. Wilding, N. E. Smeck, and G. F. Hall, eds., *Pedogenesis and Soil Taxonomy I. Concepts and Interactions,* pp. 233–251. New York: Elsevier.

Yang, J., R. W. Blanchar, R. D. Hammer, and A. L. Thompson. 1997. Soybean growth and rhizosphere pH as influenced by A horizon thickness. *Soil Science Society of America Journal* (in press).

Yates, F. 1982. Reader reaction: regression models for repeated measurements. *Biometrics* 38:850–853.

Yeakley, J. A. and W. G. Cale. 1991. Organizational levels analysis: a key to understanding processes in natural systems. *Journal of Theoretical Biology* 149:203–216.

Yodzis, P. 1984. The structure of assembled communities. II. *Journal of Theoretical Biology* 107:115–126.

Yokota, T., K. Ogawa, and A. Hagihara. 1994. Dependence of the aboveground respiration of hinoki cypress (*Chamaecyparis obtusa*) on tree size. *Tree Physiology* 14:467–479.

Yoshioka, P. M. and B. B. Yoshioka. 1989. A multispecies, multiscale analysis of spatial pattern and its application to a shallow-water gorgonian community. *Marine Ecological Progress Series* 54:257–264.

Young, T. P., N. Partridge, and A. Macrae. 1995. Long-term glades in Acacia bushland and their edge effects in Laikipia, Kenya. *Ecological Applications* 5:97–108.

Zedler, P. H., C. H. Gautier, and G. S. McMaster. 1983. Vegetation change in response to extreme events: the effect of a short interval between fires in California chaparral and coastal scrub. *Ecology* 64:809–818.

Zeng, Z. and J. J. Brown. 1987. A method for distinguishing between dispersal from death in mark-recapture studies. *Journal of Mammalogy* 68:656–665.

Zhang, Y., R. W. Blanchar, and R. D. Hammer. 1997. Pyrite oxidation in cattail rhizospheres. *Wetlands* 16 (in press).

Zhou, X. M., P. Dutilleul, A. F. Mackenzie, C. A. Madramootoo, and D. L. Smith. 1997. Effects of stem-injected sucrose on grain production, dry matter distribution, and chlorophyll fluorescence of field-grown corn plants. *Journal of Agronomy and Crop Science* 178:65–71.

Zijlstra, J. J. 1988. Fish migrations between coastal and offshore areas. In B. O. Jansson, ed., pp. 257–272. *Coastal-Offshore Ecosystem Interaction*. Berlin: Springer-Verlag.

Zobel, D. B. and J. A. Antos. 1982. Adventitious rooting of eight conifers into a volcanic tephra deposit. *Canadian Journal of Forest Research* 12:717–719.

CONTRIBUTORS

Timothy F. H. Allen
Department of Botany
University of Wisconsin
Madison, WI

Gay A. Bradshaw
Pacific Northwest Research Station
USDA Forest Service
Corvallis, OR

J. Renee Brooks
Department of Biological Sciences
University of Utah
Salt Lake City, UT

Kim J. Brown
College of Forest Resources
University of Washington
Seattle, WA

Jennifer A. Dunne
Energy and Resources Group
University of California
Berkeley, CA

Pierre Dutilleul
Department of Plant Science
McGill University
Ste-Anne-de-Bellevue, Quebec
Canada

Lenore Fahrig
Ottawa-Carleton Institute of Biology
Carleton University
Ottawa, Ontario
Canada

Joseph Fall
School of Resource and Environmental Management
Simon Fraser University
Burnaby, British Columbia
Canada

Robert H. Gardner
Appalachian Environmental Laboratory
University of Maryland
Frostburg, MD

Brett J. Goodwin
Ottawa-Carleton Institute of Biology
Carleton University
Ottawa, Ontario
Canada

R. David Hammer
School of Natural Resources
University of Missouri
Columbia, MO

Richard J. Hobbs
CSIRO, Division of Wildlife and Ecology
Midland, Western Australia
Australia

Thomas M. Hinckley
College of Forest Resources
University of Washington
Seattle, WA

John L. Innes
Swiss Federal Institute for Forest, Snow and Landscape Research
Birmensdorf, Switzerland

Anthony W. King
Environmental Sciences Division
Oak Ridge National Laboratory
Oak Ridge, TN

Jurek Kolasa
Department of Biology
McMaster University
Hamilton, Ontario
Canada

Ken Lertzman
School of Resource and Environmental Management
Simon Fraser University
Burnaby, British Columbia
Canada

Timothy A. Martin
College of Forest Resources
University of Washington
Seattle, WA

Neo D. Martinez
Tiburon Center for Environmental Studies
San Francisco State University
Tiburon, CA

Robert V. O'Neill
Environmental Sciences Division
Oak Ridge National Laboratory
Oak Ridge, TN

Claudia Pahl-Wostl
Swiss Federal Institute for Environmental Science and Technology, Dübendorf
Dübendorf, Switzerland

V. Thomas Parker
Department of Biology
San Francisco State University
San Francisco, CA

David L. Peterson
U.S. Geological Survey, Biological Resources Division
Field Station for Protected Area Research
University of Washington
Seattle, WA

Steward T. A. Pickett
Institute of Ecosystem Studies
Millbrook, NY

Dar A. Roberts
Department of Geography
University of California
Santa Barabara, CA

Edward J. Rykiel, Jr.
Battelle-Pacific Northwest Laboratory
Richland, WA

Wieger Schaap
National Reference Centre for Nature Management
Ministry of Agriculture, Nature Management and Fisheries
Wageningen, The Netherlands

David C. Schneider
Ocean Sciences Centre
Memorial University of Newfoundland
St. John's, Newfoundland
Canada

Peter K. Schoonmaker
Interrain Pacific
Portland, OR

Douglas G. Sprugel
College of Forest Resources
University of Washington
Seattle, WA

S. Jonathan Stern
Marine Mammal Research Program
Texas A&M University
Galveston, TX

John Vande Castle
LTER Network Office
University of New Mexico
Albuquerque, NM

Nigel Waltho
Department of Biology
McMaster University
Hamilton, Ontario
Canada

Deane Wang
School of Natural Resources
University of Vermont
Burlington, VT

INDEX

Printed in the USA
CPSIA information can be obtained
at www.ICGtesting.com
JSHW021435221024
72172JS00002B/2